INTERPRETING ASTRONOMICAL SPECTRA

Rev. Dr David Emerson 1943 – 1996

David Emerson was born in London, and educated at the Royal Grammar School, Newcastle upon Tyne, where he gained three distinctions at A-level. He went up to Brasenose College where he achieved a first in physics and then a doctorate. In 1969 he was appointed to a lectureship at the Institute of Astronomy at the University of Edinburgh.

His specialist field was astronomical spectroscopy: he wrote his doctoral dissertation on the 'Solar Abundance of Iron'. Subsequently he extended his research to stellar atmospheres and the spectra of galaxies, and made observations in the infrared spectrum. His skills lay in the theoretical area and he developed a deep understanding of the processes whereby radiation interacts with matter, fundamental to the understanding of astronomical phenomena. This skill was put to much use, characteristically more for other people than for himself. He was supervisor to many research students, advisor on his subject to many colleagues and annually gave a locally famous undergraduate course on the subject. His expertise in the theory of radiative transfer culminated in this book.

Throughout his life he had an unobtrusive concern for his fellow human beings, and in this he did not spare himself. In his university department he increasingly assumed a pastoral role as a director of studies. Invariably courteous and patient, he earned the lasting gratitude of his students, who brought him their academic and personal problems. With his wife, he also showed care and compassion for those in need in his local community by becoming a foster parent, and in the wider world by raising awareness of, and supporting projects for, the Third World.

Most important of all to David, throughout his life, was his Christian faith. In 1983 he was asked to undertake training for the ministry and so for the last ten years of his life, in addition to being a scientist, he was also an ordained priest in the Scottish Episcopal Church. His evident humility, intelligent preaching and gentle wit won him widespread affection.

After a fulfilling life and 24 years of happy marriage, he sadly died from heart disease. He leaves a wife, two sons and a foster daughter.

INTERPRETING ASTRONOMICAL SPECTRA

D. Emerson
Late of University of Edinburgh

JOHN WILEY AND SONS
Chichester · New York · Brisbane · Toronto · Singapore

National 01243 779777
International (+44) 1243 779777

Reprinted December 1997

Other Wiley Editorial Offices

John Wiley & Sons, Inc., 605 Third Avenue,
New York, NY 10158-0012, USA

Jacaranda Wiley Ltd, 33 Park Road, Milton,
Queensland 4064, Australia

John Wiley & Sons (Canada) Ltd, 22 Worcester Road,
Rexdale, Ontario M9W 1L1, Canada

John Wiley & Sons (SEA) Pte Ltd, 37 Jalan Pemimpin #05-04,
Block B, Union Industrial Building, Singapore 2057

British Library Cataloguing in Publication Data

A catalogue record for this book is available from the British Library

ISBN 0 471 94176 X (cloth)
ISBN 0 471 97679 2 (paper)

CONTENTS

PREFACE

This book has arisen out of the difficulties that many research students find initially in interpreting astronomical spectra, especially if they are presented with spectra from unfamiliar wavelength ranges or environments. Terminology varies, formulae are presented in diverse units, often without obvious derivation, in some areas there are textbooks, in others there are not.

The aim of this book is to present a consistent overall picture of astronomical spectroscopy, using MKS units throughout, and giving derivations wherever possible, in the hope that the jungle may appear a little less impenetrable. Of necessity, some of the material runs closely parallel with other books, and clearly it is not possible to improve upon much that appears in Mihalas, *Stellar Atmospheres*, or Osterbrock, *Gaseous Nebulae*. In other areas, the subject is much less well formed, and inevitably some errors will have crept into this text. The emphasis throughout is on analytical derivations rather than quoting the results of the latest computer simulations. For the latest data and most up-to-date corrections to formulae, the student is referred to the current literature. From the outset it was decided to exclude relativistic phenomena, and in the end it was regrettably necessary to leave out the important area of shock emission, for reasons of length. For the same reason, the treatments of polarization and of X-ray emission from very hot gases are far briefer than their importance deserves.

I wish to thank my wife, Heather, my colleagues in the Institute, and the departmental administrator, Mrs E. Gibson, for their support while this book was being written.

INTRODUCTION

In the past, someone might join as a research student a group working on stars and using visible spectroscopy, and continue dealing with stars and using that wavelength range for the rest of his/her working life. Today the situation is very different. A spectrum of a faint point object presumed to be an 'ordinary' star can reveal something much more exotic, say with strong emission lines, whose analysis may take the researcher into very different regimes of temperature and density than those to which he or she is accustomed. Multi-wavelength programmes are increasingly common, and one individual can find him or herself in an extreme case using in one year X-ray, ultraviolet, visible, infrared, millimetre and radio observations. Not only are the techniques involved very different, but so is the terminology and, at least on the surface, the approach to interpretation. In some areas, like stellar atmospheres, there are several good textbooks at various levels, but in others the newcomer is reduced to searching for review articles or summer school lectures. There is a strong temptation to leave unfamiliar wavelength bands or temperature regimes to collaborators who, it is hoped, know what they are doing!

The aim of this book is to provide a consistent and coherent introduction to the interpretation of astronomical spectra from sources ranging from coronal densities of $10^8\ \mathrm{m}^{-3}$ and less and temperatures of 10^7 K to molecular clouds with temperatures of a few tens of degrees via stars with much higher density atmospheres but intermediate temperatures and gaseous nebulae (photoionized gases) with densities of the order of $10^9\ \mathrm{m}^{-3}$ and with temperatures of the order of 10^4 K. It will, it is hoped, be found that many of the basic ideas applied to these different regimes are the same, even if they may appear to look somewhat different at first sight. The level of knowledge assumed is that of Physics undergraduates starting their final year, with a tolerant recognition of the fact that many such undergraduates seem to forget spectroscopic terminology as soon as they have finished their atomic or quantum physics courses. Most subjects are discussed first at a fairly general and qualitative level before they are considered in sufficient detail to enable the newcomer to the subject to be able to derive the basic equations and to obtain approximate results from observations. No claim is made here to have incorporated all the latest corrections and all the latest physical data such as transition probabilities—for these the reader is referred to the review articles. Units are S.I. (mks) throughout except for a few stray references to parsecs, and if a formula appears without units, S.I. units should be assumed. No discussion has been attempted here of observational techniques—there are other books dealing with these topics—but of course it is very important that the observer is fully aware of the observational constraints and problems when interpreting the data.

The first four chapters deal with fundamentals common to most regimes such as radiative transfer, statistical equilibrium and line formation and broadening. Subsequent chapters look at applications to stars, photoionized gases, the cold interstellar medium (which includes a discussion of masers), winds and circumstellar shells, ending with a brief discussion of coronal gases. We shall finish this introduction with a brief consideration of some of the concepts found in the opening chapters.

Radiative transfer is concerned with how the radiative energy flow changes as one moves through an emitting and absorbing gas. One wants in the end to be able to predict the spectrum that emerges from a given gas cloud or star as a function of direction and frequency, because this is what we observe and comparison of predictions with observations enables the composition, temperature and density of the cloud or stellar atmosphere to be ascertained. In general, the temperature and density will vary as one moves deeper into the object observed, particularly in the case of stars, so one would like to be able to model the cloud or stellar atmosphere—that is, for a given energy input to be able to predict the run of temperature through the cloud or stellar atmosphere. Any region in a cloud or stellar atmosphere must obey energy conservation with total energy input equal to total energy output. In the case of a non-extended stellar atmosphere, since there are no sources or sinks of energy in the absence of convection or other mechanical means of energy transport, the total flux (which measures the radiative energy flow) passing through successive layers of the atmosphere must be a constant, although the distribution of monochromatic flux with frequency will change. This condition of *flux constancy* expresses the law of energy conservation, and calculating the temperature gradient needed to drive a given constant flux (fixed by the luminosity and radius of the star) against the opposition of the opaqueness of the gas, together with the condition of hydrostatic equilibrium (with g fixed by the radius and mass of the star), enables a *model stellar atmosphere* to be produced giving temperature and pressure as a function of depth. In contrast, the temperature gradient is of much less importance in the case of an interstellar molecular cloud and the net radiative flow may be small with considerable non-radiative energy input in the form of cosmic rays, so that the simplest interpretations of molecular cloud spectra often assume a constant temperature. However, in more detailed work it is necessary to produce a model of the temperature distribution in molecular clouds as well.

One still has to determine the emissive and absorbing powers of the gas which enter into the equations of radiative transfer, to be more precise the *emission coefficient* and the *absorption coefficient*. These coefficients often appear in radiative transfer in the form of the ratio of emission to absorption coefficient, the *source function*, and of the *optical depth*, the integral of the absorption coefficient with respect to distance along a particular line of sight. The absorption coefficient at a particular frequency will be the sum of the absorption coefficients of all the atomic and molecular radiative processes that can take place at that frequency. The absorption coefficient for each process will be proportional to the product of the transition probability and the number of atoms or ions or molecules that are in the appropriate quantum state (lower energy level) for the transition concerned. Similarly, the emission coefficient will be proportional to the product of the transition probability and the number of atoms or ions or molecules in the appropriate upper energy level. In turn, the number of atoms or ions or molecules in the correct initial state depends not only on the abundance of the element or elements concerned, but also on the degree of excitation, ionization, and (for molecules) dissociation.

Now if we have a steady state (and in astrophysics we often do, but there are important exceptions such as shock waves and solar flares), and are given the composition, density and temperature of the gas under consideration and all the appropriate atomic and molecular constants, we can determine the degree of excitation, ionization and dissociation by using the equations of *statistical equilibrium*. For each quantum level, the rate of departure to all other levels by all processes (collisional as well as radiative) equals the rate of arrival from all other levels by all processes. An equation expressing this balance is written for each level, and the set of equations is solved simultaneously subject to the given total abundances of the elements involved and the given pressure and temperature, to obtain the populations of all the levels. In general this cannot be done because of the enormous number of levels involved which makes the matrix to be solved impossibly large. A second problem is that the rates of some of the radiative processes filling or emptying a level depend on the radiation field, which was what we were trying to find in the first place! Mathematically we need to solve simultaneously a large matrix and a set of differential equations.

The situation is made worse by the fact that many of the atomic constants involved, like transition probabilities and collisional cross-sections, are hard to measure experimentally and can only be estimated theoretically for the simplest atomic structures.

Under conditions of complete thermodynamic equilibrium the situation is greatly simplified because many of the quantities involved are related by the equations of statistical mechanics.

Thus the radiation field is given by *Planck's equation* which for frequency ν and temperature T is

$$I_\nu = \frac{2h\nu^3}{c^2} \frac{1}{e^{h\nu/kT} - 1} \tag{I.1}$$

where h is Planck's constant, c is the speed of light and k is Boltzmann's constant, here and throughout this book. The radiation field is, of course, isotropic. The source function in complete thermodynamic equilibrium is also given by Planck's equation. In complete equilibrium, the number of particles of mass M travelling with velocities between V and $V + \mathrm{d}V$, is given by *Maxwell's equation*:

$$N(V)\,\mathrm{d}V = 4\pi V^2 \left(\frac{M}{2\pi kT}\right)^{3/2} e^{-MV^2/2kT}\,\mathrm{d}V \tag{I.2}$$

which is needed in estimating collisional rates. The fraction of particles of a particular atomic, ionic or molecular species excited to level n with excitation energy E_n above the ground state (the energy difference between the level and the ground state) is given by *Boltzmann's equation*:

$$\frac{N_n}{N} = \frac{g_n}{U} e^{-E_n/kT} \tag{I.3}$$

with

$$U = \sum_i g_i e^{-E_i/kT}$$

where g_n is the statistical weight of the nth level, i.e. the number of degenerate sub-levels in the nth level and U is called the partition function and serves to ensure that the populations N_n of all the levels add up to the total population N of the species concerned. Finally in complete equilibrium the ratio of the number of particles in a particular stage of ionization to the number in the next lower stage of ionization is given for an *ionization potential I* (the energy needed to remove an electron from the lower stage of ionization) by *Saha's equation*:

$$\frac{N(\text{higher stage})N_e}{N(\text{lower stage})} = \left(\frac{2\pi m_e kT}{h^3}\right)^{3/2} \frac{2U(\text{higher stage})}{U(\text{lower stage})} e^{-I/kT} \tag{I.4}$$

where N_e is the number of electrons per unit volume and m_e is the mass of the electron. A similar equation holds for the dissociation of a molecule, where for a diatomic molecule AB one replaces N(lower) by N(AB) and N(higher) and N_e by N(A) and N(B), m_e becomes the reduced mass of the molecule $m_A m_B/m_{AB}$, and I becomes the dissociation energy D, with appropriate changes to the partition functions.

Complete thermodynamic equilibrium cannot occur in a stellar atmosphere because it would require the temperature to be the same everywhere with no temperature gradient to drive an outward flow of radiation, and because it would require an isotropic radiation field which again would imply no net flow of radiation. However, one can sometimes treat conditions locally as being characterized by a single local temperature which in Planck's equation gives the source function, in Saha's equation gives the ionization and in Boltzmann's equation gives the excitation, although the radiation field will not be precisely Planckian at the local temperature, and the 'single' temperature at which thermodynamic equilibrium holds will vary from layer to layer. In other words, what a given layer *does* to a radiation field which in general is not a thermodynamic equilibrium field is given by the equations of complete equilibrium characterized by the local kinetic temperature of the gas. This is called the approximation of *local thermodynamic equilibrium*, or LTE, as it is universally abbreviated, and its use enormously simplifies the solution of the equations of radiative transfer.

Elastic collisions in which particles exchange kinetic energy are much more frequent than collisions that produce excitation or ionization. Hence in the vast majority of astronomical circumstances Maxwell's equation for the velocity distribution holds, and in this book we shall assume that Maxwell's equation at the local kinetic temperature always gives the velocity distribution. It will be shown later that if Boltzmann's equation holds for the relative upper and lower level populations, then the source function reduces to the Planck function. Inelastic collisions producing excitation and de-excitation and ionization drive the degree of excitation towards that given by Boltzmann's equation and the degree of ionization towards that given by Saha's equation, so LTE is more likely in high density conditions. Equally if the optical depth is very large, then photons travel small distances so even if radiative processes are dominant over collisional processes, local conditions have a dominating effect and the local kinetic temperature will control the level populations via the occasional collisional process.

LTE will clearly hold in the interior of a star where the density is high, the average distance travelled by a photon is small, and the net outwards flux represents a small deviation from a nearly isotropic radiation field. Equally clearly we cannot expect LTE to hold at all in the thinnest parts of the interstellar medium where the density represents a good terrestrial vacuum, but the radiation field has major non-local contributions from the microwave background and the general star background. On the other hand it often turns out that in the lower density parts of the interstellar medium the equations of statistical equilibrium are particularly simple with only a few low-lying levels being significantly populated and the gas being optically thin so that any photon emitted is sure to escape without being reabsorbed. The latter point means that we do not need to solve an equation of radiative transfer at all. Similar considerations apply to gaseous nebulae, which also often have low densities and optical thinness at many wavelengths, with the radiation field supplied by the illuminating star. Unfortunately, most environments in stellar atmospheres and dense molecular clouds lie somewhere in between these extremes, the gas being of moderate density but optically thick and many levels being significantly populated. Some lines may be formed in LTE and some not; indeed, the contribution from deep levels in a stellar atmosphere may be in LTE but the contribution from high levels may deviate markedly from LTE, for the density is lower at higher levels and the optical depth to the edge of the atmosphere is small, so photons readily escape, reducing the level populations below their LTE values.

One has to try to estimate whether an LTE solution will be adequate for a particular purpose or what approximation to a full non-LTE solution might serve. In radio astronomy in particular one often uses the *excitation temperature* which is the temperature that would be needed to give the relative populations of the upper and lower levels of a transition *if* Boltzmann's equation held. If LTE holds, then the excitation temperature equals the kinetic temperature. A radiation field can often be characterized by a radiation temperature, which gives a Planck function approximating to the actual distribution of radiation. If the radiation field comes from a star at some distance, then the distribution of the field with frequency may be roughly Planckian at the effective temperature of the star, but the magnitude is reduced by a factor roughly proportional to the inverse square of the distance from the star called the *dilution factor*. If radiative processes are dominant and there is no dilution, then the excitation temperature will tend towards the radiation temperature.

If we now look particularly at lines, the concept of *saturation* is very important. Consider a thin cloud of gas with no continuous emission and some particular line transition which will, of course, give rise to an emission line. If we increase the number of atoms in the line of sight through the cloud, the line strength will at first increase in proportion to the number of atoms along the line of sight, but as the optical depth through the cloud increases some of the emission will be absorbed before it can escape from the cloud, and the line strength will increase more slowly than the number of atoms along the line of sight. Eventually increasing emission will be balanced by increasing absorption, the line strength will cease to grow, and we say that the line is saturated. Emission and absorption will balance to give an emission intensity equal to the Planck function evaluated at the line frequency and the excitation temperature of the line. The line is now optically thick. In general there will also be emission by continuum processes at all wavelengths, but at line wavelengths we have both continuum and line emission, so the line stands out above the continuum. In stars the continuum is optically thick, and if stellar atmospheres were isothermal both line plus continuum and continuum would radiate at the saturated strength and so the line would be invisible. However there is a temperature gradient in stellar atmospheres so the increased absorption coefficient at frequencies where we have both line and continuum opacity compared with that at neighbouring frequencies where we have continuum absorption alone means that optical depth 1 comes higher in the atmosphere at line frequencies. Since the temperature is lower higher up, the source function is usually lower, and the spectrum therefore looks less bright at the line frequency and we have an absorption line. At first the depth of such a line increases as the number of atoms capable of absorbing the line increases, but eventually we are looking at the top of the atmosphere and the line can become no deeper so again the line saturates.

We have ignored the fact that spectral lines are not monochromatic and have a natural breadth in frequency which is often swamped in astrophysical spectra by broadening due to motions in the gas (Doppler broadening) or by broadening due to interactions with other atoms, ions and electrons (collisional or pressure broadening). Not only is broadening of interest in itself, giving information on temperature, turbulence, rotation and pressure, but the line broadening affects saturation, for the centre of a line may be saturated while the wings are unsaturated and still growing with the number of atoms in the line of sight. Thus the frequency or wavelength integrated strength of the line depends on the line profile and the broadening mechanism.

In the last seven chapters we will apply some of these ideas to various astrophysical environments.

1 RADIATIVE TRANSFER AND MODEL ATMOSPHERES

In this chapter we follow the changes in intensity of the radiation field as it is absorbed and emitted in passing through a gas, with the aim of predicting the emergent intensity from a gas cloud or star with a given run of temperature and pressure, and we briefly consider the forms in which that emergent intensity is observed. As mentioned in the introduction, the run of temperature and pressure with depth in a stellar atmosphere is determined by the condition that the luminosity received from the stellar interior (after due allowance for convective energy transport) must be transmitted without overall loss or gain by each layer in the stellar atmosphere. The temperature and pressure variation with depth found by applying this condition for a given luminosity, mass and radius is called a model atmosphere. In what follows we develop the general equations needed and show how model atmospheres are constructed, leaving the question of how the source function (the ratio of emission to absorption of the gas) is estimated to the following chapter.

Radiative Transfer—Some Definitions

In the introduction, we rather vaguely talked about energy flows and absorption and emission coefficients. In this first section these quantities are defined more precisely and some related quantities are introduced.

The two fundamental measures of energy flow are (i) the energy flow in a particular direction, the *intensity* I, where $I =$ energy flow through unit area per unit time per unit solid angle with the unit area perpendicular to the chosen direction defined by the solid angle, and (ii) the net energy flow summed over all directions, the *flux* F, where $F =$ energy flow through unit area per unit time,

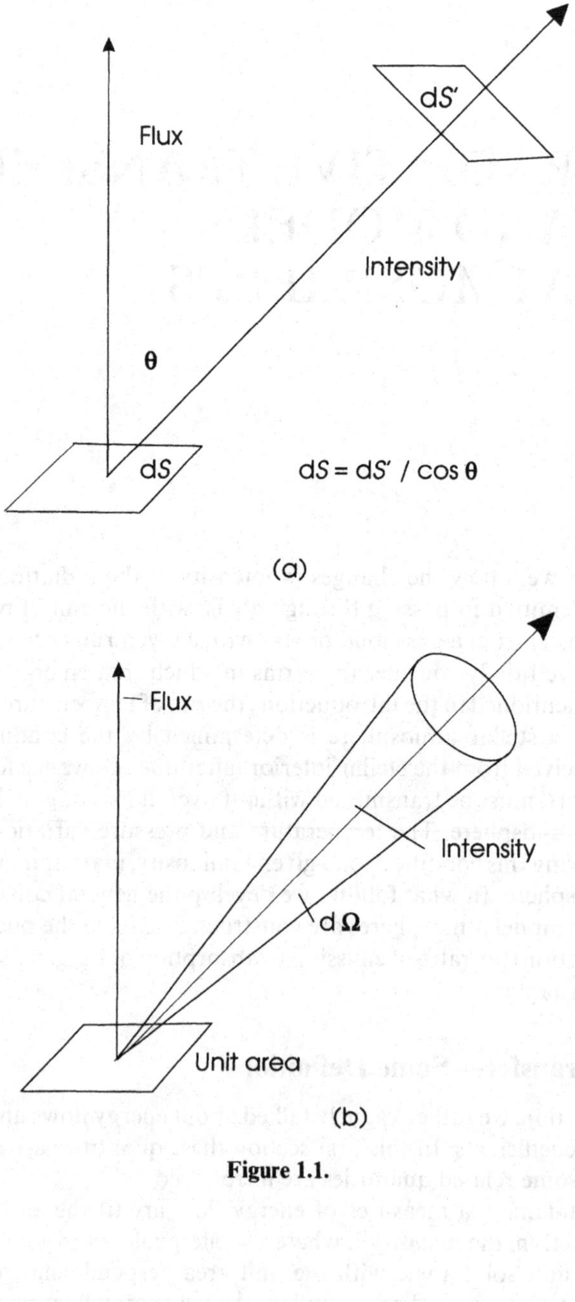
Flux
dS'
Intensity
θ
dS
dS = dS' / cos θ
(a)
Flux
Intensity
d Ω
Unit area
(b)

Figure 1.1.

where the unit area is fixed as being perpendicular to the direction in which the net flow of radiation is proceeding.

The flux is found by summing the intensity over all directions. An isotropic intensity will give a flux of zero as positive and negative contributions will cancel. The intensity is measured per unit area perpendicular to the direction being considered, whereas the unit area for the flux is in a fixed direction, so in summing intensities there is a projection factor $\cos\theta$, where θ is the angle between the direction defined by the flux and the variable direction defined by the intensity, i.e. the direction of the bundle of rays being considered when the intensity is measured (see Figure 1.1). Let solid angle be represented by Ω. Then

$$F = \int I \cos\theta \, d\Omega$$

Usually we are concerned with monochromatic intensities and fluxes such as the intensity per unit frequency interval at frequency ν, I_ν, with the total intensity given by:

$$I = \int I_\nu \, d\nu$$

In a stellar atmosphere we normally solve for the intensity as a function of direction and then integrate to obtain the flux. In particular, this operation performed at the surface gives the flux emitted by the star, F_{out}. Then the luminosity of the star, if spherical and of radius R, is given by

$$L = 4\pi R^2 F_{\text{out}}$$

and similarly for monochromatic quantities. Observations of point objects normally give an observed *flux* $= L/(4\pi d^2)$ for an object at distance d. For extended objects (resolved in angle by the telescope) we can also measure the intensity received from various parts of the object, sometimes called the *surface brightness*. For an object of uniform surface brightness, intensity received is independent of distance.

In the next chapter we will be trying to solve the equations of statistical equilibrium inside a gas cloud, and for this purpose we will need to know the radiative energy falling on an atom. The relevant quantity here is the *mean intensity* J:

$$J = \frac{1}{4\pi} \int I \, d\Omega$$

where 'all round' is of course 4π.

If one defines direction at a given point by polar angles, using the previously introduced θ (the angle between the chosen direction and the direction of net flow which runs from 0 to π), and a second angle ϕ running from 0 to 2π (Figure 1.2), then a small change in the solid angle Ω subtended at the point is given by $d\Omega = \sin\theta \, d\theta \, d\phi$, as can be seen if we construct a sphere of radius r and note that $d\Omega = dS/r^2$, where dS is an area perpendicular to the radius equal to $(r \sin\theta \, d\theta)$

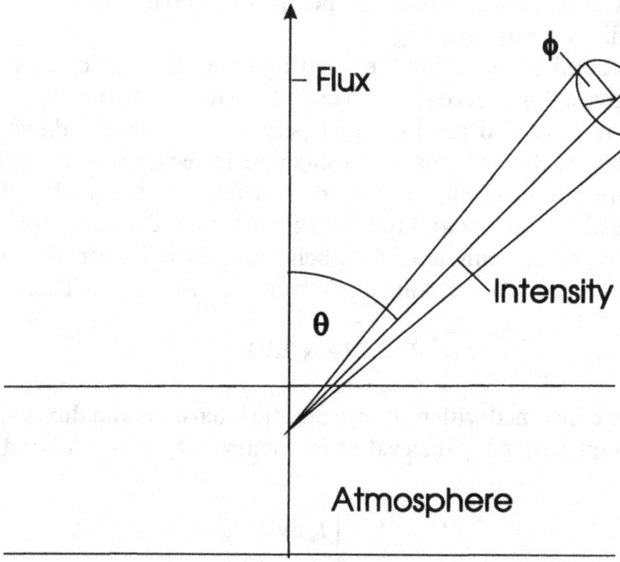

Figure 1.2.

$(r\,\mathrm{d}\phi)$. In many astrophysical situations with spherical symmetry, I is a function of θ only. Writing $\cos\theta = \mu$:

$$\int \mathrm{d}\Omega = \int_0^{2\pi}\int_0^{\pi} \sin\theta\,\mathrm{d}\theta\,\mathrm{d}\phi = 2\pi\int_{-1}^{+1} \mathrm{d}\mu$$

Hence we find

$$J = \frac{1}{2}\int_{-1}^{+1} I(\mu)\,\mathrm{d}\mu$$

It is convenient to define similarly:

$$H = \frac{1}{2}\int_{-1}^{+1} I(\mu)\mu\,\mathrm{d}\mu \tag{1.1}$$

$$K = \frac{1}{2}\int_{-1}^{+1} I(\mu)\mu^2\,\mathrm{d}\mu \tag{1.2}$$

where the flux is simply $4\pi H$. We shall see later that K measures the radiation pressure.

The absorption coefficient per unit mass, κ_ν is defined by

$$\frac{\mathrm{d}I_\nu}{I_\nu} = -\kappa_\nu\rho\,\mathrm{d}x$$

where $\mathrm{d}I$ is the change in intensity I over the distance $\mathrm{d}x$, ρ is the density and $\kappa_\nu\rho$ is the absorption coefficient per unit volume. In words, the absorption coefficient per unit volume is the fractional decrease in intensity per unit distance. Absorption is normally taken to include stimulated emission as a negative contribution, as the latter is proportional to the intensity. The absorption *cross-section* σ is given by $\kappa_\nu \rho = N\sigma$, where N is the number of absorbing atoms per unit volume.

The *emission coefficient* j_ν is defined as the emission per unit mass per unit frequency interval into a unit solid angle, so the total emission by an isotropic emitter from unit volume in all directions is $4\pi j_\nu \rho$. Finally, the *source function* is defined as the ratio of the emission to the absorption coefficient, and the *optical depth*, τ_ν, is defined by $\mathrm{d}\tau_\nu = \kappa_\nu \rho\, \mathrm{d}x$.

Radiative Transfer—The Basic Equations

We first take a simplified approach, omitting the subscript ν for the moment for clarity. Consider a 'cold, thick' cloud which absorbs but does not emit. Take some line of sight through the cloud:

$$\frac{\mathrm{d}I}{I} = -\kappa\rho\,\mathrm{d}x$$

and integrating

$$I = I_0 \exp\left(-\int \kappa\rho\,\mathrm{d}x\right)$$

$$= I_0 \mathrm{e}^{-\tau}$$

where I_0 is the incident intensity at $x = 0$.

Now consider a 'hot' cloud and first suppose that it is thin so the cloud emits but does not absorb. Then any layer of thickness $\mathrm{d}x$ at $x = x_1$ will contribute to the total intensity an amount $\mathrm{d}I = \int j\rho(x_1)\mathrm{d}x$. Now suppose the cloud is thick enough to absorb as well as emit (Figure 1.3). Then the emission from layer at x_1 will be absorbed by the overlying layers which have an optical depth $\tau(x_1)$ so the contribution from the layer at x_1 is reduced to

$$\mathrm{d}I = j\rho(x_1)\mathrm{e}^{-\tau(x_1)}\,\mathrm{d}x_1$$

Summing over all emitting layers, we obtain

$$I = \int j\rho(x_1)\mathrm{e}^{-\tau(x_1)}\,\mathrm{d}x_1$$

$$= \int \frac{j\rho}{\kappa\rho}\mathrm{e}^{-\tau(x_1)}\kappa\rho\,\mathrm{d}x_1$$

$$= \int_0^{\tau} S(\tau_1)\mathrm{e}^{-\tau_1}\,\mathrm{d}\tau_1$$

where τ is the total optical thickness of the cloud along the line of sight, S is the source function, and we have taken $I_0 = 0$.

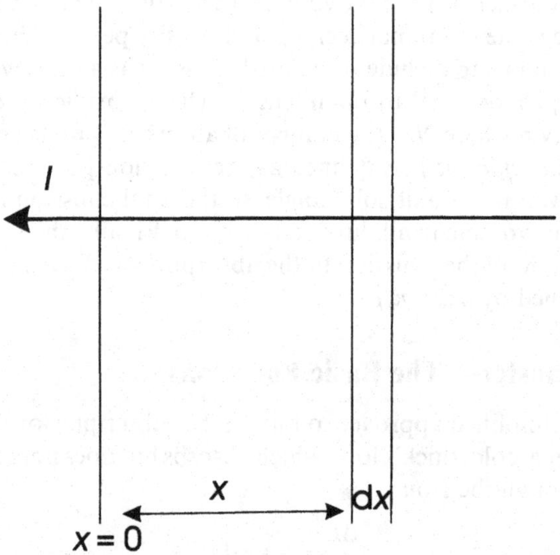

Figure 1.3.

More formally, we can obtain the same result as follows. Consider a small distance dx along some particular direction. Then in moving dx, the intensity is reduced by absorption and increased by emission: $\mathrm{d}I = -\kappa\rho I\,\mathrm{d}x + j\rho\,\mathrm{d}x$

$$\frac{\mathrm{d}I}{\kappa\rho\,\mathrm{d}x} = -I + \frac{j\rho}{\kappa\rho}$$

$$\frac{\mathrm{d}I}{\mathrm{d}\tau} = I - S(\tau) \qquad (1.3)$$

This is the basic *equation of radiative transfer*, with the sign convention of I being positive outwards (towards the observer) but τ and x being positive inwards starting at zero at the 'surface' nearest the observer.

To solve the equation, multiply both sides by $\mathrm{e}^{-\tau}$:

$$\frac{\mathrm{d}I}{\mathrm{d}\tau}\mathrm{e}^{-\tau} - I\mathrm{e}^{-\tau} = S\mathrm{e}^{-\tau}$$

$$\frac{\mathrm{d}}{\mathrm{d}\tau}(I\mathrm{e}^{-\tau}) = S\mathrm{e}^{-\tau}$$

$$I\mathrm{e}^{-\tau} = \int S\mathrm{e}^{-\tau}\,\mathrm{d}\tau$$

$$I(\tau_a)\mathrm{e}^{-\tau_a} - I(\tau_b)\mathrm{e}^{-\tau_b} = \int_{\tau_a}^{\tau_b} S(\tau)\mathrm{e}^{-\tau}\,\mathrm{d}\tau$$

Hence

$$I(\tau_a) = I(\tau_b)e^{-(\tau_b - \tau_a)} + e^{\tau_a}\int_{\tau_a}^{\tau_b} S(\tau)e^{-\tau}\,d\tau \tag{1.4}$$

which is a general solution to the equation of radiative transfer at optical depth τ_a given the intensity at optical depth τ_b as a boundary condition.

In particular, for the emergent intensity from a cloud of total optical depth τ_t we have $\tau_a = 0$, $\tau_b = \tau_t$ and $I(\tau_t) = 0$ if there is no radiation incident on the cloud, so

$$I(0) = \int_0^{\tau_t} S(\tau)e^{-\tau}\,d\tau$$

and if the cloud has a uniform source function, $S(\tau) = \text{constant}$,

$$I(0) = S(1 - e^{-\tau_t}) \tag{1.5}$$

For the emergent intensity from a stellar atmosphere, we have effectively $\tau_b = \infty$, so while $I(\tau_b)$ may be large, $I(\tau_b)\exp(-\tau_b)$ is still zero. In this case we cannot assume that $S(\tau)$ is constant, so

$$I_\nu(0) = \int_0^{\infty} S_\nu(\tau_\nu)e^{-\tau_\nu}\,d\tau_\nu \tag{1.6}$$

where we have added the subscript ν to act as a reminder that all the quantities involved are frequency dependent. $S_\nu(\tau_\nu)$ will rise with τ_ν, but not as fast as an exponential, so that $I_\nu(0) \sim S_\nu(\tau_\nu = 1)$.

The physical depth into a cloud or atmosphere (and hence the temperature, density, etc.) corresponding to a given optical depth in a particular direction will depend on the direction, so $I_\nu(\tau_\nu = 0)$ should be written in full $I_\nu(\tau_\nu = 0, \mu)$, where $\mu = \cos\theta$. It is often convenient to use optical depth measured in (the negative of) the direction of the net flow of radiation, that is radially for a spherically symmetric situation. In that case optical depth in this standard direction, $d\tau$, equals optical depth along a ray making an angle θ with the standard direction, $d\tau'$, times $\cos\theta$, so $d\tau'$ along a ray in the equation of radiative transfer can be replaced by $d\tau/\mu$ (Figure 1.4).

In a spherically symmetric situation, the angle a ray makes with the radial direction is not constant, so μ varies along the ray (see Figure 1.5). Take equation (1.3) with x measured along a ray and transform to (r, θ) as variables:

$$\frac{d}{dx} = \frac{\partial}{\partial r} \times \frac{dr}{dx} + \frac{\partial}{\partial \theta} \times \frac{d\theta}{dx}$$

Notice that $dx\cos\theta = dr$ and $dx\sin\theta = -r\,d\theta$, and replace θ with μ:

$$\frac{\partial}{\partial \theta} = \frac{\partial}{\partial \mu} \times \frac{d\mu}{d\theta}, \qquad \frac{d\mu}{d\theta} = -\sin\theta$$

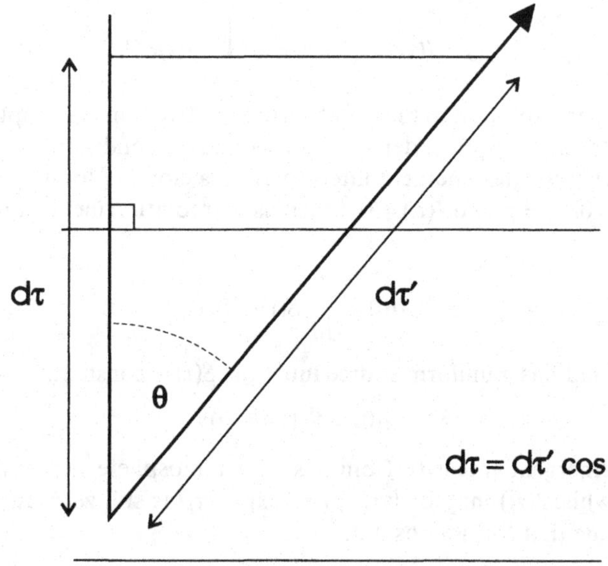

Figure 1.4.

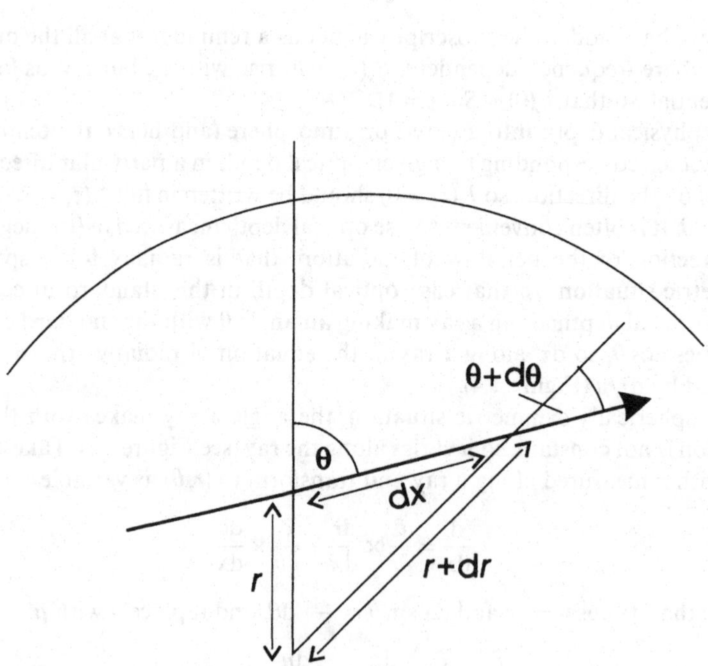

Figure 1.5.

The equation of radiative transfer with $\tau_\nu = \int \kappa_\nu \rho \, dr$ measured along the radial direction then becomes

$$\mu \frac{\partial I_\nu}{\partial \tau_\nu} + \frac{(1-\mu^2)}{\tau_\nu} \frac{\partial I_\nu}{\partial \mu} = I_\nu - S_\nu \tag{1.7}$$

However, many stellar atmospheres approximate to the plane-parallel situation, where the layers of the atmosphere are regarded as being flat and running parallel to each other and perpendicular to the net flow of radiation. The atmosphere refers to the region from which most of the radiation emitted by the star directly leaves. The plane-parallel approximation is clearly valid if the thickness of the atmosphere is small compared with its radius of curvature, effectively the radius of the star. It certainly holds for the Sun where the thickness of the photosphere is of the order of hundreds of kilometres compared with a radius of 700 000 km, but is not adequate in the case of the extended atmospheres of Wolf–Rayet stars which may extend for several stellar radii.

In the plane-parallel approximation, the equation of radiative transfer becomes

$$\mu \frac{dI_\nu}{d\tau_\nu} = I_\nu - S_\nu \tag{1.8}$$

with τ_ν measured along the perpendicular to the atmosphere. The solution for the emergent intensity (1.6) becomes

$$I_\nu(0) = \int_0^\infty S_\nu(\tau_\nu) e^{-\tau_\nu/\mu} \frac{d\tau_\nu}{\mu} \tag{1.9}$$

with similar modifications to the other equations.

This leaves the question of the value of the source function still to be discussed. In local thermodynamic equilibrium (LTE) the source function reduces to the Planck function B_ν, which is itself a function of frequency and local temperature only, so we can write $S_\nu = B_\nu(T)$. In general, however, the source function also depends on the density and on the radiation field. Estimation of the effect of the radiation field requires calculation of the mean intensity J_ν where

$$J_\nu(\tau_\nu) = \frac{1}{2} \int_{-1}^{+1} I_\nu(\mu) \, d\mu$$

$$= \frac{1}{2} \int_0^{+1} I_\nu(\mu) d\mu + \frac{1}{2} \int_{-1}^{0} I_\nu(\mu) d\mu$$

where we have split the integral into inward and outward flowing halves. Note that in the second integral, '0' is approached from the negative side.

This equation is now to be manipulated into a directly calculable form. First note that from equation (1.4) with the optical depth measured along the direction of net flow and hence with τ replaced by τ/μ and using t as a running optical depth variable we have

$$I(\tau,\mu) = e^{\tau/\mu}\int_{\tau}^{\infty} S(t)e^{-t/\mu}\frac{dt}{\mu} \quad (\mu > 0)$$

$$I(\tau,\mu) = e^{\tau/\mu}\int_{\tau}^{0} S(t)e^{-t/\mu}\frac{dt}{\mu} \quad (\mu < 0)$$

We can take the term $\exp(\tau/\mu)$ inside both integrals so the integrand becomes $S(t)\exp(-[t-\tau]/\mu)$, and then in the second integral change the signs of both μ and $(t-\tau)$ and the order of the limits so the second integral becomes

$$I(\tau,\mu) = \int_{0}^{\tau} S(t)e^{-(\tau-t)/-\mu}\frac{dt}{-\mu} \quad (\mu < 0)$$

We now substitute the two expressions for I into the equation for J, change the order of integration over μ and t, and assume the source function is independent of μ (which is usually true but not for the case of electron scattering, for instance) and hence can be taken out of the integration over μ:

$$J(\tau) = \frac{1}{2}\left[\int_{\tau}^{\infty} S(t)\int_{0}^{1} e^{-(t-\tau)/\mu}\frac{d\mu}{\mu}dt + \int_{0}^{\tau} S(t)\int_{-1}^{-0} e^{-(\tau-t)/-\mu}\frac{d\mu}{-\mu}dt\right]$$

Finally in the first half of the integral we substitute $\mu = 1/x$ and change the order of the limits, cancelling a minus sign in $dx = -x^2 d\mu$, while in the second half of the integral we substitute $\mu = -1/x$. Then

$$\begin{aligned} J(\tau) &= \frac{1}{2}\left[\int_{\tau}^{\infty} S(t)\int_{1}^{\infty} e^{-x(t-\tau)}\frac{dx}{x}dt + \int_{0}^{\tau} S(t)\int_{1}^{\infty} e^{-x(\tau-t)}\frac{dx}{x}dt\right] \\ &= \frac{1}{2}\left[\int_{0}^{\infty} S(t)\int_{1}^{\infty} e^{-x|t-\tau|}\frac{dx}{x}dt\right] \\ &= \frac{1}{2}\int_{0}^{\infty} S(t)E_1(|t-\tau|)dt \end{aligned} \quad (1.10)$$

Here E_1 is called an *exponential integral.* Exponential integrals E_n are defined by

$$E_n(z) = \int_{1}^{\infty} \frac{e^{-yz}}{y^n}dy$$

and have the properties that $E_n(0) = 1/(n-1)$, $E_n(z \to \infty) = \exp(-z)/z \to 0$ and

$$\frac{dE_n}{dz} = -E_{n-1}, \qquad \int E_n(z) = -E_{n+1}(z)$$

The flux $= 4\pi H = 2\pi\int I(\mu)\mu d\mu$ and can be evaluated directly from the source function using a similar process to the one just applied to the mean intensity to give:

$$H(\tau) = \frac{1}{2}\left[\int_{\tau}^{\infty} S(t)E_2(t-\tau)\mathrm{d}t - \int_{0}^{\tau} S(t)E_2(\tau - t)\,\mathrm{d}t\right] \tag{1.11}$$

The observed flux is related to the flux at the surface, which is given by

$$H(0) = \frac{1}{2}\int_{0}^{\infty} S(t)E_2(t)\,\mathrm{d}t \tag{1.12}$$

The Observer's Point of View

We must distinguish between the intensity or flux emergent from a stellar atmosphere or interstellar cloud (recipes for the calculation of which we discussed in the previous section) and the intensity or flux received by our detectors. A crucial question here is whether we can resolve (in angle) the astronomical source concerned. In optical observations, resolution is normally limited by the Earth's atmosphere (by 'seeing'), which under good conditions allows one arc second to be resolved. For millimetre and radio observations one is normally limited by the telescope system and one speaks of the 'beam' of the telescope which represents its resolving ability (roughly the beam such a telescope would send out if it were transmitting rather than receiving). It must be remarked that the beam is usually more complicated than a simple cone with the system 'on' inside the cone and 'off' outside the cone, and will have side lobes. The situation is further complicated by the fact that many observations are made by chopping between object and sky. These important complications will be ignored here but are crucial in the interpretation of the raw data. A source at long wavelengths is said to be resolved if its real angular size is larger than the telescope beam.

If the source can be resolved, then one can speak of the energy received from a part of the source near its western edge, say, as opposed to the energy received from a part of the source near its centre and so on. In other words, the *intensity* received, $I'(\mu')$, can be measured as a function of direction ($\mu' = \cos\theta'$), where θ' is the angle between a point on the source and the centre of the source as seen by the observer (Figure 1.6). Of course, even extended sources subtend very small angles at the observer, so μ' will always be close to 1.0. The energy received by unit area from a small area of the source perpendicular to the line of sight, ΔS, is $\Delta E = I'(\mu')$ $\Delta\Omega' = I'\Delta S/d^2$, where d is the distance to the source, $\Delta\Omega'$ is the solid angle subtended by ΔS at the observer and we have used the relation: solid angle = perpendicular area over (distance)^2. The energy emitted by the area ΔS of the source into a solid angle $\Delta\Omega$ in the direction of the observer and therefore in a direction making an angle θ with the direction of net flow (with $\cos\theta = \mu$) is $I(\mu)\Delta S\Delta\Omega$, and if we choose the solid angle to be that subtended by unit area at the observer, so $\Delta\Omega = 1/d^2(\mu' \sim 1)$, the energy emitted is $I\Delta S/d^2$. Hence, equating energy emitted with energy received, $I(\mu) = I'(\mu')$.

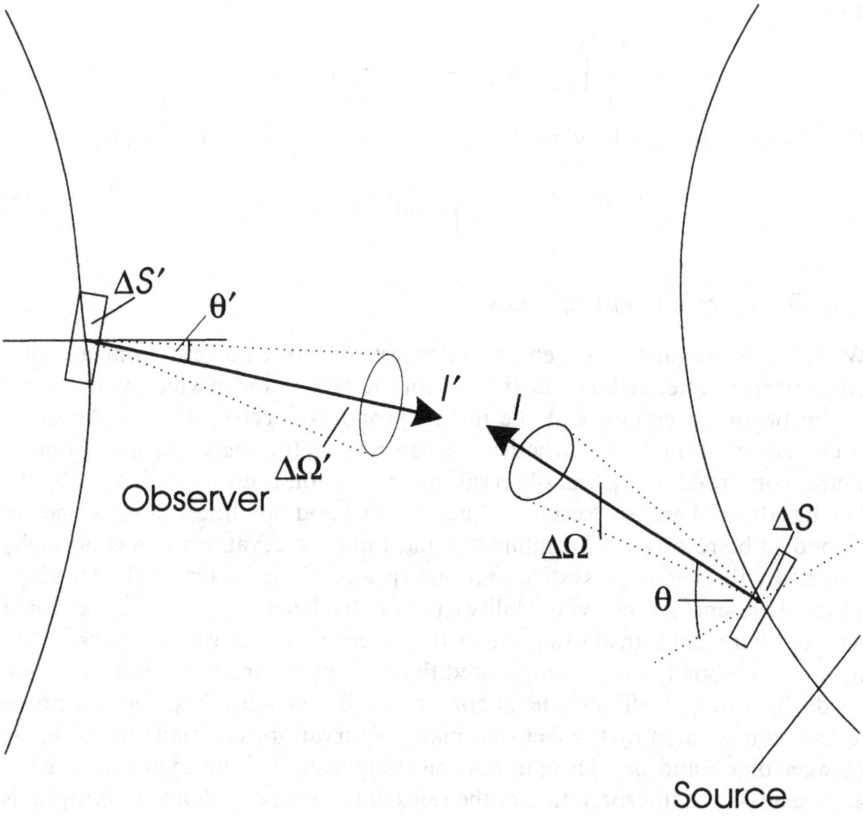

Figure 1.6.

This result appears to be independent of distance, but this only holds if the source is uniformly bright over ΔS, and since in practice $\Delta\Omega'$ is fixed by the resolution of the receiver, ΔS will get physically larger as a given source is placed farther away from the observer and the average intensity and surface brightness-change. Consider, for instance, the observed surface brightness of galaxies (usually given with energy received in the form of apparent magnitude per square arc second). Galaxies have an intrinsic surface brightness profile (say in the form of energy emitted per square parsec) that decreases steeply as one moves away from the centre, and so their surface brightness falls as the distance is increased, although not as fast as an inverse square law. Stars are normally unresolved, but of course in the case of our Sun we can measure the intensity received from different parts of the Sun's disc $I(\mu')$, the variation being called the limb-darkening. Simple geometry relates μ' to μ at the Sun's surface (Figure 1.7) and hence enables us to determine $I(\mu)$. Equation (1.6) shows that for a linear source function $S(\tau) = a + b\tau$

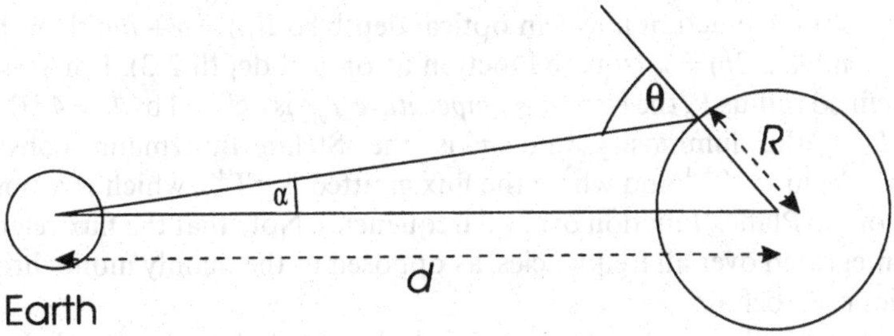

Figure 1.7. α is the angular distance of the point observed from the centre of the Sun as seen from the Earth (not to scale)

the variation of the emergent intensity should be given by $I(\mu) = a + b\mu$, so if a linear source function is assumed, the observed limb-darkening enables a and b, and hence the variation of source function with optical depth, to be found. It is straightforward to extend this analysis to more complicated depth dependencies of the source function. The absorption coefficient and the optical depth are functions of frequency, so the coefficients a and b and the limb-darkening curve also vary with frequency, although the integrated source function at a given physical depth is, of course, independent of frequency.

For an unresolved source, we can only measure the flux received, namely $\int I'(\mu')\mu' \mathrm{d}\Omega$, over the apparent source with $\mu' \sim 1.0$. Now $\int \mathrm{d}\Omega'$ = the solid angle subtended by the source, and will be proportional to $1/d^2$ unless d is very small. Hence the flux will be proportional to the inverse square of the distance to the source. Consider a spherical source with a fairly well defined surface of radius R (this will be true of most stars) and let the total luminosity of the source be L. Then the energy received by unit area at distance d is $L/(4\pi d^2)$, where $4\pi d^2$ is the surface area of an imaginary sphere of radius d through which L must flow every second. The surface area of the star is $4\pi R^2$ and if unit area emits flux F, then $L = 4\pi R^2 F$. Hence:

$$\text{flux received} = (\text{flux emitted})R^2/d^2$$

The procedure in predicting what will be observed from a given atmosphere is normally to calculate the intensity at the surface $I(\mu)$, then calculate the flux emitted:

$$F_{\text{out}} = 2\pi \int_0^{+1} I(\mu)\mu \,\mathrm{d}u$$

since we are dealing with a purely outward flow at the top of the atmosphere, and finally to estimate the flux received. Equation (1.12) can alternatively be used to find the flux emitted directly. If the star emits an intensity I_0 independent of direction but purely outwards, then the flux emitted $= \pi I_0$. More realistically, if we have a source function linear in optical depth so $I(\mu) = a + b\mu$, then the flux emitted $= \pi(a + 2/3b) = \pi$ (source function at optical depth 2/3). For stars with a well-defined radius R, the *effective temperature* T_{eff} is defined by $L = 4\pi R^2 \sigma T_{\text{eff}}^4$, where L is the luminosity and σ is the Stefan–Boltzmann constant $=$ $5.67 * 10^{-8}\ \text{W m}^{-2}\ \text{K}^{-4}$ and where the flux emitted is σT_{eff}^4, which is π times the integral of the Planck function over all frequencies. Note that the flux referred to here is integrated over all frequencies, as opposed to the mainly monochromatic quantities used before.

Now consider the case of an unresolved object that has an extended atmosphere with no definite 'surface'—examples include stellar winds. We want to estimate the flux received. Take rays parallel to the line joining the observer to the source, let z measure distance along a ray with $z = 0$ at the point of closest approach to the source, and let p be the distance to the centre of the source at the point where the ray makes its closest approach to the centre (p is the impact parameter in collision terminology). Assume the source is spherically symmetric, and divide the circular area seen by the observer (although in reality the observer cannot resolve the source!) into annuli of radius p, thickness $\mathrm{d}p$ and area $2\pi p \mathrm{d}p$ (Figure 1.8). Let the emergent intensity for a ray with distance of closest approach p be $I(p)$, so the energy emitted into the solid angle $\Delta\Omega$ in the direction of the observer by the corresponding annulus is $I(p)\, 2\pi p \mathrm{d}p \Delta\Omega$. The energy received by unit area at the observer's position from the annulus is given by putting $\Delta\Omega = 1/d^2$, and since μ' is nearly 1.0 for all rays at the observer's position, the flux received is found by summing over the source:

$$F_{\text{obs}} = \frac{1}{d^2} \int_0^{\infty} I(p) 2\pi p \mathrm{d}p \tag{1.13}$$

$$I(p) = \int_{-\infty}^{+\infty} S(z, p = \text{constant})\, \mathrm{e}^{-\tau(z)} \mathrm{d}\tau(z) \tag{1.14}$$

where the contributions from large p and z will tend to zero. The source function and the absorption coefficient will be functions of radial distance r from the centre of the source rather than of z and p, but looking at the system observer–source side on we see that $r^2 = z^2 + p^2$ so that $z\,\mathrm{d}z = r\,\mathrm{d}r$ at constant p and the integral in (1.14) can be transformed into one over r, with r going from p to infinity and the value of the integral doubled.

Intensity observations at radio and millimetre wavelengths are usually quoted as temperatures, namely the *brightness temperatures*. The brightness temperature is defined as the temperature that the source would have to have in order to give the observed temperature, *if* it radiated like a blackbody, i.e if it radiated according to Planck's law. At long wavelengths Planck's law is well approximated by the Rayleigh–Jeans approximation $I_\nu = (2kT)/c^2 \cdot \nu^2$, valid if $(h\nu)/(kT) \ll 1$ so

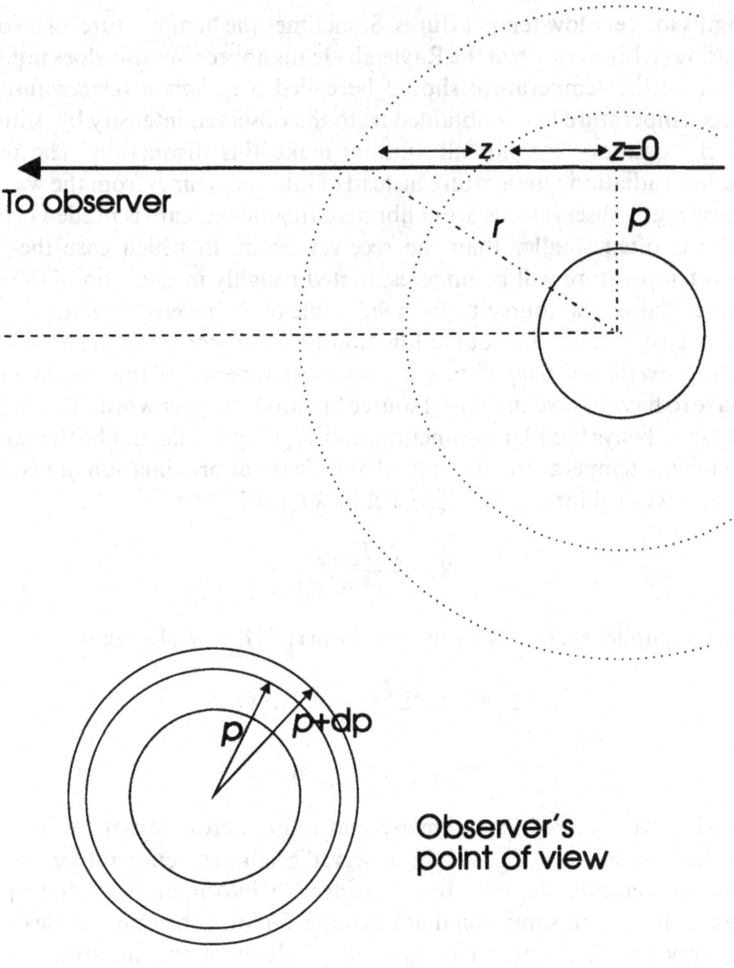

Figure 1.8.

the brightness temperature becomes

$$T_B = \frac{c^2}{2k} \frac{I_\nu(\text{observed})}{\nu^2}, \qquad \frac{h\nu}{kT} \ll 1 \tag{1.15}$$

irrespective of whether the source is a blackbody or not. Thus the brightness temperature is simply an alternative way of expressing a measurement of received intensity. The Rayleigh–Jeans approximation holds well at centimetre wavelengths and longer, but may not be a good approximation at millimetre wavelengths for very low temperatures. Sometimes the 'temperature' of a source is given using (1.15), even when the Rayleigh–Jeans approximation does not hold, in which case the temperature should be called a *radiation temperature*, the brightness temperature being obtained from the observed intensity by using the full Planck equation, but not all authors make this distinction. The use of brightness or radiation temperature instead of intensity comes from the way that radio wavelength observations are calibrated. In practice, care is needed because the source is often smaller than the receiver beam, in which case the 'true' brightness temperature will be underestimated roughly in the ratio of the solid angle subtended by the source to the solid angle of the receiver beam.

One can also replace the source function by a temperature, in this case by defining the *excitation temperature* T_{ex} as the temperature that a blackbody would have to have to give the actual source function, in other words $S_\nu = B_\nu(T_{ex})$, where B_ν is as always the Planck function, and T_{ex} may or may not be the same as the local kinetic temperature. If the Rayleigh–Jeans approximation holds at the frequency concerned for a given T_{ex}, i.e. if $h\nu/kT_{ex} \ll 1$, then

$$S_\nu = \frac{2kT_{ex}}{c^2}\nu^2$$

and since for a uniform cloud we can write from (1.5) $I_\nu = S_\nu(1 - \exp(-\tau_\nu))$, then

$$\frac{2kT_B}{c^2}\nu^2 = \frac{2kT_{ex}}{c^2}\nu^2(1 - e^{-\tau_\nu})$$

$$T_B = T_{ex}(1 - e^{-\tau_\nu}) \tag{1.16}$$

an expression widely used in millimetre and radio astronomy. If the source is optically thick, $\tau_\nu \gg 1$, $T_B = T_{ex}$ and the observed brightness temperature gives the excitation temperature directly, but provides no information on the optical depth, except for the presumption that it is large. On the other hand if the optical depth is very small, $\tau_\nu \ll 1$, then $T_B = \tau_\nu T_{ex}$. Much of the interpretation of millimetre and radio observations is concerned with deciding whether the source is optically thick or not. If $h\nu/(kT_{ex})$ is not much less than 1, we must write

$$T_B = \frac{h\nu}{k}\frac{1 - e^{-\tau_\nu}}{e^{h\nu/kT_{ex}} - 1} \tag{1.17}$$

Introduction to Model Atmospheres

As has already been indicated, a model atmosphere gives the variation of kinetic temperature with optical depth in the layers of a star from which we receive radiation and which we call the stellar atmosphere. The optical depth is the integral of the absorption coefficient with respect to geometrical depth, and so if the absorption coefficient varies with frequency, the optical depth corresponding to a given layer will also depend on frequency. Hence one normally tabulates model atmospheres against optical depth τ_0 at some standard frequency ν_0—for instance, the frequency equivalent to a wavelength of 500 nm in the middle of the visible spectrum is often chosen. The optical depth at any other frequency ν can be related to τ_0 by

$$\tau_\nu = \int_0^{\tau_0} \frac{\kappa_\nu}{\kappa_0} d\tau_0 \tag{1.18}$$

where κ_0 is the absorption coefficient at the standard frequency.

However, the absorption coefficient depends on the density and on the degree of ionization of the gas, and hence it is desirable for the model atmosphere also to tabulate related quantities. Usually the total gas pressure P_g and the electron partial pressure P_e are chosen. The gas pressure is related to the density ρ at a given temperature by the perfect gas law:

$$\begin{aligned} P_g &= NkT \\ &= \frac{\rho}{m_{av}} kT \end{aligned} \tag{1.19}$$

where N is the total number of particles per unit volume, k is (as always) Boltzmann's constant and m_{av} is the average mass of a particle. The average mass of a particle is

$$m_{av} = \frac{\sum_i m_i N_i}{N} = \mu m_H$$

where the sum is over all species (electrons, atoms, ions and molecules) of mass m_i and number density N_i in the atmosphere, m_H is the mass of a hydrogen atom and μ is the mean molecular weight (unfortunately the same symbol is customarily used for $\cos\theta$). For instance, a standard mixture of 90% hydrogen and 10% helium by number with negligible ionization would have $m_{av} = (9m_H + [1*4m_H])/10 = 1.3m_H$, since the mass of a helium atom is four times that of hydrogen. The electron pressure, which contributes to P_g, is given by $P_e = N_e kT$, where N_e is the number of free electrons per unit volume, and is a measure of the degree of ionization since free electrons come entirely from the ionization of atoms. A pure hydrogen atmosphere that was completely ionized would have an electron for each original hydrogen atom, so $P_g = P_e + P(H_{ion}) = 2P_e$.

Thus a model atmosphere consists of the tabulation of T, P_g and P_e against τ_0. The range of optical depths involved typically runs from 0 to 10—essentially no radiation escapes from optical depths greater than 10 at any wavelength. There is, of course, no edge to a star, but in most cases the density falls off fast enough to give a fairly narrow sequence of layers from which most of the radiation arises. Many stars have a temperature rise in their outermost layers where $\tau_0 \ll 1$. The regions where this happens are called the *chromosphere*, where the rise starts, and the *corona*, where very high temperatures are reached. These layers cannot be treated by normal model atmosphere techniques since there is a non-radiative energy input from acoustic and/or magnetic sources, but they have very low densities and are virtually transparent to the bulk of the radiation leaving the star. Thus they contribute very little to the star's overall spectrum except perhaps at very short wavelengths. The discussion which follows concerns the deeper layers where the temperature falls outwards, which are called the *photosphere.*

We need to specify particular models, which will depend on the luminosity L, mass M, radius R and the composition of the parent star. As has already been pointed out, the temperature–optical depth relation is determined by the flux of radiation to be carried, and the opacity (absorption) that opposes that flow—the higher the opacity, the greater the temperature gradient needed to drive a given flux out. The opacity will depend on the temperature, composition and pressure, and the pressure in turn will depend on the surface gravity of the star. The surface gravity is given by $g = GM/R^2$, where G is the gravitational constant. In practice, the masses of stars vary over a fairly small range from 1/20 solar mass to 100 solar masses, but vary over an enormous range in radius, so g is mainly determined by the radius. The flux is given by σT_{eff}^4, as has already been pointed out. Composition is usually specified by the mass fractions of hydrogen, X, helium, Y, and the other elements, often called 'metals', Z. This is because the abundances of the elements heavier than helium vary together in most cases, keeping the same proportions, although it is the exceptions to this rule that provide much of the interest in stellar atmosphere work! The misleading term 'metal' comes from the fact that iron lines can easily be detected in the spectra of the stars of most temperatures and so very often it is only the iron abundance that is measured as a representative of the elements heavier than helium. The abundances involved here are the surface abundances of the star and these abundances, almost always on the main sequence and often in other cases, represent the original composition of the star. It should be noted that measured abundances are quoted in the form of the number of atoms of the element relative to hydrogen, whereas stellar evolution models use the mass fraction of an element. The mass fraction of element i is

$$X_i = \frac{\dfrac{N_i}{N_{\text{H}}} m_{\text{H}}}{\sum_j \dfrac{N_j}{N_{\text{H}}} m_j}$$

The hydrogen and helium abundances are little different from solar values in most cases. Hence we normally specify model atmospheres by T_{eff}, g and composition, with composition usually represented by the metallicity.

Model Atmospheres—Finding the T–τ Relation

We want to find the temperature gradient that will drive the required flux through the atmosphere against the opposition of the opacity. The absorption coefficient varies with frequency, so a lack of transparency at one wavelength may be compensated for at another wavelength. However, the distribution of the flux to be transferred with respect to frequency will bear some general resemblance to a Planck function at the effective temperature. Thus for a solar type star with $T_{eff} = 5700\,K$ and a temperature range through the photosphere of 8000 to 4000 K, the absorption coefficient in the far ultraviolet will have little effect on the temperature structure of the photosphere since little energy is transported at these wavelengths.

The model atmosphere must satisfy energy conservation, and in the absence of mechanical or magnetic inputs this is expressed by the equation of *radiative equilibrium*:

emission per unit volume summed over frequency = absorption per unit volume summed over frequency

$$\int_0^\infty j_\nu \rho \, d\nu = \int_0^\infty \kappa_\nu \rho J_\nu \, d\nu \tag{1.20}$$

where of course radiation is absorbed from all directions and so the requisite radiation quantity is the mean intensity J_ν and not I_ν.

The equation of radiative transfer must now be combined with the equation of radiative equilibrium. We start with the plane-parallel case (1.5):

$$\mu \frac{dI_\nu}{d\tau_\nu} = I_\nu - S_\nu$$

Multiply both sides by 1/2 and integrate over μ:

$$\frac{1}{2}\int_{-1}^{+1} \mu \frac{dI_\nu}{d\tau_\nu} d\mu = \frac{1}{2}\int_{-1}^{+1} I_\nu \, d\mu - \frac{1}{2}\int_{-1}^{+1} S_\nu \, d\mu$$

then invert the order of integration and differentiation on the left-hand side:

$$\frac{d}{d\tau_\nu}\left(\frac{1}{2}\int_{-1}^{+1} I_\nu \mu \, d\mu\right) = \frac{1}{2}\int_{-1}^{+1} I_\nu \, d\mu - \tfrac{1}{2} S_\nu \int_{-1}^{+1} d\mu$$

where we have assumed that S_ν is independent of direction, which is often true but does not hold for scattering. Then

$$\frac{dH_\nu}{d\tau_\nu} = J_\nu - S_\nu \tag{1.21}$$

where the definition of H (1.1) has been used.

Now (1.20) can be written

$$\int_0^\infty (\kappa_\nu J_\nu - j_\nu)\rho \, \mathrm{d}\nu = 0$$

or, dividing and multiplying throughout by κ_ν and remembering that $j/\kappa = S$:

$$\int_0^\infty \kappa_\nu \rho (J_\nu - S_\nu) \mathrm{d}\nu = 0 \tag{1.22}$$

From (1.21), multiplying through by $\kappa_\nu \rho$:

$$\frac{\mathrm{d}H_\nu}{\mathrm{d}x} = \kappa_\nu \rho (J_\nu - S_\nu)$$

$$\int_0^\infty \frac{\mathrm{d}H_\nu}{\mathrm{d}x} \mathrm{d}\nu = \int_0^\infty \kappa_\nu \rho (J_\nu - S_\nu) \mathrm{d}\nu = 0$$

so

$$\frac{\mathrm{d}H}{\mathrm{d}x} = 0$$

where

$$H = \int H_\nu \, \mathrm{d}\nu$$

Hence H = constant. Remembering that flux $= 4\pi H$, we find that for a plane-parallel atmosphere:

$$\text{flux} = \text{constant}$$

is the equivalent of radiative equilibrium, as indeed is obvious without all the preceding manipulation. The condition of flux constancy, i.e. that H is independent of τ, is the fundamental one that we will use in determining the T–τ relation.

The same procedure can be applied in the spherically symmetric, non-plane-parallel case. Integrating (1.7) over μ and multiplying by 1/2, we obtain:

$$\frac{\mathrm{d}H_\nu}{\mathrm{d}\tau_\nu} + \frac{2H_\nu}{\tau_\nu} = J_\nu - S_\nu \tag{1.23}$$

where

$$\tau_\nu = \int \kappa_\nu \rho \, \mathrm{d}r$$

and the term

$$\frac{1}{2\tau} \int_{-1}^{+1} (1 - \mu^2) \frac{\partial I_\nu}{\partial \mu} \mathrm{d}\mu$$

has been integrated by parts. Then

$$\frac{1}{r^2} \frac{\mathrm{d}(r^2 H_\nu)}{\mathrm{d}\tau_{..}} = J_\nu - S_\nu$$

so

$$\frac{\mathrm{d}(r^2 H_\nu)}{\mathrm{d}r} = r^2 \kappa_\nu \rho (J_\nu - S_\nu)$$

and integrating over frequency:

$$\frac{\mathrm{d}(r^2 H)}{\mathrm{d}r} = r^2 \int \kappa_\nu \rho (J_\nu - S_\nu) \mathrm{d}\nu = 0$$

where (1.22) has been used. Thus $H = \text{constant}/r^2$, and the flux integrated over frequency obeys an inverse square law for the spherically symmetric case in the absence of sources or sinks of energy, again a fairly obvious result from general considerations.

We now return to the plane-parallel case. Take (1.5) again but this time multiply both sides by $1/2\mu$ and integrate over μ:

$$\frac{1}{2}\int_{-1}^{+1} \mu^2 \frac{\mathrm{d}I_\nu}{\mathrm{d}\tau_\nu} \mathrm{d}\tau_\nu = \frac{1}{2}\int_{-1}^{+1} I_\nu \mu \,\mathrm{d}\mu - \frac{1}{2}\int_{-1}^{+1} S_\nu \mu \,\mathrm{d}\mu$$

Invert the order of integration and differentiation on the left-hand side and assume S_ν is independent of μ to obtain:

$$\frac{\mathrm{d}K_\nu}{\mathrm{d}\tau_\nu} = H_\nu \tag{1.24}$$

where the definition of K has been used (1.2). For the spherically symmetric case, one similarly obtains, after integration by parts of the second term on the left-hand side:

$$\frac{\mathrm{d}K_\nu}{\mathrm{d}\tau_\nu} + \frac{(3K_\nu - J_\nu)}{\tau_\nu} = H_\nu \tag{1.25}$$

We now concentrate on the plane-parallel case. Equation (1.24) cannot be solved analytically, and an iterative procedure must be followed. Deep in the atmosphere, the intensity is nearly isotropic, with the net outward flow a small perturbation on a radiation field that is nearly the same in all directions. I_ν can then be expanded in a power series in μ:

$$I_\nu = I_{\nu 0} + I_{\nu 1}\mu + I_{\nu 2}\mu^2 + \cdots$$

so

$$J_\nu = I_{\nu 0} + 1/3 I_{\nu 2} + \cdots$$
$$H_\nu = 1/3 I_{\nu 1} + \cdots$$
$$K_\nu = 1/3 I_{\nu 0} + 1/5 I_{\nu 2} + \cdots$$

and to first order

$$K_\nu = 1/3 J_\nu \tag{1.26}$$

which is called the *Eddington approximation* and holds well in a stellar interior, but poorly at the top of a stellar atmosphere where the flow is far from isotropic and indeed is almost entirely outwards. The Eddington approximation with (1.24) gives:

$$\frac{dJ_\nu}{d\tau_\nu} = 3H_\nu \qquad (1.27)$$

We now also assume that κ_ν is independent of frequency, the *grey approximation*. This is certainly not correct for real stellar atmospheres, but it does enable us to obtain an analytical solution. Remove the frequency subscript from τ_ν and integrate both sides of (1.27) with respect to frequency:

$$\frac{d}{d\tau}\int J_\nu d\nu = 3\int H_\nu d\nu$$

$$\frac{dJ}{d\tau} = 3H$$

Since H = constant,

$$J = 3H\tau + \text{constant}$$

or

$$J = 3H(\tau + C) \qquad (1.28)$$

For the grey case, (1.22) becomes $\kappa\rho\int(J_\nu - S_\nu)d\nu = 0$ or $J = S$. If LTE holds, the source function is equal to the blackbody function. Equation (1.28) can then be written:

$$J = S = B = 3H(\tau + C)$$

where $B = \int B_\nu d\nu = \sigma T^4/\pi$. Noting that $H = (\text{flux})/(4\pi) = \sigma T_{eff}^4/(4\pi)$, (1.28) finally becomes:

$$T^4 = 3/4T_{eff}^4(\tau + C) \qquad (1.29)$$

The constant C can be determined from the boundary conditions. If the radiation field at $\tau = 0$ is isotropic, then $J(0) = 2H(0)$ and $C = 2/3$.

The equation $T^4 = (3/4)T_{eff}^4(\tau + 2/3)$ gives a boundary temperature of 0.841 of the effective temperature, but this has used the Eddington approximation and an approximate boundary condition. The limitations of the near-isotropy assumptions can be removed by using an iterative approach. The grey assumption with (1.10) gives

$$S(\tau) = J(\tau) = \int_0^\infty S(\tau)E_1|t - \tau|dt$$

which is an exact equation, and if we insert our first Eddington approximation solution for S inside the integral as $S(t)$, we should obtain a better estimate of $S(\tau)$.

The difference will be appreciable in the outer layers where isotropy is a very poor approximation. In fact the exact boundary temperature for a grey atmosphere comes out very close to 0.811 times the effective temperature.

The grey assumption is less easily removed. One apparently promising approach is to use a *mean* opacity. We start by combining (1.27) with the Eddington approximation:

$$\frac{\mathrm{d}J_\nu}{\mathrm{d}\tau_\nu} = 3H_\nu$$

or

$$\frac{1}{\rho}\int_0^\infty \frac{1}{\kappa_\nu}\frac{\mathrm{d}J_\nu}{\mathrm{d}x}\mathrm{d}\nu = 3\int H_\nu \mathrm{d}\nu$$

Define a mean opacity κ_R by

$$\frac{1}{\kappa_R}\int_0^\infty \frac{\mathrm{d}J_\nu}{\mathrm{d}x}\mathrm{d}\nu = \int_0^\infty \frac{1}{\kappa_\nu}\frac{\mathrm{d}J_\nu}{\mathrm{d}x}\mathrm{d}\nu$$

so

$$\frac{1}{\rho\kappa_R}\int_0^\infty \frac{\mathrm{d}J_\nu}{\mathrm{d}x}\mathrm{d}\nu = \frac{\mathrm{d}J}{\mathrm{d}\tau_R} = 3H$$

where

$$\mathrm{d}\tau_R = \kappa_R \rho \mathrm{d}x$$

and

$$J = 3H\tau_R + \text{constant}$$

This is the same solution as for the grey case, but expressed in terms of mean optical depth. This can easily be converted to real optical depth at the standard wavelength:

$$\tau_0 = \int_0^\infty \frac{\kappa_0}{\kappa_R}\mathrm{d}\tau_R$$

Finally we need to evaluate the mean opacity. We assume that $J_\nu = S_\nu$ at all frequencies and that $S_\nu = B_\nu(T)$, the Planck function. Writing $\mathrm{d}B_\nu/\mathrm{d}x = \mathrm{d}B_\nu/\mathrm{d}T \cdot \mathrm{d}T/\mathrm{d}x$ and noting that $\mathrm{d}T/\mathrm{d}x$ is independent of frequency and can be taken outside the integrals over frequency, we find

$$\frac{1}{\kappa_R} = \frac{\int_0^\infty \frac{1}{\kappa_\nu}\frac{\mathrm{d}B_\nu}{\mathrm{d}T}\mathrm{d}\nu}{\int_0^\infty \frac{\mathrm{d}B_\nu}{\mathrm{d}T}\mathrm{d}\nu} = \frac{\pi}{4\sigma T^3}\int_0^\infty \frac{1}{\kappa_\nu}\frac{\mathrm{d}B_\nu(T)}{\mathrm{d}T}\mathrm{d}\nu \qquad (1.30)$$

κ_R is called the *Rosseland mean opacity* and is used extensively in stellar work. Unfortunately the assumption that $J_\nu = S_\nu$ is a poor one in the outer layers of a star unless the degree of non-greyness is small. Hence the Rosseland mean opacity is only useful in stellar atmosphere work as a starting point for iterative procedures.

We will illustrate one such iterative procedure, namely that of Unsöld and Lucy. Once again we start with (1.27) and the Eddington approximation:

$$\frac{dJ_\nu}{d\tau_\nu} = 3H_\nu$$

integrating

$$\int_0^\infty \frac{dJ_\nu}{\rho dx} d\nu = 3 \int_0^\infty \kappa_\nu H_\nu d\nu$$

so

$$\frac{dJ}{\rho dx} = 3\kappa_F H$$

where

$$\kappa_F = \text{flux mean opacity} = \frac{\int_0^\infty \kappa_\nu H_\nu d\nu}{\int_0^\infty H_\nu d\nu}$$

If we define a Planck mean opacity κ_P and a corresponding optical depth $d\tau_P = \kappa_P \rho dx$ by

$$\kappa_P = \frac{\int_0^\infty \kappa_\nu B_\nu d\nu}{\int_0^\infty B_\nu d\nu}$$

Then

$$\frac{dJ}{d\tau_P} = \frac{3\kappa_F}{\kappa_P} H$$

$$J = \int_0^\infty \frac{3\kappa_F}{\kappa_P} H d\tau_p + 2H(0) \tag{1.31}$$

using the 'isotropic outwards at the surface' boundary condition.

Now use (1.22) with the LTE assumption that $S_\nu = B_\nu$:

$$\frac{dH_\nu}{d\tau_\nu} = J_\nu - S_\nu = J_\nu - B_\nu$$

Multiply by κ_ν and integrate over frequency:

$$\int_0^\infty \frac{dH_\nu}{\rho dx} d\nu = \int_0^\infty \kappa_\nu J_\nu d\nu - \int_0^\infty \kappa_\nu B_\nu d\nu$$

$$\frac{dH}{\rho dx} = \kappa_J J - \kappa_P B$$

where we define

$$\kappa_J = \frac{\int_0^\infty \kappa_\nu J_\nu \, d\nu}{\int_0^\infty J_\nu \, d\nu}$$

Hence dividing by κ_P:

$$\frac{dH}{d\tau_P} = \frac{\kappa_J}{\kappa_P} J - B$$

$$B = \frac{\kappa_J}{\kappa_P} J - \frac{dH}{d\tau_P}$$

$$= \frac{\kappa_J}{\kappa_P}\left[\int_0^\infty \frac{3\kappa_F}{\kappa_P} H \, d\tau_P + 2H(0)\right] - \frac{dH}{d\tau_P} \tag{1.32}$$

using (1.31).

Suppose now an approximate solution $B'(\tau)$ predicts a flux H' that varies with optical depth, but the true solution $B(\tau)$ gives a constant flux H. Write $H'(\tau) = H + \Delta\tau$ and $B'(\tau) = B(\tau) - \Delta B(\tau)$, where ΔH is the flux error and ΔB is the required correction to the source function. Then subtracting (1.32) for B from (1.32) for B':

$$-\Delta B = \frac{\kappa_J}{\kappa_P}\left[3\int_0^\infty \frac{\kappa_F}{\kappa_P} \Delta H \, d\tau_P + 2\Delta H(0)\right] - \frac{d(\Delta H)}{d\tau_P} \tag{1.33}$$

A given estimate of $T(\tau)$ enables $S_\nu = B_\nu(T)$ to be calculated and from B_ν one obtains κ_P. Equation (1.10) enables $J_\nu(\tau)$ to be found from S_ν and hence for κ_J to be calculated, and (1.31) makes possible the evaluation of $H'(\tau)$ and hence of κ_F and ΔH. Substitution in (1.33) then enables the initial guess at the source function to be corrected. Note that (1.33) uses an exact calculation of the flux error to determine the correction. When a solution is obtained that has no flux error for all τ, then that solution is exactly correct, whatever approximations were made along the way. The point is that (1.33) uses approximations to give a formula that gives corrections of the right sign and roughly the right magnitude, but when convergence is achieved the result should be exact.

Gas Pressure and Electron Pressure

The gas pressure is determined from the equation of hydrostatic equilibrium. Consider an element of gas of unit area between radii r and $r + dr$ from the centre of a star. Let the pressure at r be P and the pressure at $r + dr$ be $P + dP$, so the outward pressure force is $-dP$. The gravitational force on the element is $g * \text{mass} = g\rho dr$, where g is the acceleration due to gravity and ρ is the density.

Then balancing pressure gradient against gravity, $P = -\rho g$ and

$$\frac{dP}{dr} = -\rho g$$

$$= -\frac{Pm_{av}}{kT}g \tag{1.34}$$

where the perfect gas law has been used in the last equation. This is the equation of *hydrostatic equilibrium*. The formal solution is

$$\frac{dP}{P} = -\frac{gm_{av}}{kT}dr$$

$$P(r) = P(r_0)\exp\left(-\frac{1}{k}\int_{r_0}^{r}\frac{m_{av}g}{T}dr\right) \tag{1.35}$$

If the thickness of the atmosphere is small compared with r (the plane-parallel case), then g varies little in the atmosphere, while for an extended atmosphere round a major body (e.g. the solar corona) $g \propto 1/r^2$ approximately. If we can treat T as well as g and m_{av} as being approximately constant over an atmosphere, then

$$P(r) = P(r_0)e^{-(r-r_0)/H} \tag{1.36}$$

where $H = kT/(gm_{av})$ is called the *scale height*.

This approach is not very helpful in a stellar atmosphere where the temperature varies rapidly with depth and depth is more conveniently measured in optical depth terms. Divide both sides of (1.34) by $\kappa_0\rho$, where κ_0 is the absorption coefficient at the standard wavelength. Then

$$\frac{dP}{d\tau_0} = \frac{g}{\kappa_0} \tag{1.37}$$

with solution

$$P(\tau_0) = g\int_0^{\tau_0}\frac{d\tau_0}{\kappa_0(\tau_0)} \tag{1.38}$$

which can be evaluated numerically if κ_0 is known as a function of τ_0. The problem is that the absorption coefficient is a function of temperature and electron pressure, and the electron pressure depends on the temperature and total gas pressure, for high gas pressure will tend to suppress ionization.

If we can approximate κ_0 as a power law in temperature and pressure,

$$\kappa_0 = \kappa_s P^a T^b$$

where κ_s is a constant, then (1.37) becomes

$$\frac{dP}{d\tau_0} = \frac{g}{\kappa_s P^a T^b}$$

and

$$\int_0^P P^a \mathrm{d}P = \frac{g}{\kappa_s}\int_0^\tau T^{-b}\mathrm{d}\tau_0$$

$$P = \left[\frac{(a+1)g}{\kappa_s}\int_0^\tau T^{-b}\mathrm{d}\tau_0\right]^{1/(a+1)} \tag{1.39}$$

which can be solved if we know $T(\tau_0)$. However, a single power law is often not an adequate representation of the variation of the absorption coefficient through an atmosphere, and an iterative approach to solving (1.38) must be adopted, with a first guess at P enabling P_e and κ_0 to be found, which can then in turn be used in (1.38) to produce a better estimate of P.

Radiation can also exert a pressure, for photons carry momentum, and a photon absorbed or scattered will transfer momentum. We will designate this component of the total pressure *radiation pressure* P_r to distinguish it from the gas pressure P. The radiation pressure can be found by noting that the net pressure due to photons is equal to the rate of change of momentum perpendicular to the net flow of photons per unit area. The energy passing through unit area perpendicular to the net flow contributed by photons of intensity I in a solid angle $\mathrm{d}\Omega$, which makes an angle θ with the direction of net flow, is $I\cos\theta\mathrm{d}\Omega$ per second. Now a photon of energy E carries momentum E/c, where c is the speed of light, and a photon will therefore have momentum perpendicular to the specified area of $E\cos\theta/c$. Hence the rate of change of momentum is given by

$$\frac{2}{c}\int I\cos^2\theta\,\mathrm{d}\Omega$$

$$=\frac{4\pi}{c}\int_0^1 I(\mu)\mu^2\,\mathrm{d}\mu = \frac{2\pi}{c}\int_{-1}^{+1} I(\mu)\mu^2\,\mathrm{d}\mu$$

$$=\frac{4\pi}{c}K$$

where the definition (1.2) of the moment K has been used.

Hence the radiation pressure is

$$P_r = \frac{4\pi}{c}\int_0^\infty K_\nu\,\mathrm{d}\nu$$

$$\frac{\mathrm{d}P_r}{\mathrm{d}r} = \frac{4\pi}{c}\int_0^\infty \frac{\mathrm{d}K_\nu}{\mathrm{d}r}\mathrm{d}\nu \tag{1.40}$$

Use of equation (1.24) then shows

$$\frac{\mathrm{d}P_r}{\mathrm{d}r} = \frac{4\pi}{c}\int_0^\infty \kappa_\nu \rho H_\nu\,\mathrm{d}\nu$$

and

$$\frac{dP_r}{d\tau_0} = \frac{4\pi}{c}\int_0^\infty \frac{\kappa_\nu}{\kappa_0} H_\nu \, d\nu \tag{1.41}$$

$$= \frac{4\pi}{c}\int_0^\infty H_\nu \, d\nu, \quad \text{for a grey atmosphere}$$

$$= \frac{\sigma T_{\text{eff}}^4}{c} = \frac{L}{4\pi R^2 c}$$

One can see that (1.41) makes intuitive sense, for one would expect the radiation pressure to depend on the flux blocked, that is on the product of the flux and the absorption coefficient.

In hydrostatic equilibrium:

$$\frac{dP_r}{d\tau_0} + \frac{dP}{d\tau_0} = \frac{g}{\kappa_0}$$

However if $dP_r/d\tau_0 > g/\kappa_0$, then there is no stable solution for $dP/d\tau_0$, and the atmosphere is unbound. Radiation pressure is not important in solar type stars, but is significant for O stars, and at the highest luminosities one reaches the *Eddington limit* where the atmosphere becomes unbound and blows off. For the grey case, the Eddington limit is given by the condition

$$\frac{L}{4\pi R^2 c} = \frac{GM}{R^2 \kappa_0}$$

or

$$\frac{L}{M} = \frac{4\pi G c}{\kappa_0} \tag{1.42}$$

In fact, for very luminous stars the continuous opacity is often mainly due to electron scattering, for which the absorption is indeed grey, and we need only make the substitution $\kappa_0 = \sigma_e N_e$ in (1.42), where σ_e is the Thomson cross-section for electron scattering and N_e is the number of electrons per unit mass. On the other hand, where the question is whether radiation pressure can drive a stellar wind, the contribution of lines is important, and one has to evaluate $\int \kappa_\nu H_\nu \, d\nu$, remembering that H_ν is low in a dark line.

We now turn to the question of calculating the electron pressure. The number of electrons per unit volume is

$$N_e = \sum (\text{degree of ionization})(\text{number of ions})$$

$$= \sum_s \sum_j j N_{js}$$

where s denotes the element concerned and j the degree of ionization with $j = 0$ for a neutral atom, $j = 1$ for a once-ionized atom releasing one electron, and so on.

If we formally write the ionization fractions of the various stages of ionization of an element as $h_{js} = N_{js}/N_s$, then

$$\begin{aligned} N_e &= \sum_s N_s \left[\sum_j j h_{js} \right] \\ &= \sum_s A_s (N - N_e) [j h_{js}] \\ &= \frac{P - P_e}{kT} \sum_s A_s \left[\sum_j j h_{js}(N_e, T) \right] \end{aligned} \tag{1.43}$$

where A_s is the fractional abundance of the element s and N is the total number of particles per unit volume. To evaluate h_{js} we assume that Saha's equation holds, so that

$$N_{j+1}/N_j = f_j(T)/N_e$$

with $f_j \propto T^{3/2} \exp\{-I_j/kT)$. Suppose appreciable numbers of atoms are in the $(j-1)$th, jth and $(j+1)$th stages of ionization. Then

$$\begin{aligned} h_{js} &= \frac{N_{js}}{N_{j-1,s} + N_{js} + N_{j+1,s}} \\ &= \frac{1}{\dfrac{N_{j-1,s}}{N_{js}} + 1 + \dfrac{N_{j+1,s}}{N_{js}}} \\ &= \frac{1}{\dfrac{N_e}{f_{j-1}(T)} + 1 + \dfrac{f_j(T)}{N_e}} \end{aligned} \tag{1.44}$$

Usually, at the most three stages of ionization are appreciably populated at a given temperature, since Saha's equation is so temperature dependent.

Equations of the form (1.44) for all the elements can be inserted into (1.43), but unfortunately this leaves N_e on the right-hand side of (1.43) in a rather complicated form as well as on the left-hand side of the equation. One procedure is to make a first guess at N_e for the given T and P and then to evaluate h_{js} and use (1.43) to make a better guess at N_e and so on.

A first approximation to N_e can be made by determining which element (or small group of elements) is the principal electron donor at the temperature concerned. For instance, hydrogen is the most abundant element, and so at temperatures for which hydrogen is completely ionized (greater than 10 000 K), hydrogen will be the dominant donor of electrons, with a contribution from helium at temperatures greater than 15 000 K for which helium is singly or doubly ionized. For a pure hydrogen atmosphere, we have

$$N(H^+)/N(H^0) = f_H/N_e$$

for the ratio of hydrogen ions H^+ to neutral hydrogen atoms H^0. Now $N_e = N(H^+)$, so

$$N_e^2/N(H^0) = f_H(T)$$

We also have that $P = [N(H^+) + N(H^0) + N_e]\,kT = [2N_e + N(H^0)]kT$. Hence:

$$P = 2N_e kT + \frac{N_e^2 kT}{f_H(T)}$$

The quadratic can be solved

$$N_e = \frac{-2 + \sqrt{4 + \dfrac{4P}{f_H kT}}}{\dfrac{2}{f_H(T)}} \tag{1.45}$$

On the other hand, for solar type stars and cooler the atmospheric temperature is too low for hydrogen to be ionized at all, and the main electron donors are the much less abundant but more easily ionized heavier elements such as magnesium and silicon at the upper end of the temperature range and sodium and potassium at the lower end of the temperature range. A first approximation to the electron pressure is then obtained by selecting a typical 'metal' M to represent the group of 'metals' acting as the main electron donors at the relevant temperature, and then using an average ionization potential (remembering that the ionization potentials of members of the group are bound to be similar) and the fractional abundance A_M of the whole group. For single ionization $N_e = N(H^+)$ and

$$\frac{N_e N(M^+)}{N(M^0)} = \frac{N_e^2}{N(M^0)} = f_M(T)$$

We also have $P - P_e \sim P$ in the relevant temperature range, so

$$\frac{P}{kT} A_M = N(M^+) + N(M^0) = N_e + \frac{N_e^2}{f_M(T)}$$

Hence

$$N_e = \frac{-1 + \sqrt{1 + \dfrac{4PA_M}{f_M kT}}}{\dfrac{2}{f_M(T)}} \tag{1.46}$$

At temperatures higher than the Sun, the ionization of a typical electron-donating 'metal' is almost complete, and we can write $N_e = N(M^+) = (P/kT)A_M$ as long as hydrogen ionization is negligible. At temperatures considerably lower than those found in the Solar atmosphere, the typical electron-donating 'metal' is mainly neutral, so that $(P/kT)\,A_M = N(M^0)$ and

$$N_e = \sqrt{f_M(T)\frac{P}{kT}A_M} \propto \sqrt{PA_M}$$

Model Atmospheres—Summary

An overall procedure for producing a model atmosphere might then run something like this. We start with a given effective temperature, gravity and composition, and make some initial guess at a $T(\tau)$ relation. Then starting at $\tau = 0$, we guess the opacity, and then calculate the pressure at succeeding steps using (1.38):

$$P(\tau + \Delta\tau) = P(\tau) + (g/\kappa_0)\Delta\tau$$

with $P(0) = 0$ and $\kappa_0 = \kappa_0(P, T)$. At each step in optical depth we calculate N_e iteratively for the given P and T using some first guess as a starting point as above. Given N_e, we can calculate κ_0 exactly for the given T and P, and hence can re-estimate $P(\tau + \Delta\tau)$, then proceed to recalculate N_e and so on until convergence is achieved. When convergence has been achieved at all depths, we have P and N_e exactly for the given $T - \tau$ relation. One can then enter something like the Unsöld–Lucy method to improve $T(\tau)$, remembering that at each iteration P and P_e will have to be recalculated. This is only a sketch of a possible process, and the programmes actually used differ considerably for computational and numerical reasons.

There are still a number of difficulties. One is that in the atmospheres of stars cooler than F5, some of the energy is carried by convection rather than radiation, significantly so for M-type stars. The approach is then adopted of taking a radiative $T(\tau)$ model, estimating the convective flux and subtracting it at each depth from the total flux of energy to be carried, and then solving radiatively for the $T(\tau)$ relation required to transport the remaining flux. Convective theory is still at a very primitive stage, but fortunately the final $T(\tau)$ models do not seem to be very sensitive to the inadequacies of the convective theory used (which is far from being the case for stellar interiors), partly because the low densities of stellar atmospheres mean that only a relatively small fraction of the flux is carried convectively in many cases. A second difficulty is that we have assumed an LTE source function, but at least for continuous processes this is often an adequate approximation.

The biggest difficulty lies in calculating the opacity. Continuous processes are reasonably well understood, but lines also play a significant role in the opacity. Strictly speaking, one should integrate over every line profile in the frequency integrals in model atmosphere calculations, requiring perhaps ten extra frequency points per line. This can be done in hot stars where the spectra have a relatively small number of very strong lines, but is an impossible task in cooler stars where there may be thousands of medium-strong lines and millions of weak lines. One solution is to take a line of typical depth and make it occupy a fraction of the total frequency range corresponding to that occupied in observed spectra by the hundreds of lines it represents. For instance, one might take a typical line in the wavelength region between 500 nm and 550 nm as having in some particular case an absorption coefficient 0.30 that of the continuous opacity, and find that such lines occupy 5% of the wavelength range. These lines can then be replaced by one line with a rectangular line profile of the appropriate depth and occupying 2.5 nm of the wavelength range. A more elaborate solution would be to use several typical lines, say 'strong', 'medium' and 'weak' with widths or weights in the frequency integrals corresponding to their relative importance in the actual spectrum. This is called a *multiple picket fence* approach, and is still, of course, only an approximation to the real situation. Determining the distribution of lines of different strengths is a non-trivial task (technically one refers to this as finding the *opacity distribution function*). A further difficulty is that it is very likely that some lines will be formed in a non-LTE fashion.

The first stellar atmosphere models ignored lines, and one refers to models that take lines into account as *line-blanketed*. The line opacity has the effect of blocking the flow of radiation, so that compared with line-free models, line-blanketed models have steeper temperature gradients. This effect is often divided into a raising of the temperature of deeper layers, and a lowering of the boundary temperature. The latter component is significant in abundance work because it can lead to strong absorption lines appearing stronger (darker) than one would expect from unblanketed models.

2 MICROSCOPIC PROCESSES AND STATISTICAL EQUILIBRIUM

Introduction

If Local Thermodynamic Equilibrium (LTE) does not hold, and the source function, excitation and ionization are not given by the Planck function, Boltzmann's equation and Saha's equation, respectively, then we must calculate the populations of the levels involved directly, using statistical equilibrium. The rate of a process which causes transitions from level i to level j is defined as the number of transitions per unit time from i to j via the process concerned per particle in level i. Such transitions can occur via radiation (R_{ij}) or via collisions (C_{ij}), so the number of transitions per second from i to j per unit volume is $N_i(R_{ij} + C_{ij})$, where N_i is the number of particles per unit volume in level i, and by convention j is the higher energy level.

If the population of all levels is steady so that the number of transitions finishing at a certain level i equals the number of transitions leaving from that level, then

$$\sum_j N_i(R_{ij} + C_{ij}) = \sum_j N_j(R_{ji} + C_{ji}) \tag{2.1}$$

where the sums exclude $i = j$ of course. There will be similar equations for each level i, which can be solved simultaneously subject to the overall condition $\sum N_i = N$, the total number of particles involved. In general, 'level' includes ionized states, transitions to which must be included in the summations. In the discussion of this chapter, the ionized state compared with the neutral state or state of lower ionization will be given the designation k.

In the following section we will consider radiative processes in more detail. Here we simply note that we have the following processes:

spontaneous emission: atom(j) → atom(i) + photon

absorption: atom(i) + photon → atom (j)

stimulated emission: atom(j) + photon → atom(i) + photon + photon

where in each case the photon energy $h\nu = E_{ij}$. The process whose existence is not obvious is, of course, stimulated emission where the emission of a photon is stimulated by the presence of photons of the same frequency, the probability of the process being proportional to the number of such photons, that is to the intensity at the frequency concerned.

In a later section we will consider collisional processes of excitation:

excitation: atom(i) + colliding particle (energy E_b) → atom (j) + colliding particle (energy E_f)

where the colliding particle's energy before the collision = colliding particle's energy after the collision plus the excitation energy, so $E_b = E_f + E_{ij}$. We will also consider the reverse process of *de-excitation*, where the colliding particle leaves the collision with more kinetic energy that it started with. The colliding particles must have a centre of mass kinetic energy of at least E_{ij} for excitation, but there is no restriction in the case of de-excitation. Charged particles naturally have stronger interactions than neutral ones, and in stellar atmospheres most collisional excitation is produced by electrons.

Finally we turn to transitions between bound and free states. The process involving absorption of radiation is called

photoionization: atom + photon → ion + electron

where the photon must have an energy $h\nu$ greater than the ionization energy χ needed to ionize the atom, but where all frequencies $\nu > \chi/h$ are acceptable. The reverse radiative process, called *radiative recombination*, actually involves a collision:

radiative recombination: ion + electron → atom + photon

where the photon energy $h\nu = \chi$ + kinetic energy of the colliding electron, so that the frequency of the photon can have any value greater than χ/h. Radiative recombination at any allowed frequency can be stimulated by the presence of radiation fields at that frequency. *Collisional ionization* can also occur, but here the reverse process is

three-body recombination: ion + electron + electron → atom + electron

In stellar atmospheres the contributions of collisional ionization and three-body recombination to the ionization balance are usually small, but collisional ionization can be very important in high-temperature environments like stellar coronae.

Note that the energy required to ionize from the ground state is the *ionization potential I*, but we will sometimes need to consider ionization from excited states where the energy required is the ionization energy χ.

Radiative Rates

The basic measure of radiative rates is the *Einstein coefficient for spontaneous emission* from level j to level i, A_{ji}, which is defined as the probability per second that an atom in level j will decay to level i. Hence the number of spontaneous radiative decays from j to i per unit volume is $N_j A_{ji}$, where N_j is the number of particles in level j per unit volume. A typical value for A_{ji} for a transition allowed by the rules of quantum mechanics is $10^{-8}\,\text{s}^{-1}$.

Different authors define slightly differently the corresponding *Einstein coefficient for absorption*. Here we define this coefficient for absorption from level i to level j, B_{ij}, by equating the probability of a radiative transition from i to j per unit time in a radiation field of mean intensity J_ν to $4\pi J_\nu B_{ij}$, where $\nu = E_{ij}/h$ and E_{ij} is the energy difference between the levels. It follows that the probability of radiative absorption per second per atom in level i exposed to intensity I_ν in solid angle $d\Omega$ is $B_{ij} I_\nu \, d\Omega$. The *Einstein coefficient for stimulated emission* from level j to level i, B_{ji}, is similarly defined by equating the probability of a stimulated radiative decay from j to i per unit time in a radiation field of mean intensity J_ν at the frequency of the transition to $4\pi J_\nu B_{ji}$. It should be noted that sometimes the B_{ij} and B_{ji} coefficients are defined without the factor 4π, and sometimes they are defined in terms of the radiation energy density $U_\nu = 4\pi J_\nu/c$, so that one uses the probability of a transition per unit time per unit energy density at the appropriate frequency.

A full quantum mechanical treatment of the interaction of radiation and matter shows that the relationships between the coefficients are as follows:

$$\begin{aligned} A_{ji} &= (8\pi h\nu^3/c^2)B_{ji} \\ B_{ij} &= (g_j/g_i)B_{ji} \end{aligned} \tag{2.2}$$

where g_i and g_j are the statistical weights of the lower and upper levels. The first relationship of (2.2) will differ by a factor of 4π or $4\pi/c$ if a different definition of B is used.

We can now write down the radiative rates:

$$\begin{aligned} R_{ij} &= 4\pi B_{ij} J_\nu \\ R_{ji} &= A_{ji} + 4\pi B_{ji} J_\nu \end{aligned} \tag{2.3}$$

We can also express the emission and absorption coefficients in terms of the Einstein A and B coefficients. The emission coefficient j_ν is defined as the energy emitted into unit solid angle per unit mass. The Einstein coefficient refers to the number of transitions into 'all round' or 4π and hence must be multiplied by the photon energy $h\nu$ and divided by 4π to produce an emission coefficient:

$$j_\nu \rho = (1/4\pi)\, N_j A_{ji} h\nu \tag{2.4}$$

where ρ is the mass density and N_j is the number of emitting atoms per unit volume. The stimulated emission, like absorption, is proportional to the intensity, and so is conveniently included as a negative contribution to the absorption coefficient. The absorption coefficient refers to *energy* absorbed per unit mass so the Einstein coefficients again have to be multiplied by the photon energy to obtain the absorption coefficient, but in our definitions they already refer to a mean intensity and so no factor 4π is needed:

$$\kappa_\nu \rho = (N_i B_{ij} - N_j B_{ji}) h\nu \tag{2.5}$$

Lines have profiles and are spread over a finite range of frequencies, while the A_{ji}, B_{ij} and B_{ji} coefficients refer to the integrated effect over the line profile. If we now take (2.4) and (2.5) to give absorption and emission coefficients at a particular frequency as opposed to averages over a line profile, (2.4) can be written:

$$j_\nu \rho = (1/4\pi)(N_j A_{ji} h\nu)\phi_\nu \tag{2.4a}$$

where ϕ_ν is the normalized line profile, so $\int \phi_\nu \mathrm{d}\nu = 1$ over the line profile. Equation (2.5) can be similarly adjusted by multiplying by ϕ_ν.

The source function is defined as the ratio of the emission to the absorption coefficient. Hence

$$S_\nu = \frac{j_\nu}{\kappa_\nu} = \frac{N_j A_{ji} \dfrac{h\nu}{4\pi}}{(N_i B_{ij} - N_j B_{ji}) h\nu}$$

$$= \frac{\dfrac{A_{ji}}{4\pi B_{ji}}}{\dfrac{N_i B_{ij}}{N_j B_{ji}} - 1}$$

$$= \frac{\dfrac{2h\nu^3}{c^2}}{\dfrac{N_i g_j}{N_j g_i} - 1} \quad \text{using (2.2)} \tag{2.6}$$

$$= \frac{\dfrac{2h\nu^3}{c^2}}{\mathrm{e}^{h\nu/kT} - 1}$$

$$= B_\nu, \text{ the Planck function, if } \frac{N_j}{N_i} = \frac{g_j}{g_i}\mathrm{e}^{-E_{ij}/kT} \text{ with } E_{ij} = h\nu$$

i.e. if Boltzmann's equation holds. Thus the *source function reduces to Planck's function if Boltzmann's equation holds for the relative populations of the upper and lower levels involved.* It will be recalled that the excitation temperature is defined in such a way as to force the Planck Function to be equal to the source function if the former is evaluated at the excitation temperature.

This argument is sometimes used in the reverse direction to determine the relationships (2.2) between the Einstein coefficients. Consider a system of matter and radiation in full equilibrium, say within an enclosed cavity with the walls held at a constant fixed temperature. Then the radiation field will be given by Planck's equation and matter will be in equilibrium with radiation. Consider now a small sample of gas, whose emission will be proportional to $\int j_\nu d\nu$, and whose absorption will be proportional to $\int \kappa_\nu J_\nu d\nu$. In equilibrium, emission and absorption must balance, and if this is to be true for any sample of matter with any frequency dependence of absorption, then the same must hold at each frequency, so $j_\nu = J_\nu \kappa_\nu$. This balancing at each energy in full equilibrium is called the *principle of detailed balance.*

Now $J_\nu = B_\nu$ so $j_\nu/\kappa_\nu = S_\nu = B_\nu$. This in turn requires that the relation between A_{ji} and B_{ij} and B_{ji} is that given by (2.2). Furthermore, since the individual coefficients are determined by the properties of atomic structure, the relationships (2.2) must hold generally and not just under the very special conditions of full equilibrium. We shall use a similar line of argument later in this chapter.

Expression (2.5) for the absorption coefficient can be factorized into an absorption coefficient with no allowance for stimulated emission and a correction factor for stimulated emission:

$$\begin{aligned}\kappa_\nu \rho &= N_i B_{ij} h\nu \left(1 - \frac{N_j B_{ji}}{N_i B_{ij}}\right)\phi_\nu \\ &= N_i B_{ij} h\nu \left(1 - \frac{N_j g_i}{N_i g_j}\right)\phi_\nu \quad \text{using (2.2)} \\ &= N_i B_{ij} h\nu \left(1 - \frac{g_j}{g_i} e^{-E_{ij}/kT} \frac{g_i}{g_j}\right)\phi_\nu\end{aligned}$$

if Boltzmann's equation holds between i and j.

Hence

$$\kappa_\nu \rho = N_i B_{ij} h\nu (1 - e^{-h\nu/kT})\phi_\nu \tag{2.7}$$

The correction factor $(1 - \exp[-h\nu/kT])$ is usually fairly close to 1.0 for frequencies in the visible part of the spectrum—a typical value for the optical spectra of stars is 0.9. However, in the millimetre and radio regions the stimulated emission can become dominant, and if $h\nu \ll kT$ the exponential can be expanded in series, and taking the lowest order surviving term:

$$\kappa_\nu \rho \simeq N_i B_{ij} \frac{h^2 \nu^2}{kT} \phi_\nu$$

We can generalize to include situations where Boltzmann's equation does not hold by replacing T with the excitation temperature T_{ex}. Using (2.2), we finally find

$$\kappa_\nu \rho \simeq N_i \frac{g_j}{g_i} \frac{A_{ji}}{\nu} \frac{c^2 h}{8\pi k} \frac{\phi_\nu}{T_{ex}} \tag{2.8}$$

In ultraviolet and optical stellar astronomy, one normally finds the Einstein coefficients replaced by the *oscillator strength* f, which appears for purely historical reasons. Pre-quantum mechanical physics pictured an electromagnetic wave making an electron oscillate like a harmonic oscillator with energy being absorbed because the oscillation was damped—the fraction of energy absorbed was fixed irrespective of the transition, so an oscillator strength was introduced to allow for the fact that experimentally lines have different strengths.

The absorption coefficient can be written as $\kappa_\nu \rho = N_i \sigma_\nu$ (ignoring stimulated emission for the moment), where σ_ν is the cross-section for absorption. Integrating over the line profile and noting that the frequency changes very little from the line-centre value over a profile:

$$\int \kappa_\nu \rho \, d\nu = N_i B_{ij} h\nu = N_i \frac{\pi e^2}{4\pi\varepsilon_0 mc} f \tag{2.9}$$

where m is the mass of an electron. Equation (2.9) constitutes the definition of the oscillator strength f. Then:

$$f = \frac{4\pi\varepsilon_0 mc^3}{8\pi^2 e^2} \frac{1}{\nu^2} \frac{g_j A_{ji}}{g_i} = 1.347 \times 10^{21} \frac{1}{\nu^2} \frac{g_j A_{ji}}{g_i} \tag{2.10}$$

Quantum mechanics shows that the Einstein A coefficient can be determined from the wave functions Ψ_i and Ψ_f of the initial and final states as follows:

$$A_{ji} = \frac{64\pi^4 e^2}{3hc^3 4\pi\varepsilon_0} \nu^3 \left| \int \Psi_j^* \boldsymbol{r} \Psi_i \, dV \right|^2$$

where $\boldsymbol{r}$ is the position vector and the integral is over volume. Hence

$$A_{ji} = \frac{64\pi^4}{3hc^3 4\pi\varepsilon_0} \nu^3 |d|^2 = 1.046 \times 10^{21} \nu^3 |d|^2 \tag{2.11}$$

where $d = e\int \Psi_j^* d \Psi_j dV$ is the *dipole moment*.

If the level i is degenerate, then $|d|^2 = \sum |d_i|^2$, where the sum is over the g_i degenerate states of the final level. Molecular transition strengths are often quoted as dipole moments, sometimes in units of Debyes, where one Debye = $3.34 * 10^{-30}$ coulomb metres. An alternative way of describing the strength of a line is by the *line strength* $S(i, j)$, where $S = \sum\sum |d_i|^2$ with the sum running over the degenerate states of both upper and lower levels, so

$$A_{ji} = \frac{1}{g_j} \frac{64\pi^4}{3hc^3 4\pi\varepsilon_0} \nu^3 S(i, j) \tag{2.12}$$

There remains the vexed question of determining the values of transition probabilities or oscillator strengths, either experimentally or theoretically. In the case of hydrogen and hydrogenic ions (those with one electron), levels with the same principal quantum number n are degenerate and hence we have only to deal with the oscillator strengths for transitions n_i to n_j.

These can be calculated exactly, but the Kramers semi-classical formula gives roughly correct answers:

$$f(n_j, n_i) = \frac{32}{3\sqrt{3}\pi} \frac{1}{\left[\frac{1}{n_i^2} - \frac{1}{n_j^2}\right]^3} \frac{1}{n_j^3 n_i^5} \tag{2.13}$$

which has to be multiplied by a *Gaunt factor* (for which the symbol $g(n_i, n_j)$ is rather confusingly used) of order of magnitude unity to obtain the exact oscillator strengths.

The standard equation for the levels of a hydrogenic atom of principal quantum number n and atomic number Z is

$$E = 2.180 \times 10^{-18} \frac{Z^2}{n^2} \tag{2.14}$$

hence

$$\nu = \frac{E}{h} = 3.290 \times 10^{15} Z^2 \left[\frac{1}{n_i^2} - \frac{1}{n_j^2}\right]$$

Using this relation for ν, the relation (2.10) between A_{ji} and f, and Kramer's formula (2.13), we obtain (for hydrogenic atoms):

$$A_{ji} = 8.033 \times 10^9 Z^4 \left[\frac{1}{n_i^2} - \frac{1}{n_j^2}\right]^2 \frac{n_i^2}{n_j^2} f$$

$$= 1.575 \times 10^{10} Z^4 \frac{1}{\left[\frac{1}{n_i^2} - \frac{1}{n_j^2}\right]} \frac{1}{n_i^3 n_j^5} \tag{2.13a}$$

For large n_j the hydrogenic f values become proportional to $1/n_j^3$. It will be noted that in a series of lines with a common lower level, e.g the Lyman series from $n = 1$, the smallest jumps (those to the lowest n_j) have the largest oscillator strengths and transition probabilities, so for the Lyman series the Lyα line (to $n = 2$) has the largest A and f values. The situation becomes much more complicated when many electron atoms are considered, but it remains true that the permitted transitions from the first excited levels to the ground state usually have the largest transition probabilities.

In the simpler cases, the selection rules indicate which transitions are permitted. For a single electron making a transition it is required that the orbital angular momentum quantum number l changes by $+1$ or -1. This rule always holds. In the case of hydrogen, the levels with the same n but different l are degenerate, so for instance $n=2$ can have $l=1$ (an s state) or $l=2$ (a p state), while the ground state must be an s state (l can have integer values from 0 to n). However only $l=1$ to $l=0$ is allowed, so Lyman a is a 2p to 1s transition in emission, and the 2s to 1s transition at the same wavelength is forbidden.

In the many-electron situation, we need to make some matters of terminology clear first. Electrons in closed shells generally have total orbital angular momentum quantum number $L=0$, total spin angular momentum quantum number $S=0$, and total angular momentum quantum number $J=0$, and we are concerned with the electrons outside a closed shell. The most usual case for lighter atoms is for the individual orbital angular momenta to couple together to give a total orbital angular momentum quantum number L, for the spins to couple to give a total spin quantum number S (the individual electrons are ascribed a spin quantum number of 1/2), and then for the total orbital angular momentum and the total spin angular momentum to couple to give the total angular momentum, specified by total angular momentum quantum number J. This is called LS or Russell–Saunders coupling. The overall state is described by quantum numbers L, S, and J ($L=0$ is called an S state, $L=1$ is called a P state, $L=2$ is called a D state, and $L=3$ is called an F state) . The set of values of n and l for all the electrons is called the *configuration*, but often only the electrons outside a closed shell are specified. Thus the ground state configuration of sodium is $1s^2 2s^2 2p^6 3s$ or simply 3s, where $2p^6$ means that there are 6 (the superscript) electrons with $n=2$ and $l=1$ (p). The combination 'configuration LS' is called a *term*—for the sodium ground state we have $3s\,^2S$, where the superscript is $2S+1$ which for a single electron can only be $2\times 1/2+1=2$, and $L=0$ is again the only possibility with a single $l=0$ electron. However with two electrons outside a closed shell, a given configuration can give rise to several terms—for instance two p electrons could combine to give $L=0$, 1 or 2, and their spins could combine to give $S=0$ or $S=1$ (L must be an integer and S an integer or half-integer). Not all combinations are possible and in the case of two p electrons with the same n (equivalent electrons) the possibilities are ^1D , ^3P, and ^1S only. A given term may give rise to several *levels* distinguished by the value of the total angular momentum J—for instance, the first excited state of sodium is the term $3p\,^2P$ which can have $J=3/2$ or $J=1/2$. The set of transitions between the various levels of two given terms is called a *multiplet*.

The selection rules in LS coupling are (a) a change in L in a permitted transition is $+1$ or -1 or zero, (b) a change in S is zero. In addition it holds in general that the change in J in a permitted transition is $+1$ or -1 or zero, with $J=0$ to $J=0$ forbidden. It must be remembered that LS coupling does not hold at all for some transitions and only approximately in others, and that in complex atomic structures even the configurations may not be pure. Forbidden and molecular transitions will be discussed later. The relative strengths of the transitions within a multiplet can easily be calculated exactly if the coupling is purely of one type. The strongest transition is that involving the largest J values.

Radiative Rates—Continuous Processes

We now turn to the rates of photoionization and radiative recombination. Similar ideas to those already used for line transitions apply here, with the differences that a continuous wide range of frequencies is involved, and that recombinations involve collisions. One therefore replaces B_{ij} by $B'_{ik}(\nu)$, where the probability of photoionization by photons with frequencies between ν and $\nu + d\nu$ and mean intensity J_ν is $B'_{ik} 4\pi J_\nu \, d\nu$ per atom in level i per unit time. Recombination is proportional to the flux of electrons (number of electrons hitting unit area per unit time) $= n_e(V)\, V\, dV$ for electrons with number density $n_e(V)\, dV$ between velocities V and $V + dV$. Now on recombination one obtains a photon of energy equal to the sum of the kinetic energy of the electron and the ionization energy from the lower level concerned (χ_i) so $h\nu = (1/2)mV^2 + \chi_i$. Then $h\, d\nu = mV\, dV$ and a given small frequency interval from ν to $d\nu$ corresponds to a particular velocity range from V to $V + dV$.

We therefore define the coefficient for spontaneous recombination, $A'_{ki}(V)$, by setting the probability of emission of a recombination photon with frequency between ν and $\nu + d\nu$ per ion and per unit time equal to $A'_{ki}(V)\, V n_e(V) dV$ for a velocity range corresponding to the stated frequency range. Similarly, the coefficient for stimulated emission , $B'_{ki}(V)$, is defined by the condition that the probability per unit time that an ion is stimulated to recombine in radiation field J_ν to give a photon with frequency between ν and $\nu + d\nu$ is $B'_{ki}(V)\, V n_e(V) dV\, 4\pi J_\nu$, where again the frequency range and the velocity range correspond.

In complete equilibrium, photoionizations and recombinations balance for each frequency and corresponding velocity by the principle of complete balance. Hence

$$4\pi B'_{ik}(\nu) J_\nu d\nu\, N_i = [A'_{ki}(V) + 4\pi J_\nu B'_{ki}(V)] n_e(V) N_k V dV$$

Substituting $d\nu = (mV/h) dV$, Maxwell's equation for $n_e(V)/N_e$, where N_e is the total number of electrons per unit volume, Saha's equation for $N_k N_e / N_{\text{atom}}$ with level i belonging to the 'atom', Boltzmann's equation for N_i/N_{atom} and Planck's equation for J_ν, one obtains:

$$\frac{A'_{ki}(V)}{B'_{ki}(V)} = \frac{8\pi h \nu^3}{c^2}$$

$$\frac{B'_{ki}(V)}{B'_{ik}(\nu)} = \frac{h^2}{8\pi m^2 V^2} \frac{g_i}{U_k}$$

where U_k is the partition function for the higher state of ionization, g_i is the statistical weight of the lower level and m is the mass of an electron. These relations will always hold since they are relationships between characteristics of the atomic structure. It is customary to replace B'_{ik} by the *photoionization cross-section* α_ν, where $\alpha_\nu = B'_{ik}\, h\nu$. Then

$$A'_{ki}(V) = \frac{h^2}{c^2 m^2} \frac{g_i}{U_k} \frac{\nu^2}{V^2} \alpha_\nu$$

$$B'_{ki}(V) = \frac{h}{8\pi m^2} \frac{g_i}{U_k} \frac{1}{V^2 \nu} \alpha_\nu \tag{2.15}$$

We can now use these relationships to determine radiative rates. For photoionization, where all frequencies above a lower limit ν_0 are allowed and $h\nu_0 = \chi_i$, the energy just to ionize from level i, we have

$$R_{ik} = 4\pi \int_{\nu_0}^{\infty} \frac{\alpha_\nu J_\nu}{h\nu} \mathrm{d}\nu \tag{2.16}$$

For the hydrogenic case and level i with principal quantum number n:

$$\alpha_\nu(n) \simeq \frac{64\pi^4 m e^{10}}{3\sqrt{3}ch^6(4\pi\varepsilon_0)^5} \frac{Z^4}{n^5 \nu^3}$$

$$= \frac{2.815 \times 10^{25} Z^4}{n^5 \nu^3} m^2 \tag{2.17}$$

which needs to be multiplied by the Gaunt factor to obtain an exact answer. If $h\nu_0 > kT$, and the radiation field is Planckian, one can write $J_\nu = B_\nu \simeq (2h\nu^3/c^2)\exp(-h\nu/kT)$, so

$$R_{ik} \simeq \frac{2.815 \times 10^{25} \times 8\pi}{c^2} \frac{Z^4}{n^5} \int_{\nu_0}^{\infty} \frac{\mathrm{e}^{-h\nu/kT}}{\nu} \mathrm{d}\nu$$

$$\simeq \frac{7.87 \times 10^9 Z^4}{n^5} \int_1^{\infty} \frac{\mathrm{e}^{-ax'}}{x'} \mathrm{d}x', \qquad x' = \nu/\nu_0, \qquad a = h\nu_0/kT$$

$$\simeq \frac{7.87 \times 10^9 Z^4}{n^5} E_1(a)$$

$$\simeq \frac{7.87 \times 10^9 Z^4}{n^5} \frac{kT}{h\nu_0} \mathrm{e}^{-h\nu_0/kT} \tag{2.18}$$

since exponential integral $E_1(a) \simeq \mathrm{e}^{-a}/a$ if $a \gg 1$.

For recombination there is no lower limit to the integral and if we write out the expression for R_{ki} and then substitute for $n_e(V)/N_e$ from Maxwell's equation and convert the integral from one over velocity to one over frequency, we obtain:

$$R_{ki} = \int_0^\infty [A'_{ki}(V) + 4\pi B'_{ki}(V)J_\nu] n_e(V) V \, dV$$

$$= \frac{h}{2m^2} \frac{g_i}{U_k} \int_0^\infty \left[\frac{2h\nu^3}{c^2} + J_\nu\right] \frac{\alpha_\nu}{\nu} \frac{n_e(V)}{V^2} V \, dV$$

$$= 4\pi N_e \left(\frac{m}{2\pi kT}\right)^{3/2} \frac{h^3}{2m^3} \frac{g_i}{U_k} \int_{\nu_0}^\infty \left[\frac{2h\nu^3}{c^2} + J_\nu\right] \frac{\alpha_\nu}{h\nu} e^{-(h\nu - \chi_i)/kT} \, d\nu \tag{2.19}$$

Recombination rates are normally quoted in terms of a *recombination coefficient* α_{rec}, where $\alpha_{rec} N_e = R_{ki}$. In the hydrogenic case, substituting from (2.17) for α_ν, the photoionization cross-section, and writing for a single electron atom $g_i = 2n^2$ and $U_k = 1$, we obtain:

$$\alpha_{rec} = \frac{128\pi^3}{3}\left(\frac{2\pi}{3}\right)^{1/2} \frac{e^{10}}{(4\pi\varepsilon_0)^5 m^{1/2} c^3 h^3} \frac{Z^4}{n^3} \frac{1}{(kT)^{3/2}} e^{\chi_i/kT} \int_{\nu_0}^\infty \left[1 + \frac{c^2 J_\nu}{2h\nu^3}\right] \frac{e^{-h\nu/kT}}{\nu} \, d\nu$$

$$= 3.26 \times 10^{-12} \frac{Z^4}{n^3} \frac{1}{T^{3/2}} e^{\chi_i/kT} \int_{\nu_0}^\infty \left[1 + \frac{c^2 J_\nu}{2h\nu^3}\right] \frac{e^{-h\nu/kT}}{\nu} \, d\nu \tag{2.20}$$

The emission coefficient can be obtained by taking one frequency from the integral in (2.20) and multiplying by $N_e h\nu/(4\pi)$. The factor in square brackets is the correction for stimulated emission which can be evaluated in LTE by taking $J_\nu = B_\nu$, the Planck function. Ignoring for the moment the correction for stimulated emission, which is a good approximation if $h\nu_0 \gg kT$, the integral reduces to the exponential integral $E_1(x_0)$ if we write $x = h\nu/kT$ and $x_0 = h\nu_0/kT = \chi_i/kT$. Finally, if $h\nu_0 \gg kT$ so $x_0 \gg 1$, $E_1(x_0) \approx \exp(-x_0)/x_0$, we obtain:

$$\alpha_{rec} \simeq 3.26 \times 10^{-12} \frac{Z^4}{n^3 T^{3/2}} \frac{kT}{\chi_i} \simeq 2.06 \times 10^{-17} \frac{Z^2}{n T^{1/2}} \tag{2.21a}$$

where (2.14) has been substituted for χ_i. For the total recombination rate one must sum over all levels, which typically gives a recombination rate two to three times higher than that to $n = 1$ alone:

$$\alpha_{rec} \approx 4 * 10^{-17}/T^{1/2} \quad \text{for } T \sim 10^4 \text{ K} \tag{2.21b}$$

Lastly in this section we turn to transitions in which the electron is free both before and after the transition. A free electron cannot absorb a photon and increase its kinetic energy or emit a photon and lose some of its kinetic energy because the momentum of the photon is given by the relativistic relation $p = E/c$ and one can easily show that momentum and energy cannot both be conserved in such an interaction. However if the electron—photon interaction takes place in the presence of another particle (an atom or ion) which is interacting with the electron, then the heavy particle can take the recoil and both energy and momentum can be conserved. Thus we have the process called

bremsstrahlung: ion + electron$\rightleftharpoons$ion + electron + photon

which is sometimes regarded as an atomic transition in which the electron is free both before and after and hence is alternatively called the *free–free* process, as opposed to photoionization which is a *bound–free* process.

Note that $H^+ + e^- \rightleftarrows H^+ + e^- +$ photon is termed a free–free transition of the hydrogen atom. Bremstrahlung differs from photoionization in that there is no threshold—a very low energy photon can give a little extra kinetic energy to a photon and vice versa. An approximate expression for the cross-section for free–free absorption for hydrogenic atoms (atomic number Z, one electron) per electron travelling at velocity V and analogous to (2.17) for the photoionization cross-section is

$$a_\nu \text{ per electron} \approx \frac{4\pi}{3\sqrt{3}} \frac{e^6}{chm^2(4\pi\varepsilon_0)^3} \frac{Z^2}{\nu^3 V}$$

Integration over Maxwell's distribution of velocities gives

$$a_\nu \simeq N_e \frac{16\pi^2}{3\sqrt{3}} \frac{Z^2 e^6}{(2\pi m)^{3/2}(kT)^{1/2}(4\pi\varepsilon_0)^3 ch} \frac{1}{\nu^3}$$

$$\simeq 3.690 \times 10^{-2} \frac{N_e}{T^{1/2}} \frac{1}{\nu^3} m^2 \qquad (2.22)$$

where N_e is the total number of electrons per unit volume. This must be multiplied by the stimulated emission factor, here $(1 - \exp[-h\nu/kT])$ since free–free is purely collisional and hence LTE, and by the number of heavy particles per unit mass N^+/ρ to give the absorption coefficient κ_ν. Since free–free is an LTE process, the emission coefficient is given by

$$j_\nu = \kappa_\nu B_\nu = a_\nu N^+/\rho \cdot (2h\nu^3)/c^2 \cdot \exp(-h\nu/kT) \qquad (2.22a)$$

At long wavelengths the stimulated emission correction becomes dominant and $(1 - \exp[-h\nu/kT) \approx h\nu/kT$, so (2.22) becomes

$$\kappa_\nu \rho \simeq \frac{16\pi^2}{3\sqrt{3}} \frac{e^6 Z^2}{c(4\pi\varepsilon_0)^3(2\pi m)^{3/2}} \frac{N_e N^+}{(kT)^{3/2}} \frac{1}{\nu^2}$$

$$\simeq 1.77 \times 10^{-12} \frac{N_e N^+}{T^{3/2}} \frac{Z^2}{\nu^2} m^{-1} \qquad (2.22b)$$

However, at long wavelengths, the Gaunt factor becomes appreciably different from 1 and both emission and absorption coefficients must be multiplied by $0.5513 \ln (T^{3/2}/[Z\nu]) + 9.75$.

Free electrons can also scatter radiation without change of frequency and the cross-section for this (unless we are dealing with relativistic electrons) is given by the Thomson cross-section $= 6.65 \times 10^{-29}\,m^2$, which is independent of frequency. Needless to say, this is a completely non-LTE process.

Collisional Rates

A collisional rate can be written as the product of the flux of particles relative to the atom or ion to be excited, $N_c v$, and the cross-section for excitation, σ. However, the cross-section is a function of velocity v, and so is the flux of colliding particles. Writing the number of particles per unit volume with velocities between v and $v + dv$ as $n_c(v)dv$, usually given by Maxwell's equation, we can write for the *rate of excitation*:

$$C_{ij} = \int_{v_0}^{\infty} n_c(v) v \sigma_{ij}(v) dv \tag{2.23}$$

and for the *rate of de-excitation*:

$$C_{ji} = \int_{0}^{\infty} n_c(v) v \sigma_{ji}(v) dv \tag{2.24}$$

where v_0 is given by $(1/2)mv_0^2 = E_{ij}$ and m is the mass of colliding particle (more strictly the reduced mass, with v the relative velocity, but if the colliding particle is an electron, the distinction is unimportant). The lower limit in (2.23) is because the colliding particle must have an energy of at least the amount to be given away as excitation energy.

Equations (2.23) and (2.24) can be rewritten in terms of the kinetic energy $E = (1/2)mv^2$ with $v dv = dE/m$ and $N_c = \int n_c(v)dv = \int n_c(E)dE =$ total number of colliding particles per unit volume. Then

$$C_{ij} = N_c \int_{E_{ij}}^{\infty} \frac{n_c(E)}{N_c} \sigma_{ij}(E) \sqrt{\frac{2E}{m}}\, dE \tag{2.23a}$$

$$C_{ji} = N_c \int_{0}^{\infty} \frac{n_c(E)}{N_c} \sigma_{ji}(E) \sqrt{\frac{2E}{m}}\, dE \tag{2.24a}$$

using $n_c(v)dv = n_c(E)dE$.

Now $n_c(v)dv$ is nearly always given by Maxwell's equation, even when other aspects of LTE do not hold. Hence

$$n_c(v)dv = N_c 4\pi v^2 \left(\frac{m}{2\pi kT}\right)^{3/2} e^{-mv^2/2kT} dv$$

$$n_c(E) = N_c 4\pi \frac{2E}{m} \frac{1}{\sqrt{2mE}} \left(\frac{m}{2\pi kT}\right)^{3/2} e^{-E/kT}$$

$$C_{ij} = N_c \frac{2\pi}{(\pi kT)^{3/2}} \sqrt{\frac{2}{m}} \int_{E_{ij}}^{\infty} \sigma_{ij}(E) e^{-E/kT} E dE \tag{2.24b}$$

and similarly for C_{ji} with lower limit $E = 0$, and σ_{ji} substituted for σ_{ij}.

The expression for the collisional ionization rate C_{ik} is similar to that for C_{ij}, with the ionisation energy χ in place of the excitation energy E_{ij}. Here, however, it must be remembered that the result of a collisional ionization by an electron of energy E is energy $E-\chi$ to be shared by the ejected electron (which takes, say, energy E') and the colliding electron which will be left with $E-\chi-E'$. The cross-section $\sigma_{ik}(E)$ is therefore an average over the distribution of available kinetic energy, that is an average over E'.

What is the relation between C_{ij} and C_{ji}? In complete equilibrium, the principle of detailed balancing states that collisional excitations and collisional de-excitations are equal at *every energy*. The balance is between exciting collisions where the colliding particle comes in with kinetic energy E, loses energy E_{ij} to excitation, and departs with kinetic energy $E-E_{ij}$, and de-exciting collisions, where the colliding particle comes in with energy $E-E_{ij}$, gains energy E_{ij} from de-excitation, and leaves with kinetic energy E. If $\mathrm{d}C_{ij}(E)/\mathrm{d}E = \mathrm{d}C_{ji}(E-E_{ij})/\mathrm{d}E$, then cancelling the common factor $2\pi/(\pi kT)^{3/2}\sqrt{(2/m)}$ from both sides:

$$N_i N_c \sigma_{ij}(E) E \mathrm{e}^{-E/kT} = N_j N_c \sigma_{ji}(E-E_{ij})(E-E_{ij}) e^{-(E-E_{ij})/kT}$$

Hence

$$\frac{N_j}{N_i} = \frac{\sigma_{ij}(E)}{\sigma_{ji}(E-E_{ij})} \mathrm{e}^{-E_{ij}/kT} \frac{E}{E-E_{ij}}$$

But in complete equilibrium, $N_j/N_i = g_j/g_i \exp(-E_{ij}/kT)$ by Boltzmann's equation, so

$$\frac{\sigma_{ij}(E)}{\sigma_{ji}(E-E_{ij})} = \frac{g_j}{g_i} \frac{E-E_{ij}}{E}$$

Now complete balance is a very special case, and indeed even the balance of $N_i C_{ij}$ and $N_j C_{ji}$ does not hold in general. However the relationship between the cross-sections (which are determined individually by the atomic structure) should hold in general if it can be demonstrated in one particular case.

Substituting $E^* = E + E_{ij}$ in the de-excitation equivalent of (2.24b):

$$C_{ji} = N_c \frac{2\pi}{(\pi kT)^{3/2}} \sqrt{\frac{2}{m}} \int_{E_{ij}}^{\infty} \sigma_{ji}(E^*-E_{ij}) \mathrm{e}^{-(E^*-E_{ij})/kT} (E^*-E_{ij}) \mathrm{d}E^*$$

and using the cross-section ratio

$$C_{ji} = N_c \frac{2\pi}{(\pi kT)^{3/2}} \sqrt{\frac{2}{m}} \int_{E_{ij}}^{\infty} \sigma_{ij}(E^*) \frac{g_i}{g_j} E^* \mathrm{e}^{-E^*/kT} \mathrm{e}^{E_{ij}/kT} \mathrm{d}E^*$$

or

$$C_{ij} = C_{ij} \frac{g_i}{g_j} \mathrm{e}^{E_{ij}/kT} \qquad (2.25)$$

which should hold universally. We shall use this relation on a number of occasions later on. As one might expect, the collisional excitation rate is much less than the de-excitation rate unless the average kinetic energy of the colliding particle is of the order of magnitude of the excitation energy.

The determination of cross-sections as a function of energy is an extremely complicated matter about which there is a vast literature. However, if the exciting particle and the target are both charged (i.e. electron collisional excitation of an ion), then approximately $\sigma_{ij} \propto 1/E$. The reason is that the target attracts the colliding particle and the smaller distance of closest approach that results produces a stronger interaction and a larger cross-section. The deflection of the approaching particle will be greater if the colliding particle is moving slowly, so one would expect the cross-section to increase with decreasing kinetic energy, other things being equal. If we write $\sigma \propto \sigma_0/E_0$, then (2.24b) gives

$$C_{ij} = N_c \frac{2\pi}{(\pi kT)^{3/2}} \sqrt{\frac{2}{m}}\,\sigma_0 \int_{E_{ij}}^{\infty} e^{-E/kT}\,dE$$

$$= N_c \frac{2\sqrt{2}}{\sqrt{\pi m}} \frac{\sigma_0}{\sqrt{kT}} e^{-E_{ij}/kT} \tag{2.26}$$

In general, the energy dependence of the cross-section will differ from $1/E$, which in any case is inapplicable to neutral particles, but in nearly all cases the exponential factor in the Maxwell distribution dominates, and the main *temperature* dependence of the collisional excitation is given by $\exp(-E_{ij}/kT)$, and of collisional ionization by $\exp(-I/kT)$. In all cases, collisional rates are proportional to the density of colliding particles.

In the important case of the collisional excitation of ions by electrons for *radiatively forbidden* transitions,

$$\sigma_{ij} \sim \pi a_0^2/(E/E_0)$$
$$\sim (\pi/2m)(h/2\pi)^2\, 1/E$$

where E_0 is minus the energy of the ground state of the hydrogen atom [1 Rydberg $= 2\pi^2 me^4/(4\pi\varepsilon_0 h)^2$] and a_0 is the 'radius' of the electron orbit in the ground state of hydrogen [(1 Bohr radius $= 4\pi\varepsilon_0(h/2\pi)^2/(e^2 m)$].

Hence it is convenient to define a *collision strength*, $\Omega(i,j)$, by

$$\sigma_{ij}(E) = (\pi/2m)(h/2\pi)^2 \frac{1}{E}\frac{\Omega(i,j)}{g_i} \tag{2.27}$$

where $\Omega(i,j)$ is of order of magnitude unity (typically between 3 and 0.1). The rules of quantum mechanics that forbid a radiative transition do not, of course, forbid the corresponding collisionally induced transition. An important example is the parity rule, which requires a change in the parity of an atom on the emission of a single photon, and for a single electron produces the selection rule $\Delta l = 1$. For collisions, $\Delta l > 1$ *is* allowed.

The ratio of the cross-sections and (2.27) give for σ_{ji}:

$$\sigma_{ji}(E - E_{ji}) = \frac{\pi}{2m}\left(\frac{h}{2\pi}\right)^2 \frac{1}{E - E_{ij}} \frac{\Omega(i,j)}{g_j}$$

We now obtain from (2.26) and (2.27):

$$C_{ij} = N_e \sqrt{\frac{2\pi}{kT}} \left(\frac{h}{2\pi}\right)^2 \frac{1}{m^{3/2}} \frac{\Omega(i,j)}{g_i} e^{-E_{ij}/kT} \quad (2.28)$$

and correspondingly for C_{ji}, replacing the numerical constants by their numerical value:

$$C_{ji} = \frac{8.6 \times 10^{-12}}{T^{1/2}} N_e \cdot \frac{\Omega(i,j)}{g_j} \quad (2.28a)$$

A number of approximations have been suggested for *radiatively permitted* transitions. The collisional cross-section is roughly proportional to the oscillator strength f for a radiative transition. One example of such an approximation is the dipole approximation, with

$$\sigma_{ij}(E) \sim (8\pi/\sqrt{3}) \frac{1}{(E/E_0)} \cdot \frac{f}{(E_{ij}/E_0)} \pi a_0^2 G$$

G is adjusted to fit experimental results, and for electron excitation of positive ions, integration over the Maxwell distribution as in previous examples gives

$$C_{ij} = 3.9 \times 10^{-6} N_e \left(\frac{kT}{E_{ij}}\right) \frac{f}{T^{3/2}} \exp(-E_{ij}/kT)$$
$$\propto (1/T^{1/2}) \exp(-E_{ij}/kT) \quad (2.29)$$

For the excitation of neutral atoms, the cross-section drops to zero at $E = E_{ij}$, and the corresponding approximation to that leading to (2.29) gives

$$C_{ij} \simeq 2.16 \times 10^{-6} N_e \left[\frac{kT}{E_{ij}}\right]^{5/3} \frac{f}{T^{3/2}} \exp(-E_{ij}/kT) \quad (2.30)$$

$$C_{ji} \simeq 2.16 \times 10^{-6} N_e \left[\frac{kT}{E_{ij}}\right]^{5/3} \frac{g_i}{g_j} \frac{f}{T^{3/2}}$$
$$\simeq 1.28 \times 10^{-51} \left[\frac{kT}{E_{ij}}\right]^{5/3} \frac{1}{E_{ij}^2} \frac{N_e}{T^{3/2}} A_{ji} \quad (2.30a)$$

using equation (2.10) for the relation between oscillator strength and transition probability.

For collisional ionization, a similar approximation gives

$$C_{ik} \sim 10^{11} N_e \left(\frac{kT}{I}\right) \exp(-I/kT) \frac{\alpha_0}{T^{1/2}} \quad (2.31)$$

where α_0 is the photoionization cross-section at threshold, and the numerical factor depends on the charge on the ion. It should be emphasized that (2.29)–(2.31) are fairly crude approximations, and better approaches may exist in individual cases.

The opposite of collisional ionization is three-body recombination. However, there exists another recombination process which can become dominant in certain situations. This is *dielectronic recombination.* Normally excitation in an atom or ion involves one electron moving from the ground state to a higher level. It is, however, possible to have states in which two electrons are excited. The energies of these two electron states lie above the ionization energy for one electron excited states, and they are therefore unbound and unstable to the removal of one electron to leave a higher state of ionization. This process is called

autoionization: ion (net charge Z, two electrons excited) $\rightarrow$ ion $(Z-1)$ + electron

It requires no energy input, and hence has a very large probability.

The two electron excited state is produced by a collision of the ion $(Z-1)$ with an electron leading to recombination, and if the electron energy is high enough, to the excitation of two electrons. The equilibrium between such collisional recombinations and autoionizations leads to a small population of two electron excited states. However, there is a probability that a two electron excited state radiatively decays to a one electron excited state which is, of course, stable against autoionization. The complete dielectronic recombination process is then

ion (Z) + electron $\rightarrow$ ion $(Z-1)$ with two electrons excited

Ion $(Z-1) \rightarrow$ ion $(Z-1)$ with one excited electron + photon

with the net result:

Ion (Z) + electron $\rightarrow$ ion$(Z-1)$ + photon,

which is in the same form as a radiative recombination, and has the same dependence on electron density. The difference is that radiative recombination can happen at near zero electron energy, whereas in dielectronic recombination the electron must have sufficient energy to excite the two electron state and hence the rate involves a factor $\exp(-E/kT)$, where E is the energy of the two electron state above the ionization limit. Hence at low temperatures radiative recombination dominates, but at high temperatures like the $T \sim 10^6$ K found in the solar corona, dielectronic recombination dominates.

Statistical Equilibrium

In this section we apply the radiative and collisional rates just discussed to the determination of excitation and ionization in a few simple cases. First we consider the excitation of an atom with just two levels, in order to determine the source function for a radiative transition between those levels. Statistical equilibrium gives

$$N_i(R_{ij} + C_{ij}) = N_j(R_{ji} + C_{ji})$$

$$\frac{N_i}{N_j} = \frac{A_{ji} + 4\pi B_{ji}\int J_\nu \phi_\nu \mathrm{d}\nu + C_{ji}}{4\pi B_{ij}\int J_\nu \phi_\nu \mathrm{d}\nu + C_{ij}} \tag{2.32}$$

where we have written $B\int J_\nu \phi_\nu \mathrm{d}\nu$ instead of just BJ_ν to allow for the fact that the radiation field may be mainly due to the line itself, in which case it will vary rapidly with frequency and must be weighted by the line absorption profile ϕ_ν.

Equation (2.6) shows that the line source function can be written

$$S_\nu = \frac{2h\dfrac{\nu^3}{c^2}}{\dfrac{g_j N_i}{g_i N_j} - 1}$$

Now (2.32), with (2.2) to relate B_{ij} to B_{ji} and (2.25) to relate C_{ij} to C_{ji}, and with the substitution $E_{ij} = h\nu$, gives

$$\frac{g_j N_i}{g_i N_j} = \frac{A_{ji} + 4\pi B_{ji}\int J_\nu \phi_\nu \mathrm{d}\nu + C_{ji}}{4\pi B_{ji}\int J_\nu \phi_\nu \mathrm{d}\nu + C_{ji}\exp\left(-\dfrac{h\nu}{kT}\right)}$$

Hence

$$S_\nu = \frac{\dfrac{2h\nu^3}{c^2}\left[4\pi B_{ji}\int J_\nu \phi_\nu \mathrm{d}\nu + C_{ji}\mathrm{e}^{-h\nu/kT}\right]}{A_{ji} + C_{ji}[1 - \mathrm{e}^{-h\nu/kT}]}$$

$$= \frac{\int J_\nu \phi_\nu \mathrm{d}\nu + \dfrac{2h\nu^3}{c^2}\dfrac{C_{ji}}{A_{ji}}\mathrm{e}^{-h\nu/kT}}{1 + \dfrac{C_{ji}}{A_{ji}}[1 - \mathrm{e}^{-h\nu/kT}]}, \quad \text{using } \frac{8\pi h\nu^3}{c^2}B_{ji} = A_{ji}$$

Finally, if we substitute

$$\frac{C_{ji}}{A_{ji}} = \frac{\varepsilon}{1 - \mathrm{e}^{-h\nu/kT}}$$

we obtain

$$S_\nu = \frac{\int \phi_\nu J_\nu \mathrm{d}\nu + \varepsilon B_\nu}{1 + \varepsilon} \tag{2.33}$$

where

$$B_\nu = \frac{\dfrac{2h\nu^3}{c^2}}{e^{h\nu/kT} - 1}$$

i.e. Planck's function, with ε defined by

$$\varepsilon = \frac{C_{ji}}{A_{ji}}[1 - e^{-h\nu/kT}]$$

ε essentially represents the relative importance of collisional and radiative processes, with the factor in brackets being a small correction at visible wavelengths.

The term εB_ν represents a thermal source of photons, and the term ε a thermal sink, both at the local kinetic temperature. If $\varepsilon \gg 1$, then $S_\nu \to B_\nu$, the Planck function. In other words, if collisions dominate, with the rates C determined by the temperature through Maxwell's equation, then the source function is also thermalized at the local kinetic temperature. The term $\int \phi_\nu J_\nu d\nu$ (and the factor unity in the denominator) can be thought of as representing the reservoir of photons, scattered, but neither lost nor created. If ε were zero, there would be no solution to the equations of radiative transfer if only local conditions are specified—the radiation field would be determined by boundary conditions. In practice, $\varepsilon \ll 1$ for most lines in stellar atmospheres, but the thermal source/sink terms still ultimately decide what a particular layer does to the strength of a line. In a stellar interior, photons travel very short distances, and although they may be scattered many times before being destroyed by absorption, these scatterings will happen locally, and S_ν and J_ν will both be driven to the Planck function at the local temperature. Near the surface, photons are often able to escape from the star altogether. J_ν may be less than B_ν, and hence if $\varepsilon < 1$, the source function may be considerably less than the Planck function.

Sometimes it is convenient to write $\varepsilon' = \varepsilon/(1 + \varepsilon)$ and then

$$S_\nu = (1 - \varepsilon') \int \phi_\nu J_\nu d\nu + \varepsilon' B_\nu \tag{2.34}$$

This suggests writing the absorption coefficient as the sum of a 'true absorption' and 'scattering', $\kappa_\nu = \kappa_{abs} + \sigma$, so

$$S_\nu = \frac{\kappa_{abs} B_\nu + \sigma \int J_\nu \phi_\nu d\nu}{\kappa_\nu} \tag{2.34a}$$

although for lines, the distinction between absorption and scattering is slightly artificial and (2.34a) gives no more information than (2.34). In the case of continuum processes, scattering in Thomson electron scattering is a distinct physical process inherently involving no change in photon energy.

The simplest case occurs when J_ν is mainly due to some externally imposed radiation field. This can be the case for a radio or millimetre line from an interstellar cloud, where the line itself is weak, and the greater part of J_ν may come from the cosmic microwave background, or may be the diluted radiation field of many stars. In the previous chapter, we defined T_{ex}, the excitation temperature, by $S_\nu = B_\nu(T_{ex})$. If $h\nu/kT_{ex} \ll 1$, the Rayleigh–Jeans approximation holds, and

$$T_{ex} = \frac{c^2}{2k\nu^2} S_\nu = \frac{c^2}{2k\nu^2}\left(\frac{J_\nu + \varepsilon B_\nu}{1+\varepsilon}\right)$$

substituting from (2.33), with J_ν constant over the line profile for an external continuum, and $\int \phi_\nu d\nu = 1$.

Now if the external radiation field is blackbody, characterized by temperature T_{rad}, and $h\nu/kT_{rad} \ll 1$, then $J_\nu = (2k\nu^2/c^2)T_{rad}$ (if we are dealing with starlight, this will have to be multiplied by a dilution factor W).

Suppose the local kinetic temperature is T_e, and $h\nu/kT_e \ll 1$, so $B_\nu = (2k\nu^2/c^2)T_e$. Substituting, we find

$$T_{ex} = \frac{T_{rad} + \varepsilon T_e}{1+\varepsilon} \tag{2.35}$$

At long wavelengths

$$\varepsilon = \frac{C_{ji}}{A_{ji}}[1 - e^{-h\nu/kT}] = \frac{C_{ji}h\nu}{A_{ji}kT_e} = \frac{T_0}{T_e}, \quad \text{say}$$

so

$$T_{ex} = \frac{T_e(T_{rad} + T_0)}{T_e + T_0}$$

At frequencies adjacent to the line frequency, the brightness temperature will be equal to T_{rad}, the background. Now for the line, $T_B = T_{ex}[1 - e^{-\tau}] + T_{rad}e^{-\tau}$, where the first term represents the line emission from the cloud, and the second term represents the background, absorbed by the cloud. Then

$$T_B - T_{rad} = (T_{ex} - T_{rad})(1 - e^{-\tau}) \tag{2.36}$$

which will only give an emission line (T_B(line) $> T_{rad}$) if $T_{ex} > T_{rad}$. This requires $T_e > T_{rad}$. Now T_e and T_{rad} are typically around 5 to 50 K and since $h\nu/k$ is typically of order 1 to 10 for these wavelengths, the presence of an emission line requires $C_{ji}/A_{ji} > 1$. In other words, radiative excitation will tend to drive the source function into equilibrium with the background, and the line into invisibility. Only if collisions dominate, thermalizing the source function, will the line be seen.

Another example of the use of statistical equilibrium conditions is in the solar corona. The conditions found here are very low densities, very high kinetic temperatures of millions of degrees, and the background represented by the

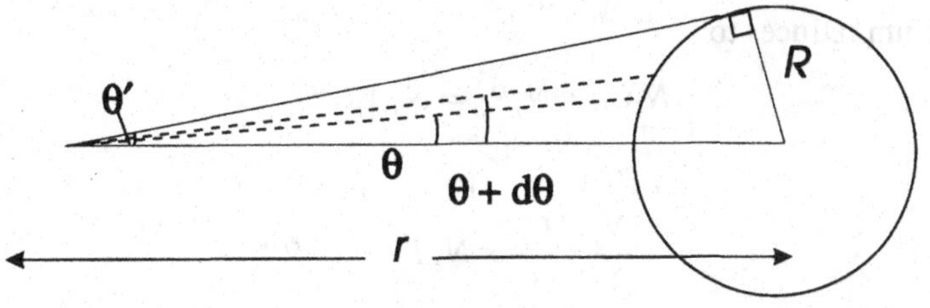

Figure 2.1.

photosphere with a T_{rad} of about 5700 K. The photospheric radiation field will be roughly blackbody in frequency distribution but will be diluted because instead of coming from all directions (solid angle $\Omega = 4\pi$), the photospheric radiation will only fall on a coronal ion from the solid angle $\Delta\Omega$ subtended by the solar disc at that point (see Figure 2.1). Let θ be the angle as seen from the corona between a point on the disc and the centre of the disc and let θ' be the value of θ for the edge of the Sun's disc, where the line of sight is tangential. Let W be the dilution factor, defined by $J_\nu = WB_\nu(T_{\text{rad}})$.

Then for R = solar radius and r = distance (coronal point to centre of Sun):

$$\Delta\Omega = 2\pi \int_0^{\theta'} \sin\theta \, d\theta, \quad \text{since } d\Omega = 2\pi \sin\theta \, d\theta$$

$$= 2\pi[1 - \cos\theta'] = 2\pi\left[1 - \sqrt{1 - \frac{R^2}{r^2}}\right]$$

$$W = \frac{\Delta\Omega}{4\pi} = \frac{1}{2}\left[1 - \sqrt{1 - \frac{R^2}{r^2}}\right] \tag{2.37}$$

If $r \gg R$, then $W \approx R^2/(4r^2)$, giving as expected an inverse square law at large distances.

Conditions in the corona are highly ionized, the levels in the ions are far apart, and the corresponding transitions between the first excited state and the ground state lie in the far ultraviolet. The radiation needed to excite such levels must also lie in the far ultraviolet, where the Sun's radiation from the photosphere is extremely weak, so radiative excitation is negligible. On the other hand, the very high coronal temperatures may mean that $kT \sim E_{ij}$ in the factor $\exp(-E_{ij}/kT)$ in the expression for C_{ij}, so collisional excitation is possible, but because of the low electron density is infrequent. De-excitation by permitted radiative transitions follows excitation almost immediately, so that the rare collisional de-excitations have virtually no chance of occurring. Stimulated emission is not important at the low J_ν involved. Hence the equation of statistical equilibrium reduces to

$$N_i C_{ij} = N_j R_{ji} = N_j A_{ji}$$

Hence

$$\frac{N_j}{N_i} = \frac{C_{ij}}{A_{ji}} \propto N_e T^{-1/2} e^{-E_{ji}/kT} \tag{2.38}$$

so

$$\begin{aligned} I_\nu &= \int j_\nu \rho \, dx \\ &= \int \frac{h\nu N_j A_{ji} \rho \, dx}{4\pi} \\ &= h\nu \int \frac{C_{ij} N_i \rho \, dx}{4\pi} \end{aligned} \tag{2.39}$$

Here $N_i \sim N_{\text{ions}}$ since nearly all the ion population will be in the ground state. Similar ideas will be involved when we consider the excitation of forbidden lines in gaseous nebulae.

Coronal ions also produce forbidden lines in the visible spectrum, from levels with the same configuration as the ground state, but different L, S, and J. These lines can be excited radiatively since photospheric radiation peaks in the visible, and in the near infrared. We then have

$$N_i(R_{ij} + C_{ij}) = N_j R_{ji}$$

with $R_{ij} = 4\pi B_{ij} J_\nu W$. Of course B_{ij} is small since we are dealing with forbidden lines, but so is C_{ij} since the density is so low. In fact the radiation field falls off as $1/r^2$, whereas it turns out that the electron density falls off more quickly, so that excitation may be by collisions at small r, and by photospheric radiation at large r.

Finally, we return to the two level atom in the optically thick case, where we have to include the term $\int \phi_\nu J_\nu d\nu$. This requires a knowledge of J_ν, which comes from other layers, and hence requires the solution of the equation of radiative transfer (which in turn involves knowing the source function, which is what we were trying to estimate in the first place!).

The other major problem is that one must take into account more than two levels, especially for stellar problems, and that the ionized state as well as bound levels must appear in the full equations of statistical equilibrium. We consider only the three level case, for two bound levels i and j, and the ionized state k. The question of the ionization equilibrium between ions and neutral atoms under non-LTE circumstances will be discussed shortly but here we are interested in the effect of transitions to and from the ionised state k on the source function for transitions between levels i and j. In stellar atmosphere conditions collisional ionization and three-body recombination are usually much less important than photoionization and radiative recombination respectively, so that we can simplify the equations of statistical equilibrium by writing $N_i(R_{ik}+C_{ik}) \simeq N_{ik}R_{ik}$ and $N_k(R_{ki}+C_{ki}) \simeq N_k R_{ki}$, and similarly for level j, to give

for level i: $N_i(R_{ij}+C_{ij}+R_{ik}) = N_j(R_{ji}+C_{ji}) + N_k R_{ki}$

for level j: $N_j(R_{ji}+C_{ji}+R_{jk}) = N_i(R_{ij}+C_{ij}) + N_k R_{kj}$

for level k: $N_k(R_{ki}+R_{kj}) = N_i R_{ik} + N_j R_{jk}$

In fact one of the equations in this set is redundant and adds no additional information. We arbitrarily choose to use the first and third equations. The third equation can be rewritten:

$$N_k = \frac{N_i R_{ik}}{R_{ki}+R_{kj}} + \frac{N_j R_{jk}}{R_{ki}+R_{kj}}$$

and substituting in the first equation:

$$N_i\left[R_{ij}+C_{ij}+\frac{R_{ik}R_{kj}}{R_{ki}+R_{kj}}\right] = N_j\left[R_{ji}+C_{ji}+\frac{R_{jk}R_{ki}}{R_{ki}+R_{kj}}\right]$$

or writing the final terms on the LHS and RHS as R^1_{ij} and R^1_{ji}, respectively:

$$N_i[(R_{ij}+C_{ij})+R^1_{ij}] = N_j[(R_{ji}+C_{ji})+R^1_{ji}]$$

This has the same form as the equation for N_j/N_i in the case of the two level atom, and can be manipulated in exactly the same way to give for the source function for levels j and i:

$$S_\nu = \frac{\int \phi_\nu J_\nu \mathrm{d}\nu + \varepsilon B_\nu + R^1_{ikj}}{1+\varepsilon+R^1_{jki}} \tag{2.40}$$

where

$$R^1_{ikj} = \frac{2h\nu^3}{c^2}\frac{g_i}{g_j}\frac{R^1_{ij}}{A_{ji}} = \frac{2h\nu^3}{c^2 A_{ji}}\frac{g_i}{g_j}\frac{R_{ik}R_{kj}}{(R_{ki}+R_{kj})}$$

and

$$R^1_{jki} = \frac{\left[R^1_{ji} - \dfrac{g_i}{g_j}R^1_{ij}\right]}{A_{ji}}$$

where it will be seen that $R^1_{ikj} \propto R_{ik}R_{kj}$, which is the rate of transference from level i to level j via photoionization to k and recombination to j, a process which of course increases the source function. Similarly R^1_{jki} is proportional to the net rate of transference from j to i via the ionized state and recombination, a process which reduces the source function.

One can write $R^1_{ikj} = R^1_{jki} \cdot B^*(T_r)$ with

$$B^* = \frac{2h\nu^3}{c^2} \frac{1}{\dfrac{g_j}{g_i}\dfrac{R_{jk}R_{ki}}{R_{ik}R_{kj}} - 1} \tag{2.41}$$

where B^* represents a pseudo-Planck function characterized by the radiation temperature T_r of the radiation field photoionizing from levels i and j to k, since the rates R_{ik} and R_{jk} are proportional to this radiation field (see the later discussion of the source function). If $\varepsilon B > R^1_{ikj} = R^1_{jki}B^*$ and $\varepsilon > R^1_{jki}$, we are essentially back to the two level situation and the line is said to be *collision-dominated.* On the other hand if $\varepsilon B < R^1_{jki}B^*$ and $\varepsilon < R^1_{jki}$, then the source function is determined by the photoionizing radiation field and T_r rather than the local kinetic temperature, and the line is said to be *photoionization-dominated.* The flux relevant to photoionization will be at a much higher frequency than the line frequency, and may originate in distant layers, so the source function becomes uncoupled from local conditions. The distinction between collisional domination and photoionization domination is important in analysing the strengths and profiles of strong lines, although it must be remembered that near the surface in both cases the 'leakage' term $\int \phi_\nu J_\nu d\nu$ will dominate and lower the source function.

Ignoring stimulated emission, we have

$$B^* \simeq \frac{\dfrac{2h\nu^3}{c^2}}{\dfrac{g_j}{g_i}\dfrac{R_{jk}R_{ki}}{R_{ik}R_{kj}}}$$

$$R^1_{jki} \simeq \frac{R_{jk}R_{ki}}{R_{ki} + R_{kj}} \frac{1}{A_{ji}}$$

$$\varepsilon \simeq \frac{C_{ji}}{A_{ji}}$$

$$\varepsilon B_\nu \simeq \frac{2h\nu^3}{c^2} \frac{C_{ji}}{A_{ji}} e^{-h\nu/kT_e}$$

$$R^1_{jki}B^* = \frac{2h\nu^3}{c^2} \frac{g_i}{g_j} \frac{R_{ik}R_{kj}}{R_{ki} + R_{kj}} \frac{1}{A_{ji}}$$

Hence, noting that $R_{ki} = N_e\alpha_{rec}(i)$ and $R_{kj} = N_e\alpha_{rec}(j)$ and that from (2.21a) $\alpha_{rec}(j)/\alpha_{rec}(i) = n(i)/n(j)$:

$$\frac{R^1_{jki}}{\varepsilon} \simeq \frac{R_{jk}}{C_{ji}} \frac{1}{\left[1 + \dfrac{\alpha_{rec}(j)}{\alpha_{rec}(i)}\right]} \simeq \frac{R_{jk}}{C_{ji}} \tag{2.42}$$

$$\frac{R^1_{jki}B^*}{\varepsilon B_\nu} \simeq \frac{g_i}{g_j} e^{h\nu/kT_e} \frac{R_{ik}}{C_{ji}} \frac{1}{\left[1 + \dfrac{\alpha_{rec}(i)}{\alpha_{rec}(j)}\right]} \simeq \frac{R_{jk}}{C_{ji}} e^{h\nu/kT_e}$$

Substitution in these equations from (2.18) and from (2.30a) then enables an estimate to be made as to whether photoionization domination or collisional domination holds in any particular case.

Ionization Balance

Ionization may be by photoionization or collisional ionization, and is balanced by recombination via radiative recombination, three-body recombination, or dielectronic recombination. The rate of photoionization (per atom) is independent of the density, but the rates of collisional ionization, radiative recombination, and dielectronic recombination are all proportional to the electron density, and the rate of three-body recombination (involving two electrons and an ion) is proportional to the density squared. At the densities appropriate to stellar atmospheres, three-body recombination can be neglected compared with the other processes.

Strictly we should sum over the ionization rates from all levels to obtain the total ionization rate, but the populations of excited levels are usually small, so in what follows we consider only ionization from $n = 1$. For a hydrogenic atom or ion the photoionization rate can be taken from (2.18), multiplied by a dilution factor W for those cases outside stellar atmospheres. Now taking $Z = 1$ in the approximate expression for collisional ionization (2.31) and remembering to use the kinetic electron temperature since this is a collisional process:

$$C_{ik} \simeq \frac{1.55 \times 10^{11}}{T_e^{1/2}} \alpha_0 \frac{\exp(-h\nu_0/kT_e)}{(h\nu_0/kT_e)} N_e \tag{2.43}$$

Substituting for α_ν at the threshold frequency, $\alpha_0 = 2.8 * 10^{25}/\nu_0^3$:

$$C_{ik} \simeq \frac{kT_e}{h\nu_0} \frac{N_e}{T_e^{1/2}} \frac{\exp(-h\nu_0/kT_e)}{\nu_0^3} \times 4.3 \times 10^{36}$$

So

$$\frac{R_{ik}}{C_{ik}} \simeq 1.8 \times 10^{-27} \nu_0^3 \frac{WT_r}{T_e^{1/2} N_e} \exp\left[-\frac{h\nu_0}{k}\left(\frac{1}{T_r} - \frac{1}{T_e}\right)\right] \tag{2.44}$$

so with $\nu_0 = 3.28 * 10^{15}$ for the case of hydrogen, $W = 1$, and $T_e \sim T_r$, we obtain $R_{ik}/C_{ik} \sim 6 * 10^{19} T^{1/2}/N_e$.

For typical stellar atmosphere temperatures and densities photoionization dominates collisional ionization, and this will also be true for circumstellar shells where the dilution factor will be small, but the density will be very small. It should be noted that we are dealing here with the ionization balance and hence with ground states, but that if we were interested in highly excited state populations the situation might become rather more favourable to collisional ionization in some circumstances.

In the solar corona, $T_r \sim 5000$ K(photospheric), but $T_e \sim 10^6$ K, so despite the low density, collisional ionization is dominant. As has been pointed out already, dielectronic recombination is dependent on the kinetic temperature via an exponential factor, whereas radiative recombination is only weakly temperature dependent, so under high temperature coronal conditions dielectronic recombination can dominate radiative recombination, and the ionization balance is essentially one between collisional ionization and dielectronic recombination. Both these rates are proportional to the electron density, so for stages of ionization with net charges Z and $Z+1$ we have

$$N(Z)C_{1k}(N_e, T) = N(Z+1)\sum_i D_{ki}(N_e, T)$$

In this equation we have taken $i = 1$ for the ionization rate since only the ground state is appreciably populated, but we have summed over all the dielectronic recombination rates D_{ki}, assuming that all single electron excited states i thus formed will decay to the ground state. $N(Z+1)/N(Z)$ will be a function of temperature only since both C_{1k} and D_{ki} are directly proportional to the electron density. Similar equations can be written for other stages of ionization with the constraint that the sum of the number densities in all stages of ionization must equal the total number density of the element concerned, but in practice the temperature dependence is very steep, so that a particular stage of ionization is only found in appreciable quantities over a limited temperature range, and for most of that temperature range that particular stage of ionization is dominant.

Now consider the case of stellar atmospheres and circumstellar envelopes,where the balance is between photoionization and radiative recombination. In what follows we take N_k as fixed and calculate N_i relative to it. It is convenient to use the quantity N_i^*, which is the value N_i would have if LTE held. Thus for the ionization potential I, the excitation potential E_i of level i, and the ionization energy from level i, $\chi_i = I - E_i$, Saha's and Boltzmann's equations give

$$\frac{N_k}{N_i^*} = \frac{1}{N_e}\left[\frac{2\pi mkT}{h^2}\right]^{3/2}\left[\frac{2U_k}{g_i}\right]\exp(-\chi_i/kT)$$

where U_k is the partition function of the ion. The radiative recombination rate is given by (2.19)

$$R_{ki} = 4\pi\left[\frac{m}{2\pi kT}\right]^{3/2}\frac{h^3 g_i}{2m^3 U_k}N_e\int_{\nu_0}^{\infty}\left(\frac{2h\nu^3}{c^2} + J_\nu\right)\frac{\alpha_\nu}{h\nu}\exp\left[-\frac{h\nu - \chi_i}{kT}\right]d\nu$$

$$= \left[\frac{N_i^*}{N_k}\right]4\pi\int_{\nu_0}^{\infty}\frac{\alpha_\nu}{h\nu}\left[\frac{2h\nu^3}{c^2} + J_\nu\right]\exp\left(-\frac{h\nu}{kT}\right)d\nu \qquad (2.45)$$

Suppose we want to find the departure from LTE of the population of level i by determining the departure coefficient d_i defined by $N_i = (1 + d_i)N_i^*$ (note that $b_i = 1 + d_i$ is sometimes used in nebular studies). Suppose also that we can ignore inputs to level i and losses from level i through transitions with other levels of the lower stage of ionization. Then

$$N_i R_{ik} = N_k R_{ki}$$

$$N_i^* 4\pi\left[\int_{\nu_0}^{\infty} \frac{\alpha_\nu J_\nu}{h\nu} \mathrm{d}\nu + d_i \int_{\nu_0}^{\infty} \frac{\alpha_\nu J_\nu}{h\nu} \mathrm{d}\nu\right] = N_i^* 4\pi\left[\int_{\nu_0}^{\infty} \left(\frac{\alpha_\nu}{h\nu}\right)\left(\frac{2h\nu^3}{c^2} + J_\nu\right) \mathrm{e}^{-h\nu/kT} \mathrm{d}\nu\right]$$

Hence

$$d_i = \frac{\displaystyle\int_{\nu_0}^{\infty} \frac{\alpha_\nu}{h\nu}\left[\frac{2h\nu^3}{c^2} \mathrm{e}^{-h\nu/kT} + J_\nu(\mathrm{e}^{-h\nu/kT} - 1)\right] \mathrm{d}\nu}{\displaystyle\int_{\nu_0}^{\infty} \frac{\alpha_\alpha J_\nu}{h\nu} \mathrm{d}\nu}$$

This finally gives for the departure coefficient, using $B_\nu[1 - \exp(-h\nu/kT)] = 2h\nu^3/c^2 \exp(-h\nu/kT)$:

$$d_i = \frac{\displaystyle\int_{\nu_0}^{\infty} [B_\nu(T_e) - J_\nu](1 - \mathrm{e}^{-h\nu/kT}) \frac{\alpha_\nu}{h\nu} \mathrm{d}\nu}{\displaystyle\int_{\nu_0}^{\infty} \frac{\alpha_\alpha J_\nu}{h\nu} \mathrm{d}\nu} \tag{2.46}$$

This equation has sometimes been used in stellar atmosphere work to make a rough estimate of departures from LTE in the lower levels of bound–free transitions, but this raises the question of whether the neglect of levels other than i in the lower stage of ionization is justifiable. Consider the case of hydrogen and the population of the $n = 2$ level. The radiative line transitions to and from $n = 2$ (Lyman α downwards and the Balmer series upwards) are all very strong lines which will be optically thick in the continuum forming layers of a stellar atmosphere, and hence upward and downward radiative transitions in these lines balance and can be removed from the equations of statistical equilibrium. If collisional transitions to other bound levels can be neglected in comparison with R_{ik} and R_{ki}, then (2.46) will be valid as a first approximation.

Equation (2.46) then indicates that there will be departures from LTE if the radiation field J_ν deviates from the Planck function at the local kinetic temperature, with level i being underpopulated if $J_\nu > B_\nu(T_e)$ and vice versa. The frequencies of importance are those for which the probability of photoionization from level i is large, that is frequencies close to the threshold frequency ν_0 for photoionization since the cross-section $\propto \nu^{-3}$. For a grey atmosphere with a linear source function, $J(\tau = 0) = (1/2) B(T_{\text{eff}})$ in the Eddington approximation and $B(\tau = 0) \sim B(T = 0.81 T_{\text{eff}})$, so very approximately

$$\frac{J_\nu}{B_\nu} \simeq \frac{1}{2} \frac{B_\nu(\nu_0, T_{\text{eff}})}{B_\nu(\nu_0, 0.81 T_{\text{eff}})} \simeq \frac{1}{2} \exp\left[0.23 \frac{h\nu_0}{kT_{\text{eff}}}\right]$$

if the frequency is such that Wien's law, $B_\nu \sim 2h\nu^3/c^2 \exp(-h\nu/kT)$, holds. Hence if $h\nu_0/kT_{\rm eff} > 3$, $J > B$ at ν_0 and we can expect underpopulation, with $h\nu_0 = \chi(n=2)$ for the case being considered. This gives underpopulation for $T_{\rm eff} = 10\,000$ K, but overpopulation for $T_{\rm eff} = 35\,000$ K. These rough results have to be checked with a proper model atmosphere calculation, but in general the deviations from LTE are small for hydrogen levels in the continuum forming regions in most stars.

We now turn to the case of the ionization balance in extended stellar atmospheres, circumstellar shells, and stellar winds. The radiation field can be taken as externally imposed and given by the diluted radiation field $J_\nu = W\,B_\nu(T_{\rm rad})$, the local kinetic temperature is $T_{\rm e}$, and we again consider a single level i in the lower stage of ionization and take N_k as being fixed, since we are only interested in relative populations. The ionization balance gives $N_i R_{ik} = N_k R_{ki}$. If LTE held one would have $N_i^* R_{ik}^* = N_k^* R_{ki}^* = N_k R_{ki}$ since the recombination process is a collisional one and always proceeds at the LTE rate at the local kinetic temperature. Hence $N_i R_{ik} = N_i^* R_{ik}^*$, with R_{ik}^* given by the photoionization rate when exposed to a Planckian radiation field at the local kinetic temperature, $J_\nu = B_\nu(T_{\rm e})$.

Now

$$\frac{N_k}{N_i} = \frac{N_k N_{\rm e}}{N_i}\frac{1}{N_{\rm e}} = \frac{N_k N_{\rm e}}{N_i^*}\frac{N_i^*}{N_i}\frac{1}{N_{\rm e}} = \left[\frac{N_k N_{\rm e}}{N_i}\right]^* \frac{R_{ik}}{R_{ik}^*}\frac{1}{N_{\rm e}}$$

Substituting Saha's equation for $(N_k N_{\rm e}/N_i)^*$ and assuming i is the ground state, with

$$R_{ik} = 4\pi \int_{\nu_0}^{\infty} \frac{W\alpha_\nu B_\nu(T_{\rm rad})}{h\nu}\,{\rm d}\nu \quad \text{and} \quad R_{ik}^* = 4\pi \int_{\nu_0}^{\infty} \frac{\alpha_\nu B_\nu(T_{\rm e})}{h\nu}\,{\rm d}\nu$$

then

$$\frac{N_k}{N_i} = \left[\frac{2\pi m k T_{\rm e}}{h^2}\right]^{3/2} \frac{2U_k}{U_i}\,{\rm e}^{-I/kT_{\rm e}}\,\frac{W}{N_{\rm e}}\,\frac{\displaystyle\int_{\nu_0}^{\infty}\frac{\alpha_\nu B_\nu(T_{\rm rad})}{h\nu}\,{\rm d}\nu}{\displaystyle\int_{\nu_0}^{\infty}\frac{\alpha_\nu B_\nu(T_{\rm e})}{h\nu}\,{\rm d}\nu} \tag{2.47}$$

If Wien's law holds for $\nu > \nu_0$, i.e. if $h\nu_0/kT > 1$, and $\alpha \propto 1/\nu^3$, then

$$\int_{\nu_0}^{\infty} \frac{\alpha_\nu B_\nu(T)}{h\nu} \propto \int_{\nu_0}^{\infty} \frac{1}{\nu^4}\frac{2h\nu^3}{c^2}\exp\left(-\frac{h\nu}{kT}\right){\rm d}\nu$$

$$\propto \int_{x_0}^{\infty} \frac{{\rm e}^{-x}}{x}\,{\rm d}x \text{ where } x = h\nu/kT$$

$$\propto E_1(x_0) \simeq \frac{{\rm e}^{-x_0}}{x_0} \quad \text{since } x_0 = h\nu_0/kT \gg 1$$

$$= \frac{kT}{h\nu_0}\exp\left(-\frac{h\nu_0}{kT}\right)$$

Substituting in (2.47) with $T = T_{\rm rad}$ in the integral in the numerator and $T = T_{\rm e}$ in the integral in the denominator, and writing $W \approx R^2/4r^2$ for a distance to the star, r, much greater than the radius of star R, we have with $I = h\nu_0$:

$$\frac{N_k}{N_i} = \frac{R^2}{4r^2}\frac{1}{N_{\rm e}}\left[\frac{2\pi mk}{h^2}\right]^{3/2}\frac{2U_k}{U_i}T_{\rm e}^{1/2}\,T_{\rm rad}\exp\left(-\frac{I}{kT_{\rm rad}}\right) \tag{2.48}$$

This is Stromgren's ionization equation.

Under what circumstances is it applicable? In a thin gas cloud most atoms will be in the ground state, so $N_i \sim N_{\rm atom}$, where 'atom' here means the lower stage of ionization being considered. Line transitions will be optically thin in many cases, so one can ignore radiative line transitions upward from i, and collisional excitations and de-excitations will usually be unimportant in the low density conditions, except to low lying levels, which can be considered for present purposes as part of the ground state. However there will be recombinations to levels other than i, and these recombinations will cascade rapidly down to the ground state, so our original balance should be modified to $N_i R_{ik} = \sum_i N_k R_{ki}$, and in (2.47) the integral in the denominator should therefore be a sum over integrals with differing ν_0.

Thus the simple formula will underestimate the recombination rate by factors of 2 to 3, and correspondingly overestimate the degree of ionization. In the case of hydrogen, the line radiation connecting $n = 1$ and $n = 2$ is optically thick (Lyman α) and cannot be neglected; indeed the recombination cascade must be treated as effectively ending on $n = 2$, which serves as a pseudo ground state. One must be watchful for other effects of line optical thickness. Note also that the simplicity of (2.48) results from the assumption that the radiation field can be treated as an externally imposed given quantity, and hence this approach is not appropriate deep in a stellar atmosphere, where finding J_ν is part of the problem.

3 LINE PROFILES

Introduction

Line formation, the subject of the next chapter, can only be understood on the basis of an understanding of line broadening, for while the wings of a line will be weak, the core of a line may be saturated. Line broadening can be divided into *natural line broadening*, which is always present, *Doppler line broadening*, due to the motion of the observed atoms in different directions with different velocities with the resulting Doppler shifts producing a spread in the frequency of the observed line, and *collisional* or *pressure line broadening*, due to the effects of other particles on the radiating atom. Doppler broadening has the property that it is always proportional, in wavelength units, to the wavelength of the line centre, whereas natural and collisional line broadening vary in their effect from line to line, with no systematic trend with wavelength. On the other hand, collisional broadening is always proportional to the number of colliding particles per unit volume, and hence to the pressure.

There is, however, another important division, namely between broadening which occurs locally, on the small scale, and broadening that is only produced on the large scale. An example of the latter is the rotation of a star. Because stars are unresolved, our spectra include light from parts of the star that are moving away from us, and light from parts of the star that are moving towards us, and the difference in Doppler shift results in a broadened line. However this superimposition due to our lack of resolution has no effect on line formation. It is only the small scale local effects that influence the formation of the line and saturation.

Doppler Line Broadening

First let us consider line broadening due to the motion of individual radiating atoms in a hot gas. The frequency ν of a line emitted from an atom moving with velocity v_r in the line of sight is given by

$$\Delta\nu = \nu - \nu_0 = -(v_r/c)\cdot\nu_0$$

where ν_0 is the rest frequency of the line and c is the speed of light. The sign of $\Delta\nu$ is negative (a redshift in wavelength terms) if v_r is positive, i.e. away from the observer. The same is, of course, true of the line absorbed by such an atom. The normalized profile of the line ϕ_ν will be given by the distribution of line of sight velocities v_r, since the radiation that we observe is normally the result of the radiation of many atoms. The number of atoms travelling with velocities between v_r and $v_r + dv_r$ in a particular direction is given by the 'one-dimensional' form of Maxwell's equation

$$n(v_r)\mathrm{d}v_r = N\left(\frac{m}{2\pi kT}\right)^{1/2} \mathrm{e}^{-mv_r^2/2kT}\mathrm{d}v_r$$

where N is the total number of atoms of mass m. Substituting for v_r in terms of the Doppler shift, we obtain the resulting line profile:

$$\phi(\Delta\nu) \propto n[\Delta v_r] = (\text{constant})\cdot\exp[-(\Delta\nu/\Delta\nu_D)^2]$$

where the *Doppler width* $\Delta\nu_D = \nu_0/c\sqrt{([2kT]/m)}$.

Now $\phi(\Delta\nu)$ is normalized by

$$\int_0^\infty \phi_\nu \mathrm{d}\nu = \int_{-\infty}^{+\infty} \phi(\Delta\nu)\mathrm{d}(\Delta\nu) = 1$$

so

$$(\text{constant})\,\Delta\nu_D \int_{-\infty}^{+\infty} \mathrm{e}^{-x^2}\mathrm{d}x = 1, \quad \text{with } x = \frac{\Delta\nu}{\Delta\nu_D}$$

Hence the constant has the value $1/(\sqrt{\pi}\cdot\Delta\nu_D)$ since the value of the integral is $\sqrt{\pi}$. Thus we finally have for thermal Doppler broadening:

$$\phi_\nu = \frac{1}{\sqrt{\pi}\Delta\nu_D}\mathrm{e}^{-(\Delta\nu/\Delta\nu_D)^2} \tag{3.1}$$

with

$$\Delta\nu_D = \frac{\nu_0}{c}\sqrt{\frac{2kT}{m}}$$

or, in wavelength terms:

$$\phi_\lambda = \frac{1}{\sqrt{\pi}\Delta\lambda_D}\mathrm{e}^{-(\Delta\lambda/\Delta\lambda_D)^2} \tag{3.1a}$$

$$\Delta\lambda_D = \frac{\lambda_0}{c}\sqrt{\frac{2kT}{m}}$$

where λ_0 is the wavelength of the line centre, $\Delta\lambda = \lambda - \lambda_0$, and the profile is now normalized with respect to $\Delta\lambda$.

The bell-shaped profile of the form $\exp[-(\Delta x)^2]$ is called *Gaussian*. The profile falls to half its maximum height/depth (which is reached when $\Delta\lambda = 0$) when

$\exp[-(\Delta\lambda/\Delta\lambda_D)^2] = 1/2$, i.e. when $\Delta\lambda = \surd(\log_e 2)\,\Delta\lambda_D$. Thus the *full width at half height* (FWHH) of the line profile, that is the wavelength separation between the points where the profile drops to half its maximum height/depth on either side of the line centre, is given by

$$\text{full width at half height} = 2\surd(\log_e 2)\Delta\lambda_D = 1.667\,\Delta\lambda_D$$

Thus for an iron atom (atomic weight 55) in a solar-type atmosphere ($T = 5700$ K, say) and for a line in the visible ($\lambda_0 = 500$ nm, say), the FWHH of the profile is $3.63 * 10^{-3}$ nm. This refers to both the absorption and emission coefficient profiles—the actual line will be subject to radiative transfer effects, with saturation depressing the central portion of the observed line (reducing the dip in an absorption line) compared with points further out and hence increasing the observed FWHH compared with the absorption or emission coefficient profile. Note that the Doppler width is proportional to the rest wavelength, is inversely proportional to the square root of the mass of the atom concerned (and so is much greater for a hydrogen atom than for an iron atom), and is otherwise the same for all lines.

The motions of individual atoms are not the only contributors to Doppler line broadening, for there are also larger scale collective motions of groups of atoms. These motions are usually called *turbulence*, although there are often contributions from oscillations and streaming motions, which are not 'turbulence' in the fluid dynamicist's meaning of the word . If the size of the moving elements is considerably less than the thickness of the line forming region, then the movements will affect line formation in much the same way as the movements of individual atoms and we call the phenomenon *microturbulence.*

It is usually assumed that the velocities of the microturbulent elements have a Gaussian distribution, that is that the probability of velocity v in the line of sight is proportional to $\exp(-v^2/V_t^2)$ where V_t is a constant called the *microturbulent velocity*. The resulting line profile will be the same as for thermal Doppler broadening, since the assumed velocity distribution is the same as that for thermal broadening. In general, both thermal and microturbulent Doppler broadening will be present and the resulting line profile will correspond to the convolution of the two profiles. The convolution of two Gaussians is a Gaussian with width equal to the square root of the sum of the squares of the widths of the contributing Gaussians. Hence the resulting profile is given by (3.1) but with Doppler width:

$$\Delta\lambda_D = \frac{\lambda_0}{c}\sqrt{\frac{2kT}{m} + V_t^2}$$

$$\Delta\nu_D = \frac{\nu_0}{c}\sqrt{\frac{2kT}{m} + V_t^2} \qquad (3.2)$$

But is the distribution of microturbulent velocities Gaussian? Indeed, do we expect there to be microturbulent velocities in stellar atmospheres or interstellar clouds, and if so, what do we expect those velocities to be? Theory at the moment does not allow us to answer any of these questions convincingly, but we can observe that there are certainly large scale motions in cool stars, and such motions will generate smaller scale motions and we can also note that the overall result of random processes often approximates to a Gaussian distribution. Observationally, microturbulent Doppler broadening behaves like thermal Doppler broadening in that it is proportional to the wavelength, but unlike it in that the broadening is independent of the mass of the atom concerned. Hence, in principle, the two contributions can be distinguished. One finds for nearly all stars a Doppler broadening that affects line formation and saturation and so must be small scale, but which is considerably larger than could be produced by thermal motions alone. The microturbulent velocity thus inferred is typically 1–2 km s^{-1} for solar-type stars and considerably more for giants, and is comparable with or greater than the mean thermal velocity. In general, the microturbulent velocity will be anisotropic and depth dependent, as certainly happens in the Sun.

Motions involving gaseous elements larger than the line-forming regions also exist. These are called *macroturbulence*. Some of these motions will be in the direction of the observer, giving a line that is Doppler shifted to higher frequencies, and some will be away from the observer, giving rise to a line that is Doppler shifted to lower frequencies. The observed line will result from many such elements, and as a result we will see a broadened line. If the distribution of these large scale motions is Gaussian, then the line absorption coefficient will have a Gaussian profile, with a FWHH of $1.667\,\lambda/c\,v_{\mathrm{mt}}$ in wavelength units, where v_{mt} is the root mean square velocity and is called the *macroturbulent velocity*. Again we have no general theory at the moment that says that the distribution of velocities must be Gaussian, and the Gaussian assumption must remain a best guess. The important point here is that line formation takes place *within* a moving element and hence is unaffected by the motion. The effect is as if we were observing the sum of the spectra of a set of nearly identical stars with different radial velocities. The strength of a line is unaffected by macroturbulence, in contrast with the cases of thermal and microturbulent Doppler velocities.

We know theoretically that main sequence stars with spectral types later than about F5, together with red giants and supergiants, have convective envelopes. In these envelopes energy is transferred by hot rising and cold falling streams of gas. The convective motions are predicted to extend outwards into the deeper layers of the atmosphere, where we would expect them to give rise to macroturbulent broadening. This can be observed directly in the case of the Sun, where under conditions of excellent angular resolution the smooth photosphere can be seen to be divided up into many bright *granules* separated by darker lanes. If a spectrograph with a long slit is used under these conditions, the resulting spectrum shows

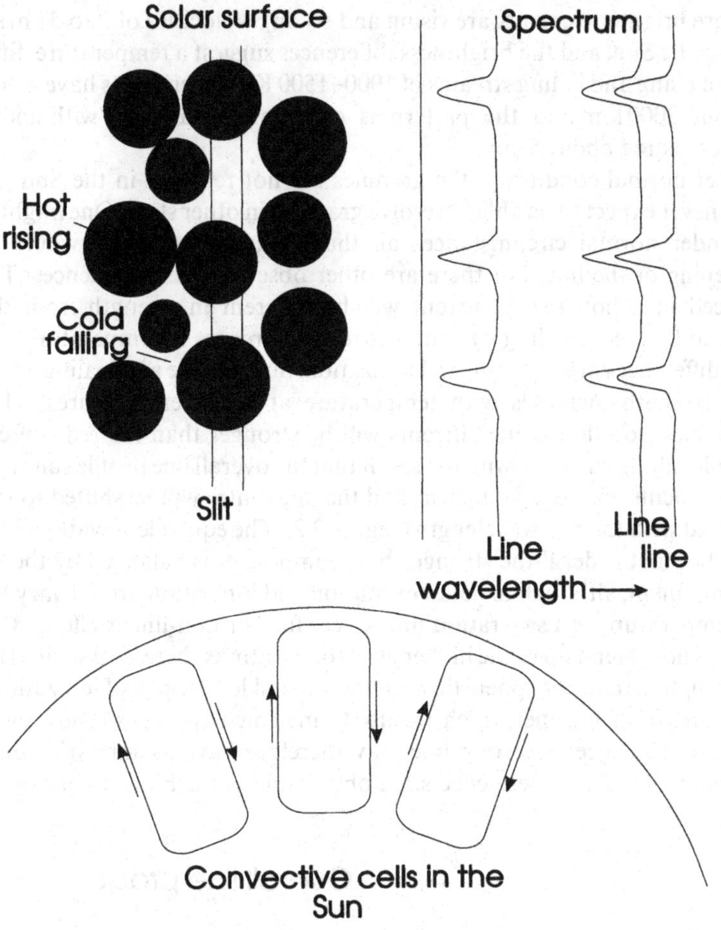

Figure 3.1.

'wiggly lines' because elements moving away from the observer will produce a line shifted to the red in the spectrum at the corresponding points along the slit and vice versa for elements moving towards the observer (Figure 3.1). It is found that the bright granules give blue shifted lines and the darker lanes give rise to redshifted lines, just as would be predicted for convection (the hotter and therefore brighter elements are rising and so on). Velocities of 2 to 3 km s^{-1} are found in the Sun, and the brightness differences suggest a temperature difference between rising and falling streams of 1000–1500 K. The granules have a diameter of about 2000 km and the pattern is constantly changing, with individual granules lasting about 5 min.

Under normal conditions, the granules are not resolved in the Sun, and we would never expect to be able to resolve granules in other stars. One might expect that under normal circumstances all that could be observed would be the broadening of the line, but there are other observable consequences. The line produced in a hot, rising current will be different in strength from the line produced by a cold falling current, because differences in temperature will give rise to differences in excitation and ionization. Suppose we are dealing with a line whose strength increases with temperature at solar temperatures. The blue shifted lines from hot rising currents will be stronger than the red shifted lines from cold falling currents, with the result that the overall line profile summed over many elements will be asymmetric, and the line centre will be shifted to the blue compared with the rest wavelength (Figure 3.2). The equivalent width will not be affected to first order if the stronger blue component is balanced by the weaker red component, although in detail excitation and ionization do not vary linearly with temperature and saturation introduces further non-linear effects. Convection does not extend up to the higher levels of the atmosphere, and so lines formed mainly high in the atmosphere (low excitation and low degree of ionization lines) or the cores of strong lines which are also formed high up will not show the effects of macroturbulence. A strong line may therefore have asymmetric wings but a symmetric core. All these effects are observable in the Sun , where we can go

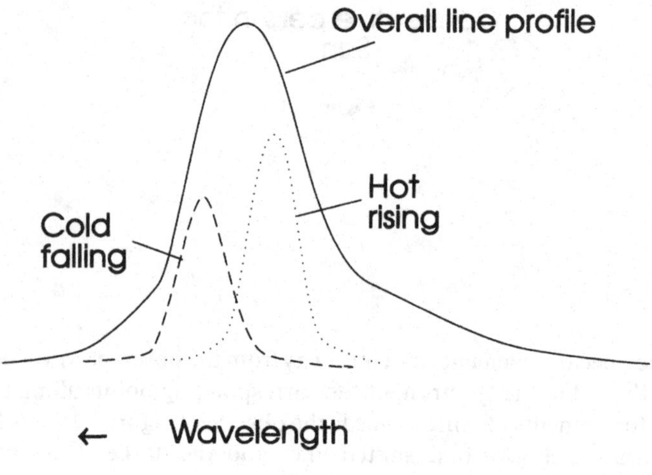

Figure 3.2.

further and observe the situation 'side-on' by looking towards the edge or limb. We will then see the tangential component of the macroturbulent velocity which will now be in the line of sight, and which turns out to be generally smaller than in the radial direction. Recently it has become possible to observe in the Sun (and in some other stars) a rich variety of oscillations on various scales, some of which contribute to the macroturbulence. In the Sun, observations of lines from near the limb of the Sun can measure the tangential component of macroturbulence.

When we turn to other stars, very little of this detail is accessible, for such stars are not resolved, and although asymmetry is in principle observable, often the spectral resolution or the quality of the spectra are not high enough for macroturbulent asymmetry to be detected and distinguished from other effects so that all we can normally detect is a macroturbulent velocity from the broadening. Even this presents difficulties. We are looking for a Doppler component of line broadening that does not affect the line strength, which still leaves the possibility of confusion with rotational broadening. The line profile predicted for rotation is well established, and differs somewhat from the Gaussian shape assumed for macroturbulence in being more rectangular. More helpfully, we can assume that rotating stars will sometimes be seen pole-on, with no rotational broadening in the observed spectrum. If a given type of star never shows zero 'macro' Doppler line broadening, then we can infer that the pole-on examples have macroturbulence, and hence that probably all examples of this type of star have appreciable macroturbulence. Macroturbulence of at least the same order as the microturbulent and thermal velocities appears to be present in stars with convective envelopes. Cool supergiants appear to have large macroturbulent velocities of the order of $10\,\mathrm{km\,s^{-1}}$. Convective theory suggests that supergiants with their low surface gravities and large scale heights should have very large, long-lived convective cells, and it could be that the surfaces of such stars are covered by small numbers of cells, leading to variations of an irregular kind in the magnitude of the star, depending on whether the side facing us has n or $n-1$ cells at a given time.

Rotational line broadening occurs because if we observe a rotating star perpendicular to the rotation axis, one edge or limb will be rotating away from us and will give rise to a line shifted to longer wavelengths, while the other limb will be approaching and give rise to a line shifted to shorter wavelengths. As the star will not be resolved, we will see a line resulting from the superimposition of the limb and intermediate zones and hence a broadened line. If the rotation axis is not perpendicular to the line of sight, the broadening will be less. Indeed, if we happen to see the star pole-on, all motions will be across the line of sight, and so no rotational Doppler broadening will be seen at all. Clearly, the rotation will not affect the formation of a line, for again we are seeing the superimposition of line-forming regions of different Doppler shifts, and so rotation affects the profile but not the strength of a line. Indirectly, there is an effect, for the centrifugal terms in the equation of hydrostatic equilibrium alter the effective surface gravity, and lead to stellar models in which the effective temperature varies over the surface of a star between the equator and the poles. This in turn leads to a variation of line strength over the surface of the star, whose effect on the observed line integrated over the whole disc depends on the orientation of the star.

Consider then a rotating star as seen by the observer, and to start with assume the rotation axis is perpendicular to the line of sight. Let distance from the rotation axis seen from this point of view be p, and let distance parallel to the rotation axis be x (Figure 3.3), with $x = 0$ at the equator. Thus a strip at fixed p extends from $x = h = \sqrt{(R^2 - p^2)}$ to $x = -h = -\sqrt{(R^2 - p^2)}$, where R is the radius of the star. Now imagine looking down on the star along the rotation axis (Figure 3.4), and suppose the line of sight to the earth for some particular point (p, x) on the star's projected disc intersects the spherical surface at a point O, where in the plane containing the line of sight O lies a distance r from the rotation axis, the plane intersecting the axis at point C. Let the angle between CO and the line of sight from C on the rotation axis to the earth be θ. Then $p = r\sin\theta$, as can be seen if we imagine the plane containing CO intersecting the star in a circle of radius r. The angle between the tangent to the surface at O, which is perpendicular to CO, and the line of sight to O is $90 - \theta$, and if the surface is rotating at velocity $v(r)$, the radial velocity (that is the velocity component in the line of sight) is $v(r)\sin\theta$.

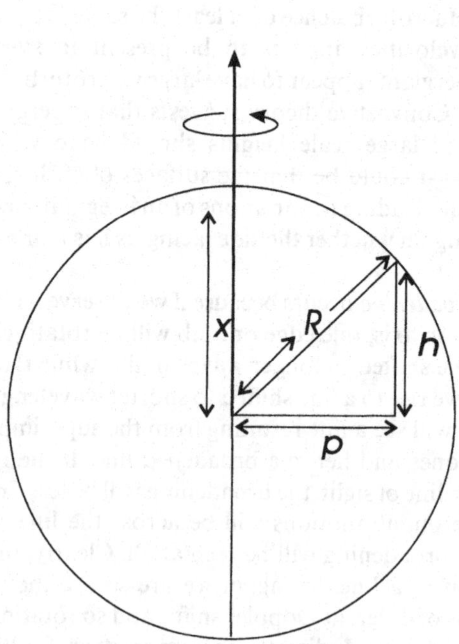

Figure 3.3. Star as seen by the observer

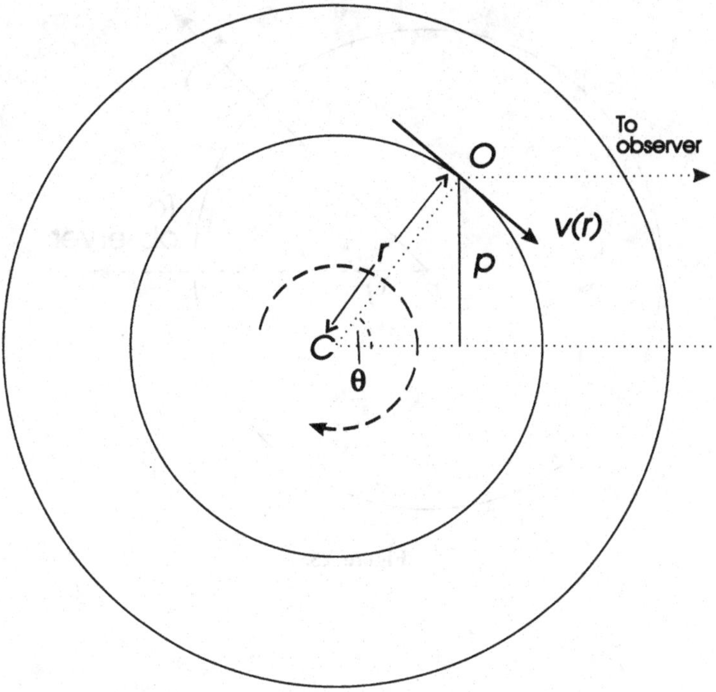

Figure 3.4. Star viewed along the rotation axis

Now suppose the star rotates like a solid body (i.e. all points on the surface have the same period of rotation) so $v(r) \propto r$ since period $= 2\pi r/v(r)$. Then if the velocity of the equator ($r = R$) is v_R, we have that $v(r) = (r/R)v_R$ and that the velocity in the line of sight $= (r/R)v_R \sin\theta = (p/R)v_R$. Hence all points on a strip of given p will have the same velocity in the line of sight, and will give the same Doppler shift for a line of rest wavelength λ_0, namely $\Delta\lambda = \lambda_0(p/R)v_R/c$. In general, of course, the rotation axis will not be perpendicular to the line of sight but will be inclined at an angle i to it (Figure 3.5), in which case we will measure only the radial component of the velocity $v(r)\sin i$, so we obtain $\Delta\lambda = (\lambda_0/c)(v_R \sin i)(p/R)$.

The profile of the line will be given by the relative contributions from strips at different p (say between p and dp), each giving its own characteristic $\Delta\lambda$, with the corresponding strip at $-p$ giving $-\Delta\lambda$ on the other side of the profile (Figure 3.6). For a uniformly bright disc:

$$I(\Delta\lambda)d(\Delta\lambda) = I(p)dp\,2h$$

Hence

$$\frac{I(p)}{I(0)} = \frac{h}{R} = \sqrt{1 - \frac{p^2}{R^2}}$$

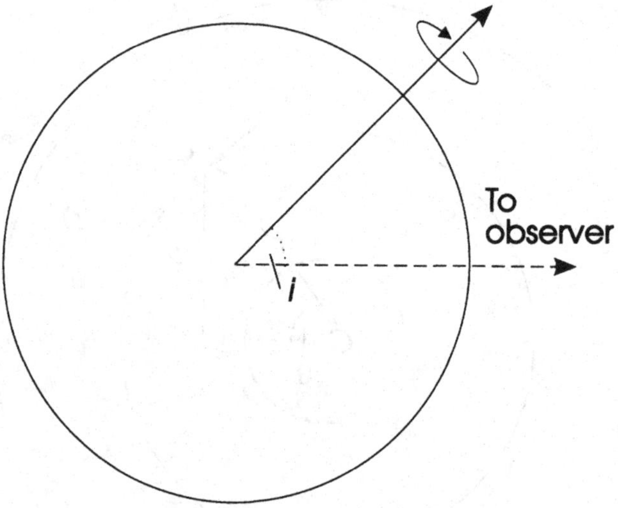

Figure 3.5.

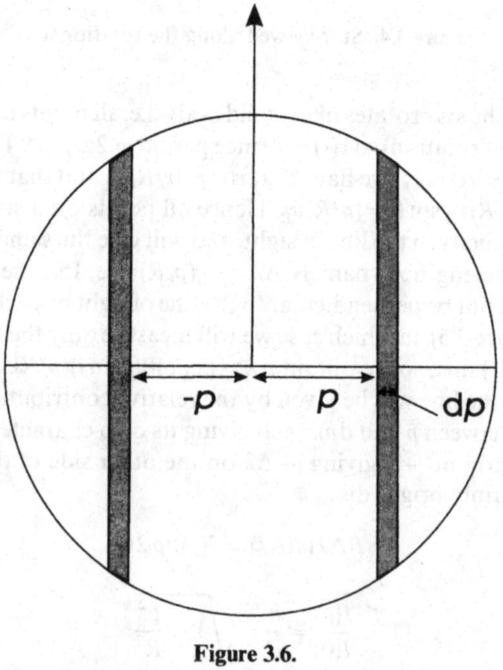

Figure 3.6.

Thus

$$\frac{I(p)}{I(0)}=\sqrt{1-\left(\frac{\Delta\lambda}{\Delta\lambda_{\max}}\right)^2} \tag{3.3}$$

with $\Delta\lambda_{\max}=\lambda_0/c\cdot v_R \sin i$.

This can be normalized to give a normalized profile $\phi(\Delta\lambda)$ such that $\int\phi d(\Delta\lambda)=1$. Thus writing $\phi=\text{constant}\sqrt{(1-[\Delta\lambda/\Delta\lambda_{\max}]^2)}$:

$$\Delta\lambda_{\max}\int_{-1}^{+1}(\text{constant})\sqrt{1-\left(\frac{\Delta\lambda}{\Delta\lambda_{\max}}\right)^2}\,d\left(\frac{\Delta\lambda}{\Delta\lambda_{\max}}\right)=1$$

which gives the value of the constant to be $2/(\pi\Delta\lambda_{\max})$ and hence

$$\phi(\Delta\lambda)=\frac{2}{\pi\Delta\lambda_{\max}}\sqrt{1-\left(\frac{\Delta\lambda}{\Delta\lambda_{\max}}\right)^2} \tag{3.3a}$$

The full width at half height is $\sqrt{3}\Delta\lambda_{\max}$ and the profile is much more rectangular than the Gaussian shape found (or assumed) for other kinds of Doppler broadening. Rotational broadening is relatively easily distinguished from thermal or microturbulent broadening by its lack of effect on line strengths. Distinguishing rotational broadening from macroturbulent broadening is harder, as has already been discussed, although it must be remembered that macroturbulence does not normally exceed 10 km s^{-1}, whereas rotational velocities can reach several hundred kilometres per second.

In general the disc will not be uniformly bright, but will be affected by limb darkening, so that in the simplest case $I=a+b\mu'$, where $\mu'=\cos\theta'$ and θ' is the angle between the line of sight and the radius vector from the centre of the star C_0 (not C) to O (Figure 3.7). Consider a plane through O perpendicular to the line of sight and suppose that the line of sight to the centre of the star intersects this plane at B. Then $OB=\sqrt{(p^2+x^2)}$. Consider now the angle $C_0OB=\alpha$ in the right-angled triangle C_0OB. We find that $\cos\alpha=OB/R$ and noting that $\theta'=90-\alpha$, we finally obtain

$$\mu'=\cos\theta'=\sqrt{1-\frac{p^2+x^2}{R^2}}$$

Hence the energy received from a strip on the star running from p to $p+\mathrm{d}p$ and from $+h$ to $-h$ is

$$\frac{\mathrm{d}p}{(\text{distance of star})^2}\int_{-h}^{+h}(a+b\mu')\,\mathrm{d}x$$

which can be rewritten as being proportional to

$$\propto 2ah+bR\int_{-h}^{+h}\sqrt{\left(1-\frac{p^2}{R^2}\right)-\frac{x^2}{R^2}}\frac{\mathrm{d}x}{R}$$

$$\propto 2aR\sqrt{1-\frac{p^2}{R^2}}+2bR\left(1-\frac{p^2}{R^2}\right)\frac{\pi}{4}$$

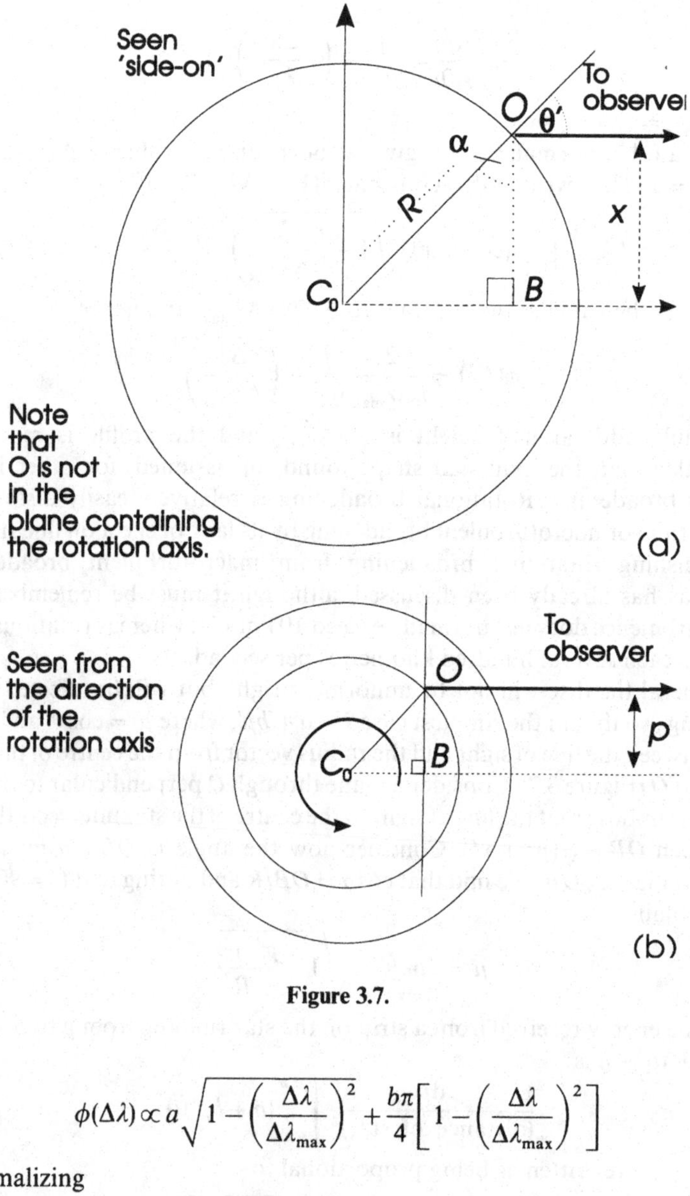

Figure 3.7.

Hence

$$\phi(\Delta\lambda) \propto a\sqrt{1-\left(\frac{\Delta\lambda}{\Delta\lambda_{\max}}\right)^2}+\frac{b\pi}{4}\left[1-\left(\frac{\Delta\lambda}{\Delta\lambda_{\max}}\right)^2\right]$$

and normalizing

$$\phi(\Delta\lambda)=\frac{1}{\Delta\lambda_{\max}}\frac{\dfrac{2}{\pi}\sqrt{1-\left(\dfrac{\Delta\lambda}{\Delta\lambda_{\max}}\right)^2}+\dfrac{b}{2a}\left[1-\left(\dfrac{\Delta\lambda}{\Delta\lambda_{\max}}\right)^2\right]}{1+\dfrac{2b}{3a}} \tag{3.4}$$

We have the remaining problem that there is no way of determining i, the inclination, for any individual star, and hence no way of doing more than setting a lower limit to the equatorial velocity of the star. However if, for a particular type of star, we assume that all such stars have the same v_R and are orientated at random, v_R can be estimated. The average measured equatorial velocity will be given by

$$\begin{aligned}(v_R \sin i)_{\text{av}} &= v_R(\sin i)_{\text{av}} \\ &= v_R \frac{1}{4\pi}\int_0^{4\pi} \sin i \, \mathrm{d}\Omega \\ &= v_R \frac{2\pi}{4\pi}\int_0^{\pi} \sin^2 i \, \mathrm{d}i \\ &= v_R \frac{\pi}{4}\end{aligned}$$

Hence

$$v_R = \frac{4}{\pi}(v_R \sin i)_{\text{av}} \tag{3.5}$$

It is found, for instance, that main sequence stars hotter than spectral type F5 are all fast rotators with equatorial velocities of several hundred km s^{-1}, while cooler main sequence stars are slow rotators with equatorial velocities less than 10 km s^{-1}, if they can be measured at all.

Natural and Pressure Line Broadening

Spectral lines have an intrinsic width even in the absence of all external effects of relative motion or interaction with other particles. If the excited states of an atom have very long lifetimes, each state would have a very small spread in energy and hence give rise to a very sharp line. In practice, decay is usually rapid, so lifetimes are very short, and by Heisenberg's uncertainty principle, $\Delta E \Delta t \sim h/(2\pi)$, so there will be an uncertainty in energy or a spread in energy level, giving rise to a correspondingly broadened line. The magnitude of the broadening is $\Delta\nu \sim \Delta E/h \sim 1/(2\pi\Delta t) \sim A_{ji}/(2\pi)$ since the lifetime is the reciprocal of the decay probability. This is called *natural line broadening*, which we include here with pressure broadening since the profile is similar.

We shall take ΔE to be the difference in energy from the mean energy of the upper level of the transition j to i, E_j, with the line central frequency $\nu_0 = E_{ij}/h$ and we take level i to be the ground state and hence unbroadened (Figure 3.8). Now the wave function describing the upper state in quantum mechanics can be separated into a part depending only on the spatial coordinates and a part depending only on the time. For a non-decaying state the latter can be written $\exp(-i2\pi E_{ij}/h \cdot t)$ so the full wave function becomes $\Psi_j = \psi_j(\mathbf{r})\exp(-2\pi i E_{ij}/h \cdot t)$.

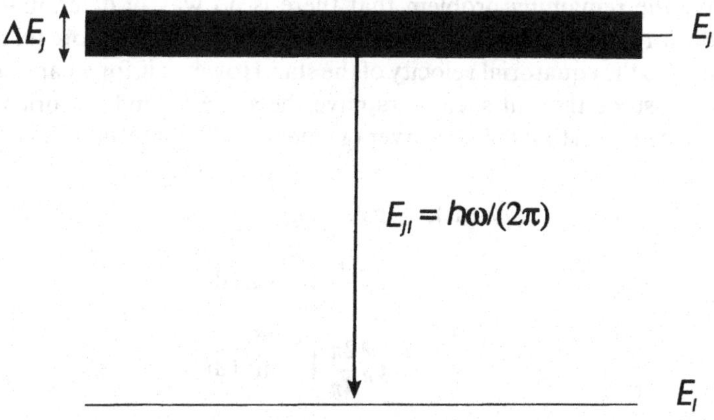

Figure 3.8.

The probability of finding an atom in level j is proportional to $\Psi_j^*\Psi_j$ and integrated over all space this is proportional to $\psi_j^*\psi_j$, which is independent of time, as it should be.

Now if we take into account the interaction of radiation and matter, we find that level j spontaneously decays, with decay probability per unit time A_{ji}. A constant decay probability means that $dN_j/dt = -A_{ji}N_j$, which has the solution $N_j = N_j(t=0)\exp(-A_{ji}t)$. Since $N_j \propto$ probability of finding an atom in the state j and is proportional to $\Psi_j^*\Psi_j$, we must modify our expression for Ψ_j to

$$\Psi_j = \psi_j(r)\exp[-(2\pi i E_{ij}/h)t]\exp[-A_{ji}t/2)$$

Suppose we excite j at $t = 0$ and then let it decay. The time dependence of the wave function is given by

$$f(t) = \exp[-2\pi i\nu_0 t]\exp\{-A_{ji}t/2\}, \quad \text{from } t = 0 \text{ to } \infty$$

This can be Fourier analysed into a set of non-decaying frequencies:

$$f(t) = 2\pi\int F(\nu)\exp\{-2\pi i\nu t)\,d\nu$$

where the amplitude of each frequency is given by

$$F(\nu) = \frac{1}{2\pi}\int f(t)e^{+2\pi i\nu t}\,dt$$

$$\propto \int_0^\infty \exp\left(-2\pi i(\nu_0 - \nu) - \frac{A_{ji}}{2}t\right)dt$$

$$\propto \frac{1}{2\pi i(\nu_0 - \nu) + \dfrac{A_{ji}}{2}}$$

Now the energy emitted at each frequency is proportional to the complex square of the amplitude, i.e.

$$I \propto F^*(\nu)F(\nu)$$

Hence

$$I \propto \frac{1}{4\pi^2(\nu_0 - \nu)^2 + A_{ji}^2/4}$$

Or

$$I \propto \frac{1}{(\nu - \nu_0)^2 + \left(\frac{A_{ji}}{4\pi}\right)^2}$$

The normalized profile is then found from

$$(\text{constant}) \int_{-\infty}^{+\infty} \frac{d(\Delta\nu)}{(\nu - \nu_0)^2 + \left(\frac{A_{ji}}{4\pi}\right)^2} = (\text{constant}) \frac{4\pi^2}{A_{ji}} = 1$$

Hence

$$\phi_\nu = \frac{A_{ji}}{4\pi^2} \frac{1}{(\nu - \nu_0)^2 + \left(\frac{A_{ji}}{4\pi}\right)^2} \tag{3.6}$$

Of course there may be more than one downward transition from level j, in which case we replace A_{ji} by $\Gamma_j = \Sigma_l A_{jl}$. Finally, if the lower level of the transition is not the ground state, then this level too will be broadened, and we must replace Γ by $\Gamma_j + \Gamma_i$ in the expression for the profile.

We now turn to *pressure* or *collisional broadening*, caused by interactions with other particles, and hence proportional to their number density and to the pressure. We first discuss a classical approach to such interactions, which views their effect as due to the interruption of the radiation process. This is called the *impact* picture. We imagine the radiating atom as producing an effectively infinite train of waves of frequency ν_0, starting at time $t = 0$, which is then interrupted after time T, so the amplitude of the wave is given by

$$f(t) \propto \exp(-2\pi i\nu_0 t), \quad t = 0 \text{ to } t = T$$

This can be Fourier analysed into a set of infinite monochromatic waves

$$f(t) \propto \int F(\nu)\exp(-2\pi i\nu t)d\nu$$

where the amplitude at each frequency is given by

$$F(\nu) = \frac{1}{2\pi}\int_0^{\infty} f(t)e^{+2\pi i\nu t}\,dt$$

$$\propto \int_0^{T} e^{-2\pi i(\nu_0-\nu)t}\,dt$$

$$\propto \frac{e^{-2\pi i(\nu_0-\nu)T} - 1}{-2\pi i(\nu_0-\nu)}$$

The energy emitted at each frequency is $\propto F^*(\nu)F(\nu)$. Hence

$$I(\nu) = \frac{2 - e^{2\pi i(\nu_0-\nu)T} - e^{-2\pi i(\nu_0-\nu)T}}{4\pi^2(\nu_0-\nu)^2} \tag{3.7}$$

However, there will not just be one fixed interval T between collisions. Suppose collisions occur on average at intervals T_0, so the probability per unit time of a collision is $1/T_0$. The number of particles remaining 'unhit' after time t, $N(t)$, is given by

$$dN/dt = -1/T_0 N$$

so

$$N = N_0 \exp(-t/T_0)$$

Hence the probability that a particle remains 'unhit' for a time t and then suffers its first collision between t and $t + dt$ is $1/T_0 \exp(-t/T_0)dt$ and the probability that a time T elapses between collisions is $1/T_0 \exp(-T/T_0)$. Equation (3.7) must now be averaged over T with each T value given the weighting just calculated:

$$I(\nu) \propto \frac{1}{4\pi^2(\nu_0-\nu)^2 T_0}\int_0^{\infty} (2 - e^{2\pi i(\nu_0-\nu)T} - e^{-2\pi i(\nu_0-\nu)T})e^{-T/T_0}\,dT$$

$$\propto \frac{1}{4\pi^2(\nu_0-\nu)^2 T_0}\left(2T_0 - \frac{1}{\left[-2\pi i(\nu_0-\nu) + \frac{1}{T_0}\right]} - \frac{1}{\left[2\pi i(\nu_0-\nu) + \frac{1}{T_0}\right]}\right)$$

$$\propto \frac{1}{4\pi^2(\nu_0-\nu)^2 T_0}\left[2T_0 - \frac{\frac{2}{T_0}}{4\pi^2(\nu_0-\nu)^2 + \frac{1}{T_0^2}}\right)$$

$$\propto \frac{1}{4\pi^2(\nu_0-\nu)^2 + \frac{1}{T_0^2}}$$

Thus the normalized profile is

$$\phi_\nu = \frac{\Gamma}{4\pi^2}\frac{1}{(\nu-\nu_0)^2+\left(\dfrac{\Gamma}{4\pi}\right)^2} \tag{3.8}$$

where the Lorentzian width $\Gamma = 2/T_0$.

This line shape is called *Lorentzian* and is identical to that found for natural line broadening. The 'Full Width at Half Height' of the profile in frequency units is $\Gamma/(2\pi)$.

But how can Γ, or equivalently T_0, be calculated? If the cross-section for a collision that interrupts the radiation is σ, and the flux of colliding particles with respect to some radiating atom is $N_c v$ per unit area and per unit time, where N_c is the number density of colliding particles, then the number of effective collisions per second is $\sigma N_c V$ and the average interval between collisions is $T_0 = 1/(N_c \sigma V)$, so $\Gamma = 2N_c\sigma V$. Here V is the *average relative* velocity of colliding and radiating particles (as opposed to the root mean square velocity) so

$$V = \sqrt{\frac{8kT}{\pi}\left(\frac{1}{m_r}+\frac{1}{m_c}\right)} \tag{3.9}$$

where m_r and m_c are the masses of the radiating and colliding particles, respectively.

Early attempts to visualize the collision as one between solid billiard balls with radii of R_r and R_c, respectively, both being of the order of 10^{-10} m, so $\sigma = \pi(R_r^2 + R_c^2)$, and which predicted that all lines of a given atom would have the same broadening, were of course doomed to failure. The collision is actually an interaction produced by the effect of the electrical field of the colliding particle on the radiating particle. Suppose the effect of this interaction has the form

$$\Delta\nu = (\Delta E_{ij}/h) = 1/(2\pi)K_{ij}/r^n$$

where ΔE_{ij} is the shift in energy produced by the interaction, n gives the form of the interaction, and K_{ij} is a constant depending on atomic structure and varying from line to line (i and j representing as usual the lower and upper levels of the transition). The total *phase shift* produced by a passing particle is

$$\eta = 2\pi\int \Delta\nu\,\mathrm{d}t$$

$$= \int \frac{K_{ij}}{r^n}\mathrm{d}t$$

$$\eta(\rho) = \int_{-\infty}^{\infty}\frac{K_{ij}}{(\rho^2+[Vt]^2)^{n/2}}\mathrm{d}t$$

where ρ is the distance of closest approach or impact parameter (see Figure 3.9)

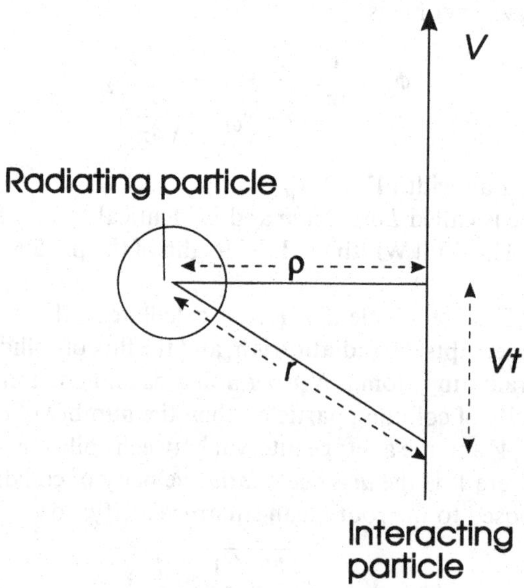

Figure 3.9.

with $t = 0$ at $r = \rho$. It has been assumed that the colliding particle is hardly deflected and pursues a straight line course, so $r^2 = \rho^2 + (Vt)^2$ since the distances along the particles track from the point of closest approach are $\pm Vt$. Then

$$\eta(\rho) = \frac{K_{ij}}{\rho^{n-1}V}\int_{-\infty}^{\infty}\frac{\frac{V}{\rho}}{\left(1+\left[\frac{Vt}{\rho}\right]^2\right)^{n/2}}\,\mathrm{d}t = \frac{K_{ij}}{\rho^{n-1}V}I_n \tag{3.10}$$

where the integral $I_n = \pi$ for $n = 2$, $\pi/2$ for $n = 4$, and $3\pi/8$ for $n = 6$.

Hence if we define a collision as happening when $\eta \geqslant \eta_0$, then the maximum impact parameter ρ_0 is given by

$$\rho_0 = \left[\frac{K_{ij}I_n}{\eta_0 V}\right]^{1/(n-1)}$$

Taking $\eta_0 = 1$, a value of the phase shift that would certainly disrupt the radiation, we obtain $\rho_0 = \rho_W$, the *Weisskopf radius*, and hence find for the Lorentz width:

$$\Gamma = 2\pi N_c V\left[\frac{K_{ij}I_n}{V}\right]^{2/(n-1)} \tag{3.11}$$

Now $N_c \propto P/T$ by the gas law, and $V \propto \sqrt{(T/m)}$, so

$$\Gamma \propto P \times T^{(-n-1)/2(n-1)}$$

giving, as always, a broadening proportional to the pressure, and a weak dependence on the temperature. This approach can only be an approximation, for it ignores the effects of weak, distant encounters with $\eta < 1$, and treats all strong encounters as being equally effective. The Lindholm approach takes into account the differing effects of collisions with various values of ρ and hence varying phase shifts η. We now take $\eta(t)$ to mean the phase shift at time t as a result of interactions up to that time with the radiating atom. The intensity at frequency ν now becomes

$$I(\nu) = \int e^{-2\pi i(\nu_0-\nu)t - i\eta(t)} dt \times \int e^{2\pi i(\nu_0-\nu)t + i\eta(t)} dt$$

$$= \int e^{-2\pi i(\nu_0-\nu)t' - i\eta(t')} dt' \times \int e^{2\pi i(\nu_0-\nu)t'' + i\eta(t'')} dt''$$

$$= \int_{-\infty}^{\infty}\int_{-\infty}^{\infty} e^{2\pi i(\nu-\nu_0)(t''-t') + i[\eta(t'')-\eta(t')]} dt'\, dt''$$

$$= \int_{-\infty}^{\infty} e^{2\pi i(\nu-\nu_0)t} dt \int_{-\infty}^{\infty} e^{i[\eta(t+t')-\eta(t')]} dt' \tag{3.12}$$

where in the last equation we have written $t'' = t + t'$, and the second integral $I_1(t)$, the autocorrelation function times $\exp(-2\pi i\nu_0 t)$, represents a time average. Consider the effect of increasing t by Δt which changes the phase shift by $\Delta\eta$ (due to a collision):

$$\Delta I_1 = \int_{-\infty}^{\infty} e^{i[\eta(t+t')-\eta(t')]}[e^{i\Delta\eta} - 1] dt'$$

The two factors in the integrand are independent if Δt is much shorter than the interval between collisions and much longer than the time of a collision so the time average represented by the integral can be written as the product of the averages of the two factors, giving

$$\Delta I_1 = I_1 \langle e^{i\Delta\eta} - 1 \rangle$$

The time average represented by the second factor can be replaced by an average over impact parameters, that is by considering at one time the many collisions with different values of ρ. The number of collisions per unit volume in time Δt with the impact parameter between ρ and $\rho + d\rho$ is $N_c V \Delta t\, 2\pi\rho\, d\rho$, so we obtain:

$$\Delta I_1 = I_1 N_c V \Delta t 2\pi \int_0^{\infty} [e^{i\Delta\eta} - 1]\rho\, d\rho$$

which integrates to give $I_1 = \exp\{-N_c VSt)$, with

$$S = S(\text{real}) - iS(\text{imag}) = \int_0^\infty [1 - e^{i\Delta\eta(\rho)}] 2\pi\rho \, d\rho$$

since the shift $\Delta\eta$ will be a function of the impact parameter alone for a given velocity.

This value of I_1 can then be inserted into (3.12) to give the line profile in much the same way as for (3.8), with the real part of S giving the line broadening so $N_c VS$(real) replaces $1/T_0$. Writing $\exp(i\Delta\eta) = \cos(\Delta\eta) + i\sin(\Delta\eta)$, we find

$$\Gamma = 2N_c V 2\pi \int [1 - \cos(\eta[\rho])] \rho \, d\rho \tag{3.13}$$

In addition there is a small shift of the centre of the line profile (effectively a change in ν_0 in the Lorentz formula), $1/(2\pi)N_c VS(\text{imag})$, which is again proportional to the pressure (a *pressure shift*). It should be noted that classical theory has been used here, and quantum mechanics gives similar but not identical results. It should also be noted that it has been assumed that no change of energy level is involved in the collisions, which is not a good approximation for broadening by electrons.

We now consider the other approach to line broadening, the *quasi-static* approximation. Here we picture the radiating atom as having its energy levels perturbed by static perturbers at various distances, the line being shifted by the perturbation, and the overall line profile being given by the distribution of perturber distances. We shall consider here only the effect of one perturber, the nearest neighbour. Let a perturber at distance r produce a shift $\Delta\nu = K'_{ij}/r^n$, which is as before except for convenience we have defined K'_{ij} in terms of frequency ν instead of circular frequency $\omega = 2\pi\nu$. Let the probability of the nearest neighbour being at a distance between r and $r + dr$ from the radiating atom be $P(r)dr$. Then $P(\Delta\nu)$, the probability that absorption or emission is shifted by $\Delta\nu$, gives the normalized line profile, with

$$P(\Delta\nu)d(\Delta\nu) = P(\Delta\nu)d\nu = -P(r)dr$$

Now

$$d(\Delta\nu) = d\nu = -nK_{ij}/r^{n+1} dr$$

and so

$$P(\Delta\nu) = -P(r)dr/d\nu = P(r)r^{n+1}/(nK_{ij})$$

Now the probability that the nearest neighbour lies between r and $r + dr$ is given by the product of (probability that there is a neighbour between r and $r + dr$) and (probability that there are no neighbours at less than r), or

$$P(r)dr = 4\pi r^2 N dr \left[1 - \int_0^r P(x)\,dx\right]$$

where N is the number of perturbing particles per unit volume, and $4\pi r^2 \mathrm{d}r$ is the volume of the shell between r and $r + \mathrm{d}r$.
Hence

$$\frac{P(r)}{4\pi r^2 N} = 1 - \int_0^r P(x)\mathrm{d}x$$

$$\frac{\mathrm{d}}{\mathrm{d}r}\left[\frac{P(r)}{4\pi r^2}\right] = -P(r)$$

$$= -4\pi r^2 N\left[\frac{P(r)}{4\pi r^2 N}\right]$$

$$P(r) = 4\pi r^2 N \mathrm{e}^{-(4/3)\pi r^3 N}$$

It is convenient now to define a standard distance r_0 by $4/3\pi r_0^3 = 1/N$, where $1/N$ is the average volume occupied by a perturbing particle and so $r_0 \sim$ average distance of a perturbing particle from a radiating atom. Then the standard shift $\Delta\nu_0$ produced by a perturber at the standard distance is $\Delta\nu_0 = K'_{ij}/r_0^n$, so $(r/r_0)^n = (\Delta\nu_0/\Delta\nu)$. Substituting for N in $P(r)$ we obtain:

$$P(r) = \frac{3}{r}\left(\frac{r}{r_0}\right)^3 \mathrm{e}^{-(r/r_0)^3}$$

$$P(\Delta\nu) = \frac{3}{r}\left(\frac{r}{r_0}\right)^3 \mathrm{e}^{-(r/r_0)^3}\frac{r^{n+1}}{nK'_{ij}}$$

$$= \frac{3}{n}\left(\frac{r}{r_0}\right)^{3+n} \mathrm{e}^{-(r/r_0)^3}\frac{r_0^n}{K'_{ij}}$$

$$= \frac{3}{n}\left(\frac{\Delta\nu}{\Delta\nu_0}\right)^{-(3/n+1)} \mathrm{e}^{-(\Delta\nu/\Delta\nu_0)^{-3/n}}\frac{1}{\Delta\nu_0} \tag{3.14}$$

$$\simeq \frac{3}{n}\left(\frac{1}{\Delta\nu}\right)^{(3/n+1)} (\Delta\nu_0)^{3/n}, \quad \Delta\nu \gg \Delta\nu_0 \tag{3.15}$$

where the last expression holds in the wings of the line.

In the literature, one often finds this shape written in terms of a parameter β, where $\beta = F/F_0$, the field of a perturbing particle at distance r is F and the field of a perturbing particle at distance r_0, the normal field, is F_0. The probability distribution for the nearest neighbour then becomes

$$P(\Delta\nu) = \frac{3}{n}\beta^{-(3+n)/2}\mathrm{e}^{-\beta^{-3/2}}\frac{1}{\Delta\nu_0} \tag{3.16}$$

which is often written as $W(\beta)$, where $W(\beta)\,\mathrm{d}\beta = P(\Delta\nu)\mathrm{d}(\Delta\nu)$, with $W(\beta)$ independent of pressure and the nature and magnitude of the interaction, so

$$P(\Delta\nu) = W(\beta)2/n\, 1/(\Delta\nu_0)\beta^{1-n/2}$$

Then

$$W(\beta) = P(\Delta\nu)\frac{\mathrm{d}\nu}{\mathrm{d}\beta} = \tfrac{3}{2}\beta^{-5/2}\mathrm{e}^{-\beta^{-3/2}} \tag{3.17}$$

For the linear case with $n = 2$, the perturbation is proportional to the field, so $\Delta\nu = k_{ij}F$ say, and

$$P(\Delta\nu) = \frac{W(\beta)}{\Delta\nu_0} = \frac{W(\beta)}{k_{ij}F_0} \tag{3.18}$$

where

$$\begin{aligned} F_0 &= \frac{e}{4\pi\varepsilon_0 r_0^2} \\ &= \frac{e}{4\pi\varepsilon_0}\left[\frac{4}{3}\pi N\right]^{2/3} \\ &= 3.74 \times 10^{-9} N^{2/3} \mathrm{V\,m^{-1}}, \quad \text{with } N \text{ in m}^{-3} \end{aligned} \tag{3.19}$$

Alternatively, since $\beta = \Delta\nu/(k_{ij}F_0)$ and hence for a particular $\Delta\nu$ depends on the value of k_{ij} for the sub-component considered, one sometimes finds the probability distribution quoted as $S(\alpha)$, where the parameter $\alpha = k_{ij}\,\beta = (\Delta\nu)/F_0$ with $P(\Delta\nu) = S(\alpha)/F_0$ or the equivalent in wavelength units.

The profile for broadening in the quasi-static approximation so far deduced assumes that only a single perturber acts at a time. This is adequate when $\Delta\nu$ is large, for it is unlikely that a second perturber will be close enough to significantly alter the large effect of the first very close perturber (remembering that large shifts require a nearby perturber). Hence (3.15) will still hold for the wings of a line even when multiple perturbers are considered. On the other hand, for small shifts multiple distant simultaneous perturbers are quite likely, and one has to use the more elaborate Holtsmark theory for the core of the line. A positively charged particle tends to be surrounded by a slightly higher than average density of negatively charged particles, and this *shields* the positively charged particle, an effect that has to be taken into account at high densities. If the density is low enough for shielding to be neglected and $\beta < 1$, Holtsmark theory gives

$$W(\beta) \simeq \frac{4\beta^2}{3\pi}[1 - 0.463\beta^2] \tag{3.20}$$

using the β form of the profile.

The quasi-static profile is asymmetric, entirely on one side or other of $\Delta\nu = 0$, and rises from zero at $\Delta\nu = 0$ to a maximum at $\Delta\nu \sim 0.711\Delta\nu_0$ for the nearest neighbour approximation and about twice as far out using the full Holtsmark theory (Figure 3.10). The whole profile scales as $(\Delta\nu_0)^{3/n}$ along the frequency axis (3.14), and this quantity is proportional to r_0^{-3} and hence to N, so again the

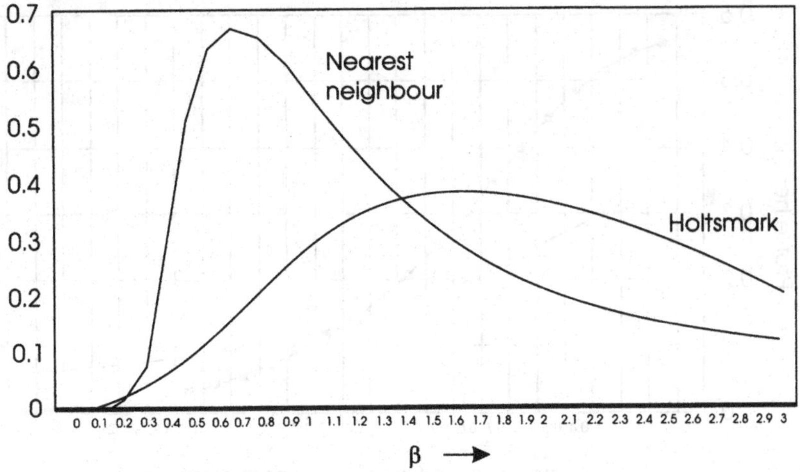

Figure 3.10. Quasi-static profiles

broadening is proportional to pressure. In the wings the profile is proportional to $(\Delta\nu)^{-7/4}$ for $n=4$ and to $(\Delta\nu)^{-5/2}$ for $n=2$, to be compared with $(\Delta\nu)^{-2}$ for the wings of a Lorentzian profile. Thus all collisional profiles are characterised by slowly falling wings, compared with the sharper fall-off of the Gaussian profiles produced by Doppler broadening (Figure 3.11). Collision broadening (or natural broadening if the pressure is very low) will therefore dominate in the wings and Doppler broadening in the core, and the details of the collision broadened profile near the centre of a line are not seen in the observed profiles.

Most atomic levels are degenerate, possessing a number of sub-levels with the same energy. Collisional interactions will remove some of this degeneracy, and so the above considerations should be applied separately to each sub-level, which will have its own 'susceptibility to interaction', K_{ij}, and hence its own $\Delta\nu_0$, and then the results should be summed over to obtain the total line profile. It must also be remembered that both the upper and the lower levels of a line are collision broadened, the line broadening corresponding to the sum of the spreads in energy of the upper and lower levels.

We have still to answer the question of when the impact theory gives the best approximation, and when the quasi-static theory is likely to be nearest to the true situation. The characteristic time for the establishment of a frequency shifted by $\Delta\nu$ from the line centre is $1/(\Delta\nu)$. Impact theory assumes (very roughly speaking) that the effect of a collision depends only on the overall phase shift η, which will be true if the length of a collision is much less than the characteristic time for the shift concerned. On the other hand, if the length of a collision is much greater than the characteristic time for the frequency shift, one can treat the collision as a series of

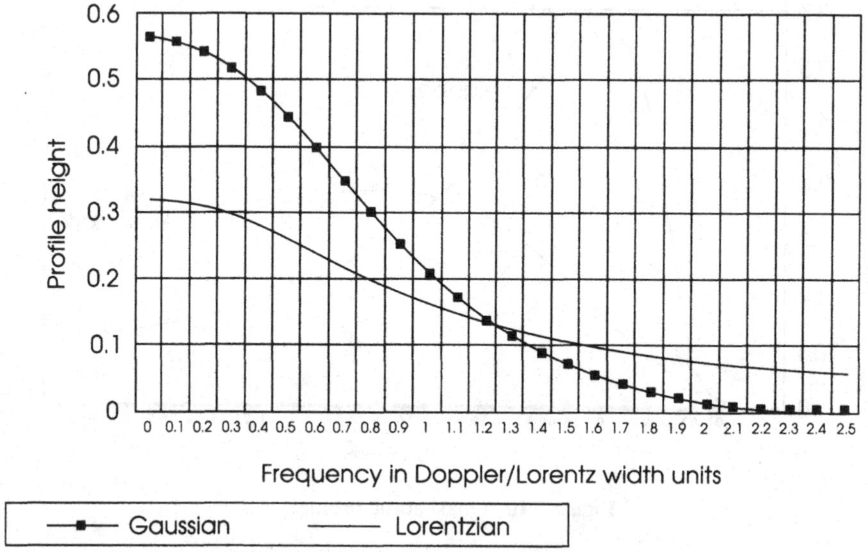

Figure 3.11. Gaussian v. Lorentzian profiles

stationary snapshots, with the perturber at first far away, producing a small shift in the energy levels of the radiating atom, then nearer, producing a larger shift, then farther away again. We can then replace collisions by a static picture of interactions with the perturbing particles distributed according to probability theory, in other words we can use the quasi-static approach. Now the length of a collision $\sim \rho/V$, where ρ is, as before, the impact parameter (this is the time for r to vary from ρ to $\sqrt{2}\cdot\rho$). Thus according to this very rough argument, the dividing line between the regime of the impact theory and that of the quasi-static theory comes when

$$\frac{\rho}{V} \simeq \frac{1}{\Delta\nu}$$

Now the instantaneous shift $\Delta\nu = K_{ij}/\rho^n$ for $r = \rho$ so the dividing line comes at $V/\rho \sim K_{ij}/\rho^n$ and the impact parameter on the dividing line is $\rho_{\text{crit}} \sim (K_{ij}/V)^{1/(n-1)}$. The frequency shift on the dividing line is then

$$(\Delta\nu)_{\text{crit}} \simeq \frac{V^{n/(n-1)}}{K_{ij}^{1/(n-1)}}$$

$$\propto \left(\frac{T}{m}\right)^{n/2(n-1)} \tag{3.21}$$

using the fact that $V^2 \propto T/m$ and omitting factors of the order of 2π.

For $\Delta\nu > (\Delta\nu)_{crit}$, in the wings of the line, the quasi-static theory will be most appropriate, whereas for $\Delta\nu < (\Delta\nu)_{crit}$ in the core of the line, the impact theory will apply. The dividing line depends on the temperature, but more critically on the mass of the particle. If the colliding particles are electrons as opposed to heavy ions, $(\Delta\nu)_{crit}$ will be large, and so for electrons impact theory will apply far out into the wings, essentially for the whole of the profile, whereas for broadening by ions quasi-static theory will be appropriate for the wings and the core will in any case be dominated by Doppler broadening.

What of the form of the interaction itself? First consider the case where the interacting particle is charged. Then the interaction corresponds to the application of an external electric field to the radiating atom, and since the field F of a point charge is inversely proportional to the square of the distance, we might expect $n = 2$ to be appropriate. However we have to consider quantum mechanical perturbation theory here, with perturbation in potential energy of eFz, where z is displacement in the direction of the field. The shift in energy level in first-order perturbation theory is $\int \Psi^* eFz\Psi \mathrm{d}V$, where the Ψ are the wave functions describing the radiating atom and the integral is over space. Now the wave functions for a particular atomic state will have a definite parity (depending for a single electron on l) so that changing z to $-z$ either leaves the wavefunction unchanged or multiplies it by -1. Hence $\Psi^*(z)\Psi(z) = \Psi^*(-z)\Psi(-z)$ and the positive z and negative z contributions to the integrand over volume cancel and the first-order perturbation vanishes. One has then to go to second-order perturbation theory, where the shift in energy levels is proportional to the square of the perturbation, that is to the field squared. This is called the *quadratic Stark effect*, and corresponds to $n = 4$. The quadratic Stark effect is one-sided, that is it moves all the energy sub-levels (which are degenerate in the absence of a field) in the same direction so that a line is not only split and broadened, but the average frequency of the line is changed.

The exception occurs for one electron atoms and ions, most importantly for hydrogen. States with a given principle quantum number but different orbital angular momentum quantum numbers and hence different parities are degenerate in such atoms and ions, and so their wavefunctions in the unperturbed state must be written as a mixture of two wavefunctions of different parity:

$$\Psi = a\Psi_+ + b\Psi_-.$$

When first-order perturbation theory is applied to such wavefunctions, the integrand contains terms like $a^*\Psi_+^* eFz\, b\Psi_-$, which do not vanish on integration. Hence one has a first-order perturbation proportional to the field which dominates second-order effects. This is called the ***linear Stark effect***, and corresponds to $n = 2$. The linear Stark effect splits a level symmetrically into $2n_p^2$ sub-levels for principle quantum number n_p, so that for every sub-level that decreases in energy there is a sub-level that increases in energy (Figure 3.12). The broadening is therefore symmetric, and the profile is found by summing over the

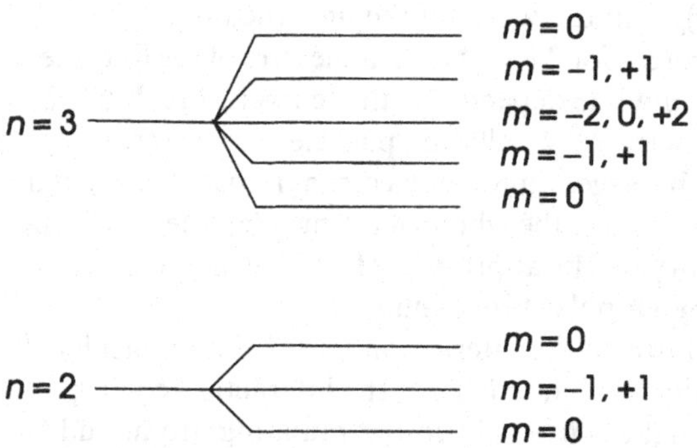

Figure 3.12. Linear Stark effect

profiles of each sub-component s with its own particular weight w_s in the sum $S(\alpha) = \sum(w_s/k_s)W(\alpha/k_s)$.

In cool stars, there are too few electrons and ions present to make a large contribution to line broadening, despite the strong nature of the interactions involved. Neutral particles interact by *Van der Waals* forces as follows. Atoms contain central positive electrical charges surrounded by negative charge clouds, and can mutually induce in each other dipole moments. The induced moment is proportional to $1/r^3$ and the interaction between dipoles is proportional to $1/r^3$ so $\Delta E \propto r^{-6}$ and $n = 6$. The induced moments are also proportional to the polarizabilities of the radiating atom and the radiating atom. The most significant broadening agent is nearly always hydrogen. Calculation of the constants k_{ij} is hard for complex atoms, and the experimental determination of the constants is also hard for many substances since the hot vapour of an element such as iron reacts violently with hydrogen gas. Inert gases like argon and neon with polarizabilities of the same order as that of hydrogen can be used and the results interpolated to find the value of k_{ij} that would be found for a gas of the polarizability of hydrogen. An additional complication comes from the fact that the interaction also has an appreciable $1/r^{12}$ term.

The Line Absorption Coefficient

The line profile enters the line formation problem through the line absorption coefficient. We have already seen that an absorption coefficient can be written as the product of the cross-section σ_ν per atom, with $N_{\rm abs}$ the number of atoms per unit mass capable of absorbing radiation at the appropriate frequency:

$$\kappa_\nu = \sigma_\nu N_{\rm abs} \tag{3.22}$$

where

$$N_{abs} = \frac{N_i}{N_{ion}} \frac{N_{ion}}{N_{el}} \frac{N_{el}}{N_H} N_H \qquad (3.23)$$

Here N_i is the number of atoms of the appropriate element in the right state of ionization and excitation to absorb the line concerned, N_{ion} is the number of atoms of the appropriate element in the right state of ionization, N_{el} is the number of atoms of the appropriate element, and N_H is the number of atoms of hydrogen per unit mass. 'Atom' is taken as usual to include both atoms and ions, and the number densities in the ratios can, of course, both be per unit mass or both be per unit volume. N_H could be replaced by the total number of atoms of all sorts per unit mass, but since the composition of the majority of astronomical environments (excluding planetary systems) is close to 90% hydrogen by number of atoms, it is convenient to use hydrogen as the reference. Thus N_{el}/N_H is the abundance of the element concerned with respect to hydrogen, N_{ion}/N_{el} is the fraction of the element that is in the right state of ionization (neutral, once ionized, etc.) to produce the line concerned, and N_i/N_{ion} is the fraction of the appropriate ion that is in the right state of excitation to absorb the line, that is, in the right initial level of the transition concerned. In LTE, N_i/N_{ion} will be given directly by Boltzmann's equation, and N_{ion}/N_{el} can be found from Saha's equation which gives N_{ion}/N_{ion+1} and N_{ion}/N_{ion-1}, where N_{ion+1} refers to the next higher stage of ionization and N_{ion-1} refers to the next lower stage of ionization. Then

$$N_{ion}/N_{el} = \frac{N_{ion}}{N_{ion-1} + N_{ion} + N_{ion+1}}$$

$$= \frac{1}{\dfrac{N_{ion-1}}{N_{ion}} + 1 + \dfrac{N_{ion+1}}{N_{ion}}} \qquad (3.24)$$

since usually only three stages of ionization will be significantly populated at a given temperature.

The cross-section can now be expressed as the product of the line profile and a constant for a particular line:

$$\sigma_\nu = \sigma_{\nu 0} \phi(\nu)$$

where the line profile is normalized so $\int \phi(\nu)\, d\nu = 1$ and hence $\int \sigma_\nu\, d\nu = \sigma_{\nu 0}$. As we saw in equation (2.9), the oscillator strength f, used in stellar work as a measure of line strength, is defined by

$$\sigma_{\nu 0} = \frac{\pi e^2}{4\pi\varepsilon_0 mc} f$$

where it will be recalled that m and e are the mass and charge of the electron, respectively. Equivalently:

$$\sigma_{\nu 0} = B_{ij}h\nu$$

$$= \frac{g_j}{g_i}\frac{c^2}{8\pi\nu^2}A_{ji}$$

The profile is determined by the combined effects of Doppler broadening (thermal and microturbulent), natural line broadening and collisional broadening. Large scale motions like rotation do not come into consideration as far as the absorption coefficient is concerned, since they do not affect the process of line formation but only result from the fact that observationally our spatial resolution is limited and we therefore observe simultaneously elements of the object's surface moving at different velocities. If we wish to predict an observed line profile, one calculates the emergent line profile at the surface of the object concerned, and then convolves this with a suitably weighted distribution of large scale velocities.

Often the line profile of the absorption coefficient is mainly determined by Doppler effects or is mainly determined by collisional and natural line broadening. In the first case the shape is Gaussian:

$$\phi(\nu) = \frac{1}{\sqrt{\pi}\Delta\nu_D} e^{-(\Delta\nu/\Delta\nu_D)^2} \tag{3.25}$$

for Doppler width $\Delta\nu_D$ and distance from the line centre $\Delta\nu$. This is sometimes written

$$\phi(x) = \frac{1}{\sqrt{\pi}} e^{-x^2} \tag{3.25a}$$

where $x = \Delta\nu/\Delta\nu_D$ and $\phi(\nu) = \phi(x)/\Delta\nu_D$. In the second case:

$$\phi(\nu) = \frac{\Gamma}{4\pi^2} \frac{1}{(\Delta\nu)^2 + \left(\frac{\Gamma}{4\pi}\right)^2} \tag{3.26}$$

where the Lorentzian width Γ is the sum of the collisional and natural line widths. This is sometimes written

$$\phi(x) = \frac{1}{\pi}\frac{1}{x^2+1} \tag{3.26a}$$

where x is now defined somewhat differently by $x = 4\pi\Delta\nu/\Gamma$ and $\phi(\nu) = \phi(x)$ $(4\pi/\Gamma)$.

The overall line profile always contains both Doppler and collisional/natural line broadening components, with the Doppler component dominating near the line centre (core) and the collisional/natural component dominating far from the line centre (the wings of the line). Equation (3.25a) represents the case where the transition from Doppler to collisional domination takes place so far out in the line profile that the latter can be ignored and the whole profile approximated as a Gaussian, and vice versa for equation (3.26a).

Often, however, both components need to be taken into account, and this requires the convolution of the Gaussian and Lorentzian forms. The convolution of normalized functions $G(x)$ and $H(x)$ is given by

$$K(y) = \int G(y - x)H(x)\mathrm{d}x$$

If we adopt $x = u = \Delta\nu/\Delta\nu_D$ and define $a = \Gamma/(4\pi\Delta\nu_D)$, then the Lorentzian profile becomes $\phi(u) = \phi(\nu)\,\mathrm{d}\nu/\mathrm{d}u = \phi(\nu)\,\Delta\nu_D$, and hence

$$\phi(u) = \frac{a}{\pi}\frac{1}{u^2 + a^2}$$

The convoluted profile $\phi(\nu) = \phi(u)/\Delta\nu_D$ is given by

$$\phi(\nu) = \frac{a}{\pi^{3/2}\Delta\nu_D}\int_{-\infty}^{\infty}\frac{\mathrm{e}^{-x^2}}{(u - x)^2 + a^2}\mathrm{d}x \tag{3.27}$$

The same result can be obtained by taking a Lorentz profile but shifting it according to the velocity V_x of the radiating atom in the line of sight so the line centre becomes $\nu_0 + V_x/c \cdot \nu_0$ and a point on the profile shifted by pressure effects by $\Delta\nu'$ will be observed at frequency $\nu = \nu_0 + \Delta\nu' + V_x/c \cdot \nu_0$, where ν_0 is the rest central frequency of the line. The shifted profile is then

$$\phi(\nu) = \frac{\Gamma}{4\pi^2}\frac{1}{\left(\nu_0 + \frac{V_x}{c}\nu_0 - \nu\right)^2 + \left(\frac{\Gamma}{4\pi}\right)^2}$$

This form is then averaged over different V_x weighted according to the one-dimensional Maxwell equation in the same way as was done in deriving the Doppler profile. The result is again (3.27).

The convolved profile is called the *Voigt* profile. The integral has to be evaluated numerically, and a/π times the integral in (3.27) is called the Voigt function $V'(a, u)$, tabulations of which are available. In terms of the Voigt function, the profile is given by

$$\phi(\nu) = 1/(\pi\Delta\nu_D)V'(a, u) \tag{3.28}$$

In these expressions, u is the frequency distance from the line centre in Doppler widths and a is the ratio of the Lorentz width to the Doppler width and measures the relative importance of pressure and Doppler broadening. It should, however, be noted in more precise work that the pressure broadened profile is not exactly Lorentzian.

The line absorption coefficient can now be written in a number of equivalent forms. Writing out the cross-section in terms of oscillator strength:

$$\kappa_\nu = \frac{\pi e^2}{4\pi\varepsilon_0 mc} f N_{\text{abs}} \phi_\nu$$

$$= \kappa'_0 \phi_\nu, \text{ say} \qquad (3.29)$$

where κ'_0 represents a frequency integrated absorption coefficient.

Alternatively, the absorption coefficient can be written with the Voigt profile:

$$\kappa_\nu = \frac{\pi e^2}{4\pi\varepsilon_0 mc} f N_{\text{abs}} \frac{1}{\sqrt{\pi}\Delta\nu_{\text{D}}} V'(a, u)$$

$$= \alpha_0 V'(a, u), \quad \text{say} \qquad (3.30)$$

where α_0 is the line centre absorption coefficient for $a = 0$, that is for the pure Doppler case with $V' = \exp(-u^2)$. Then

$$\alpha_0 = \frac{\pi e^2}{4\pi\varepsilon_0 mc} \frac{1}{\sqrt{\pi}} \frac{1}{\Delta\nu_{\text{D}}} f N_{\text{abs}} \qquad (3.31)$$

For stellar atmospheres, $a \ll 1$, and the line centre absorption coefficient is only slightly less (by a factor $\sim a$) than α_0.

One sometimes finds a mean absorption coefficient κ_0 used with

$$\kappa_0 = \frac{\int \kappa_\nu \, d\nu}{\Delta\nu_{\text{D}}}$$

$$= \frac{\pi e^2}{4\pi\varepsilon_0 mc} \frac{f}{\Delta\nu_{\text{D}}} N_{\text{abs}} \qquad (3.32)$$

Thus $\kappa_0 = \kappa'_0/\Delta\nu_{\text{D}} = \alpha_0\sqrt{\pi}$ and would give the absorption coefficient for the whole line if the line had a rectangular profile with width $\Delta\nu_{\text{D}}$.

Finally, all these different forms for the absorption coefficient should be corrected for stimulated emission, since stimulated emission is normally taken as a negative contribution to the absorption coefficient. The correction factor is $(1 - g_{\text{L}}/g_{\text{U}} \cdot N_{\text{U}}/N_{\text{L}})$ for statistical weights g and populations N of the upper and lower levels of the transition concerned. If Boltzmann's equation holds for the relative populations of the upper and lower levels of the transition, then the correction factor can be written $(1 - \exp[-h\nu/kT])$, as in equation (2.7).

4 LINE FORMATION

Introduction

We first expand on the thought experiment in the introduction in which one imagines observations of a gas cloud which is initially optically thin but which is steadily increased in column density through the cloud either by increasing the density ρ or preferably by increasing the physical thickness R. For simplicity we shall assume that we are observing in a direction perpendicular to a plane parallel cloud. Thus initially the optical depth $\tau_\nu = \kappa_\nu \rho R$ is very small at all wavelengths and the intensity emitted is $I_\nu = j_\nu \rho R$. At most wavelengths there will just be continuous processes operating giving rise to a very weak continuum, but at certain wavelengths there will also be line processes, $j_\nu = j_{\text{line}} + j_{\text{cont}}$ will be higher than at neighbouring wavelengths, and so the intensity will be higher than in the continuum and we will see an emission line (Figure 4.1(a)). If we now increase the optical thickness of the cloud, at first both line and continuum intensities will grow proportionally. A stage will be reached at which the optical depth at the centre of the strongest lines becomes significant, and a further increase in R will lead to absorption as well as emission. The line strength will now grow more slowly than R and the optical depth. Eventually, the strong line being considered will saturate so that further increases in emission are fully balanced by increases in absorption (Figure 4.1(b)). The intensity emitted for $\tau_\nu \gg 1$ will be given by the source function, which will be equal to the Planck function if LTE holds. As R is increased further still, not only will the line centre saturate but so will the line at frequencies adjacent to the central frequency, when the optical depth at these frequencies (although less than at the centre) becomes greater than 1.

We have so far assumed that the source function of the cloud is the same all over the cloud. Even if the cloud is homogeneous (and of course it will not be if the cloud is a star), the source function will in general drop towards the edge because of leakage of radiation. At frequencies where the cloud is optically thick, the intensity emitted will correspond to the source function evaluated at an optical

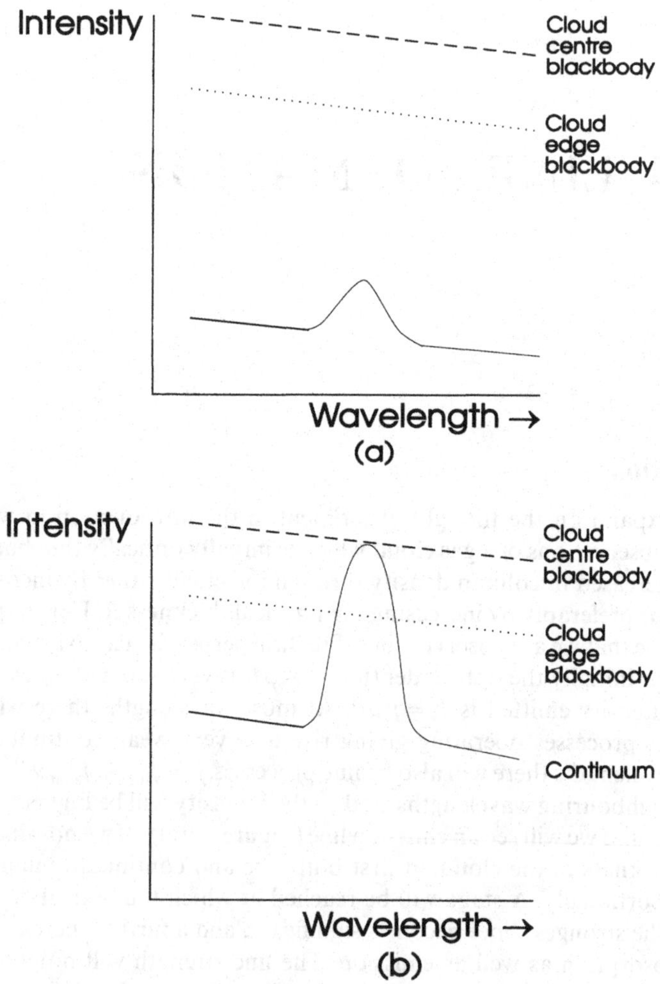

Figure 4.1. (a) Weak emission line; (b) saturated emission line; (c) saturated emission line—two-temperature cloud; (d) absorption line; (e) strong absorption line in atmosphere with chromosphere

depth of about one at that frequency. Thus we 'see' radiation originating from nearer the edge of the cloud at the line centre than at adjacent (but still optically thick) frequencies where the optical depth is less. Hence the line centre will look less bright (lower source function) than neighbouring frequencies in the line if the source function falls outwards. We will see an emission line with a dip at the centre (Figure 4.1(c)); in other words, an emission line with an absorption core usually described as a *self-reversed* line.

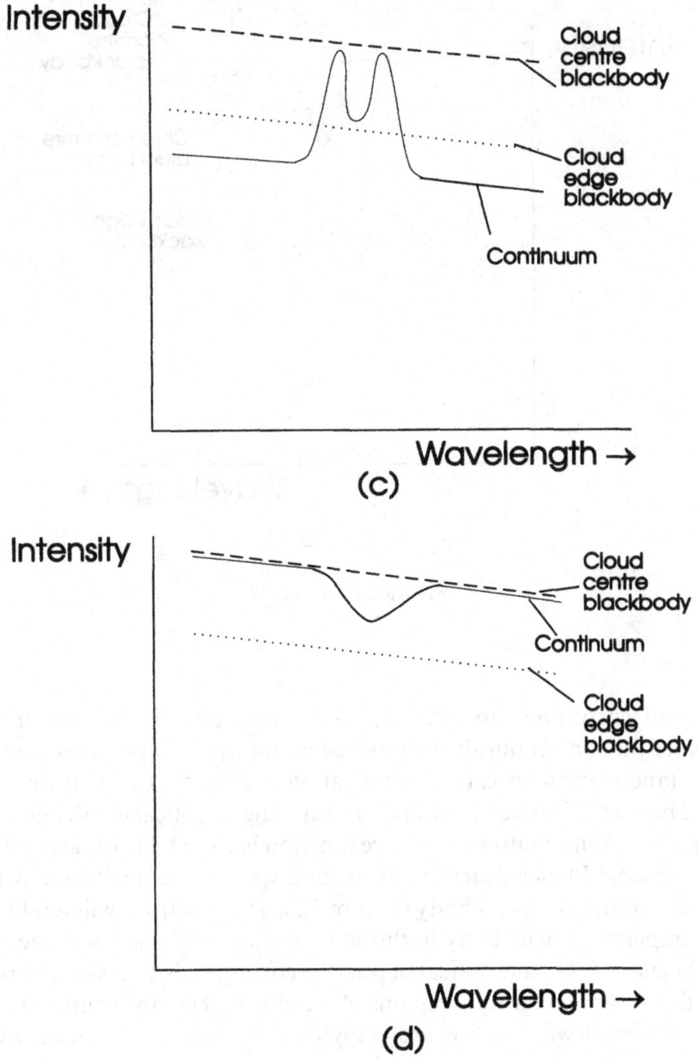

Figure 4.1. (*Contd*)

If R is increased further still, the optical depth will become greater than one at continuum frequencies, with the result that the continuum intensity also becomes equal to the source function. If the source function were really independent of depth, the whole spectrum would eventually become featureless, with no lines. In the more realistic case that the source function varies with depth, the continuum

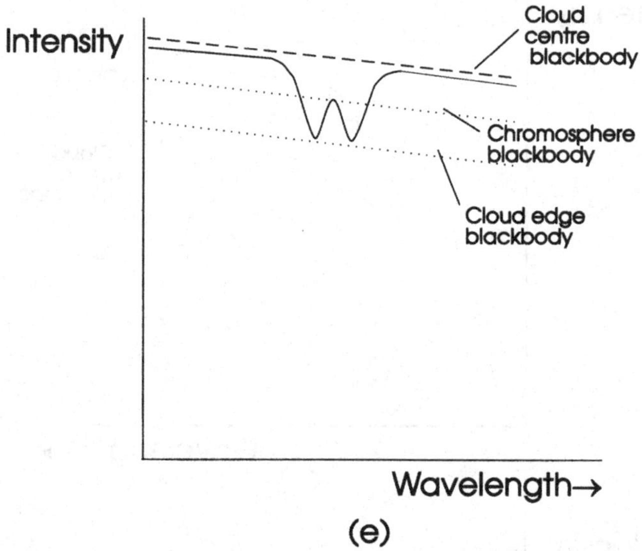

Figure 4.1. *(Contd)*

intensity will correspond to different source functions in different frequency regions since the continuum absorption coefficient will certainly vary slowly with frequency (and sometimes vary abruptly at discontinuities), so that the physical depth and hence the source function corresponding to optical depth one will vary with frequency. If the continuum source function has the LTE value and hence is equal to the local Planck function, the overall spectrum of the 'cloud' will look rather like a composite blackbody (Figure 4.2). The spectrum will tend towards a lower temperature blackbody in those wavelength regions where the absorption coefficient is high and optical depth one corresponds to a shallow physical depth in the lower temperature regions of the cloud where the source function is low, and conversely will tend towards a higher temperature blackbody where the absorption coefficient is low. Thus the absorption coefficient in a solar type atmosphere is at a minimum around a wavelength of 1500 nm, but rises both to longer wavelengths in the near infrared and to shorter wavelengths, with corresponding effects on the observed intensity at those wavelengths. One must be careful in extrapolating too far along these lines, for if one goes far into the infrared or ultraviolet, one is observing the thin chromosphere of a solar type star, where the temperature actually rises outwards and LTE may not hold.

What happens to the lines in the case of an optically thick continuum and depth-dependent source function, as would be the case in a stellar atmosphere? The absorption coefficient of the line will be additional to the continuum

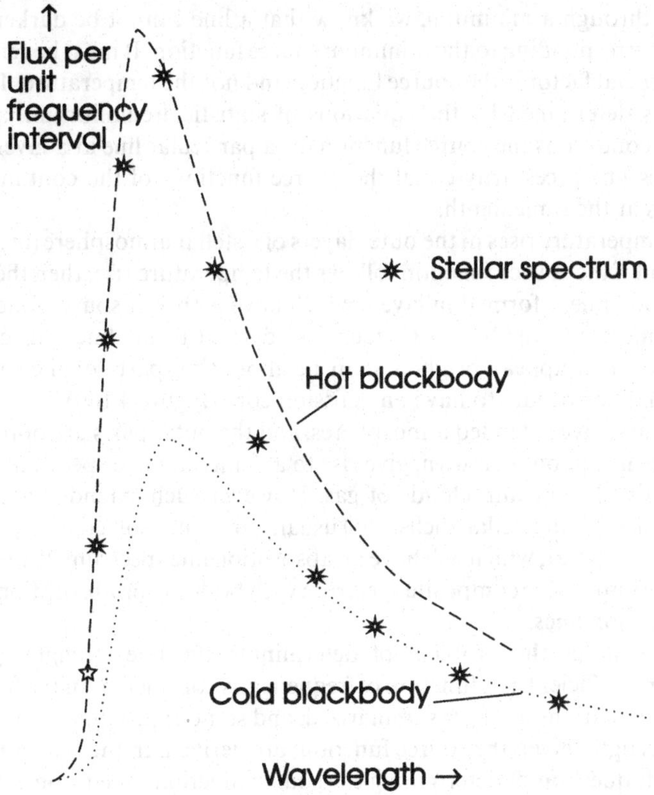

Figure 4.2. A hypotheticatical case where a stellar spectrum corresponds to a hot blackbody at short wavelengths and to a cold blackbody at long wavelengths

absorption coefficient at line frequencies and so at line frequencies one will be observing shallower layers than at neighbouring continuum frequencies, and the source function in those shallower layers will in general be lower than deeper down. Hence the intensity in the line (corresponding to the source function at line optical depth one) is lower than at neighbouring continuum frequencies where it will correspond to continuum optical depth one. One will therefore see a 'less bright' line, that is an *absorption* line (Figure 4.1(d)). For weak lines, where the additional absorption caused by the line is a small fraction of the continuous absorption coefficient, the depth of the line as a fraction of the intensity of the neighbouring continuum is proportional to the line absorption coefficient as a fraction of the continuum absorption coefficient, but strong lines grow more slowly than this and eventually *saturate*. The temperature reaches a minimum in solar type stars in the upper layers of the atmosphere and if the source function also goes through a minimum, we know that a line cannot be darker than the intensity corresponding to the minimum source function. It must be remembered that the crucial factor is the source function and not the temperature. The source function is determined by the equations of statistical equilibrium, and under non-LTE conditions the source function for a particular line at a given physical depth does not necessarily equal the source functions of the continuum or of other lines at the same depth.

If the temperature rises in the outer layers of a stellar atmosphere (in a chromosphere) and if the source function follows the temperature rise, then the centre of a very strong line is formed in layers which have a greater source function than those giving rise to the line at frequencies adjacent to the line centre. The line centre may then appear brighter than neighbouring parts of the line, so the absorption line appears to have an emission core (Figure 4.1(e)).

Some stars have extended atmospheres, and the outer parts are optically thin, and would, if seen on their own, give rise to an emission line spectrum similar to those obtained from thin clouds of gas. However such extended atmospheres (stellar winds, circumstellar shells, etc.) usually are not resolved in angular terms from the central star, which will have an absorption line spectrum. We then see an apparently impossible composite spectrum with both strong absorption lines and strong emission lines.

In this chapter the problem of determining the line strength, given the absorption coefficient as a function of frequency, is considered. In the first section the formal equations used are summarized, and some approximate equations for the line strength (given the source function) are derived. In the following section the difficult question of determining the source function under non-LTE conditions is discussed. In the final section, the escape probability approach is outlined. In this approach, deviations of the source function from the LTE value are taken as being due to the leakage of photons from the atmosphere.

Line Formation—Predicting the Emergent Spectrum

We start by gathering together some of the expressions derived in Chapter 1 for the emergent intensity and flux. In general, for optical depth measured along the line of sight:

$$I_\nu(\tau_\nu = 0) = \int_0^\infty S_\nu(\tau_\nu) e^{-\tau_\nu} \, d\tau_\nu$$

and for a plane parallel stellar atmosphere we have equation (1.9):

$$I_\nu(\tau_\nu = 0, \mu) = \int_0^\infty S_\nu(\tau_\nu) e^{-\tau_\nu/\mu} \frac{d\tau_\nu}{\mu}$$

where τ_ν is now measured perpendicular to the atmosphere.

The flux emitted is given by

$$F_\nu = 2\pi \int_0^{+1} I_\nu(\mu)\mu\,d\mu$$

$$= 2\pi \int_0^{\infty} S_\nu(t_\nu)E_2(t_\nu)\,dt_\nu$$

from (1.12), and the flux received from a spherical star of radius R at distance d is

$$F_\nu(\text{received}) = \frac{R^2}{d^2} 2\pi \int_0^{+1} I_\nu(\mu)\mu\,d\mu$$

If the source is an extended atmosphere or wind, the plane parallel assumption no longer applies, and it may be more convenient to use (1.13) and (1.14), where the integration is along the line of sight, to give an intensity which is then summed over various apparent distances from the centre of the source to give the flux.

The inputs into these various forms for the prediction of the observed spectrum are the source function (or equivalently the ratio of the upper to the lower level populations) and the absorption coefficient.The absorption coefficient at a particular frequency is the sum of the absorption coefficients $\kappa_{L\nu}$ of all the lines whose profiles have a significant strength at that frequency added to the continuous absorption coefficient κ_C (which will vary very slowly with frequency):

$$\kappa_\nu = \sum_{ij} \kappa_{L\nu}(ij) + \kappa_C$$

$$= \sum_{ij} \phi_{ij}(\nu)\sigma_{\nu 0}(ij)N_{abs}(i) + \kappa_C$$

$$= \frac{\pi e^2}{4\pi\varepsilon_0 mc} \sum_{ij} f_{ij}\phi_{ij}(\nu)N_{abs}(i) + \kappa_C \tag{4.1}$$

where use has been made of equations (2.9) and (3.22).

Each line is defined by upper level j and lower level i and has oscillator strength f_{ij}. We shall only consider one line in what follows, but it should be borne in mind that at short wavelengths nearly all lines overlap with others. In LTE, N_{abs} depends only on the composition, temperature and pressure, but usually it cannot be assumed in calculating N_{abs} that Boltzmann's equation and Saha's equation hold.

For a stellar atmosphere, one normally specifies conditions such as temperature and pressure in terms of continuum optical depth perpendicular to the atmosphere at some standard wavelength or frequency. In the case of solar type stars, for instance, a wavelength of 500 nm is often used. Let the absorption coefficient at the standard frequency be κ_s with corresponding optical depth τ_s.

Then

$$\kappa_\nu = \frac{\kappa_\nu}{\kappa_s}\kappa_s = \frac{\kappa_{L\nu} + \kappa_C}{\kappa_s}\kappa_s$$

so

$$\tau_\nu = \int \frac{\kappa_{L\nu} + \kappa_C}{\kappa_s}\,d\tau_s$$

Hence

$$I_\nu(\mu) = \int_0^\infty S_\nu(\tau_\nu[\tau_s])\frac{\kappa_{L\nu} + \kappa_C}{\kappa_s}\,e^{-\tau_\nu(\tau_s)}\,d\tau_s \tag{4.2}$$

where, for simplicity, the optical depth has been taken along the line of sight.

The major problem in predicting line strengths lies in the estimation of the source function. If LTE holds, then $S_\nu = B_\nu(T)$, the Planck function, for all lines and the continuum. In general this is not the case, and the source functions for the line, S_L, and for the continuum, S_C, are different. This can be allowed for in the equations above if one recalls the definition of the source function as the emission coefficient divided by the absorption coefficient, so

$$S_\nu = \frac{j_\nu}{\kappa_\nu} = \frac{\kappa_{L\nu}S_L + \kappa_C S_C}{\kappa_{L\nu} + \kappa_C} \tag{4.3}$$

Discussion of how the source function can be calculated in non-LTE situations is left to the final sections of this chapter.

Absorption lines are observed against a continuous background. For each point in the profile, observations give the *depth* or dip of the line:

$$r = \text{depth at frequency } \nu = (F_C - F_\nu)/F_C$$

where F_ν is the flux in the line at frequency ν and F_C is the flux at a neighbouring continuum point, so the depth can vary between 0.0 and 1.0. The overall line strength is often given in the form of the *equivalent width* W:

$$W = \int_0^\infty \text{depth}\,d\lambda$$

$$= \int_0^\infty \frac{F_C - F_\nu}{F_C}\,d\lambda = \frac{\lambda^2}{c}\int_0^\infty \frac{F_C - F_\nu}{F_C}\,d\nu$$

since for wavelength λ and speed of light c, $d\nu = -c/\lambda^2\,d\lambda$, and the line occupies such a small proportion of the frequency range from 0 to ∞ that λ^2 can be taken out of the integral.

In addition to the equivalent width of a line, the depth r_0 at the line centre is sometimes given, together with the full width at half height, $\Delta\lambda_{1/2}$, which is the wavelength separation of the points at which the profile falls to half of its maximum depth. For a triangular or rectangular profile, $W = r_0\Delta\lambda_{1/2}$. Hence the equivalent width can also be defined as the width in wavelength of a rectangular-profiled line 100% deep which has the same area in a flux-wavelength plot as the actual line.

We now attempt to predict the equivalent widths of lines using various approximations. First, consider the case of a cloud of uniform source function S_1, thickness along the line of sight h, and density ρ, that interacts with radiation only at line wavelengths, and is illuminated by a small continuum source of intensity I_C, with $I_C > S_1$. This is the situation found when observing a star through interstellar or circumstellar matter, or indeed in laboratory experiments where a bright lamp is viewed through a heated vapour. For this case $\mu = 0$, so

$$I_\nu = \int_0^{\tau_{L\nu}} S_1 e^{-t_{L\nu}} dt_{L\nu} + I_C e^{-\tau_{L\nu}}$$

$$= S_1(1 - e^{-\tau_{L\nu}}) + I_C e^{-\tau_{L\nu}}$$

where $\tau_{L\nu}$ is the line of sight optical depth through the cloud at frequency ν. The equivalent width in the intensity spectrum is then:

$$W = \frac{\lambda^2}{c} \int \frac{I_C - I_\nu}{I_C} d\nu$$

$$= \frac{\lambda^2}{c}\left(1 - \frac{S_1}{I_C}\right) \int (1 - e^{-\tau_{L\nu}}) d\nu \qquad (4.4)$$

It can be seen that the central depth of the line r_C is given by

$$r_C = \left(1 - \frac{S_1}{I_C}\right)(1 - e^{-\tau_{L\nu}(\text{centre})})$$

which for a very strong line becomes:

$$r_C = R_C = (1 - S_1/I_C)$$

Thus R_C is the maximum possible line depth.

Equation (4.4) can be rewritten so that the left-hand side is in dimensionless form:

$$\frac{W}{\lambda} = \frac{\lambda}{c} R_C \int (1 - e^{-\tau_{L\nu}}) d\nu \qquad (4.4a)$$

For a weak line $1 - \exp(-\tau) \sim \tau$ if $\tau \ll 1$ even at the line centre, and so

$$\left(\frac{W}{\lambda}\right)_{\text{weak}} = \frac{\lambda}{c} R_C \int \tau_{L\nu} d\nu \qquad (4.5)$$

Now $\tau_{L\nu} = \kappa_{L\nu}\rho h = \kappa'_0 \phi_\nu \rho h$ with κ'_0 from (3.29). Thus:

$$\left(\frac{W}{\lambda}\right)_{\text{weak}} = \frac{\lambda}{c} R_C \kappa'_0 \rho h \int \phi_\nu \, d\nu$$

$$= \frac{\pi e^2}{4\pi\varepsilon_0 mc^2} R_C \lambda f \rho h N_{\text{abs}} \tag{4.5a}$$

since $\int \phi_\nu \, d\nu = 1$.

Equation (4.4) also represents a crude approach to line formation in a stellar atmosphere. It has already been pointed out that the essential feature of line formation is that at line wavelengths the observed radiation comes from higher in the atmosphere, and that if the source function falls outwards, the emitted intensity will be less than at neighbouring wavelengths, giving rise to an absorption line. Suppose we approximate a stellar atmosphere by two layers: a deeper layer with source function S_2 which produces the continuum, and a higher layer with source function S_1 which extends from $\tau = 0$ to $\tau = \tau_{L\nu}$ in which only line absorption is significant. Then:

$$I_C = \int_0^\infty S_\nu e^{-t/\mu} \frac{dt}{\mu} = S_2 \int_0^\infty e^{-t/\mu} \frac{dt}{\mu} = S_2$$

$$I_{L\nu} = \int_0^{\tau_{L\nu}} S_1 e^{-t/\mu} \frac{dt}{\mu} + \int_{\tau_{L\nu}}^\infty S_2 e^{-t/\mu} \frac{dt}{\mu}$$

$$= S_1(1 - e^{-\tau_{L\nu}/\mu}) + S_2 e^{-\tau_{L\nu}/\mu}$$

where τ is assumed measured perpendicular to a plane parallel atmosphere. Then the line depth is:

$$r = \left(1 - \frac{S_1}{S_2}\right)(1 - e^{-\tau_{L\nu}/\mu})$$

and we obtain the same expression as found in (4.4) for the depth and equivalent width if we put $I_C = S_2$.

In effect, the continuous run of source function with optical depth has been approximated by two layers, with the dividing point in geometrical depth, x_1, chosen such that $\tau_C = \kappa_C \rho x_1$ is considerably less than 1. Suppose $\kappa_C \rho x_1 = q$, then $\tau_{L\nu} = \kappa_{L\nu} \rho x_1 = \kappa_{L\nu}/\kappa_C q$ so the optical depth represents the ratio of line to continuous absorption, bringing out the point that it is the contrast between line and continuum that is important in line formation. For a weak line we obtain as before:

$$\left(\frac{W}{\lambda}\right)_{\text{weak}} = \frac{\pi e^2}{4\pi\varepsilon_0 mc^2} R_C \lambda f \frac{q}{\kappa_C} N_{\text{abs}} \tag{4.5b}$$

This sort of approach is sometimes called a Schuster–Schwarzschild model.

The result for the line depth can also be expressed in flux terms with

$$F_C = 2\pi \int_0^{+1} S_2 \mu d\mu = \pi S_2$$

$$F_L = 2\pi S_1 \int_0^{+1} (1 - e^{-\tau_{L\nu}/\mu})\mu d\mu + 2\pi S_2 \int_0^{+1} e^{-\tau_{L\nu}/\mu}\mu d\mu$$

$$= \pi S_1 - 2\pi S_1 E_3(\tau_{L\nu}) + 2\pi S_2 E_3(\tau_{L\nu})$$

where in the last line we have substituted $z = 1/\mu$ and used the definitions and properties of exponential integrals $E_n(t)$ given in Chapter 1 after equation (1.10). Hence

$$1 - \frac{F_L}{F_C} = \left(1 - \frac{S_1}{S_2}\right)[1 - 2E_3(\tau_{L\nu})] \tag{4.6}$$

The Schuster–Schwarzschild model can only represent a very rough approximation to a real stellar atmosphere where both line and continuum absorption occur at all levels and there is a continuous variation of temperature and source function with depth. A more realistic approximation is given by the *Milne–Eddington* model in which a constant ratio of line to continuous absorption coefficients independent of depth, $\kappa_L/\kappa_C = \eta_\nu$, is assumed. It is also assumed that the source function is linear in the continuum optical depth so $S = a + b\tau_C$. Then in the line

$$\tau_\nu = \int \frac{\kappa_{L\nu} + \kappa_C}{\kappa_C} d\tau_C = (\eta_\nu + 1)\tau_C$$

At continuum wavelengths we have

$$I_C = \int_0^\infty (a + b\tau_C)e^{-\tau_C/\mu}\frac{d\tau_C}{\mu} = a + b\mu$$

$$= a + b\mu$$

$$F_C = 2\pi \int_0^{+1} (a + b\mu)\mu d\mu = \pi\left(a + \frac{2}{3}b\right)$$

At line wavelengths we have:

$$I_\nu = \int_0^\infty (a + b\tau_C)e^{-\tau_\nu/\mu}\frac{d\tau_\nu}{\mu}$$

Hence

$$I_\nu = \int_0^\infty \left[a + \frac{b\tau_\nu}{1 + \eta_\nu}\right] e^{-\tau_\nu/\mu}\frac{d\tau_\nu}{\mu}$$

$$= a + \frac{b}{1 + \eta_\nu}\mu$$

$$F_\nu = \pi\left[a + \frac{2}{3}\frac{b}{1 + \eta_\nu}\right]$$

Thus the equivalent width of the line (in flux) is

$$W = \frac{\lambda^2}{c} \int \frac{a + \frac{2}{3}b - a - \frac{2}{3}\frac{b}{1+\eta_\nu}}{a + \frac{2}{3}b} \mathrm{d}\nu$$

$$= \frac{\lambda^2}{c} \left[\frac{2b}{3a + 2b}\right] \int \frac{\eta_\nu}{1 + \eta_\nu} \mathrm{d}\nu \tag{4.7}$$

The maximum possible central depth of a strong line is $2b/(3a + 2b) = R_C$ say, and again writing the left-hand side of the expression for the equivalent width in dimensionless form, we obtain:

$$\frac{W}{\lambda} = \frac{\lambda}{c} R_C \int \frac{\eta_\nu}{1 + \eta_\nu} \mathrm{d}\nu \tag{4.7a}$$

which is very similar to (4.5). In the case of weak lines with $\eta \ll 1$:

$$\left(\frac{W}{\lambda}\right)_{\text{weak}} = \frac{\lambda}{c} R_C \int \eta_\nu \mathrm{d}\nu$$

$$= \frac{\lambda}{c} R_C \int \frac{\kappa_{L\nu}}{\kappa_C} \mathrm{d}\nu$$

$$= \frac{\pi e^2}{4\pi\varepsilon_0 mc^2} R_C \lambda f \frac{N_{\text{abs}}}{\kappa_C} \tag{4.8}$$

which is the same as the result obtained using the Schuster–Schwarzschild model apart from the rather arbitrary factor q introduced in the latter. Real stellar atmospheres do not have source functions linear in optical depth, and in general the line and continuum source functions vary differently with temperature and pressure and hence with depth, but for weak lines, line formation and continuum formation take place in neighbouring layers and hence (4.8) should hold reasonably accurately.

In stellar work it is customary to expand N_{abs} as was done in (3.23) but to remove a factor g (the statistical weight of the lower level of the transition) from the excitation factor N_i/N_{ion} and to combine g with the oscillator strength f and wavelength λ in a factor λgf. This is done because a number of lines may have similar excitation potentials and hence the same $\exp(-E/kT)$ factor, but each line will have a different statistical weight, wavelength and oscillator strength, so it is convenient to group these parameters together. It is also usual to write the continuum absorption coefficient κ_C as the product of the continuous absorption cross-section per hydrogen atom, α_C, and the number of hydrogen atoms per unit mass, N_H, since most important absorption processes are directly or indirectly proportional to the number of hydrogen atoms. We obtain:

$$\left(\frac{W}{\lambda}\right)_{\text{weak}}=\frac{\pi e^2}{4\pi\varepsilon_0 mc^2}\frac{R_C}{\alpha_C}\lambda gf\frac{N_i}{gN_{\text{ion}}}\frac{N_{\text{ion}}}{N_{\text{el}}}\frac{N_{\text{el}}}{N_H} \tag{4.9}$$

The first term in this equation is made up of physical constants and has the value $8.85 * 10^{-15}$ m. If LTE holds, R_C is fixed for a given star, and α_C is a function of temperature and pressure in the continuum forming layer that varies slowly with wavelength, so that R_C/α_C can be taken as constant for a given star and a given wavelength region. The terms λgf and N_i/gN_{ion} (the fraction excited over g, given by Boltzmann's equation in LTE) will in general be different for different lines, and it will be recalled that $N_{\text{ion}}/N_{\text{el}}$ is the fraction ionized to the appropriate degree and is given by Saha's equation in LTE, while N_{el}/N_H is the abundance of the element with respect to hydrogen.

R_C is nominally the depth of the strongest possible line, and one might think that it should be possible to determine R_C empirically from the observed spectrum. In some stars the strongest lines have central depths of 0.8 or even 0.9. However, the cores of such strong lines are almost certainly not formed in LTE. Alternatively, one may use models of the atmosphere for a given effective temperature to determine, say, a and b in the Milne–Eddington model, giving typically rather smaller values of R_C of around 0.5. These should be applicable to weak or medium strength lines, provided that they are formed in LTE. For interstellar or circumstellar lines, $R_C = 1.0$, since the interstellar or circumstellar cloud emits a negligible amount of radiation compared with the background.

It can be seen from equation (4.9) that as long as a line remains weak, W/λ will increase linearly for a given atmosphere with $\lambda gf(N_i/N_H)$ as the latter quantity is increased either by looking at lines with greater oscillator strengths or by looking at a single line and increasing the abundance of the element (note that $N_i/N_H \propto N_{\text{el}}/N_H$ for fixed temperature and pressure). Such a plot of W/λ versus $\lambda gf\,N_i/N_H$ is called a *curve of growth.* The curve of growth is linear for weak lines, but once the lines become even moderately strong, the curve of growth starts to saturate, and the variation of the equivalent width with $\lambda gf\,N_i/N_H$ becomes very slow. Unfortunately, the degree of saturation depends on the profile of the line. Suppose the profile of the line is Gaussian:

$$\phi_\nu=\frac{1}{\sqrt{\pi}\Delta\nu_D}\,e^{-(\Delta\nu/\Delta\nu_D)^2}$$

and write $\eta=\tau_{L\nu}=\kappa_0'/\kappa_C\cdot\phi_\nu$. Then (4.4) becomes (for the disc centre intensity with $\mu=1$):

$$\frac{W}{\lambda}=\frac{\lambda}{c}R_C\int[1-e^{-\kappa_0'/\kappa_C\phi_\nu}]\,d\nu$$

Expanding the exponential as a power series ($\tau_{L\nu} < 1$):

$$\frac{W}{\lambda} = \frac{\lambda}{c} R_C \int \left[1 - 1 + \frac{\kappa'_0}{\kappa_C} \phi_\nu - \frac{1}{2} \left(\frac{\kappa'_0}{\kappa_C} \phi_\nu \right)^2 + \cdots \right] d\nu$$

$$= \frac{\lambda}{c} R_C \frac{\kappa'_0}{\kappa_C} \left[\int \phi_\nu \, d\nu - \frac{1}{2} \frac{\kappa'_0}{\kappa_C} \int \phi_\nu^2 \, d\nu \right]$$

to first order. Substituting for ϕ_ν in the second term inside the square brackets and changing the integration from one over ν to one over $\Delta\nu$, the term becomes

$$\frac{1}{2} \frac{\kappa'_0}{\kappa_C} \frac{1}{\pi(\Delta\nu_D)^2} \int_{-\infty}^{\infty} e^{-2(\Delta\nu/\Delta\nu_D)^2} \, d(\Delta\nu)$$

If $u = \Delta\nu/\Delta\nu_D$ is substituted, the integral becomes:

$$\Delta\nu_D \int_{-\infty}^{\infty} e^{-2u^2} \, du = \Delta\nu_D \sqrt{\frac{\pi}{2}}$$

Hence

$$\frac{W}{\lambda} = \frac{\lambda}{c} R_C \frac{\kappa'_0}{\kappa_C} \left[1 - \frac{1}{2} \frac{\kappa'_0}{\kappa_C} \frac{1}{\sqrt{2\pi} \Delta\nu_D} \right] \tag{4.10}$$

Here the first term represents a linear curve of growth, so that departures from linearity become significant (reduce the observed equivalent width by 10%) when

$$\frac{\kappa'_0}{\kappa_C} \frac{1}{2\sqrt{2\pi}} = 0.1 \Delta\nu_D$$

i.e. when

$$W_{\text{linear}} = \frac{\lambda^2}{c} R_C 0.2 \sqrt{2\pi} \Delta\nu_D = 0.501 \, \Delta\lambda_D R_C \tag{4.11}$$

If we use the Milne–Eddington model, $(1 - \exp(-\tau_{L\nu}))$ must be replaced by $\eta_\nu/(1 + \eta_\nu)$ in the integrand:

$$\int \frac{\eta_\nu}{1 + \eta_\nu} d\nu = \int \eta_\nu (1 - \eta_\nu + \tfrac{1}{2}\eta_\nu^2 - \cdots) \, d\nu$$

$$= \int \eta_\nu \, d\nu - \int \eta_\nu^2 \, d\nu \text{ to first order}$$

$$= \frac{\kappa'_0}{\kappa_C} \left[1 - \frac{\kappa'_0}{\kappa_C} \int \phi_\nu^2 \, d\nu \right]$$

Hence a 10% deviation from linearity occurs as above when $W_{\text{linear}} = 0.250 \Delta\lambda_D R_C$.

For a solar-type star at a wavelength of 500 nm and with $\Delta\lambda_D = 3*10^{-3}$ nm, $R_C = 0.5$, even a line with equivalent width of $3.7*10^{-4}$ nm shows incipient saturation. Such a line with central depth less than 0.1 would be so weak that it could only be observed accurately in an unblended region of a high resolution spectrum. Hence much stellar abundance work is performed using lines that at best are on the 'bend' of the curve of growth. Equation (4.10) and the equivalent for the Milne–Eddington model are only valid if $\tau_{L\nu}$ or η_ν are less than 1 at the line centre. For stronger lines with line centre optical depths much greater than one, we can write

$$\frac{\kappa_0'}{\kappa_C}\frac{1}{\sqrt{\pi}\Delta\nu_D} = e^{u_0^2}$$

so

$$\tau_{L\nu} = \eta_\nu = e^{u_0^2 - u^2}$$

If we substitute in the Schuster–Schwarzschild intensity at the centre of the disc expression:

$$\frac{W}{\lambda} = 2R_C\frac{\lambda}{c}\Delta\nu_D\int_0^\infty [1 - \exp(-\exp(u_0^2 - u^2))]du$$

giving $W \approx 2R_C\lambda^2/c\Delta\nu_D u_0 \approx 2R_C\Delta\lambda_D u_0$ since the integrand is close to 1 from $u = 0$ to $u = u_0$ if u_0 is large, and the contribution to the integral from $u > u_0$ is small, again if u_0 is large. A similar result is obtained from the Milne–Eddington model. Substituting for u_0:

$$\left(\frac{W}{\lambda}\right)_{sat} \simeq 2R_C\frac{\Delta\lambda_D}{\lambda}\sqrt{\ln_e\left[\frac{\kappa_0'}{\kappa_C}\frac{1}{\sqrt{\pi}}\frac{\lambda^2}{c}\frac{1}{\Delta\lambda_D}\right]} \tag{4.12}$$

The curve of growth (Figure 4.3) thus saturates at an equivalent width of the order of $2R_C\Delta\lambda_D$ (the square root factor is of order 1), which is the equivalent width of a line of rectangular profile with depth R_C and width $2\Delta\lambda_D$, as might be expected. The equivalent width grows as the square root of the logarithm of $\lambda gf N_i/N_{el}$, which is, of course, an extremely slow growth–hence the description 'flat' sometimes used for this portion of the curve of growth.

We have so far taken the line as having a Gaussian profile, since the ratio $\Gamma/(4\pi\Delta\nu_D)$ is usually much less than one so the Gaussian element dominates except in the wings, and the latter contribute little to the total equivalent width in weak lines or lines of medium strength. However, for a strong line, the core will be saturated while the wings can still grow, and so as the strength increases, the wings will eventually come to supply the major part of the equivalent width. In that case the equivalent width can be found approximately by assuming that the

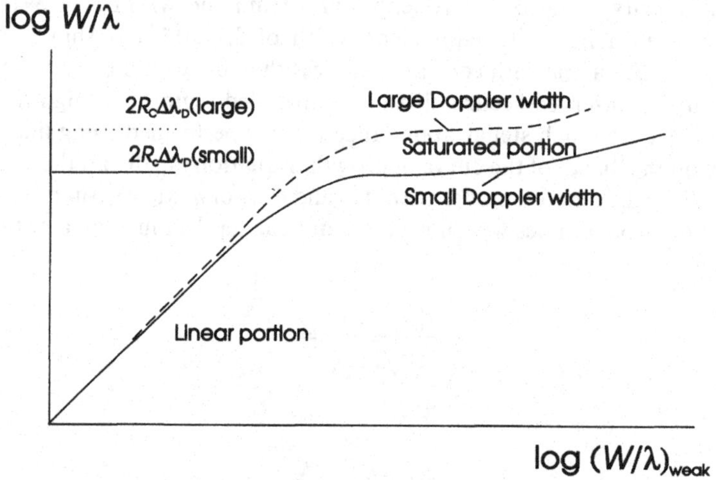

Figure 4.3. Curve of growth—saturation of Doppler profile

line has a pure Lorentzian profile:

$$\phi_\nu = \frac{\Gamma}{4\pi^2} \frac{1}{(\Delta\nu)^2 + \left(\frac{\Gamma}{4\pi}\right)^2}$$

$$\simeq \frac{\Gamma}{4\pi^2} \frac{1}{(\Delta\nu)^2} \text{ in the wings}$$

The Schuster–Schwarzschild model for intensity at the centre of the stellar disc then gives:

$$\left(\frac{W}{\lambda}\right)_{\text{strong}} = 2R_C \frac{\lambda}{c} \int_0^\infty (1 - e^{-(\kappa_0'/\kappa_C)\phi_\nu}) d(\Delta\nu)$$

Let

$$\frac{\kappa_0'}{\kappa_C} \phi_\nu = \frac{\kappa_0'}{\kappa_C} \frac{\Gamma}{4\pi^2} \frac{1}{(\Delta\nu)^2} = z$$

Then

$$\left(\frac{W}{\lambda}\right)_{\text{strong}} = 2R_C \frac{\lambda}{c} \sqrt{\frac{\kappa_0'}{\kappa_C} \frac{\Gamma}{4\pi^2}} \int_\infty^0 (1 - e^{-z}) d\left(\frac{1}{z^{1/2}}\right)$$

The integral can be written

$$\int_\infty^0 (1 - e^{-z}) d\left(\frac{1}{z^{1/2}}\right) = \frac{1 - e^{-z}}{z^{1/2}}\bigg|_\infty^0 + \int_0^\infty \frac{1}{z^{1/2}} e^{-z} dz$$

$$= 2\int_0^\infty e^{-h^2} dh = \sqrt{\pi} \quad \text{using } z = h^2$$

Hence

$$\left(\frac{W}{\lambda}\right)_{\text{strong}} = R_C\frac{\lambda}{c}\sqrt{\frac{\kappa_0'}{\kappa_C}\frac{\Gamma}{\pi}} \tag{4.13a}$$

The Milne–Eddington model similarly gives

$$\left(\frac{W}{\lambda}\right)_{\text{strong}} = R_C\frac{\lambda}{c}\int_0^\infty \frac{\frac{\kappa_0'}{\kappa_C}\phi_\nu}{1+\frac{\kappa_0'}{\kappa_C}\phi_\nu}\,d\nu$$

$$= 2R_C\frac{\lambda}{c}\int_0^\infty \frac{d(\Delta\nu)1}{1+\dfrac{(\Delta\nu)^2}{\frac{\kappa_0'}{\kappa_C}\frac{\Gamma}{4\pi^2}}}$$

$$= R_C\frac{\lambda}{c}\sqrt{\frac{\kappa_0'}{\kappa_C}\frac{\Gamma}{4}} \tag{4.13b}$$

This gives rise to what is called the *damping* portion of the curve of growth (from the classical idea of natural line broadening). The equivalent width now rises as the *square root* of $\lambda gf\, N_i/N_{el}$ but also depends on the square root of Γ. The latter will vary from line to line, and the damping portion of the curve of growth will branch off sooner for lines of large Γ (Figure 4.4).

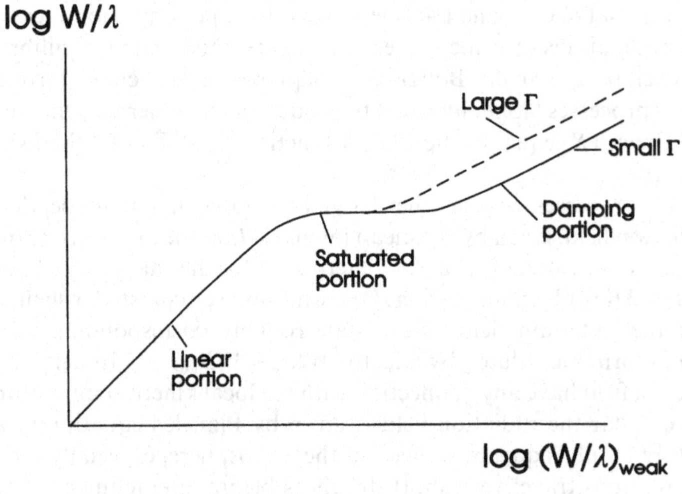

Figure 4.4. Curve of growth showing damping portion for different values of the Lorentz width (damping constant) G

The intensity from an emission-line-producing cloud at radio wavelengths is expressed in brightness temperature terms, as we saw earlier, by $T_B = T_{ex}(1 - \exp[-\tau_\nu])$. Integrated over a line, we have $\int T_B d\nu = T_{ex}\int(1 - \exp[-\tau_\nu])d\nu$, where the integral is the same as in a Schuster–Schwarzschild model so that all the features of the absorption line curve of growth are carried through, with κ'_0/κ_C replaced by $\kappa'_0\rho R$ for cloud thickness R.

The Source Function

The most difficult problem in line formation is the estimation of the source function. We start by reviewing the results of Chapter 2. The source function, as was seen there, is an expression of the ratio of upper to lower level population N_j/N_i:

$$S_\nu = \frac{\dfrac{2h\nu^3}{c^2}}{\dfrac{g_j}{g_i}\dfrac{N_i}{N_j} - 1} \tag{2.6}$$

These populations are determined by both collisional and radiative transitions between the levels i and j, and also by radiative transitions to other levels.

Collisional transitions involve an integral over the relative velocities of the colliding particles. The distribution of velocities is determined by frequent elastic collisions (which only exchange energy between colliding particles) and so is nearly always given by Maxwell's equation, characterized by the *local kinetic temperature*, T_k. Hence the much rarer elastic collisions which change the internal state of the colliding particles (excitation and ionization) tend to bring the level populations into the corresponding thermodynamic equilibrium relation for excitation, namely Boltzmann's equation, again characterized by T_K. Collisional processes alone thus tend to produce a local thermodynamic equilibrium (LTE) with S_ν equal to the Planck function B_ν at $T = T_k$, the *local* kinetic temperature.

Radiative processes, on the other hand, have rates that are dependent on the local radiation field, given by the mean intensity J_ν at the appropriate frequency. In an interstellar cloud, J_ν may be determined by the microwave background $J_\nu = B_\nu(T = 3\text{ K})$ or by illuminating stars, while for a circumstellar shell, J_ν will be given by the radiation field of the star (roughly corresponding to Planck's equation in form but diluted by a factor W) $J_\nu \sim WB_\nu(T_{star})$. In neither case will the radiation field have any connection with the local kinetic temperature. In the interior of a star the radiation field is given by Planck's equation at the local temperature at all frequencies. Deep in the atmosphere, especially for a strong line, line photons travel very short distances before interacting so J_ν is locally determined, ultimately by the rare collisional interactions, and is again given by $B_\nu(T_k)$. In the upper layers of the atmosphere the radiation field will be diluted because there will be little inward radiation flow. Photons in the upper layers will travel large distances, so the radiation field will no longer be locally determined. Even if the line centre absorption coefficient is large, the wings will be transparent, and if shuffling between the centre and the wings takes place, non-local effects can be introduced to the whole source function through the wings. One can always determine a radiation temperature $T_{\rm rad}$ such that $J_\nu = B_\nu(T_{\rm rad})$, at least for a small frequency range, but $T_{\rm rad}$ may well be quite different from $T_{\rm K}$, and in general radiative processes will pull the source function away from its LTE value.

This immediately raises a point not hitherto discussed, namely the question as to whether photons on absorption and re-emission change frequency within the line profile. The process where a line photon is absorbed and then re-emitted as a photon of the same line is called *scattering*, but may not have the same properties as pure continuum scattering processes like Thomson electron scattering, where the scattered photon must have exactly the same frequency as the incoming photon in the rest frame of the electron, and where the distribution of emission directions relative to the direction of the incoming photon takes a particular form, as does the polarization of the emitted photon. Note that in conventional terminology the line 'absorption' coefficient includes both a scattering component and an absorption followed by a collisional de-excitation component, with the latter component of the absorption coefficient called 'absorption' to distinguish it from absorption followed by re-emission, called 'scattering'!

If no change in frequency takes place in line scattering (*coherent scattering*), the probability of a photon being emitted at a particular point in the line profile is proportional to the probability of a photon being absorbed at the same point in the line profile, namely $4\pi\phi_\nu(\text{abs})J_\nu$. The emission coefficient is then proportional to $\phi_\nu J_\nu$ which, for a case in which the radiation field is strongly affected by the line itself and so J_ν varies rapidly with frequency, is not the same as ϕ_ν. Thus for coherent scattering, the emission coefficient line profile is not the same as the absorption coefficient line profile, and the scattering source function is J_ν and varies across the line. If atoms did not move, a resonance line, that is one with the ground state as lower level, which suffered no collisional broadening would be a case of pure coherent scattering, for the lower level has no natural broadening since it has no downward transitions, and so the emitted photon must have the same frequency as the absorbed photon. On the other hand, if the upper level is strongly broadened by collisions, then between the time of absorption and the time of emission of a photon collisions may bring or take away energy, and so emission may take place anywhere in the line profile, with a probability given by the absorption profile ϕ_ν. The emission coefficient in this case is proportional to $\phi_\nu \int \phi_\nu J_\nu \,d\nu$, since the probability of absorption somewhere in the line profile is proportional to $\int \phi_\nu J_\nu \,d\nu$. Thus $S_\nu = \int \phi_\nu J_\nu \,d\nu$ and is independent of frequency. This is called the case of *complete redistribution*. The natural line width $\Gamma_{\rm n} \sim 1/(\text{lifetime of the level})$ and the collisional width $\Gamma_{\rm c} \sim 1/(\text{time between collisions})$ so if $\Gamma_{\rm c} > \Gamma_{\rm n}$ (which is the most common case in stellar atmospheres), collisions will on the average take place before re-emission of the scattered photon, and one might expect redistribution to be the better approximation.

However, the effect of the Doppler shift due to the movement of the scattering atom still has to be taken into account. Consider an atom moving radially outwards scattering a line centre photon also moving radially outwards. The atom will 'see' the photon redshifted, and hence if the scattering is coherent the re-emitted photon will be in the red wing of the line as seen by the atom. If the re-emission is without change of direction and the scattered photon is absorbed again by a stationary atom, then that atom will see the photon as being at the line centre again since the scattering atom will appear to be blue-shifted. If the scattered photon is re-emitted with a change of direction, it will be 'seen' by the stationary atom with less of a blue shift, and so will no longer be received at the line centre frequency in the stationary frame of reference. Thus a combination of Doppler shift and change of direction on scattering leads to at least partial redistribution.

The study of the combination of the effects of natural and collisional broadening of both levels of a transition with Doppler shift and change of direction on scattering is an involved one, but the overall result is that *complete redistribution* is a reasonable approximation in most cases of interest, although the opposite extreme of coherent scattering can sometimes be approached in the wings of a line profile. Complete redistribution will be assumed in what follows.

Consider the case of a two level atom. In Chapter 2 we showed by solving the equations of statistical equilibrium that the source function can be written

$$S_\nu = (1 - \varepsilon') \int \phi_\nu J_\nu \, d\nu + \varepsilon' B_\nu \tag{2.34}$$

with

$$\varepsilon' = \frac{\varepsilon}{\varepsilon + 1}$$

and

$$\varepsilon = \frac{C_{ji}}{A_{ji}} [1 - e^{-h\nu/kT}]$$

This expression can be alternatively derived by following the progress of an individual photon through an atmosphere. Such a photon, on meeting an atom in a suitable state, can either be scattered or it can be absorbed and the resulting excited atom de-excited by a collision. The latter case, and the reverse process of collisional excitation followed by the emission of a photon, are called *thermal* processes. Thermal processes involve the creation and destruction of photons and are characterized by a source function $B_\nu(T_k)$ of Planckian form. Photon scattering will depend only on the frequency averaged radiation field (for complete redistribution) with source function $\int \phi_\nu J_\nu \, d\nu$. An absorbed photon can either be re-emitted or collisionally de-excited with relative probabilities A_{ji} and

C_{ji}, so the probability of thermal destruction is $C_{ji}/(A_{ji}+C_{ji})$ which equals ε' if the stimulated emission term $(1-\exp[-h\nu/kT])$ is ignored. Hence we arrive at equation (2.34.) The crucial fact to remember is that $\varepsilon \ll 1$ in stellar atmospheres, so $\varepsilon' \sim \varepsilon \sim C_{ji}/A_{ji}$ at optical wavelengths, and a typical line photon is scattered many times before it is thermalized and destroyed. Imagine the photon zig-zagging through the atmosphere, being scattered $1/\varepsilon$ times on average before being absorbed and destroyed. Since each step will on average have a length in optical depth units of one, and since the distance moved in any particular direction in a random walk is $\sqrt{(\text{Number of steps})} * (\text{step size})$, the photon will drift in optical depth a distance $1/\sqrt{\varepsilon}$ in any particular direction before being thermalized. The average distance moved before thermalization is called the *thermalization distance*, and is a measure of the non-localization of the source function.

Unfortunately, the value for the thermalization distance derived above assumes a constant average value of the optical depth for each step, which will only be the case for coherent scattering. Complete redistribution means that photons will be reshuffled in the line profile, and of course optical depth one means different geometrical distances at different frequencies. A line centre photon can be transferred to the transparent wings where it will travel much farther in geometrical distance depth terms, whereas redistributing a wing photon to line centre simply means that it will rapidly be scattered again, with a fresh possibility of being redistributed. Hence we would expect the net effect of redistribution to be an increase in the thermalization distance, and we would also expect this increase to be greater in the case of a profile of Lorentz form, where very transparent wings are a major part of the profile, than in the case of a line of Gaussian profile.

We would like to know how much non-LTE effects alter the source function, and how deep into the atmosphere they extend. One can formally take equation (2.34) and insert it into the equation of radiative transfer:

$$\mu \frac{dI_\nu}{d\tau_\nu} = I_\nu - S_\nu \tag{4.14}$$

or writing $\tau_\nu = \phi_\nu \tau'$

$$\mu \frac{dI_\nu}{d\tau'} = \phi_\nu (I_\nu - S_\nu)$$

$$= \phi_\nu \left[I_\nu - (1-\varepsilon') \int J_\nu \phi_\nu \, d\nu - \varepsilon' B_\nu \right] \tag{4.15}$$

$$= \phi_\nu \left[I_\nu - (1-\varepsilon') \frac{1}{4\pi} \iint I_\nu \phi_\nu \, d\Omega \, d\nu - \varepsilon' B_\nu \right] \tag{4.15a}$$

This equation involves on the right-hand side a double integral of the intensity over frequency and angle, which makes the solution difficult. The reader who is not interested in the details should move directly to equation (4.21).

The most obvious approach notes that for a plane parallel atmosphere $J_\nu(\tau_\nu)$ is given by equation (1.10):

$$\int_0^\infty \phi_\nu J_\nu(\tau_\nu)\mathrm{d}\nu = \frac{1}{2}\int_0^\infty \phi_\nu \int_{-1}^{+1} I_\nu \mathrm{d}\mu \mathrm{d}\nu$$
$$= \frac{1}{2}\int_0^\infty \phi_\nu \int_0^\infty S(t_\nu)E_1(|t_\nu - \tau_\nu|)\mathrm{d}t_\nu \mathrm{d}\nu$$

Writing $\tau_\nu = \tau'\phi_\nu$, we obtain at the mean optical depth τ':

$$\int_0^\infty \phi_\nu J_\nu(\tau')\mathrm{d}\nu = \int_0^\infty K_1(t', \tau')\, S(t')\mathrm{d}t' \qquad (4.16)$$

with

$$K_1 = \frac{1}{2}\int_0^\infty \phi_\nu^2 E_1(\phi_\nu|t' - \tau'|)\mathrm{d}\nu$$

where K_1 is the *kernel* (and has nothing to do with the intensity moment K). If we use frequency in a dimensionless form so that x is the frequency displacement from the line centre in units of Doppler width and replace ϕ_ν by ϕ_x, and τ' is similarly replaced by τ'' with $\tau_\nu = \phi_x\tau''$, then for a given form of the line profile the kernel function depends only on $|t'' - \tau''|$. One might then have expected that (4.16) substituted in (4.15) would give an equation that could be solved iteratively for the source function. Unfortunately convergence is so slow that this direct method usually cannot be employed.

In order to understand the general nature of the solution, we will use the Eddington approximation introduced in Chapter 1 to remove the integral over angle. Equation (1.27) of that chapter can be written

$$\frac{1}{3}\frac{\mathrm{d}J_\nu}{\mathrm{d}\tau_\nu} = H_\nu$$

and differentiating:

$$\frac{1}{3}\frac{\mathrm{d}^2 J_\nu}{\mathrm{d}\tau_\nu^2} = \frac{\mathrm{d}H_\nu}{\mathrm{d}\tau_\nu}$$
$$= J_\nu - S_\nu \qquad (4.17)$$

by (1.21). The simplest boundary condition of isotropic (outwards only) radiation requires $J_\nu(0) = 2H_\nu(0)$, as was seen in Chapter 1, which with (1.27) gives $\mathrm{d}J_\nu/\mathrm{d}\tau_\nu(\tau_\nu = 0) = 3/2J_\nu(0)$. A rather better approximation for the boundary condition is

$$\frac{\mathrm{d}J_\nu}{\mathrm{d}\tau_\nu}(\tau_\nu = 0) = \sqrt{3}J_\nu(0)$$

and this will be used in what follows.

Putting $\tau_\nu = \tau'\phi_\nu$ again and substituting (2.34) for the source function in (4.17):

$$\frac{d^2J_\nu}{d(\tau')^2} = 3\phi_\nu^2\left[J_\nu - (1-\varepsilon')\int J_\nu\phi_\nu d\nu - \varepsilon' B\right] \tag{4.17a}$$

with the boundary condition

$$\frac{dJ_\nu}{d\tau'}(\tau'=0) = \sqrt{3}\phi_\nu J_\nu(\tau'=0) \tag{4.18}$$

These equations become soluble analytically if we also assume coherent scattering which removes the integral over frequency, so $\int \phi_\nu J_\nu d\nu = J_\nu$. Equation (4.17a) then becomes

$$\frac{d^2J_\nu}{d(\tau')^2} = 3\phi_\nu^2\varepsilon'(J_\nu - B_\nu) \tag{4.17b}$$

For instance, for an atmosphere where the thermal part of the source function, B, is linear in optical depth so $B = a + b\tau'$, it can be verified at once by substitution that the solution is

$$J_\nu = B + J_{0\nu}\exp[-\sqrt{(3\varepsilon')}\phi_\nu\tau']$$

with the constant $J_{0\nu}$ given by the boundary condition as

$$J_{0\nu} = \frac{\dfrac{b}{\sqrt{3}\phi_\nu} - a}{1+\sqrt{\varepsilon'}}$$

The source function then becomes

$$S_\nu = (1-\varepsilon')J_\nu + \varepsilon' B_\nu$$
$$= B_\nu + (1-\varepsilon')J_{0\nu}e^{-\sqrt{3\varepsilon'}\phi_\nu\tau'} \tag{4.19}$$

which reduces to the Planck function B_ν if the exponential term can be neglected, that is when

$$\tau_\nu > \frac{1}{\sqrt{3\varepsilon'}}$$

Thus the order of magnitude of the thermalization distance is $\sqrt{(3\varepsilon')}$, as was found in the random walk argument. If we specialize further to the case where the thermal source function is isothermal so $B = \text{constant} = a$, $b = 0$, and $J_{0\nu} = -B_\nu/(1+\sqrt{\varepsilon'})$, then the source function at the surface is

$$S_\nu(\tau'=0) = B_\nu\left[1 - \frac{1-\varepsilon'}{1+\sqrt{\varepsilon'}}\right]$$
$$= \sqrt{\varepsilon'}B_\nu \tag{4.19a}$$

If b is not zero but $\varepsilon' \ll 1$, which is usually the case, then the source function at the surface becomes

$$S_\nu(\tau'=0) \simeq \sqrt{\varepsilon'}\, B\left(\tau' = \frac{1}{\sqrt{3\varepsilon'}\phi_\nu}\right) \tag{4.19b}$$

i.e. the source function at the surface is $\sqrt{(\varepsilon')}$ the thermal source function evaluated at the thermalization depth.

Thus far we have ignored the presence of continuum absorption, coefficient κ_C, which should be added to the line absorption, coefficient $\kappa_{L\nu}$, discussed above. Since $\kappa_\nu = \kappa_{L\nu} + \kappa_C$, $d\tau_\nu = (\kappa_{L\nu} + \kappa_C)\rho\, ds$ for displacement ds, and $d\tau' = \kappa_{L\nu}/\phi_\nu \rho\, ds$, we have

$$d\tau_\nu = (\kappa_C/\kappa_{L\nu} + 1)\phi_\nu d\tau'$$

The source function (assuming LTE for continuum processes) then becomes:

$$S_\nu = \frac{\kappa_{L\nu}}{\kappa_C + \kappa_{L\nu}}\left[(1-\varepsilon')\int \phi_\nu J_\nu\, d\nu + \varepsilon' B\right] + \frac{\kappa_C}{\kappa_C + \kappa_{L\nu}} B$$

so that (4.17) in the Eddington approximation becomes

$$\frac{d^2 J_\nu}{d(\tau')^2} = 3\left(\frac{\kappa_C}{\kappa_{L\nu}} + 1\right)^2 \phi_\nu^2 \left[J_\nu - \frac{(1-\varepsilon')}{\left(1 + \frac{\kappa_C}{\kappa_{L\nu}}\right)}\int \phi_\nu J_\nu - \left(1 - \frac{1-\varepsilon'}{\left(1+\frac{\kappa_C}{\kappa_{L\nu}}\right)}\right) B\right]$$

which reduces to (4.17a) if that equation is written with ϕ_ν outside the square brackets replaced by

$$\phi'_\nu = \left(\frac{\kappa_C}{\kappa_{L\nu}} + 1\right)\phi_\nu \tag{4.20}$$

and ε' replaced throughout by

$$\varepsilon'' = \frac{\frac{\kappa_{L\nu}}{\kappa_C}\varepsilon' + 1}{\frac{\kappa_{L\nu}}{\kappa_C} + 1} \tag{4.20a}$$

Hence we obtain the same result as for line absorption alone but with ε'' instead of ε' in the expression for the thermalization depth. For a weak line $\kappa_{L\nu}/\kappa_C \ll 1$ so $\varepsilon'' = 1$ and the line is formed in LTE. Only for a strong line with $\kappa_{L\nu} \gg \kappa_C$ will ε'' reduce to $\varepsilon' \ll 1$. Thus, as one might expect, the effect of continuum opacity is to reduce thermalization depths and drive the source towards LTE.

The discussion so far (after equation (4.18) has been limited to the case of coherent scattering. The more common case of complete redistribution is harder to deal with. One approach is to replace the integral over frequency by a sum.

Divide the frequency range covered by the line profile into N discrete frequencies labelled $i = 1, \ldots, N$ and then write:

$$\int \phi_\nu J_\nu \mathrm{d}\nu = \sum_{i=1}^{N} p_i J_i$$

with the weights p_i normalized by $\sum p_i = 1$. The rather involved solution of the resulting set of equations is discussed in appendix 1, where it is shown that at the boundary:

$$S(0) = \sqrt{(\varepsilon')} B \tag{4.21}$$

for a constant thermal source function B. The discrete frequency method can be modified to include the continuum absorption in a similar way to that used for coherent scattering above, although in the complete redistribution case it needs to be remembered that ε'', unlike ε', is frequency dependent, so that if continuous absorption is present, ε' in (4.21) must be replaced by a frequency averaged value of ε'', namely $\int \phi_\nu \varepsilon''(\nu) \mathrm{d}\nu$. The discrete frequency method does not lead to a neat analytical expression for the thermalization depth, and it is more profitable to

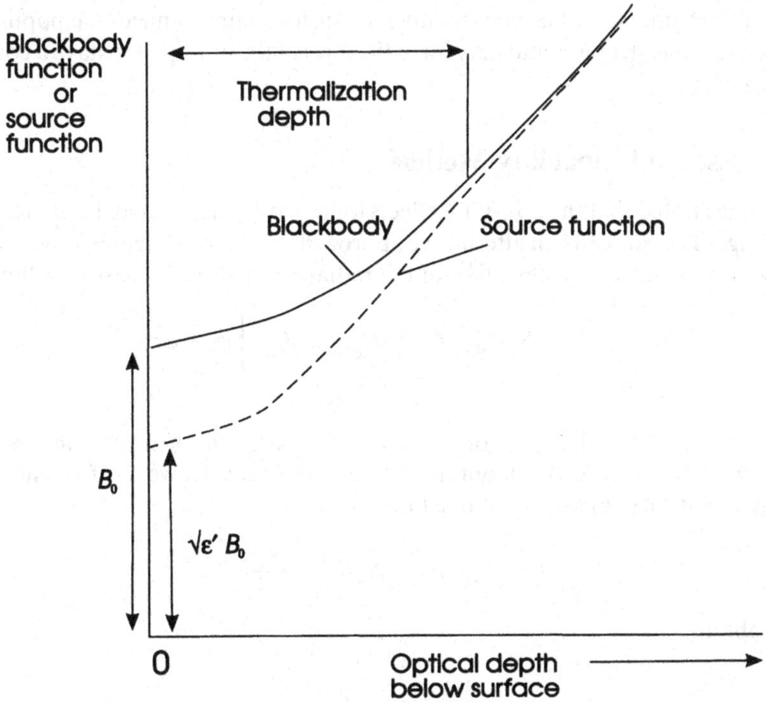

Figure 4.5.

pursue the question of the thermalization depth in the case of complete redistribution using a different approach, the escape probability method, which will be considered in the next section. However, it is clear that redistribution from the centre of a line into the wings will enable a photon to escape, thus increasing the non-LTE influence on the source function and increasing the depth to which leakage effects are significant, i.e. increasing the thermalization depth.

Thus the overall conclusion is that the source function drops by a factor $\sqrt{(\varepsilon')}$ at the edge of an atmosphere or cloud due to the leakage of photons reducing the pool of photons near that edge (Figure 4.5). This leads to dark absorption lines, even from isothermal atmospheres, and for atmospheres with a temperature gradient produces lines that are darker than one might expect from arguments assuming a thermal source function. The value of ε' is generally small, but for weak lines is brought close to the LTE value of 1 by the contribution of LTE continuum absorption. However, complete redistribution and line broadening tend to drive the source function away from LTE values down to depths depending on the line profile. Finally, it should be noted that only the case of a two-level atom has been considered here. Where lines share a common level, the source functions of the lines will be linked. An important case is that of a multiplet, where lines originate from closely spaced levels only differing in total angular momentum quantum number. Collisions often connect the populations of these levels strongly, and may force the lines of the multiplet to have a common source function.

The Escape Probability Method

We have noted that the non-LTE effects for a two-level atom can be attributed to leakage. This suggests an alternative approach. We start by defining the quantity β, which expresses the deviation of the radiation field from the source function:

$$\beta = \frac{N_j A_{ji} + 4\pi(N_j B_{ji} - N_i B_{ij}) \int \phi_\nu J_\nu \, d\nu}{N_j A_{ji}} \tag{4.22}$$

Equation (4.22) is the ratio of the net downward rate of transitions (per unit volume and time) to the spontaneous emission rate. However if we substitute from equation (2.6) for the source function

$$4\pi(N_i B_{ij} - N_j B_{ji}) = \frac{N_j A_{ji}}{S_\nu}$$

we obtain

$$\beta = 1 - \frac{\int \phi_\nu J_\nu \, d\nu}{S_\nu} \tag{4.23}$$

where in full thermal equilibrium $J_\nu = S_\nu$ and so it can be seen that β expresses the importance of leakage. Now

$$S_\nu = (1-\varepsilon')\int \phi_\nu J_\nu \,\mathrm{d}\nu + \varepsilon' B$$

$$= (1-\varepsilon')(1-\beta)S_\nu + \varepsilon' B$$

$$S_\nu = \frac{\varepsilon' B}{\varepsilon' + (1-\varepsilon')\beta} \tag{4.24}$$

$$\simeq \frac{\varepsilon' B}{\beta} \tag{4.24a}$$

if $\varepsilon' \ll 1$ and $\varepsilon' \ll \beta$, with β tending to zero at large depths where leakage is negligible.

Now consider the escape probability for a photon. Let the probability that a photon of frequency ν travelling in a certain direction escapes altogether from the atmosphere without further interaction, the *escape probability*, be p_ν. Then if the optical depth from the point under consideration to the edge of the atmosphere in the specified direction is τ_ν, $p_\nu = \exp(-\tau_\nu)$. This can be averaged over frequency, weighted by the line emission profile, to give

$$p_e = \int_0^\infty \phi_\nu p_\nu \,\mathrm{d}\nu$$

Finally one can average p_e over angle to give the probability that photons emitted from a given spot in a particular line but in all directions will escape:

$$P_e = \frac{1}{4\pi}\int p_e \,\mathrm{d}\Omega$$

Now the equation of radiative transfer can be formally integrated along a particular direction to give the emergent intensity I_{out}:

$$-\frac{\mathrm{d}I_\nu}{\mathrm{d}\tau_\nu} = S_\nu - I_\nu$$

$$I_{\text{out}} = \int (S_\nu - I_\nu)\,\mathrm{d}\tau_\nu$$

$$= \int S_\nu\left(1 - \frac{I_\nu}{S_\nu}\right)\mathrm{d}\tau_\nu \tag{4.25a}$$

But also

$$I_{\text{out}} = \int S_\nu \mathrm{e}^{-\tau_\nu}\,\mathrm{d}\tau_\nu = \int p_\nu S_\nu \,\mathrm{d}\tau_\nu \tag{4.25b}$$

The equality of the left-hand sides of these equations suggest that $(1 - I_\nu/S_\nu)$ averaged over a ray with weighting proportional to the emission per unit optical depth (i.e. to the source function) is equal to p_ν averaged in the same way. An average over a ray is not a convenient quantity to use, but if one writes $d\tau_\nu = \phi_\nu \kappa' \rho \, ds$ and averages both equations over frequency so $\phi_\nu p_\nu$ in (4.25b) becomes p_e (assuming complete redistribution and hence a frequency independent source function) and then over angle, I_ν in (4.25a) becomes $\int \phi_\nu J_\nu$ which can be replaced by $S_\nu(1 - \beta)$. Finally, averaging over all rays as well as along each ray which is the same as an average over volume:

$$\frac{\int \beta(\mathbf{r}) S(\mathbf{r}) \kappa'(\mathbf{r}) \rho \, dV}{\int S(\mathbf{r}) \kappa'(\mathbf{r}) \rho \, dV} = \frac{\int P_e S(\mathbf{r}) \kappa'(\mathbf{r}) \rho \, dV}{\int S(\mathbf{r}) \kappa'(\mathbf{r}) \rho \, dV}$$

$$\langle \beta \rangle = \langle P_e \rangle \tag{4.26}$$

where in (4.26) the average is over the whole volume, emission weighted (remembering that $S\kappa\rho$ is just the emission per unit volume), $\mathbf{r}$ being a positional vector. The significance of this result is that β is what we would like to know in order to determine the source function, and P_e can in principle be calculated. Unfortunately it is β at a particular point rather than the volume average $\langle \beta \rangle$ that is required if we want to know how β increases from zero as non-LTE effects become important near the edge of an atmosphere (see the review by Rybicki [1] for a clear discussion of the relation between β and P_e).

The basic approximation in escape probability theory lies in assuming $\beta(\mathbf{r}) = P_e(\mathbf{r})$, i.e. that the leakage effect on the source function is determined entirely by the *local* escape probability. One case in which this holds is when photons *either* travel for very short distances before interacting again *or* escape completely. For those photons travelling very short distances, $\beta = P_e = 0$ and

$$N_j A_{ji} = 4\pi (N_i B_{ij} - N_j B_j) \int \phi_\nu J_\nu \, d\nu$$

The net transfer rate from upper to lower level locally is therefore equal to the number of photons escaping completely. Hence:

$$N_j A_{ji} - 4\pi (N_i B_{ij} - N_j B_{ji}) \int \phi_\nu J_\nu \, d\nu = P_e N_j A_{ji}$$

and therefore $\beta = P_e$. This situation certainly holds quite often in clouds or atmospheres with mass movements. An emitted photon can be absorbed locally, but if it travels any distance it will reach layers that are moving relative to the layer which emitted the photon, so that the absorption profile is Doppler shifted with respect to the emission profile and the photon will not be absorbed but will escape.

In gas clouds with no large scale internal movements or in static stellar atmospheres the situation is less clear, but it can be argued that $P_e = 0$ for line centre photons of strong lines, and that such a line centre photon will stay near the line centre as it zig-zags through a cloud, being scattered many times until a rare scattering event redistributes it to the wings of the line, when it will escape. If we can therefore put $\beta \approx P_e$, equation (4.24) becomes

$$S_\nu = \frac{\varepsilon' B}{\varepsilon' + (1 - \varepsilon')P_e} \tag{4.27}$$

which reduces for $\varepsilon' \ll 1$ to $S_\nu = 0.5B$ when $P_e \sim \varepsilon'$. Thus the depth at which $P_e \sim \varepsilon'$ is the depth at which the leakage of photons starts to become important, and so is the thermalization depth. One can argue alternatively that an emitted photon either escapes or interacts, and if it interacts it has a chance $C_{ji}/(C_{ji} + A_{ji})$ of being thermalized, failing which it will be emitted again. When leakage, represented by P_e, becomes as likely as thermalization, there will be departures from LTE, and this will happen when $P_e = C_{ji}/(C_{ji} + A_{ji}) \sim \varepsilon'$. For a static stellar atmosphere, the approximation $P_e(\mathbf{r}) = \beta(\mathbf{r})$ is unlikely to be a good one at the edge of the atmosphere, and indeed (4.26) does not give the exact result (4.27) there. On the other hand, the approximation is likely to improve deeper into the atmosphere, and so may be a useful way of calculating the thermalization depth.

We must now consider how escape probabilities are actually calculated. The answer depends on whether we are considering a gas 'cloud' from which escape is possible in all directions (such as an H II region) or a cloud in which escape is possible in just one direction (such as a stellar atmosphere). It also depends on whether we are dealing with plane-parallel geometry or spherical geometry, on whether we are concerned with a static cloud or one in which there are large scale movements, and on whether complete redistribution in the line profile is an adequate approximation or not. Our approach will also differ depending on whether we are interested only in the mean escape probability from the cloud or whether we want to know the escape probability from a particular depth.

Consider first the static case in a plane-parallel atmosphere when we are trying to find the one-sided escape probability from a particular depth assuming complete redistribution. Suppose that optical depth perpendicular to the atmosphere is τ_P so that optical depth in a direction making an angle θ with the perpendicular is τ_P/μ with, as usual, $\mu = \cos\theta$, and suppose that depth at frequency ν is $\tau_{P\nu} = \tau'_P \phi_\nu$. The escape probability is

$$P_e(\tau_P) = \frac{1}{2}\int_0^{+1}\int_0^{\infty} \phi_\nu p_\nu \, d\nu \, d\mu$$

where we have taken a semi-infinite atmosphere with escape in one direction only so the lower limit on angle is $\mu = 0$.

Hence:

$$P_e = \frac{1}{2}\int_0^\infty \phi_\nu \int_0^{+1} e^{-\tau'_P\phi_\nu/\mu}\,d\mu\,d\nu$$

$$= \frac{1}{2}\int_0^\infty \phi_\nu \int_1^\infty e^{-\tau'_P\phi_\nu z}\frac{dz}{z^2}\,d\nu, \quad \text{with } z = \frac{1}{\mu}$$

$$= \frac{1}{2}\int_0^\infty \phi_\nu E_2(\tau'_P\phi_\nu)\,d\nu$$

Rewrite in terms of the mean optical depth $\tau_{OP} = \tau'_P/\Delta\nu_V$ and dimensionless frequency x, where $d(\Delta\nu) = \Delta\nu_V\,dx$ and $\Delta\nu_V$ is the Doppler or Lorentz width, and note that since $\phi_x dx = \phi_\nu d\nu$, $\phi_x = \phi_\nu \Delta\nu_V$. Then

$$P_e(\tau_{OP}) = \frac{1}{2}\int_{-\infty}^{\infty} \phi_x E_2(\tau_{OP}\phi_x)\,dx \tag{4.28}$$

Now when $\tau_{OP}\phi_x \gg 1$, $E_2(\tau_{OP}) \sim 0$ and the contribution from this frequency range is therefore small. This corresponds to the case of photons that are absorbed or emitted in the core of the line. On the other hand, when $\tau_{OP}\phi_x \ll 1$, $E_2(\tau_{OP}) \sim 1$, we are in the wings of the line and most photons escape. This situation can be approximated by taking x_1 as the dimensionless frequency at which $\tau_{OP}\phi_x(x_1) = 1$, and by assuming that no photons escape for $|x| < x_1$. Hence:

$$P_e(\tau_{OP}) \simeq \int_{x_1}^{\infty} \phi_x\,dx \tag{4.29}$$

For Doppler broadening:

$$\phi_x = \frac{1}{\sqrt{\pi}} e^{-x^2}$$

so

$$e^{-x_1^2} = \frac{\sqrt{\pi}}{\tau_{OP}} \quad \text{and} \quad x_1 = \sqrt{\ln_e\left(\frac{\tau_{OP}}{\sqrt{\pi}}\right)}$$

$$P_e(\tau_{OP}) \simeq \frac{1}{\sqrt{\pi}}\int_{x_1}^{\infty} e^{-x^2}\,dx \simeq \frac{e^{-x_1^2}}{2\sqrt{\pi}x_1}$$

where the last step follows if $x_1 \gg 1$, i.e. if $\tau_{OP} \gg 1$. Substituting for x_1:

$$P_e(\tau_{OP}) \simeq \frac{1}{2\tau_{OP}}\frac{1}{\sqrt{\ln_e\left(\frac{\tau_{OP}}{\sqrt{\pi}}\right)}} \tag{4.30}$$

Writing $P_e \sim \varepsilon'$ at the thermalization depth, we obtain:

$$\tau_{OP}\,(\text{Doppler, thermalization}) \sim 1/(2\varepsilon') \tag{4.31}$$

since $\sqrt{(\ln_e[\tau_{OP}/\sqrt{\pi}])}$ is between 1 and 2 for τ_{OP} between 4 and 1000.

For a Lorentz profile:

$$\phi_x = \frac{1}{\pi}\frac{1}{1+x^2} \simeq \frac{1}{\pi x^2}, \quad \text{for } x > 1$$

$$\frac{1}{\pi x_1^2} = \frac{1}{\tau_{OP}}, \qquad x_1 = \sqrt{\frac{\tau_{OP}}{\pi}}$$

$$P_e(\tau_{OP}) \simeq \frac{1}{\pi}\int_{x_1}^{\infty} \frac{dx}{x^2}$$

$$\simeq \frac{1}{\pi x_1} \simeq \frac{1}{\sqrt{\pi \tau_{OP}}} \tag{4.32}$$

Hence, putting $P_e \sim \varepsilon'$ at the thermalization depth:

$$\tau_{OP}(\text{Lorentz, thermalization}) \sim /(\pi\varepsilon'^2) \tag{4.33}$$

Thus the thermalization depth, which was $1/\sqrt{\varepsilon'}$ in terms of the optical depth at a particular frequency in the line profile for the case of coherent scattering, has the much larger but frequency independent values of $1/\varepsilon'$ and $1/\varepsilon'^2$ for Doppler and Lorentz profiles in the case of complete redistribution, and is now given in terms of mean optical depth.

In the limit of large line centre optical depth τ_{LCP}, it is possible, following Ivanov, to write down an exact asymptotic expression for the escape probability (see, for example, Canfield *et al.* [2]), which for the Doppler case gives

$$P_e(\text{Doppler}) = \frac{1}{4\sqrt{\pi}\tau_{LCP}}\frac{1}{\sqrt{\ln_e(\tau_{LCP})}} \tag{4.34}$$

$$= \frac{1}{4\tau_{OP}\sqrt{\ln_e\left(\frac{\tau_{OP}}{\sqrt{\pi}}\right)}} \tag{4.34a}$$

since $\tau_{LCP} = 1/\sqrt{\pi}\tau_{OP}$ for Doppler broadening where $\phi_x(0) = 1/\sqrt{\pi}$ and $\tau_{\nu P} = \tau_{OP}\phi_x$.

The case of a pure Lorentz profile is rarely met, and more commonly one is dealing with a Voigt profile which is Doppler dominated at its centre, but whose equivalent width comes mainly from strong Lorentzian wings. If $a = \Gamma/(4\pi\Delta\nu_D) \ll 1$, the Voigt function can be expanded :

$$\phi_x = \frac{1}{\sqrt{\pi}}(e^{-x^2} + aH_1(a,x) + \cdots)$$

with H_1 tending to $1/(\sqrt{\pi}x^2)$ at large x. Since the escape probability is determined by the wings of the line, we can approximate the normalized profile by $\phi_x = a/(\pi x^2)$.

Then the same argument as was used in the Doppler case gives for the escape probability in the limit of the large line centre optical depth:

$$P_e = \frac{1}{3}\sqrt{\frac{a}{\sqrt{\pi}\tau_{LCP}}} \tag{4.35}$$

$$= \frac{1}{3}\sqrt{\frac{a}{\tau_{OP}}} \tag{4.35a}$$

If the line profile has Stark broadened wings in the quasi-static approximation with normalized profile $\phi_x = \alpha/x^{5/2}$, then (Puetter [3])

$$P_e = \frac{0.3723}{\pi^{3/10}} \frac{\alpha^{2/5}}{\tau_{LCP}^{3/5}} \tag{4.36}$$

For a derivation of the formula leading to (4.34) and (4.35) see Appendix 2.

It has been assumed so far in deriving escape probabilities that the emission profile is the same as the absorption profile. This may not be the case in the Lorentzian wings of a very strong resonance line, particularly if the line is not strongly interlocked with another resonance line. The most important example is Lyman α. If the pressure is low enough for collisional broadening to be less significant than natural line broadening, then the upper level of the transition will be naturally broadened and the lower level unbroadened (it will have no natural broadening since it is the ground state). The scattering of a photon in the wings of such a line will be coherent in the rest frame of the scattering atom, that is the emitted photon will have the same frequency as the absorbed photon. However, the fact that the scattering atom is moving will lead to Doppler shifts of the absorbed and emitted photons, with values depending on the relative directions of the photon and atom. This will lead to a partial redistribution in frequency, on average of about one Doppler width. The shifts can be to higher or lower frequency, with a slight bias in favour of shifts towards the line centre, so that a photon being scattered x Doppler widths from the line centre will on average be emitted $1/x$ Doppler widths nearer the line centre, and hence after x^2 scatterings will, on average, reach the line centre *if* it has not managed to escape by then.

Now in a very strong line, even in the wings, the optical depth will be large, and so our wing photon may be scattered many times, moving through the cloud in a random walk and escaping if this random walk takes it close to the surface but otherwise becoming trapped in the very opaque core of the line. Adams [4] argues that the distance travelled on average by a photon in any particular direction in a random walk of N scatterings is $\sqrt{N}*$(distance travelled between scatterings), and since the step size in optical depth is one or $1/\phi_x$ in mean optical depth units (mean optical depth = optical depth at frequency x divided by ϕ_x) and the number of scatterings before return to the core is about x^2 on average, then the distance travelled on average in a particular direction while still in the wings is about x/ϕ_x in mean optical depth units. The profile is $\phi_x = a/(\pi x^2)$ and the distance to the surface in mean optical depth units is τ_{OP}, so the distance travelled in any direction equals the distance to the surface when $x = x_e = (a\tau_{OP}/\pi)^{1/3}$. Photons which are redistributed from the core to frequency x_e or greater will on average escape. The source function turns out to be constant quite far from the centre of the line, so to a first approximation one can still estimate the probability of a photon being emitted at x after a scattering as being given roughly by ϕ_x. It will stay in the wings for roughly x^2 scatterings, as we have seen, so the probability that it is redistributed out to x_e for the first time is approximately ϕ_x/x^2.

Hence the escape probability is given by:

$$P_e = 2\int_{x_e}^{\infty} \frac{\phi_x}{x^2}\mathrm{d}x = 2\int_{x_e}^{\infty} \frac{a}{\pi x^4}\mathrm{d}x$$

$$= \frac{2}{3\tau_{OP}} \tag{4.37}$$

substituting for x_e. Numerical calculations bear out this result to a factor 2 or so.

All these estimates of the escape probability refer to large optical depths. Near the surface, the one-sided escape probability must become 1/2 so approximate formulae valid at all depths have the form:

$$P_e \simeq \frac{0.5}{1 + \text{constant}\,\tau^n}, \quad n \simeq 1$$

We now turn to the case of a medium with mass movement, say of expansion. The optical depth to the surface from a particular point in some particular direction at some frequency is given by the distance travelled before mass motions Doppler shift the absorption profile of the gas away from the emission profile of the gas at the chosen point. The corresponding geometrical distance will be of the order of L, given by:

$$\Delta V = \mathrm{d}V/\mathrm{d}s \cdot L = \Delta/\nu_0 c$$

where the velocity gradient along the chosen direction is $\mathrm{d}V/\mathrm{d}s$, and the line has a rest frequency of ν_0 and a characteristic width in frequency of Δ (Doppler or Lorentz width). If the emission frequency is $x = (\Delta\nu)/\Delta$ from the centre of the line in the usual dimensionless units, then the optical depth is:

$$\tau = \frac{\kappa'}{\Delta}\rho\int_{0}^{\infty} \phi_x(x - s/L)\mathrm{d}s$$

$$= \tau_S\int_{-\infty}^{x} \phi_x(x')\mathrm{d}x', \quad \text{where } \tau_S = \kappa'\rho\frac{L}{\Delta}$$

and $x' = x - s/L$ so $\mathrm{d}s = -L\mathrm{d}x'$. The escape probability is then:

$$p_\nu = \exp\left(-\tau_s\int_{-\infty}^{\infty} \phi(x')\mathrm{d}x'\right)$$

Averaging over frequency:

$$p_e = \int_{-\infty}^{\infty} \phi_x \exp\left(-\tau_s \int_{-\infty}^{\infty} \phi(x')\mathrm{d}x' \right)\mathrm{d}x$$

Hence:

$$p_e = \int_0^1 e^{-\tau_s \eta}\mathrm{d}\eta, \quad \text{where } \eta = \int_{-\infty}^{x} \phi_x(x')\mathrm{d}x'$$

$$= \frac{1 - e^{-\tau_s}}{\tau_s} \tag{4.38}$$

and using the definition of L

$$\tau_s = \kappa' \rho \frac{1}{\nu_0} \frac{c}{\dfrac{\mathrm{d}V}{\mathrm{d}s}} \tag{4.39}$$

This result has still to be averaged over direction to give P_e. Suppose we have a spherically symmetric situation with position given in polar coordinates by r and θ, and the local direction of a ray with respect to the radial direction given by angle θ'. Consider a small displacement along such a ray with $\mathrm{d}\theta = -\mathrm{d}\theta'$, $\mathrm{d}r = \mathrm{d}s \cos\theta'$ and $\sin\theta' = -r\mathrm{d}\theta/\mathrm{d}s$ (Figure 4.6). Then

$$\frac{\mathrm{d}V}{\mathrm{d}s} = \frac{\partial V}{\partial r}\frac{\mathrm{d}r}{\mathrm{d}s} + \frac{\partial V}{\partial \theta'}\frac{\mathrm{d}\theta'}{\mathrm{d}s}$$

If the velocity field is purely radial, $V_r(r)$, along a ray $V = V_r \cos\theta'$ and

$$\frac{\mathrm{d}V}{\mathrm{d}s} = \frac{\mathrm{d}V_r}{\mathrm{d}r}\cos^2\theta' + \frac{V_r(-\sin\theta')(-\sin\theta')}{r}$$

$$= \frac{\mathrm{d}V_r}{\mathrm{d}r}\mu^2 + \frac{V_r}{r}(1 - \mu^2)$$

If the velocity is linear, $V_r = kr$, then $\mathrm{d}V/\mathrm{d}s = k$ and $\mathrm{d}V/\mathrm{d}s$ and τ_s are independent of μ. Hence

$$P_e = \frac{1}{2}\int_{-1}^{+1} \frac{1 - e^{-\tau_s}}{\tau_s}\mathrm{d}\mu$$

$$= \frac{1 - e^{-\tau_s}}{\tau_s} \tag{4.40}$$

The plane-parallel atmosphere gives a similar result, taking z as the outward direction and $V = V_z(z)\cos\theta'$, $\mathrm{d}V/\mathrm{d}s = \mathrm{d}V_z/\mathrm{d}z\, \mu'^2$ (Figure 4.7). Hence $\tau_s = \tau'_s/\mu'^2$,

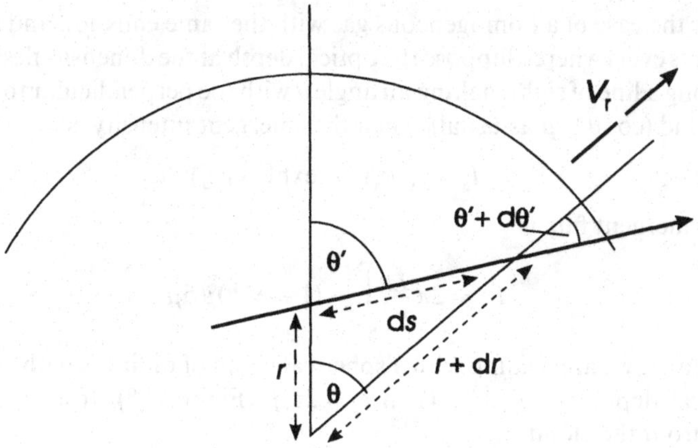

Figure 4.6.

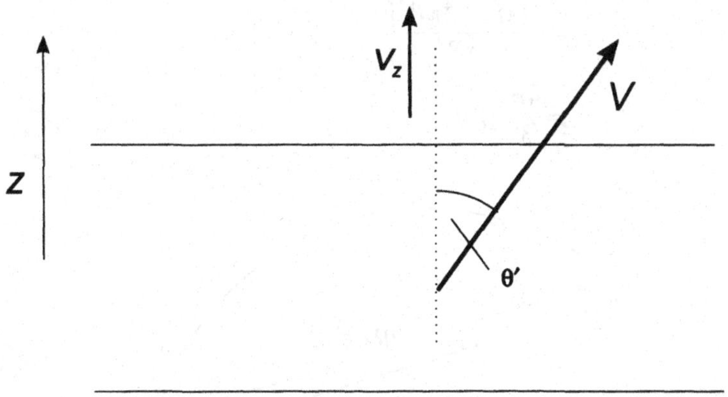

Figure 4.7.

where $\tau_s' = \kappa'\rho/\nu_0 \cdot c/(\mathrm{d}V_z/\mathrm{d}z)$. Hence

$$P_e = \frac{1}{2}\int_{-1}^{+1} \frac{1 - e^{-\tau_s'/(\mu')^2}}{\tau_s'} (\mu')^2 \, \mathrm{d}\mu'$$

$$\simeq \frac{1}{3\tau_s'}(1 - e^{-3\tau_s'}) \tag{4.41}$$

noting that the first expression tends to $1/(3\tau_s')$ for large τ_s'.

Finally we discuss the volume-averaged escape probability, which is not useful in predicting line profiles but can be helpful in estimating level populations.

Consider the case of a homogeneous gas with the same emission and absorption coefficients everywhere. Suppose the optical depth at the dimensionless frequency x is t_x along a line of sight making an angle θ with the perpendicular to the surface of the cloud ($\cos\theta = \mu$ as usual). Then the emergent intensity is

$$I_x = j_x/\kappa_x(1 - \exp[-t_x])$$

and the emergent flux is

$$F_x = 2\pi \frac{j_x}{\kappa_x} \int_0^{+1} (1 - e^{-t_x})\mu \, d\mu$$

Now suppose we are dealing with a spherical cloud of radius R with corresponding optical depth $\tau_x = \kappa_x \rho R$. Then $t_x = 2\tau_x \mu$ (Figure 4.8). If all the radiation escaped from the cloud:

$$F_x = \frac{(\text{emission per unit volume})(\text{volume})}{\text{surface area}}$$

$$= \frac{(4\pi j_x \rho)(\frac{4}{3}\pi R^3)}{4\pi R^2}$$

$$= \frac{4\pi}{3} j_x \rho R$$

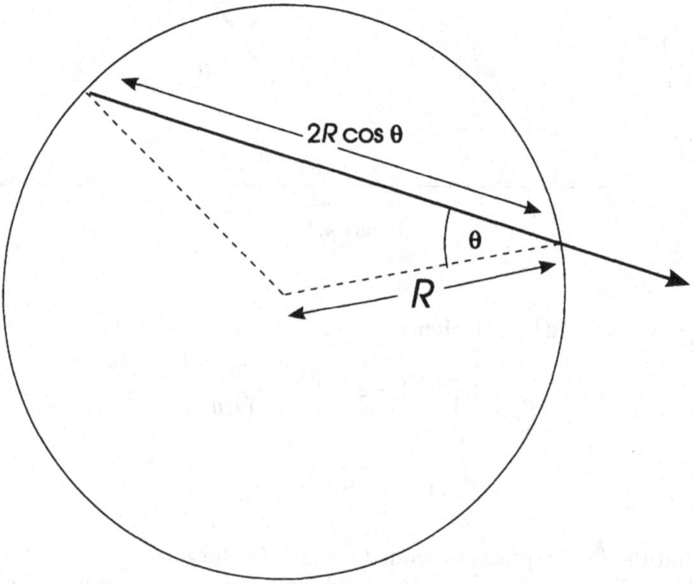

Figure 4.8.

Thus the escape probability averaged over position becomes

$$\langle\beta_x\rangle = \frac{\text{flux}}{\text{flux with no absorption}}$$

$$= \frac{3}{2j_x\rho R}\frac{j_x}{\kappa_x}\int_0^{+1}(1 - e^{-t_x})\mu\, d\mu$$

$$= \frac{3}{2\tau_x}\int_0^{+1}(1 - e^{-2\tau_x\mu})\mu\, d\mu \tag{4.42}$$

$$= \frac{3}{4\tau_x}\left[1 - \frac{1}{2\tau_x^2} + e^{-2\tau_x}\left(\frac{1}{\tau_x} + \frac{1}{2\tau_x^2}\right)\right] \tag{4.42a}$$

The mean escape probability is then found by averaging β_x over the line profile:

$$\beta = \int_{-\infty}^{\infty}\langle\beta_x\rangle\phi_x\, dx \tag{4.43}$$

where strictly speaking we should replace ϕ_x by the emission profile.

If we assume Doppler broadening, then we should have complete redistribution with absorption and emission profiles identical and a profile

$$\phi_x = 1/\sqrt{\pi}\exp(-x^2)$$
$$\tau_x = \tau_{LC}\exp(-x^2)$$

so $x = \sqrt{(\ln_e[\tau_{LC}/\tau_x])}$.

Dividing the range of integration in (4.43) into 2 for the Doppler case, we obtain:

$$\beta = \frac{2}{\sqrt{\pi}}\int_0^{\infty}\langle\beta_x\rangle e^{-x^2}dx$$

$$= \frac{1}{\sqrt{\pi}}\int_0^{\tau_{LC}}\frac{\langle\beta_x\rangle}{\sqrt{\ln_e\left(\frac{\tau_{LC}}{\tau_x}\right)}\,\tau_x}d\tau_x$$

where the variable of integration has been changed. This integral with (4.42) substituted for β_x must be evaluated numerically. Cox and Matthews [5] give an approximate fit to the result, valid for $\tau_{LC} < 40$

$$\beta = \frac{1.72}{1.72 + \tau_{LC}} \tag{4.44}$$

Alternatively, Capriotti [6] has substituted (4.42) directly in (4.43) and reversed the order of integration over μ and x to find an approximate formula, valid for

$\tau_{LC} > 2.5$:

$$\beta = \frac{3}{2\sqrt{\pi}\tau_{LC}}\left[\sqrt{\ln_e(2\tau_{LC})} - \frac{0.25}{\sqrt{\ln_e(2\tau_{LC})}} + 0.15\right] \tag{4.45}$$

References

[1] G. B. Rybicki, in W. Kalkofen (ed.), *Methods in Radiative Transfer*, Cambridge University Press, 1985.
[2] R. C. Canfield, A. N. McClymont, and R. C. Puetter, in W. Kalkofen (ed.), *Methods in Radiative Transfer*, Cambridge University Press, 1985.
[3] R. C. Puetter, *Astrophysical Journal*, **251**, 446, 1981.
[4] T. F. Adams, *Astrophysical Journal*, **174**, 439, 1972.
[5] D. P. Cox and W. G. Mathews, *Astrophysical Journal*, **155**, 859, 1969.
[6] E. R. Capriotti, *Astrophysical Journal*, **142**, 1101, 1965.

5 STELLAR SPECTRA

Introduction

The main stellar parameters that we would like to know are a star's luminosity L, mass M, radius R and composition (we would of course also like to know the star's age, but that cannot be observed directly). Most of this chapter will be concerned with the determination of abundances, but the determination of other parameters is briefly discussed in this introduction. The effective temperature $T_{\rm eff}$ can replace either luminosity or radius since $L = 4\pi R^2 \sigma T_{\rm eff}^4$ by definition of the effective temperature. The surface gravity g can replace either R or M since $g = GM/R^2$, but g is more sensitive to R than M, and stellar radii cover a far larger range than stellar masses. Secondary quantities that we would like to know include stellar rotation velocities and magnetic fields.

Masses can be found for certain binary systems, and radii can be found for some eclipsing binary systems as well as from occultation and interferometer measurements of single stars. These measurements are not the concern of this book, but it is worth noting that in both cases the number of stars for which accurate results can be obtained is very limited. Luminosities can be found for stars of known distance from the received flux $F = L/4\pi d^2$, but the flux must be integrated over all frequencies. There are inevitably frequency regions where there are no observations, and some interpolation and extrapolation is required. A temperature can be found by taking the ratio of fluxes at two wavelengths and assuming a blackbody spectrum (provided that the wavelengths are not so long that one is working in the Rayleigh–Jeans region where the shape of the spectrum does not depend on temperature). This is called a *colour temperature*. Similarly, ratios of the strengths of lines of a given atom or ion originating from levels of different excitation potential and ratios of the strengths of lines from different ions (ratios which essentially define spectral classes) give *excitation* and *ionization* temperatures when fitted to Boltzmann's and Saha's equations.

If the stellar continuous spectrum were a blackbody spectrum, then any colour temperature would give the effective temperature, and if the lines defining spectral class came from the continuum-forming layer and were formed in LTE, then the excitation temperature would be the same as the effective temperature. However, the spectra of stars deviate markedly from those of blackbodies because of the variation of the continuous absorption coefficient with wavelength and because of the cumulative effect of many absorption lines in certain spectral regions. Colour temperatures may then be very different from effective temperatures. Lines are formed in different layers of a stellar atmosphere from the continuum, and the assumption of LTE is rarely exactly true, so that in general excitation temperatures are not equal to the effective temperature. We need to use either a series of model atmospheres or the set of empirical temperatures found from the measured luminosities and radii for the relatively few stars where this is possible to calibrate colour temperatures and excitation temperatures in terms of effective temperature.

In cool stars knowledge of model atmospheres can suggest a wavelength range in the infrared at which we should be seeing a layer close to the effective temperature. If a star is at a distance d, the observed flux $F_1(\lambda)$ at this wavelength can be equated to the flux at the surface of the star times $(R/d)^2$, with the flux at the surface equalling the blackbody flux at the effective temperature, $\pi B_\nu(T_{\text{eff}})$. The observed flux summed over all wavelengths (with some extrapolation and interpolation), F_2, is equal to the total flux at the surface of the star times $(R/d)^2$ and this total flux at the surface must equal σT^4_{eff}. The two conditions enable R/d and T_{eff} to be found, for F_1/F_2 is a function of T_{eff} alone, and if d is known, R can be found. Another approach applicable to cool stars where the opacity varies non-monotonically in the infrared is to find two wavelengths at which we would expect to see down to the same depth, i.e. two wavelengths where the absorption coefficient for the appropriate temperature is the same. If the continuum is formed in LTE, the ratio of the fluxes at these two wavelengths should be equal to the ratio of blackbody fluxes at the two wavelengths for some temperature T_1, so this temperature can be determined. The flux received at one of the wavelengths is then $(R/d)^2\,\pi B_\nu(T_1)$, enabling R/d to be found. The total observed flux, F_2, which is $(R/d)^2\sigma T^4_{\text{eff}}$, then supplies T_{eff}. The wavelengths at which the optical depth is the same are temperature dependent, so an iterative procedure must be followed (see, for example, Jones *et al.* [1]).

Model atmospheres are functions of T_{eff}, composition, and g. The value of g affects the pressure in line-forming regions and hence the profiles and equivalent widths of pressure broadened strong lines. For instance the profile of the $H\gamma$ line, and in particular the pressure broadened wings, are used as a pressure and hence surface gravity indicator in hot stars. The pressure controls the degree of ionization for fixed temperature and hence the ratio of lines coming from different stages of ionization. The trick here is to find a line ratio (preferably of lines that are close in wavelength and that are strong enough to be easily detected but not so strong as to be saturated) that depends strongly on the electron pressure and hence on the surface gravity but depends only weakly on the temperature. An example often used in stars of spectral types F to K is the ratio of Sr II 407.7 nm to Fe I 404.6 nm, which increases with decreasing pressure and surface gravity as the strontium becomes more strongly ionized and the iron become less neutral. As we shall see later, the pressure also affects the continuous absorption coefficient and the size of the Balmer jump. All these dependencies can be used to find $g = GM/R^2$, although it must be recognized that they are nearly all also affected to some extent by temperature and hence temperature and pressure must be solved for simultaneously.

A word or two more must be said on the general subject of composition. The quantity normally derived from observations of stellar spectra is the number of atoms of some element as a fraction of the number of atoms of hydrogen. Abundances are expressed relative to hydrogen because in the vast majority of stars hydrogen atoms are close to 90% of all atoms, and spectral lines are measured relative to the continuum, whose intensity is usually directly or indirectly related to the hydrogen abundance. It is also found that in the vast majority of cases the abundances of elements heavier than helium relative to each other are very similar from star to star—this relative abundance distribution (iron:magnesium:carbon, etc.) is sometimes called the *cosmic abundance distribution.* However the total abundance of all the elements heavier than helium relative to hydrogen does vary considerably between stars. This quantity is sometimes called the *metallicity* because lines of iron are often the easiest to detect in faint stars. The metallicity or the iron abundance is commonly quoted relative to its value in the Sun. The terminology employed is [Fe/H] which means $\log(N_{Fe}/N_H) - \log(N_{Fe}/N_H)_{SUN}$, and stars with a negative value of this quantity are described as being *metal poor.*

All these observed abundances refer to the surface of the star. Research workers concerned with stellar interiors or stellar evolution use the mass fraction of an element X_j, that is the fraction of the total mass that is composed of that element. The composition is usually parameterized by X (with no subscript), the mass fraction of hydrogen, Y, the mass fraction of helium, and Z, the mass fraction of the 'metals', that is everything else, so $X + Y + Z = 1$. Y is around 0.30 for most stars, and its exact value is of considerable importance in cosmology, although very difficult to determine observationally. Z is around 0.02 for stars of solar composition, as opposed to metal-rich or metal-poor stars. The mass fraction of an element j can be determined from $X_j = (N_j/N_H)m_j / \sum_i (N_i/N_H)m_i$.

It must be remembered that the observed N_j/N_H are the surface abundances in the star, which in most cases are the interstellar medium abundances at the time that the star was formed. These should then give the initial X_j for a model of the evolution of the star, the interior of course becoming inhomogeneous with different abundances in the core and envelope as evolution proceeds. Thus the study of the surface abundances of the main sequence stars is really a study of the composition of the interstellar medium as a function of time, the time in a particular case being the time of formation of that star. For most stars we simply

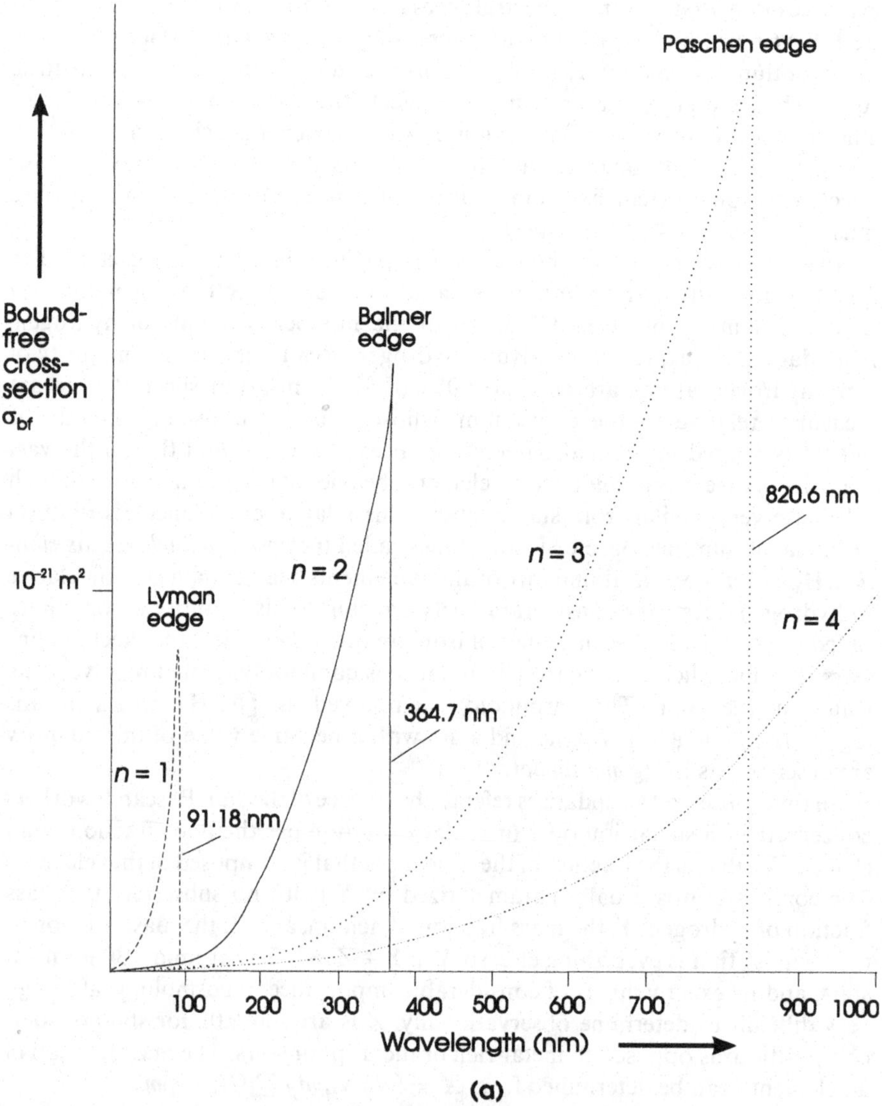

Figure 5.1. (a) Bound–free cross-section for hydrogen; from lower levels 1, 2, 3, and 4. (b) Hydrogen absorption coefficient for $T = 10\,000$ K, $P = 170$ N m^{-2}, $N_e = 3.9 * 10^{20}$ m^{-3}, no stimulated emission correction

wish to establish the metallicity, but much more detailed analyses are required for some stars to establish the cosmic abundance distribution and its uniformity, and to look for subtle changes in ratios such as N(O)/N(Fe) or N(N)/N(Fe) with time which might be expected since nitrogen, oxygen and iron are probably produced

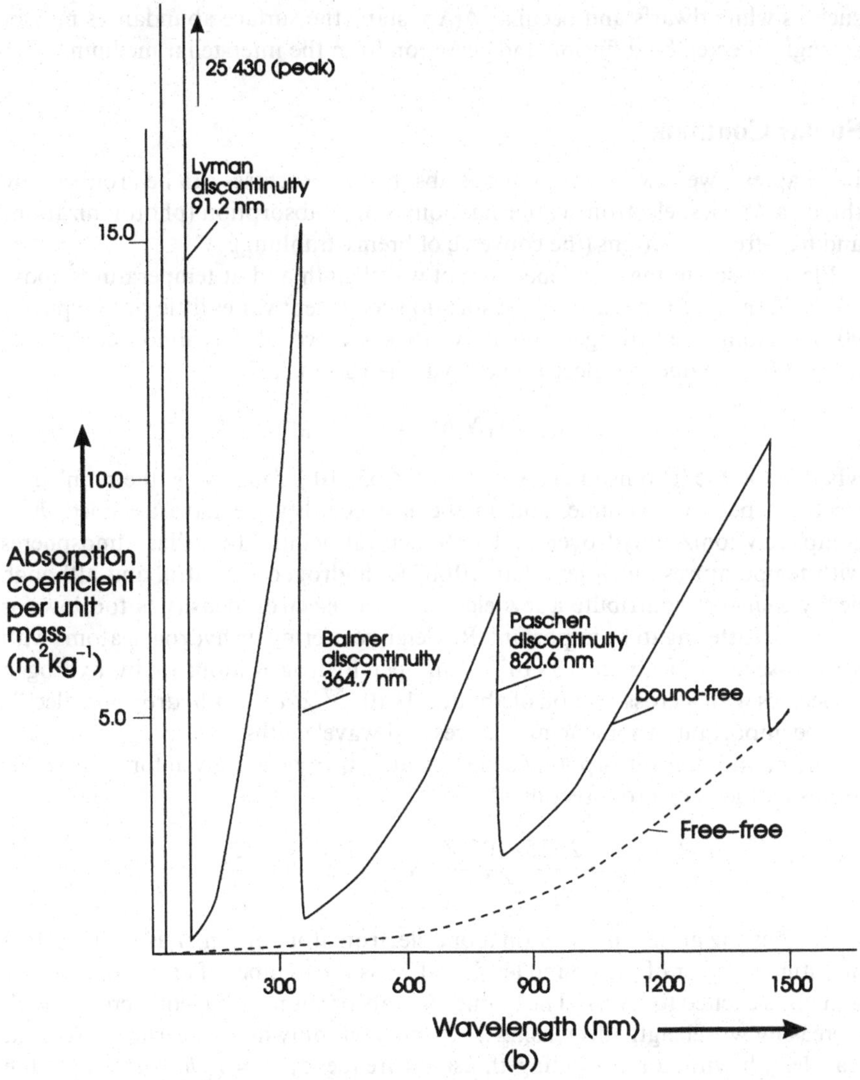

Figure 5.1. (*Contd*)

at different sites during stellar evolution. However, evolved stars like red giants sometimes show surface abundances affected by material convected from the interior and hence also affected by nucleosynthetic processes in the interior. Some stars have lost their envelopes wholly or partially, such as Wolf–Rayet stars, and in these instances we may see nearly pure interior abundances. In other cases, such as white dwarfs and peculiar A(Ap) stars, the surface abundances may be strongly affected by diffusion and accretion from the interstellar medium.

Stellar Continua

In Chapter 2 we saw that continuous absorption processes can be grouped into three categories: electron scattering, bound–free absorption (photoionization), and free–free transitions (the converse of bremsstrahlung).

Electron scattering is independent of wavelength and at temperatures above 10 000 K the electron scattering absorption coefficient varies little since typically 90% of atoms are hydrogen and above 10 000 K hydrogen is almost completely ionized to give one free electron per hydrogen atom. Thus:

$$\kappa_{es} = \sigma_T N_e/\rho = \sigma_T/(1.4m_H)$$

where σ_T is the Thomson cross-section $= 6.65*10^{-29}\,\mathrm{m}^2$, N_e is the number of electrons per unit volume, and in the last equality we have assumed 90% completely ionized hydrogen and 10% neutral helium. In stellar atmospheres with temperatures much less than 10 000 K, hydrogen is neutral and although heavy elements contribute a few electrons, the electron density is too low for electron scattering to be important. Rayleigh scattering by hydrogen atoms with a cross-section of about $5.8*10^{57}/\lambda^4\,\mathrm{m}^2$ per hydrogen atom or by hydrogen molecules with a cross-section of about $8.4*10^{-57}/\lambda^4\,\mathrm{m}^2$ per hydrogen molecule can be important at some temperatures and wavelengths.

The bound–free absorption coefficient at frequency ν is given for hydrogenic atoms and ions by cross-section (2.17)

$$\kappa_{bf} = \frac{2.815*10^{25} Z^4}{n^5 \nu^3} N_n, \qquad \nu < \nu_0 = \frac{I_n}{h}$$

where Z is the atomic number (in a one electron atom or ion), n is the principal quantum number of the bound level, and N_n is the number of atoms or ions per unit mass excited to level n. The bound–free absorption coefficient increases with increasing wavelength (decreasing frequency) but only up to a certain threshold wavelength with corresponding threshold frequency $\nu_0 = I_n/h$, where I_n is the ionization energy from level n and $I_n = 2.18*10^{-18} Z^2/n^2$. At any frequency the main contribution to κ_{bf} from a particular ion will be that from the lowest allowed value of n.

Thus κ_{bf} plotted against frequency (or wavelength) has a characteristic saw-tooth shape, so that starting at the highest frequencies κ_{bf} is dominated by $n = 1$ and rises with $1/\nu^3$ until the threshold for $n = 1$ is reached. At lower frequencies beyond this threshold, κ_{bf} is dominated by $n = 2$, so there is an abrupt drop at the threshold, with a $1/\nu^3$ rise until the threshold for $n = 2$ is reached, and so on (Figure 5.1(a)). The value of κ_{bf} in any of these frequency ranges is mainly determined by N_n for the lowest allowed value of n at that frequency. We shall assume for simplicity that N_n is given by Boltzmann's equation.

Consider the bound–free opacity due to hydrogen. For wavelengths less than the *Lyman* limit, $\lambda = (hc)/I_1 = 91.2\,\text{nm}$, photoionizations from the ground state are possible so κ_{bf} is dominated by transitions from $n = 1$. Since most hydrogen atoms are in the ground state, κ_{bf} is very large in the region on the short wavelength side of the Lyman limit, and in most stars bound–free absorptions from the ground state of hydrogen dominate the absorption coefficient below 91.2 nm, with helium bound–free making an important contribution below 50.4 nm (the threshold for photoionizations from the ground state of neutral helium). In stars of spectral type O, helium is ionized and He II has a hydrogen-like spectrum with the energies multiplied by $Z^2 = 4$ so the He II threshold from the ground state at 22.8 nm marks an abrupt increase in the absorption coefficient. Below this wavelength He II absorption is dominant for the bound–free absorption coefficient is proportional to Z^2 which would outweigh the higher abundance of hydrogen even without making allowance for the fact that hydrogen is largely ionized at these temperatures. It should be remembered that observations in the plane of the galaxy below the Lyman limit are difficult because of the same bound–free absorption by neutral hydrogen but now by interstellar gas.

At the Lyman limit there is an abrupt drop in the absorption coefficient in going from the short to the long wavelength side, and above the limit, bound–free absorption from $n = 2$ rises with increasing wavelength to the *Balmer* limit at $\lambda = (hc)/I_2 = 365\,\text{nm}$, where there is again an abrupt drop in the absorption coefficient as the $n = 3$ level takes over the major role in the visible spectrum.

The crucial point here is that the $n = 2$ and $n = 3$ levels contain only a small fraction of all hydrogen atoms. For hydrogen and hydrogenic ions, the statistical weight of the nth level, g_n, equals $2n^2$ and the partition function $= 2$ so

$$N_n = n^2 \exp(-E_n/kT) N(H^0)/N(\text{H})\, N(\text{H})/\rho$$

where the last factor is the number of hydrogens per unit mass, and the second last factor the fraction of all hydrogen that is neutral. At $T = 10\,000\,\text{K}$, $N_n/N(\text{H}^0) = 2.9 * 10^{-5}$ for $n = 2$ and $7.2 * 10^{-6}$ for $n = 3$ (Figure 5.1(b)), while the corresponding figures for $T = 5000\,\text{K}$ are $2 * 10^{-10}$ and $5.7 * 10^{-12}$. This means that absorption processes from the ground states of species that are much less abundant than neutral hydrogen can compete with hydrogen bound–free in certain wavelength ranges and at certain temperatures.

If hydrogen bound–free processes dominate, then the ionization edges will be reflected in changes in the continuum flux. Longwards in wavelength of the Balmer limit, the drop in the absorption coefficient resulting from the fact that $\text{n} = 2$ bound–free transitions are no longer possible means that the atmosphere is more transparent and so the flux observed comes from deeper and hotter layers and hence is larger. The resulting abrupt step-up in the continuum level in going to longer wavelengths is called the *Balmer discontinuity* or *Balmer jump*. There are similar Lyman and Paschen discontinuities at 91.2 nm and 820 nm, respectively, although the latter is usually fairly small.

The ratio of the hydrogen bound–free absorption coefficients just longward and just shortward of the Balmer discontinuity is given by

$$\frac{\kappa_{bf}(>365\,\text{nm})}{\kappa_{bf}(<365\,\text{nm})} = \frac{\kappa_{bf}(n=3)+\cdots}{\kappa_{bf}(n=2)+\kappa_{bf}(n=3)+\cdots}$$

$$\simeq \frac{\kappa_{bf}(n=3)}{\kappa_{bf}(n=2)}$$

$$= \frac{8}{27}\exp[-(E_3-E_2)/kT]$$

$$= 0.0037 \text{ at } 5000\,\text{K and } 0.033 \text{ at } 10000\,\text{K}$$

Thus we expect the Balmer discontinuity to increase in size as one goes to lower temperatures if hydrogen bound–free transitions are the main contributors to the continuous absorption.

However, in cooler stars the continuous absorption in the visible is dominated by bound–free transitions of the H^- ion. The H^- ion (i.e. hydrogen with two electrons) only exists in small quantities in stellar atmospheres, having an ionization potential of just $1.208*10^{-19}$ J, but since it possesses one state, the ground state, photoionizations from the ground state are possible right through the visible to the limit at 1645 nm. When Saha's equation is applied to H^-, H^- is the 'atom' and H^0 is the 'ion'

$$\frac{N(H^0)N_e}{N(H^-)} = 4 \times 2.411 \times 10^{21}\, T^{3/2} e^{-8753/T}, \qquad U(H^-) = 1$$

For

$$T = 5700, \qquad N_e = 3.8 \times 10^{19}\,\text{m}^{-3}, \qquad \frac{N(H^-)}{N(H^0)} \simeq 4 \times 10^{-8}$$

so that under solar conditions the low abundance of hydrogen atoms excited to $n = 3$ more than outweighs the low abundance of hydrogen in the form of H^- ions.

The H^- ion does not behave like a single electron atom, of course, and the cross-section for bound–free absorption does not have the sawtooth form of H^0, but rises to a maximum of $4*10^{-21}\,\text{m}^{-2}$ at a wavelength of 850 nm, and then falls smoothly to zero at the threshold at 1650 nm. At solar temperatures, H^- bound–free absorption dominates at all wavelengths greater than the Balmer limit up to the H^- threshold (Figure 5.2(a)). Neutral hydrogen dominates in the visible for temperatures greater than about 7500 K

One consequence of this is that the Balmer jump is smaller than would be expected if all the absorption was due to neutral hydrogen, and instead of increasing as one goes to low temperatures, decreases as the relative importance of H^- increases. Thus typically the Balmer jump reaches its maximum size in stars with effective temperatures of around 10 000 K, which corresponds to spectral type A0. The H^- absorption coefficient is proportional to the electron density, and so is higher in dwarf stars with high surface gravities and gas pressures in the continuum-forming layers than in supergiant stars.

Other elements with much lower abundances than hydrogen can have bound–free absorption competing with that of hydrogen in certain wavelength ranges if at those wavelengths they can have bound–free transitions from their ground states while hydrogen bound–free transitions must come from an excited state. Thus we are concerned with these elements only for wavelengths greater than the Lyman limit. The most abundant elements after hydrogen have ionization potentials greater than that of hydrogen (helium and neon) or similar to that of hydrogen (oxygen, carbon and nitrogen), but silicon, magnesium and iron have abundances $\sim 2\text{–}4 * 10^{-5}$ times that of hydrogen and ionization potentials corresponding to ground state ionization edges with wavelengths at 152 nm, 162 nm and 157 nm, respectively, and can play an important role in cooler stars between these wavelengths and the Lyman limit. The situation is complicated by the fact that some of these atoms have low-lying excited levels that are appreciably populated at moderate temperatures and which can therefore also contribute to bound–free absorption at somewhat longer wavelengths. An example is Mg I where the threshold for absorption from the first excited state comes at 251.7 nm, and gives rise to an appreciable discontinuity in the overall absorption coefficient in very cool stars.

In very hot stars the bound–free absorptions of ions of various light elements like C III and C IV, N III and N IV, O III, IV and V and Ne IV and V with edges between 46 and 10 nm may be significant for certain surface gravity and temperature ranges, and the O VI bound-free absorption with an edge at 9 nm may play a significant role in O supergiants.

Before leaving bound–free absorption we need to consider in a little more detail the question of the electron density in cool stars, for the absorption coefficient is proportional to the density of H^- ions, which in turn is proportional to the electron density. Where do the electrons come from in cool stars? For a first estimate for a given temperature and pressure, we can take one at a time the various elements that might be principal *electron donors* and calculate the resulting electron density, taken as equal to the density of ions of that element, so $N_e = N_X^+$. The largest value of the electron density indicates the main electron donor, and this value can then be used in an iterative procedure taking into account all possible electron donors (see Chapter 1).

For hydrogen, the total gas pressure $P \sim P(H^0) + P(H^+) + P_e$, ignoring the 10% contribution from helium, so $N(H^0) = P/kT - 2N_e$. Saha's equation is

$$N(H^+)N_e/N(H^0) = N_e^2/[P/kT - 2N_e] = \text{function of temperature, } f_H(T)$$

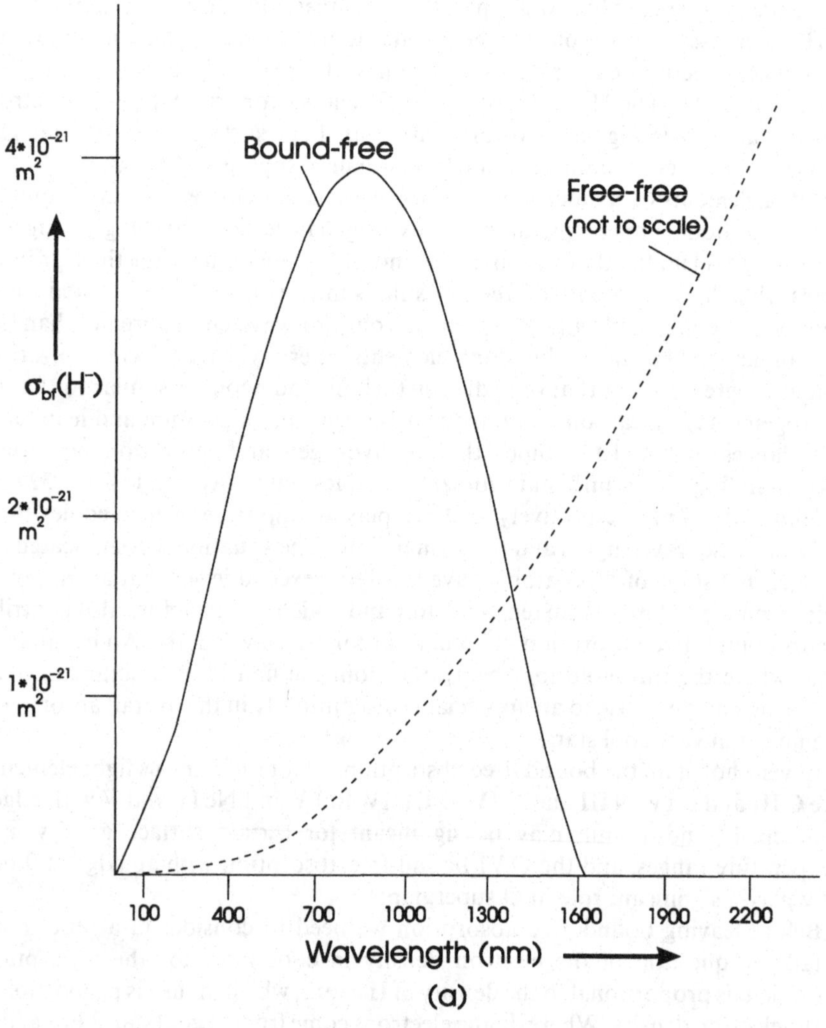

Figure 5.2. (a) Bound–free cross-section of H^- and the form of free–free cross-section. (b) H^- absorption coefficient (proportional to electron pressure) for $T = 5700\,\mathrm{K}$, $P_e = 3\,\mathrm{N\,m^{-2}}$, no stimulated emission correction

which gives a quadratic to be solved for N_e. We are concerned here with temperatures at which the degree of ionization of hydrogen is very low. If the main electron donor is a metal X with abundance A with respect to hydrogen, then to a first approximation $P = N(\mathrm{H})kT$, $N(\mathrm{X}^+)N_e/N(\mathrm{X}^0) = f_\mathrm{X}(T)$, where f_X is

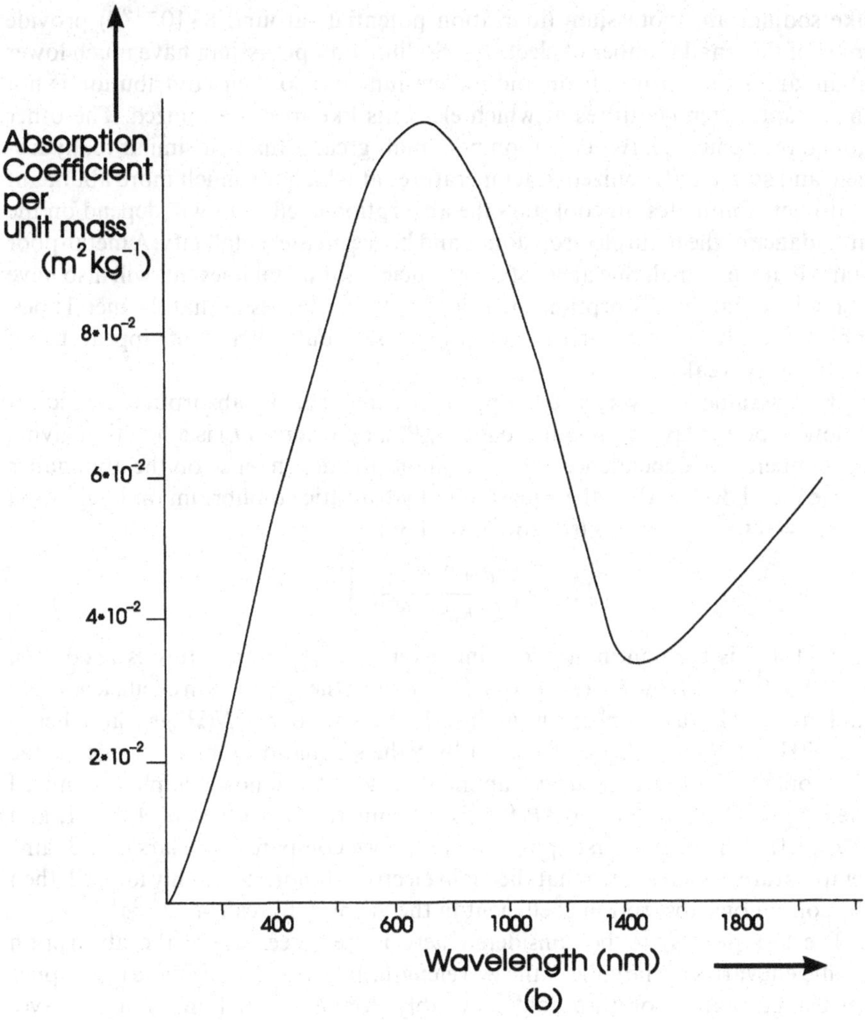

Figure 5.2. (*Contd*)

the Saha function for X, and $A = [N(X^+) + N(X^0)]/N(H)$ so

$$AP/kT = N(X^+)[1 + N(X^0)/N(X^+)] = N_e[1 + N_e/f_X]$$

Thus solving $N_e^2 + f_X N_e - AP/(kT) f_X(T) = 0$ will give the electron density.

At $T = 10^4$ K, hydrogen is the dominant electron donor, while at 6000 K elements with ionization potentials $\sim 1.2 * 10^{-18}$ J, like iron, silicon and magnesium, dominate, and at still lower temperatures very easily ionized elements like sodium and potassium (ionization potentials around $8*10^{-19}$ J) provide most of the small number of electrons. Sodium and potassium have much lower abundances than iron, silicon and magnesium and so their contribution is not important at temperatures at which elements like iron are ionized. The other abundant elements have ionization potentials greater than, or similar to, hydrogen, and so are only ionized at temperatures at which the much more abundant hydrogen dominates. In cool stars the absorption coefficient will depend on the abundance of the main electron donor and hence on the metallicity. A metal-poor star will have a small line absorption coefficient for metal lines but will also have a small continuous absorption coefficient, with the net result that the metal lines, which are always measured against the continuum, may not appear to be particularly weak.

If we assume as a very rough approximation that the absorption coefficient follows a power law in pressure so $\kappa = \kappa_0 P^a h(T)$, where $h(T)$ is a function giving the temperature dependence and κ_0 contains the dependence on the abundance of the metal donor, then the equation of hydrostatic equilibrium on integration from $\tau = 0$ to $\tau = \tau'$ yields (see equation (1.39))

$$P(\tau') = \left[\frac{(a+1)g}{\kappa_0}\frac{1}{h(T)}\right]^{1/(a+1)}$$

so that if τ' is the continuum-forming layer and the temperature is fixed then $P \propto (g/\kappa_0)^{1/(a+1)}$. Now for H^-, $\kappa \propto N(H^-)/\rho$ since the absorption coefficient is per unit mass. The atmosphere is mainly hydrogen so $\rho \sim N(H^0)m_H$ and hence $\kappa \propto N(H^-)/N(H^0) \propto N_e$ (for *fixed* T) by Saha's equation for H^-. Hence if the electrons come from a metal with abundance A that is almost completely ionized then $N_e = AN(H^0)$ and $\kappa \propto AP$ for fixed temperature with $\kappa_0 \propto A$, $a = 1$, and $P \propto (g/A)^{1/2}$. Thus to a first approximation if we compare cool stars of the same temperature, which is such that the main electron donors are highly ionized, then the continuous absorption coefficient in the visible will scale as $A^{1/2}g^{1/2}$.

The last process to be considered here is free–free, where the absorption coefficient varies as the cube of the wavelength, and which, since there is no upper wavelength cut-off or threshold inevitably dominates at long enough wavelengths. At temperatures high enough for hydrogen to be ionized, the free–free absorption of H^0 (that is, the interaction of a free electron with an H^+ ion) will be the most important contributor, owing to the great abundance of hydrogen. The free–free absorption coefficient for hydrogen-like ions is proportional to Z^2, where Z is the charge on the nucleus, so in stellar interiors where most atoms are completely ionized an element like oxygen with $Z^2 = 256$ will be able to make an appreciable contribution despite having an abundance relative to hydrogen of less than 1/1000. However, even in very hot stellar atmospheres atoms are only a few times ionized, and only helium is likely to make an appreciable contribution to the free–free absorption. In cool stars, H^- free–free absorption (that is, the interaction of a free electron with a neutral hydrogen atom), will be dominant, while in very cool stars H_2^- free–free will be most important (the cross-section for the latter is about half that for H^- at a given wavelength).

In stars cooler than 7000 K, H^- bound–free dominates the continuous absorption for most of the visible spectrum but drops towards its cut-off at 1650 nm, and the rising H^- free–free takes over at about 1500 nm and longer wavelengths. The result is that in most cool stars the minimum absorption coefficient, where the atmosphere is most transparent, is found at around 1600 nm (Figure 5.2(b)).

Lines

The purpose of this brief section is to give a general account of the spectral lines most prominent in observed spectra before turning to their detailed analysis.

It will be recalled that the energy level structure of hydrogen has levels which become closer together as one moves to higher quantum numbers (Figure 5.3). The shortest wavelength transitions, the Lyman series, are from the ground state and hence can be absorbed by nearly all hydrogen atoms. The transition probabilities and oscillator strengths decrease as one goes to higher upper quantum number levels, so the Lyman α line from $n = 1$ to $n = 2$ is the strongest. The transition between the ground state and the first excited state permitted to radiatively decay to ground is called the *resonance line*.

The energy level structures of other atoms and ions are more complex than that of hydrogen, but the strongest line is often the resonance line, and the stronger lines tend to be found at the shortest wavelengths. The ionization potential and the excitation potentials both depend on the nuclear charge reduced by the shielding effect of electrons other than the outermost, and so as one proceeds to higher stages of ionization of a particular element, the ionization potential, the excitation potentials and the separation of levels all increase, and the strong lines tend to be found at shorter wavelengths. Atoms where the 'outermost' electron is in the same shell as many others will have incomplete shielding of the nucleus and hence large ionization potentials and strongest lines at very short wavelengths, as can be seen in the case of the inert gases like helium and neon. In discussing these general trends we have ignored the inter-electron interactions that give the detailed structure of the energy levels—in the case of the true 'metals' like iron, nickel, cobalt, titanium, vanadium and manganese the structures are so complex that a simple description is impossible, resulting in stellar spectra with many lines of appreciable strength spread over a wide range of wavelengths.

The lines that we actually observe in a particular stellar spectrum depend on the temperature and, to a lesser extent, on the pressure through Boltzmann's and Saha's equations (which leads to the system of spectral classification) and much less sensitively on the abundances. However the elements observed also depend on the wavelength region used. Most observations have been made in the 'visible' which starts at the short wavelength end at the Earth's atmospheric cut-off around 330 nm. Many of the earlier photographic spectra stopped at around

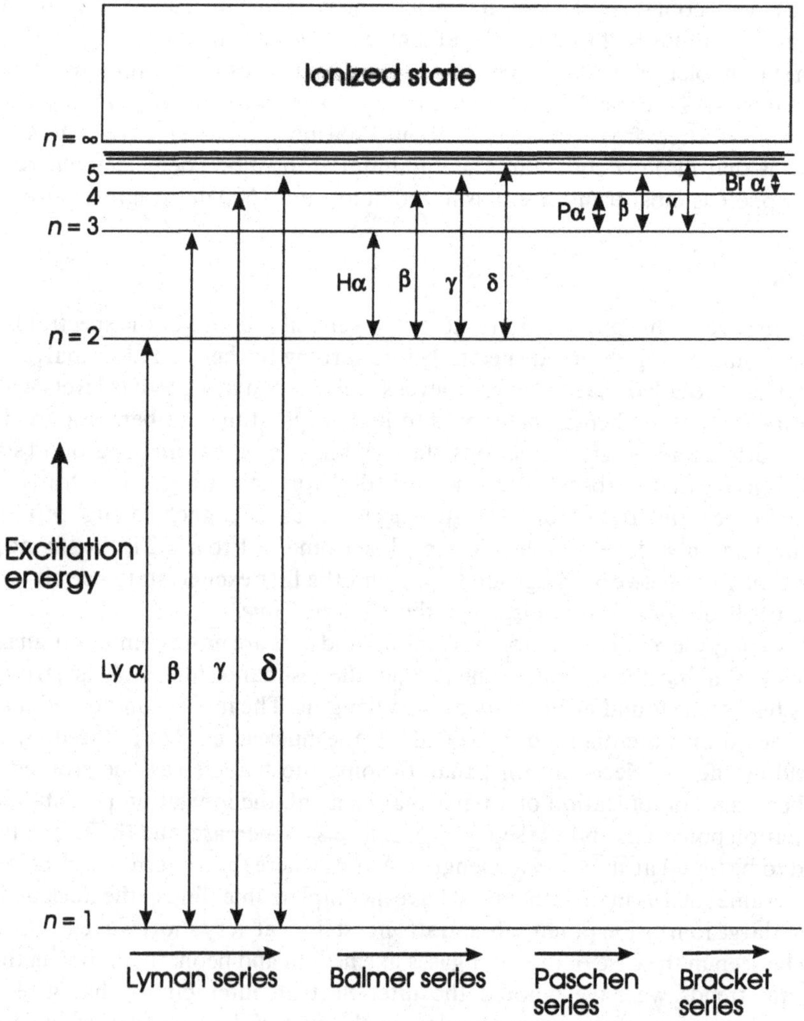

Figure 5.3. Lines and levels of hydrogen

500 nm at the long wavelength end, but the classical 'visible' ran up to about 700 nm, and modern detectors easily take us up to 900 nm, beyond which atmospheric absorption starts to become a problem. Infra-red spectra, particularly in the atmospheric windows at 1600 nm and 2300 nm, are now common and for many bright stars ultraviolet spectra are available at least down to the Lyman limit at 121.6 nm.

The most abundant elements (hydrogen, helium, oxygen, carbon, neon, and nitrogen in that order) all have relatively large ionization potentials and resonance lines in the ultraviolet and the same holds even more strongly for their ions in hotter stars. At solar temperatures helium and neon are unobservable in the visible, and there are only a few forbidden or high excitation lines of carbon, oxygen and nitrogen. The first excited level of hydrogen gives rise to the Balmer series of lines in the visible, in order of decreasing strength running from Hα ($n = 2$ to $n = 3$) at 656.5 nm through Hβ ($n = 3$ to $n = 4$) at 486.1 nm to Hγ ($n = 2$ to $n = 5$) at 434.0 nm and so on. The high abundance of hydrogen means that these lines are fairly strong even in cool stars with a low degree of excitation. The Balmer lines increase in strength with increasing temperature, reaching a maximum at a temperature of about 10 000 K and decreasing at higher temperatures as hydrogen becomes ionized, although non-LTE effects mean that hydrogen lines are not as weak in the hottest stars as one might expect.

The second most abundant element, helium, has a more complicated energy level structure split into singlet levels with spin quantum number $S = 0$ and triplet levels with $S = 1$, depending on whether the electron spins are opposed or aligned. The ground state of helium is $1s^2\ ^1S$, but the lowest excited level is $1s2s\ ^3S$ in the triplet system, $3.18*10^{-18}$ J above the ground state, with the lowest excited singlet level $1s2s\ ^1S$ slightly higher still (Figure 5.4). Radiative transitions between the singlet and triplet systems are (weakly) forbidden, so the lowest triplet state is called metastable since it cannot decay radiatively to the ground state, a fact that leads the two systems to behave somewhat differently under non-LTE conditions. The transition $1s^2 \rightarrow 1s2s$ is also forbidden by $\Delta l = 1$, so the longest wavelength line from the ground state is $1s^2 \rightarrow 1s2p$ at 58.4 nm. For visible lines one has transitions like those from the first and second levels of the triplet system: $1s2s\ ^3S \rightarrow 1s3p\ ^3P$ 388.8 nm, $1s2p\ ^3P \rightarrow 1s3d\ ^3D$ 587.5 nm, $1s2p\ ^3P \rightarrow 1s3s\ ^3S$ 706.5 nm and $1s2p\ ^3P \rightarrow 1s4d\ ^3D$ 447.1 nm.

In addition there are the transitions from the first excited levels of the singlet system: $1s2s\ ^1S \rightarrow 1s3p\ ^1P$ 501.5 nm and $1s2s\ ^1S \rightarrow 1s4p\ ^1P$ 396.4 nm, etc. The high excitation energies of the lower levels of these transitions mean that they are not observed until temperatures of the order of 15 000 K are reached. At still higher temperatures, above 25 000 K, helium becomes ionized. The energy level structure of once ionized helium is identical to that of hydrogen except for the fact that all energies must be multiplied by $Z^2 = 4$, with the result that Hα coincides with He II $n = 4$ to $n = 6$, and indeed alternate lines of the He II series with lower level $n = 4$, called the Pickering series, coincide with theBalmer series. The He II series from $n = 3$ has a prominent line in the visible at 468.6 nm ($n = 3$ to $n = 4$).

In the visible we have the temperature sequence running from coolest to hottest stars with the hydrogen lines strengthening to 10 000 K and then weakening, with He I lines then appearing, to be replaced in the hottest stars with He II lines. In the visible, these lines dominate the spectra of stars of spectral types O and B but lines of carbon, oxygen, nitrogen, silicon and neon ions are also present and the change from Si II to Si III to Si IV, for instance, indicates the changing degree of ionization with rising temperature. In the ultraviolet, the lines of some of these ions are

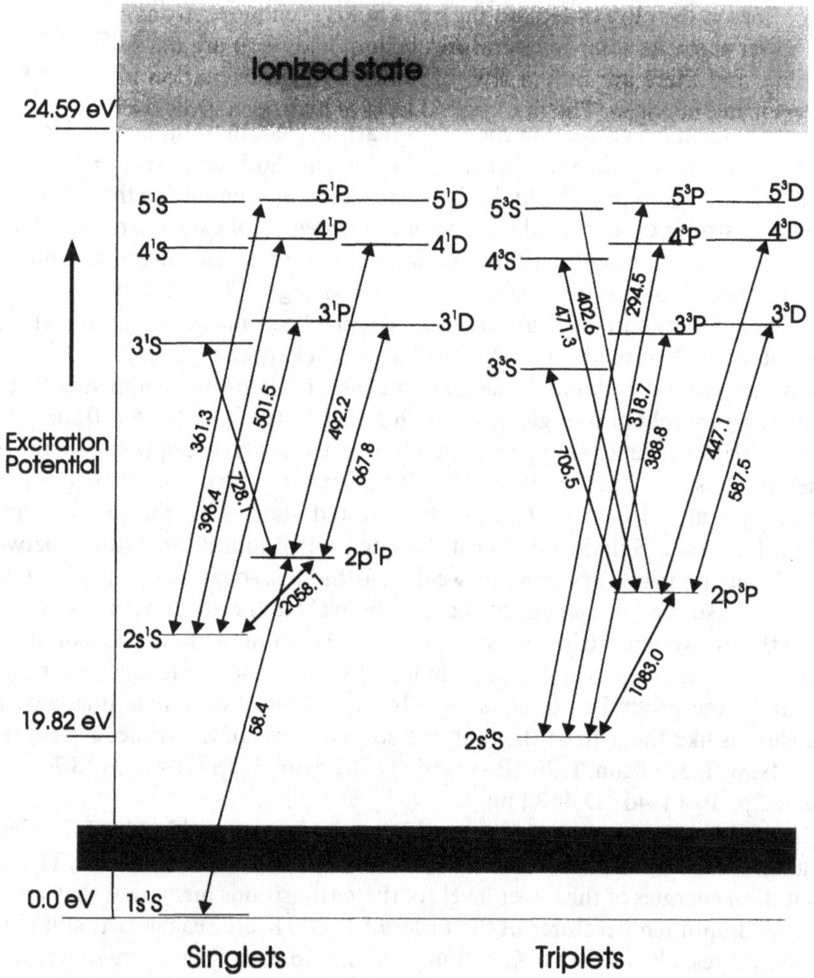

Figure 5.4. Helium energy levels; wavelengths in nanometres

very strong, for instance the resonance lines of C IV at 154.8 nm which appear in stars of spectral type O, and the Si IV resonance lines at 139.3 nm which are again prominent in the hottest stars but are replaced by lower stages of ionization in cooler stars, e.g. the C II resonance lines at 133.5 nm in the cooler B stars.

Thus in O stars temperature classification (O9 to O3) is based on the He II/H I line strength ratio, with the strength of the Si IV lines being used as a luminosity classification indicator (dwarf V to supergiant I), the visible Si IV lines at 408.9 and 411.6 nm and the ultraviolet resonance lines at 139.4 and 140.3 nm strengthening as one moves from dwarf to supergiant. O star spectra are strongly influenced by stellar winds and departures from LTE, and many lines show emission features as a result, emission increasing relative to absorption as one moves to higher luminosities. An important example is the He II line at 468.6 nm, which changes from an absorption to an emission line as the luminosity increases. In B stars light ion line strengths relative to He I lines can be used as a temperature indicator, the Si II line at 412.9 nm and the Mg II line at 448.1 nm both weakening as the temperature is increased in the cooler B stars and these ions become further ionized, whereas in hotter B stars the Si IV lines have appeared together with C III lines, which strengthen with increasing temperature. In both B and A type stars, the hydrogen lines weaken with increasing luminosity, owing to the pressure broadening being less in a low gravity supergiant, and this can be used as a luminosity indicator (there are a number of other line ratios that can be used in B stars). A star spectra are dominated by hydrogen lines, and can be divided into temperature sub-classes by the weakening of the Ca II lines relative to hydrogen with rising temperature as the calcium becomes triply ionized. A schematic B star spectrum is sketched in Figure 5.5(a), with the variations with spectral type and luminosity in the ultraviolet and visible sketched in Figure 5.5(b).

In stars of solar type, as well as hydrogen lines there are strong lines of sodium, magnesium and calcium in the visible, and a number of strong iron lines. These are all elements with relatively modest ionization potentials. Sodium has a low abundance (typically $2*10^{-6}$ that of hydrogen), but the resonance lines of neutral sodium (the D lines at 589.2 and 588.9 nm) lie in the visible. Even at solar temperatures sodium is fairly strongly ionized and at higher temperatures the D lines weaken. Ionized sodium has an inert gas structure and no lines in the visible. Calcium has nearly the same abundance as sodium and is slightly harder to ionize, but is still partially ionized at solar temperatures. It has the unusual feature that the resonance lines of both neutral calcium at 422.6 nm and of ionized calcium at 393.3 nm and 396.9 nm (K and H, the latter coinciding with a hydrogen line) lie in the visible. Both are strong in the Sun (particularly the Ca II lines) with Ca I increasing in importance in cooler stars and Ca II persisting in hotter stars.

Magnesium is an order of magnitude more abundant than sodium and calcium, but has a very similar energy level structure to calcium with a rather higher ionization potential and slightly farther apart energy levels. The Mg II H and K lines lie at 279.5 and 280.2 nm in the ultraviolet and had to await the advent of UV astronomy before their very considerable strengths in medium temperature stars were observed. The Mg I line corresponding to Ca I 422.6 nm is at 285.2 nm, again in the ultraviolet, but the first excited level of Mg I, 3s3p ^3P, lies only $4.3*10^{-19}$ J above the singlet ground state $3s^2\ ^1S$ (the line joining these two states, 457.1 nm, is weak because it breaks the $\Delta S = 0$ rule), and the triplet 3s3p $^3P \rightarrow 3s4s\ ^3S$: 518.37 nm, 517.27 nm and 516.74 nm in order of decreasing

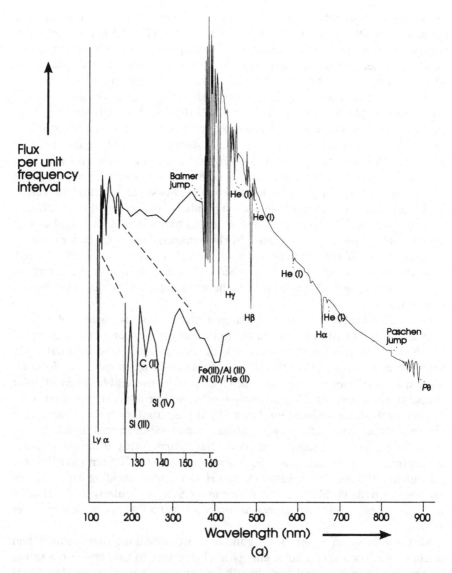

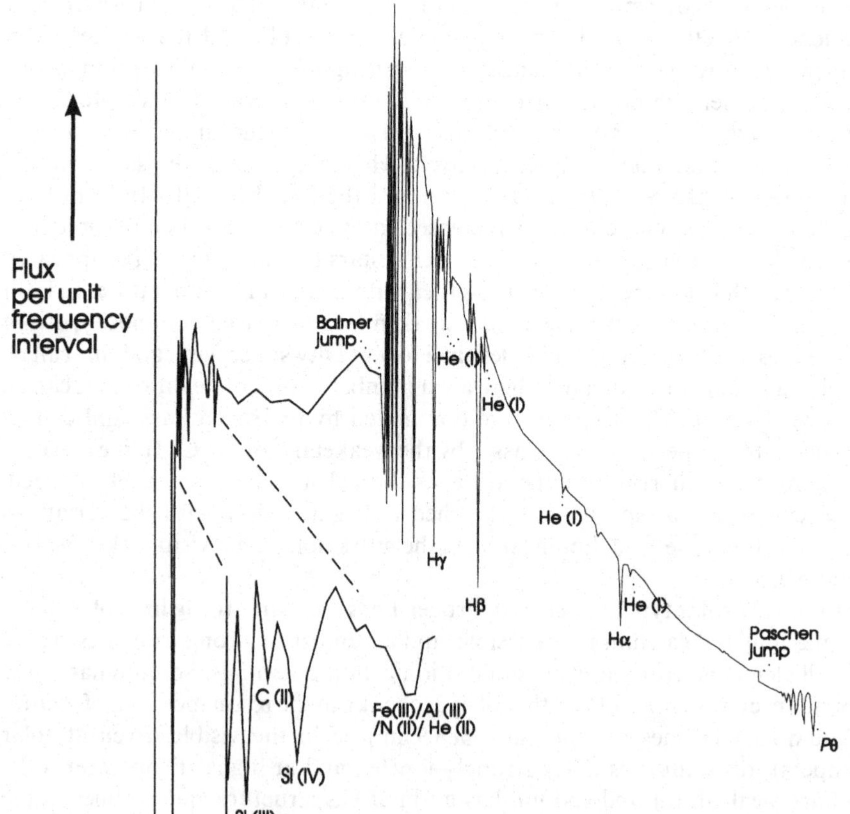

Figure 5.5. (a) B star spectrum (B4 dwarf—the appearence will depend on resolution). (b) O and B stars in the ultraviolet. Note the luminosity effects in the two lower spectra (after Panek and Savage, Astrophysical Journal, **206**, 1976.)

strength, is prominent in solar type spectra. It should be noted that the 448.1 nm line of Mg II (an ion which is harder to ionize than Ca II) comes from an excited level and hence appears at higher temperatures (in F stars) than the Ca II lines and persists to higher temperatures than the Ca II lines.

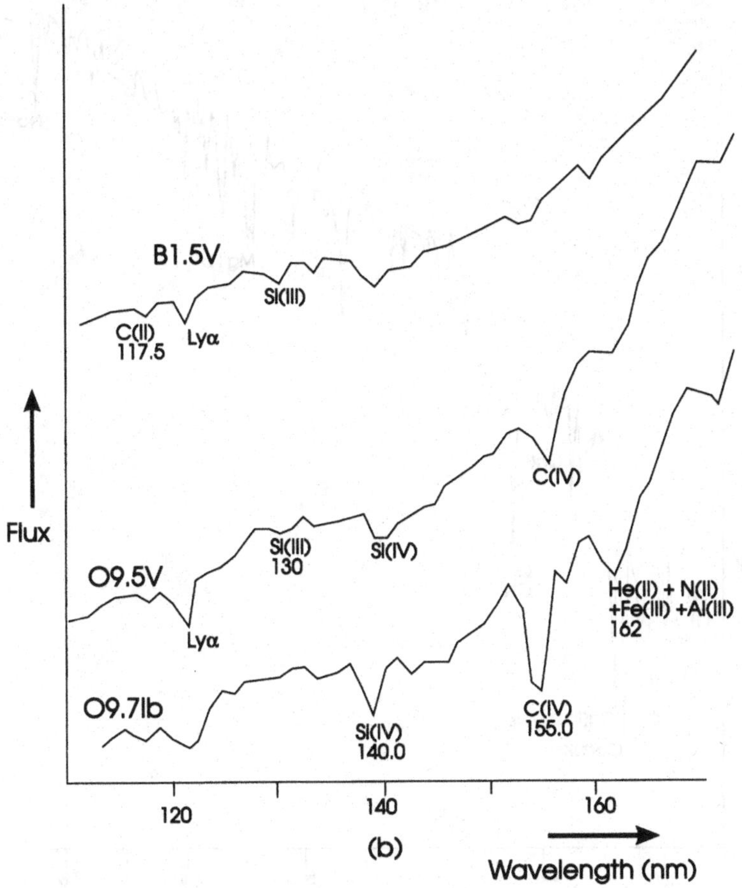

Figure 5.5. (*Contd*)

The 397.0 nm Ca II plus hydrogen line is very strong in cool stars and is followed at shorter wavelengths by many iron and other lines and the strong Ca II K line (Figure 5.6). At low dispersion the spectra of G and K stars therefore appear to show a discontinuity at around 400 nm followed at shorter wavelengths by an apparently lowered continuum, in reality made up of many blended lines. This is sometimes called the '4000 A break' and is used in low dispersion spectroscopy, particularly of faint elliptical galaxies whose spectrum is dominated by cool stars and for which the wavelength of the break can give an approximate redshift.

Cool stars show prominent features owing to molecular bands. In molecules the electrons can change their state in much the same way as electrons can in

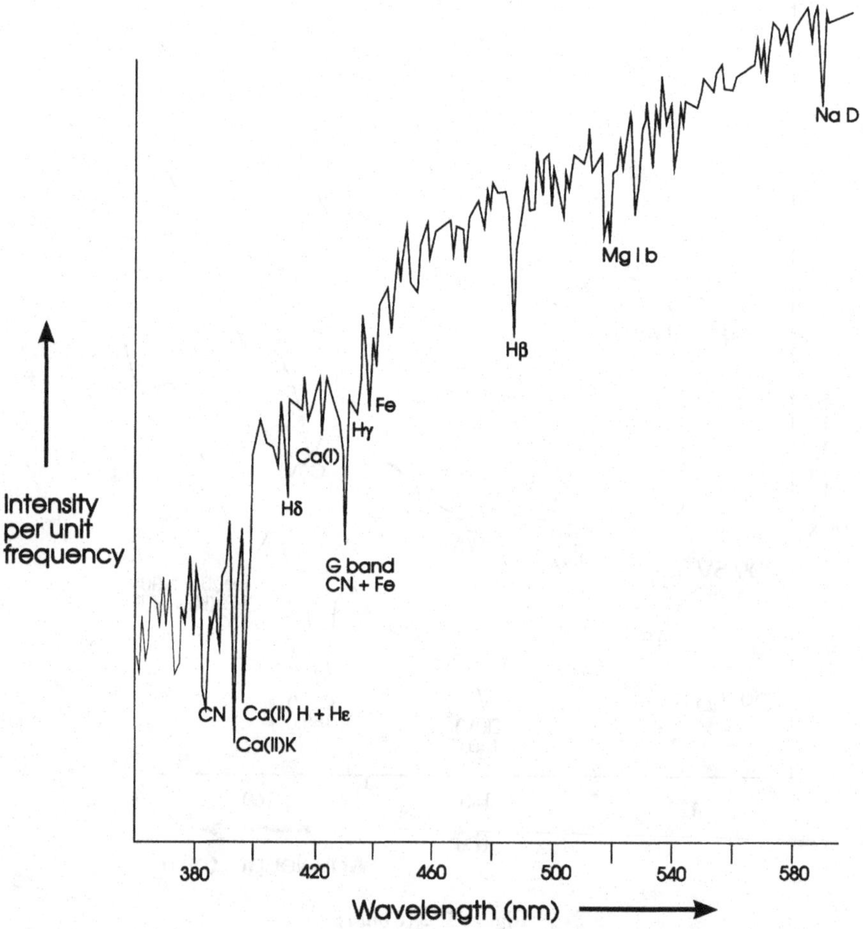

Figure 5.6. Solar spectrum; resolution = 1 nm

atoms. However the nuclei can also change their vibrational energy, where the vibrations are about the mean internuclear separation, and in addition can change their rotational energy. Electronic energies are much greater than vibrational energies, which in turn are much greater than rotational energies (Figure 5.7). We discuss here only the case of diatomic molecules. Transitions in which only the rotational energy is changed are possible, and emit or absorb radiation in the millimetre region of the spectrum. Such *pure rotational transitions* are observed from the interstellar medium, but are not seen in stellar spectra.

Transitions in which both vibrational and rotational energies change, but the electronic state stays the same, are also possible. Such *vibrational–rotational*

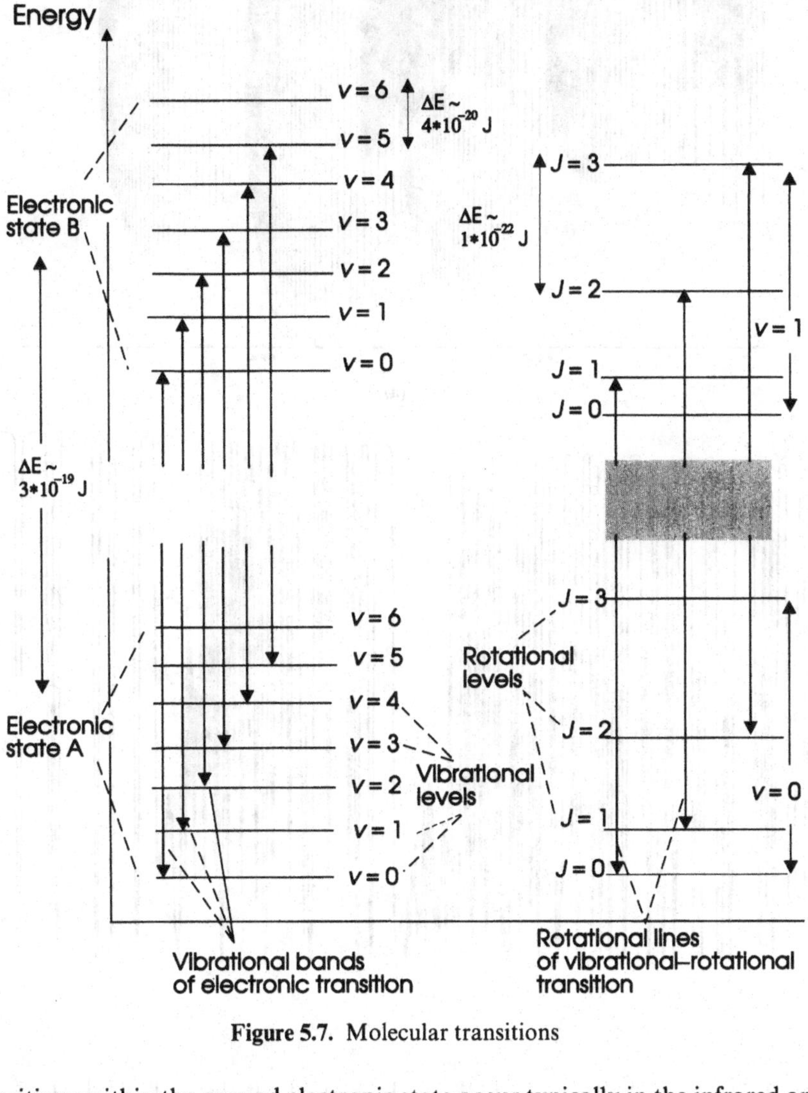

Figure 5.7. Molecular transitions

transitions within the ground electronic state occur typically in the infrared and are observed in cool stars. The strongest transitions occur when the change in the vibrational quantum number v is 1, giving rise to a series of *fundamental* bands, $0 \rightarrow 1$, $1 \rightarrow 2$, etc. Each band is composed of many lines which differ in the rotational levels involved, obeying the selection rule for the rotational quantum number J that $\Delta J = +1$ to give the R branch or $\Delta J = -1$ to give the P branch—thus R(13) of the (1,0) band involves a transition in absorption from $v = 0, J = 13$ to $v = 1. J = 14$.

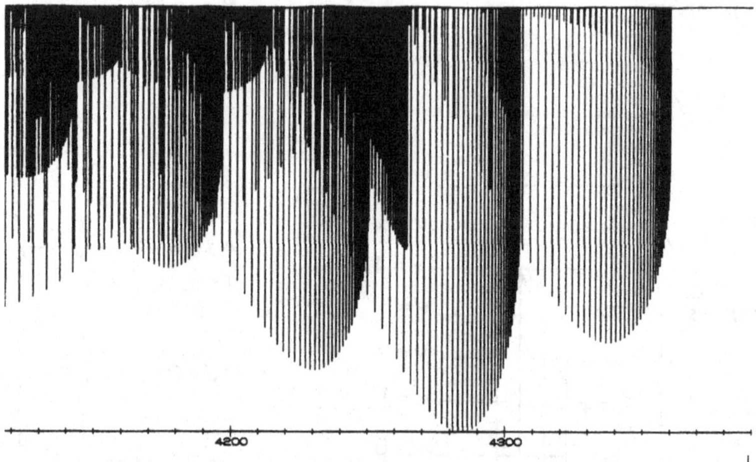

(a)

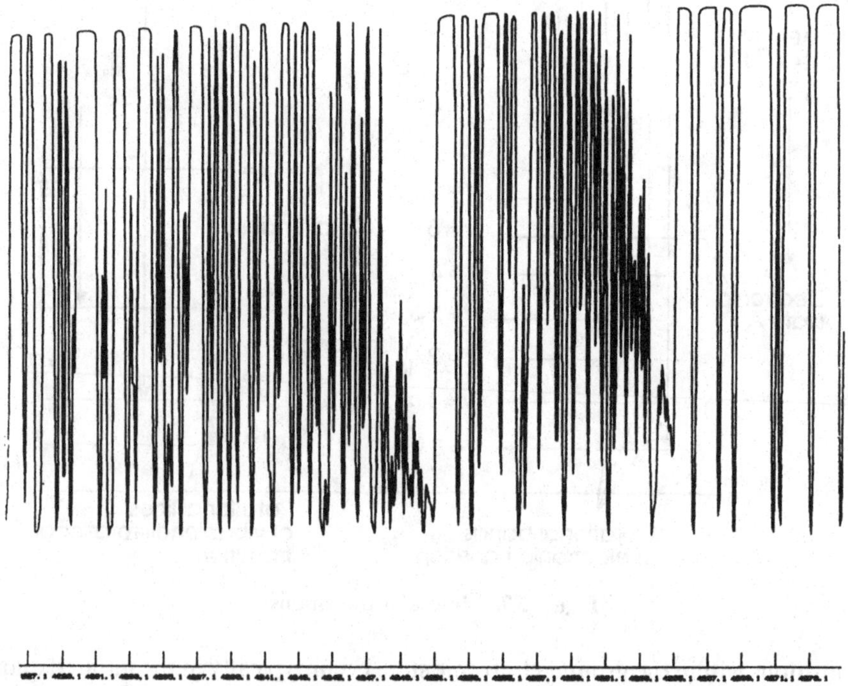

(b)

Figure 5.8. CO vibtational–rotational bands in infrared. First overtone at 2.3 μm—horizontal scale in cm^{-1}. (a) Line strengths at 3500 K, 2–0 band to right, then 3–1, then 2–0 ^{13}CO, then 4–2, etc. (b) Saturated lines of 4–2 band and 2–0 ^{13}CO bands from simulated atmosphere. (c) 5.8b as convolved to a resolution of 20 cm^{-1}. (d) 2.3 μm band system as might be observed

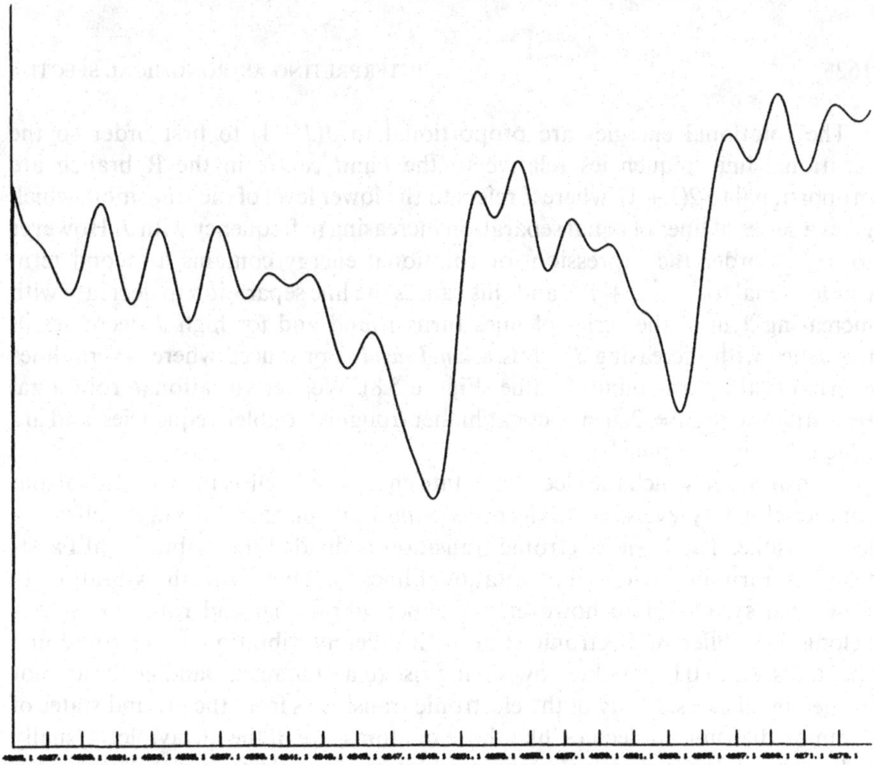

(c)

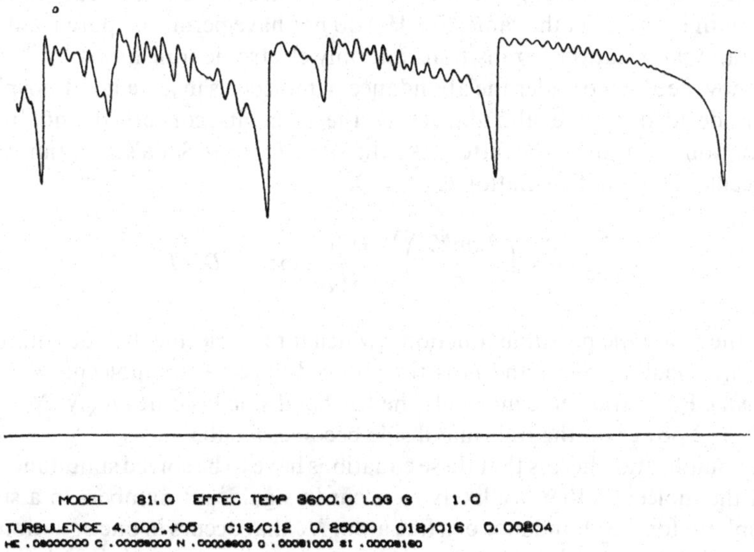

(d)

Figure 5.8. (*Contd*)

The rotational energies are proportional to $J(J+1)$ to first order so the rotational line frequencies relative to the band centre in the R branch are proportional to $2(J+1)$, where J refers to the lower level of the transition, which gives a series of lines of equal separation increasing in frequency with J. However to higher order the expression for rotational energy contains a second term proportional to $-J^2(J+1)^2$, and this causes the line separation to decrease with increasing J until the series of lines turns round and for high J decreases in frequency with increasing J. Thus a *band head* is produced where several lines overlap at the turn-round J value (Figure 5.8). Weaker vibrational–rotational transitions with $\Delta v = 2$ also occur at higher (roughly double) frequencies, and are called the *first overtone* bands.

Transitions in which the electronic state changes as well as the vibrational and rotational energy levels, called *electronic transitions*, produce lines in the ultraviolet or visible. The basic electronic transition is divided into vibrational bands which in turn are divided into rotational lines, much as with the vibrational–rotational system. Here however the upper vibrational and rotational levels belong to a different electronic state with different vibrational and rotational constants. $\Delta J = 0$ is possible now, giving rise to a Q branch. Band heads are not formed in all cases. Many of the electronic transitions from the ground states of common diatomic molecules, like those of atoms, are in the ultraviolet. Usually only at the most one or two electronic transitions from electronic levels excited in stars occur in the visible spectrum, but the band system for one electronic transition can spread over most of the visible spectrum. Homonuclear molecules (those with both nuclei the same, like H_2) do not have permitted pure rotational or vibrational–rotational transitions but only electronic transitions.

We now need to consider the abundance of molecules in a stellar atmosphere. This is decided by the abundances of the elements concerned and by the dissociation equations. The latter take the same form as Saha's equation so for a molecule XY with dissociation energy D:

$$\frac{N_X N_Y}{N_{XY}} = \left(\frac{2\pi mkT}{h^2}\right)^{3/2} \frac{U_X U_Y}{U_{XY}} \exp(-D/kT)$$

where the U are the partition functions (which for a molecule include rotational and vibrational energies) and m is the reduced mass of the molecule $= m_X m_Y/(m_X + m_Y)$. For a triatomic molecule the left-hand side becomes $N_X N_Y N_Z/N_{XYZ}$ and the 3/2 power on the right-hand side becomes a cube.

The complication here is that these equations have to be solved simultaneously for all the molecules that are likely to occur in significant numbers in a stellar atmosphere, for the abundance of a free atom N_X that occurs in one equation will be affected by how much of that atom is incorporated in some other molecule. Of course the end result is that the molecules with the largest dissociation energies, that is the most stable molecules, tend to occur in the largest numbers, but the results of the competition for free atoms between the various molecules is not always obvious. Carbon monoxide has the largest dissociation energy ($1.8*10^{-18}$ J) of any of the molecules found in stellar atmospheres, followed by N_2 at $1.6*10^{-19}$ J and CN at $1.3*10^{-18}$ J. In cool atmospheres nearly all the carbon and oxygen atoms that can find a partner are associated into CO. If the oxygen abundance is greater than that of carbon, surplus oxygen atoms are left over from the pairing and can then make less stable oxides like OH and TiO and other metal oxides. If the carbon abundance is greater than that of oxygen, surplus carbon atoms are left over which can then make carbides like C_2, CN and CH.

Nitrogen is less abundant than carbon, so although much nitrogen is in the form of N_2, considerable amounts go into the stable molecule CN in carbon stars. What is not so obvious is that even in oxygen-rich stars, if the temperature is high enough for some CO to be dissociated, appreciable quantities of CN are predicted by the dissociation equations. H_2 and OH have similar modest dissociation energies of around $7*10^{-19}$ J, and NH and CH are rather less stable, so that in the cooler stars most of the hydrogen is in the form of H_2, with appreciable amounts of the other hydrides. At very low temperatures H_2O is formed.

H_2, N_2, and CO have their electronic transitions in the ultraviolet and therefore are inaccessible for most cool star observations. H_2, C_2, and N_2 are homonuclear and therefore have no vibrational–rotational transitions, with the result that H_2 and N_2, despite their considerable abundances, are not normally observed in stellar spectra. Vibrational–rotational transitions are weaker than electronic ones, but the great abundance of carbon monoxide means that the CO fundamental at around 5 μm and the first overtone at around 2.3 μm are both strong in cool stars and detectable at low resolution (from above the atmosphere in the case of the fundamental), and at high resolution the second overtone at around 1.5 μm is also detectable. The vibrational–rotational transitions of OH and NH are also observed. CN has a ground state electronic transition $A^2\Pi$—$X^2\Pi$ which is observed as the CN red system from 580 nm longwards, and a second electronic transition from a lower level of modest excitation potential, $B^2\Sigma$—$X^2\Sigma$, which is observed as the CN violet system from 385 nm to 422 nm. Indeed it is difficult to find any stretch of a cool star spectrum in the visible and the near infrared in which there are no CN lines. C_2 also has an electronic transition from the ground state in the visible, $A^3\Pi$—$X^3\Pi$, the Swan system from 470 to 560 nm, as has CH, the $A^2\Delta$—$X^2\Pi$ system around 430 nm. TiO and VO have strong electronic transitions in the visible. Thus the visible spectra of cool (effective temperature less than 3500 K) oxygen-rich stars are dominated by TiO, and of carbon-rich stars by C_2 and CN. Finally, very cool dwarfs are dominated by H_2O bands in the infrared, which usually appear as a bumpy pseudo-continuum. Some schematic spectra of cool stars are given in Figure 5.9.

Some stellar spectra show emission lines or absorption lines that turn to emission lines at the centre of the line. There are a variety of reasons for these emission lines. In some cases we are dealing with lines in emission from

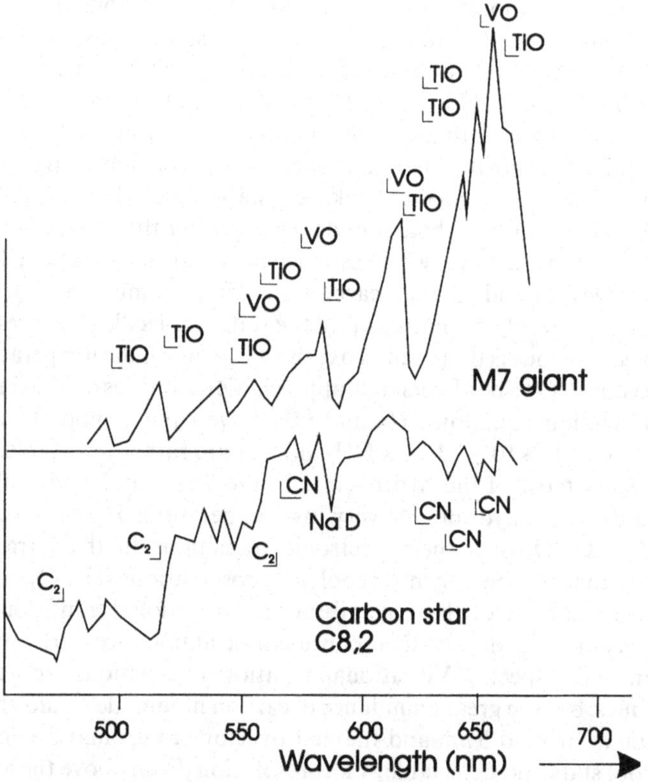

Figure 5.9. Cool giant spectra

a circumstellar envelope that is superimposed on the stellar spectrum because the shell is not resolved from the star by the telescope. In other cases in hot stars we are dealing with an extreme non-LTE situation, for example the N III lines in O stars. The Sun, and indeed all stars with effective temperatures less than 6500 K, have outer atmospheres, called *chromospheres*, in which the temperature rises as one moves outwards beyond a temperature minimum at the top of the photosphere. The density in the chromosphere is low, so that only the centres of strong lines originate in the chromosphere.

If strong lines were formed in LTE, one would expect the higher source function in the chromosphere to result in a greater emergent intensity, and hence in a line profile that reversed towards the centre, giving the appearance of an emission line at the centre of the strong absorption line. We saw in Chapter 4 that the source function falls below the blackbody value near the surface, and this means that very near the centre of a strong line the profile will cease to reflect the rising chromospheric temperature. Thus we would expect a strong absorption

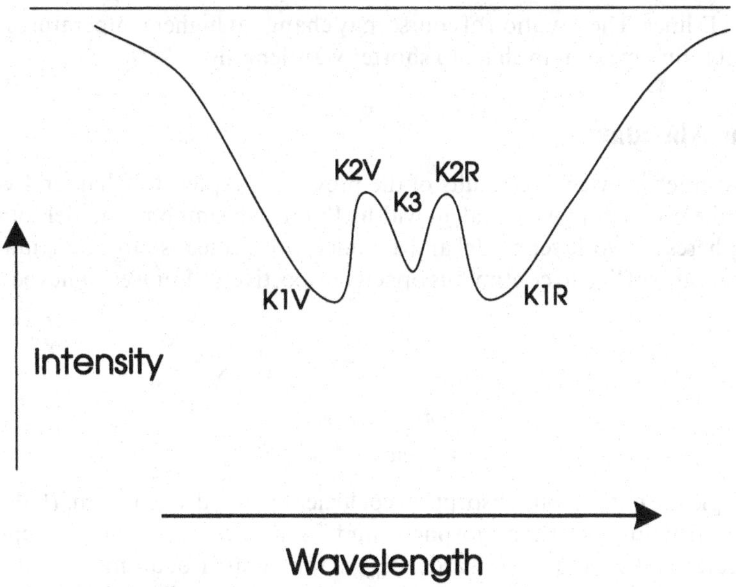

Figure 5.10. Ca II line profile in a cool star (schematic)

line to have an emission line in the central regions which again changes to an absorption line at the very centre. This sort of profile is seen in the H and K lines of Ca II, and not unexpectedly also in the similar ultraviolet pair of Mg II lines (Figure 5.10). However the strong Balmer lines show a simple absorption profile.

The reason for the difference, as was seen in Chapter 4, is that when transitions to the ionized state are taken into account, the calcium lines are collision-dominated and hence follow the local temperature until 'edge' non-LTE effects take over, while the Balmer lines are photoionization dominated with a source function that is not linked to the local kinetic temperature in the upper photosphere but which is controlled by the radiation temperature at the wavelengths at which photoionization takes place. The calcium lines have the ground state as the lower level and a state of low excitation ($5*10^{-19}$ J) as the upper level so collisional excitation rates are large, while the ionization edges are at 105 nm and 142 nm where the flux from the solar photosphere is low. On the other hand the excitation energies of the lower and upper levels of the Balmer lines are large ($1.6*10^{-18}$ J and $1.9*10^{-18}$ J) so collisional excitation rates are low, while the photoionization edges for the two levels lie at 365 nm and 820 nm where the solar flux is high so photoionizations will be important. In other cases the question of collision domination versus photoionization domination is not so obvious and a more detailed calculation on the lines of Chapter 4 needs to be performed. For solar temperatures the Lyman α line is collision-dominated while the resonance lines of neutral metals are photoionization-dominated with the exception of the sodium D lines. The situation of course may change at higher temperatures where the stellar flux maximum shifts to shorter wavelengths.

Stellar Abundances

We first briefly review the results of the previous chapter. In Chapter 4 we discussed the prediction of equivalent widths W for two simplified models of stellar atmospheres (a two-layer model and a model with a linear source function and a constant ratio of line to continuous opacity respectively). For weak lines equation (4.9) gave

$$\left(\frac{W}{\lambda}\right)_{\text{weak}} = \frac{\pi e^2}{4\pi\varepsilon_0 mc^2}\frac{R_c}{\alpha_c}\lambda g f \frac{N_i}{gN_{\text{ion}}}\frac{N_{\text{ion}}}{N_{\text{el}}}\frac{N_{\text{el}}}{N_{\text{H}}}$$
$$= \frac{\pi e^2}{4\pi\varepsilon_0 mc^2}\frac{R_c\lambda}{\alpha_c N_{\text{H}}} f N_{\text{abs}}$$

where α_c is the continuous absorption coefficient per hydrogen atom, R_c depends on the structure of the atmosphere and in particular on how steeply the temperature falls moving outwards, N_{abs} is the number of atoms per unit mass capable of absorbing the line of wavelength λ and oscillator strength f, and $N_{\text{el}}/N_{\text{H}}$ is the quantity we are trying to determine, namely the abundance of the element relative to hydrogen. In simple models R_c is the central depth of a strong line and can be determined empirically, but the centres of such lines are not formed in LTE, and R_c is better estimated using models of the atmosphere.

If we are given the temperature and pressure in the appropriate layers of the atmosphere and LTE holds, then we can calculate α_c for the appropriate wavelength range, and the Boltzmann and Saha factors N_i/N_{atom} and $N_{\text{ion}}/N_{\text{el}}$. The equivalent width of any weak line should then give the abundance of the element concerned. However, for a line to be truly 'weak' in the solar spectrum requires that the equivalent width is less than 2 nm. Such lines may not occur at all in the visible spectrum for some elements, particularly those with simple atomic structures which have relatively few lines. When a weak line is present, very high quality spectra both in resolution and in signal-to-noise terms are required if an accurate measurement of the equivalent width is to be made. Furthermore, a stellar spectrum contains many weak lines, any particular weak line is likely to be blended with some other line, and if wavelengths are closely coincident the blend may not even be detected.

Thus it is often necessary to use lines that are not strictly weak in the sense that equation (4.9) can be used. In Chapter 4 we saw that as fN_{abs} increases, the central parts of the line saturate and the equivalent width settles at a value of about twice the Doppler width $\Delta\lambda_D$, only growing very slowly with increasing fN_{abs}. Fully saturated lines give very little information about abundances, but lines 'on the bend' between the linear weak line region and the medium-strong saturated region form the basis of most abundance determinations. It was seen

(4.10)–(4.12) that in this transition region the equivalent width depends not only on fN_{abs} and R_c, but also on the Doppler width. For even larger values of fN_{abs} the wings of the line, which are unsaturated and therefore still growing, begin to contribute a major share of the equivalent width, and (4.13) shows that W/λ is proportional to $\sqrt{(fN_{abs}\Gamma)}$ for damping or Lorentz width Γ. If Γ is known, then an abundance can be found from such a strong line, if the line is formed in LTE. Now Γ can be due to natural damping, in which case it depends only on transition probabilities, but in many cases Γ is dominated by pressure broadening and although the pressure can be estimated, the constants of interaction needed to predict Γ often cannot be found accurately from either experiment or theory.

A plot of $\log(W/\lambda)$ versus $\log(fN_{abs})$ is called a *curve of growth.* A curve of growth shows a *linear* portion for *weak* lines, bending over into a nearly horizontal portion called the *saturated* portion for medium-strong lines and then rising again but more slowly for very strong lines on the *damping* portion of the curve. Each line has its own value of Γ so different lines will turn off the saturated portion onto the damping portion at different values of fN_{abs}, the more strongly broadened lines turning off at lower values of fN_{abs}. The curve of growth is a convenient way of showing graphically the interplay of saturation and line profile, but more importantly it provides a means whereby the equivalent widths of *all* lines of a given atom or ion can contribute to determining the abundance.

A plot of $\log(W/\lambda)$ as observed versus $\log(fN_{abs}) - \log(N_{el}/N_H)$, i.e. one that omits the term N_{el}/N_H in calculating the abscissa, is called an *empirical curve of growth.* A plot of $\log(W/\lambda)$ calculated as in Chapter 4 (equations (4.5), (4.9)–(4.11), (4.13)) versus $\log(fN_{abs})$ is called a *theoretical curve of growth.* If the two curves are superimposed (Figure 5.11), the horizontal displacement between the two will be $\log(N_{el}/N_H)$. Initially, we may not know the temperatures and pressures needed to calculate the Saha and Boltzmann factors. If we leave out N_{ion}/N_{el}, the Saha factor, from the abscissa of the empirical curve of growth, a comparison of empirical and theoretical curves of growth will determine $\log(N_{ion}/N_H)$, and if the element is mainly in two stages of ionization and we can observe both, simple addition will give us N_{el}/N_H. However, we still need the temperature to calculate the Boltzmann factors for individual lines. An empirical *excitation temperature* can be found by plotting simple empirical curves of growth of $\log(W/\lambda)$ versus $\log(\lambda gf)$ separately for lines with similar lower excitation potentials, so that all the lines on any particular curve have the same (unknown) Boltzmann factor. If the curves are plotted on the same graph (Figure 5.12), the horizontal shifts between them will be due only to the differences in Boltzmann factors. Thus the shift between curves with lower excitation potentials E_1 and E_2 will be $\log[\exp(-E_1/kT)] - \log[\exp(-E_2/kT)] = 5040(E_1 - E_2)/T_{ex}$ if the energies are measured in electron volts. Thus the excitation temperature can be found. If the abundances of two ions of a given element are found, their ratio should be

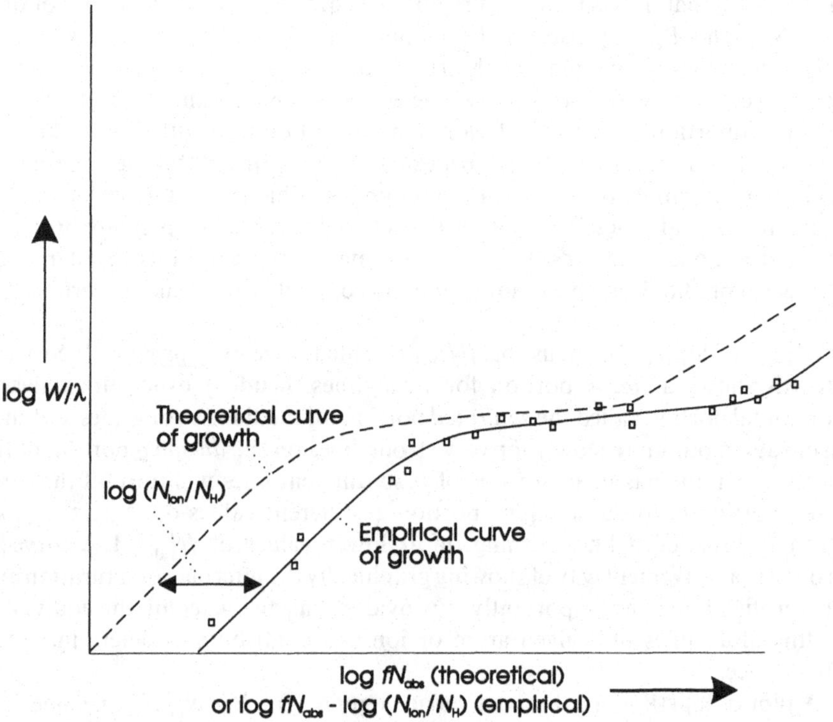

Figure 5.11.

given by Saha's equation, and hence allow us to find the electron pressure if the temperature is known.

It is possible to determine the abundance of an element in one star relative to that in another fairly similar star much more accurately than it is possible to determine the absolute value of that abundance in either star. This fact is exploited in *differential curve of growth* analyses. The basic assumption is that the shape of the curve of growth is the same in the programme star and some standard star with which it is to be compared. 'Similar' here means that the effective temperatures of the stars are not too different, and preferably the surface gravity and overall metallicity are not too different either. The Sun is often used as the standard star in spectral type G differential curve of growth analyses because the Sun has accurately determined abundances for many elements.

Differential curve of growth analyses are carried out in a variety of different ways, but one approach would proceed as follows. Log(W/λ) is plotted against $\log(N_{\mathrm{abs}}f)$ for the standard star, and a mean curve is fitted through the observed points. The observed $\log(W/\lambda)$ values for the programme star are now plotted

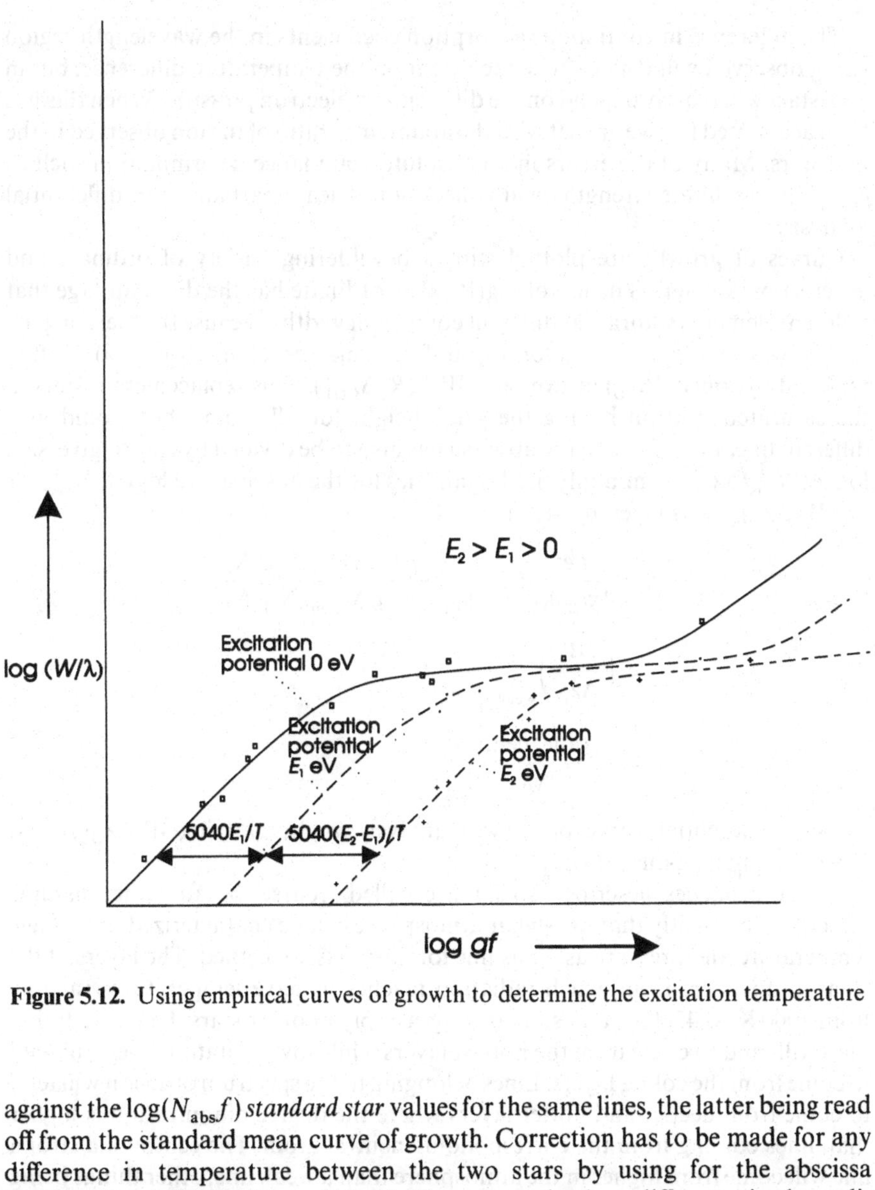

Figure 5.12. Using empirical curves of growth to determine the excitation temperature

against the $\log(N_{abs}f)$ *standard star* values for the same lines, the latter being read off from the standard mean curve of growth. Correction has to be made for any difference in temperature between the two stars by using for the abscissa $\log(N_{abs}f)_{std} + 5040(1/T_{std} - 1/T_{prog})$. There may also be a difference in the ordinate of the saturated portions of the standard and programme star curves of growth of $\log[\Delta\lambda_D(\text{prog}) - \Delta\lambda_D(\text{std})]$, which will also need to be corrected for. We should then be left with a horizontal shift between the programme star and standard star curves of growth which will equal $\log(N_{ion}/N_H)_{prog} - \log(N_{ion}/N_H)_{std} - [\log \alpha_C(\text{prog}) - \log \alpha C(\text{std})]$, Figure 5.11.

The difference in continuous absorption coefficients in the wavelength region being observed will depend to some extent on the temperature difference, but in cool stars will mainly depend on the difference in electron pressure. When this has been accounted for, we are left with the abundance ratio of the ion observed in the two stars. Many of the errors in an absolute abundance determination such as errors in oscillator strengths will cancel out at least partially in a differential analysis.

Curves of growth are plotted with a bewildering variety of ordinate and abscissa parameters. The use of $\log(W/\lambda)$ as ordinate has the disadvantage that different elements saturate at different equivalent widths because the thermal part of $\Delta\lambda_D$ will be different for atoms of different masses. Hence $\log(W/\lambda)$ is often replaced by $\log(W/\Delta\lambda_D)$ or even $\log(W/\{2R_C\Delta\lambda_D\})$. This replacement results in the saturated portion having the same height for all atoms, but would give different linear portions, so the abscissa has also to be divided by $\Delta\lambda_D$ to give, say, $\log(\lambda f N_{abs}/\Delta\lambda_D)$. Commonly used quantities for the abscissa are $\log(W/\lambda)_{weak}$ or $\log(W/\Delta\lambda_D)_{weak}$ as given by (4.9) or

$$\eta = \frac{\pi e^2}{4\pi\varepsilon_0 mc^2}\frac{\lambda}{\sqrt{\pi}\Delta\lambda_D}\frac{\lambda g f}{\alpha_C}\frac{N_i}{gN_{atom}}\frac{N_{iob}}{N_{el}}\frac{N_{el}}{N_H}$$

$$= \left(\frac{W}{\Delta\lambda_D}\right)_{weak}\frac{1}{\sqrt{\pi}R_C}$$

$$= \frac{\kappa_L(\text{line centre})}{\kappa_C}$$

Thus a differential curve of growth analysis might plot $\log(W/\Delta\lambda_D)_{std}$ and $\log(W/\Delta\lambda_D)_{prog}$ against $\log\eta_{std}$.

The approaches described so far are called 'coarse' because they assume directly or implicitly that the stellar atmosphere can be characterized by a single temperature and pressure as far as line formation is concerned. The layers of the solar atmosphere from which radiation reaches us have temperatures ranging from 7000 K to 4300 K, and similar ranges apply in other stars. High excitation lines will tend to come from the hottest layers while low excitation lines will tend to come from the colder layers. Lines belonging to the spectrum of an ion will tend to come from deeper and hotter layers where the degree of ionization is higher than lines coming from the corresponding neutral atom. The centre of a strong line will come from higher in the atmosphere than a weak line, other things being equal. Thus the excitation temperature best describing the formation of a line will depend on the lower excitation potential of the line, and the ionization temperature and electron pressure to be used in Saha's equation will also depend on the species involved.

Various approximate methods of allowing for these effects have been developed. However, a *fine analysis* takes full account of the temperature and pressure structure of the atmosphere. One might start by picking a model atmosphere (T_{eff}, $\log g$, metallicity), perhaps using de-reddened colours to find T_{eff}, hydrogen line profiles to estimate the pressure and hence $\log g$, and a line ratio of an ion to an atom which is sensitive to both T_{eff} and $\log g$ to further tie down the values of these parameters. The model atmosphere enables Boltzmann and Saha factors, and hence the line absorption coefficient, to be calculated for any particular line at each depth. The continuous absorption coefficient at the line wavelength can also be found for each depth, to give the total absorption coefficient t_λ at any depth and any wavelength in the line profile. If LTE is assumed to hold, the source function $S(t_\lambda)$ can also be found as the blackbody function at the temperature of the layer corresponding to total optical depth t_λ.

Then the emergent flux from the top of the atmosphere is

$$F_\lambda = 2\pi \int_0^\infty S(t_\lambda)E_2(t_\lambda)\mathrm{d}t_\lambda$$

F_λ integrated over the line profile gives the equivalent width. If the whole process is repeated for various values of the abundance, we have constructed a curve of growth for the particular line concerned. If we are given the observed equivalent width of the line, the abundance can be read off.

One can then look at the abundances found from individual lines of a particular atom or ion. If the abundance appears to increase with lower excitation potential of the line, then the model atmosphere used has been underestimating the Boltzmann factor, and hence has a too low effective temperature. Similarly if the abundance derived from medium-strong lines is higher than that found from weak lines, then we have probably underestimated the microturbulent contribution to the Doppler width. If high abundances are found for strong lines only, then the Lorentz widths Γ used may have been too small, implying for the case of pressure broadening that the model atmosphere used had a too low value of $\log g$. Similarly, if the relative abundance found in two stages of ionization does not agree with that predicted by Saha's equation, then the model atmosphere used may have values of T_{eff} and $\log g$ that are in error. Considerations such as these lead to the selection of a different model atmosphere and a repeat of the abundance analysis.

Difficulties may arise if we are finding the abundance of an element represented only by a single weak line which is blended (lithium in solar type stars is an example), or where we are working in a crowded region where most lines are blended (particularly near the head of a molecular band). In these instances we may use the model atmosphere to predict a whole stretch of spectrum, with several overlapping lines often contributing to the absorption coefficient at any particular wavelength. The abundances are adjusted until the predicted spectrum matches observations. This is called the *method of spectrum synthesis*—it is, of course dependent on having available oscillator strengths and line broadening parameters for all the lines involved (Figure 5.12).

We now need to consider some of the problems that arise in abundance analyses. Up to this point we have assumed that line formation is in local thermodynamic equilibrium (LTE). This means the following assumptions

(1) The source function is given by Planck's function, which is equivalent to requiring that the ratio of upper to lower level populations is given by Boltzmann's equation.
(2) The excitation of the lower level is given by Boltzmann's equation.
(3) The degree of ionization of the element concerned is given by Saha's equation.

If the relevant collisional rates were greater than the radiative rates, then, as we saw in Chapter 4, LTE would certainly hold, but such a clear-cut situation rarely applies in stellar atmospheres. Only a fully coupled statistical equilibrium-radiative transfer solution can decide whether LTE holds. Such solutions can be attempted for moderate numbers of levels, but become very difficult for an element like iron where there may be hundreds of significantly populated levels.

There are, however, certain general comments that can be made. We have seen that the source function drops at the highest layers in an atmosphere to $\sqrt{\varepsilon'}$ of the blackbody value. One therefore cannot expect the centres of strong lines to be formed in LTE, and indeed strong lines sometimes have greater central depths than the value of R_C predicted by model atmospheres and LTE.

On the other hand a weak line with the ground state as the lower level originating from the state of ionization that contains most of the element concerned is likely to be represented quite well by LTE calculations. For instance if the line comes from a neutral atom, and according to Saha's equation 10% of the element is ionized in a particular star, then even if departures from LTE mean that 20% of the element is ionized, the error in calculating the abundance from the neutral line strength using Saha is only 10%. Under typical stellar conditions nearly all atoms and ions are in the ground state, and deviations from Boltzmann's equation will not greatly alter this conclusion. In most cases continuum processes are close to LTE, and since in a weak line the line absorption coefficient is considerably less than the continuum absorption coefficient, one would expect that continuum absorption and emission should drive weak line source functions towards LTE values. A weak line from the ground state will have a small downwards radiative transition probability compared with the collisional de-excitation probability and this will also favour an LTE source function. An example is the oxygen doublet 630.04 nm and 636.39 nm from the ground state $2s^22p^4\ {}^3P$ to the first excited state ${}^1D$, which is of course a forbidden transition and a weak line in cool stars, but satisfies our requirements for LTE.

For main sequence stars it seems that abundances derived from LTE calculations are reasonably accurate for the cooler stars, with significant deviations appearing in stars of spectral type B, and LTE calculations being totally inappropriate for effective temperature characteristics of O stars (that is above 28 000 K). This does not, however, mean that LTE holds for all individual lines in cool stars or that LTE calculations will give the correct line profiles and central depths in those stars.

The other main cause of difficulty in determining stellar abundances is the non-availability of accurate oscillator strengths f, or equivalently of gf values or transition probabilities. Transition probabilities for hydrogen and helium and other cases where the atomic structure is relatively simple can be calculated theoretically, but for most atoms and ions we are dependent on experimental results. It must be remembered that ideally the lines used in stellar abundance estimates are weak, but are produced by large numbers of atoms or ions in the line of sight, and may come from levels that require high stellar temperatures to excite or belong to states of ionization which are again only produced at high temperatures. These conditions may be hard to simulate in a laboratory source with well determined conditions of temperature distribution.

An obvious approach is to use a laboratory source with a known number of atoms of the species concerned, and to measure the strength of the resulting emission lines, or to shine continuum radiation through the source and measure the strength of the resulting absorption lines. The latter approach constitutes using a mini 'stellar atmosphere' in the laboratory, the source of the absorption lines being a column of vapour in a furnace through which a beam from a much 'hotter' continuum-producing lamp is shone. Emission line sources include arcs, sparks and shocks. Furnaces are limited to temperatures of less than 3000 K, so lines from high excitation levels and lines from highly ionized species must be obtained from emission line sources.

Determination of the number of atoms in the source is difficult and so most determinations of oscillator strengths find the *relative* f values of a given atom or ion. An *absolute* oscillator strength is found in a separate experiment for the strongest line or lines, with a measurement of the number of atoms producing a line of the observed strength, or alternatively a measurement of the lifetime of the upper level of the transition. This absolute oscillator strength is then used to put the comprehensive sets of relative oscillator strengths on an absolute scale. It is still necessary in determining relative oscillator strengths to know the Boltzmann factors for the upper levels (emission lines) or lower levels (absorption lines) and this means knowing the temperature of the source. Usually temperatures cannot be measured directly for emission line sources like arcs, and one has to use lines of known oscillator strength but different excitation potential from some element to determine an empirical excitation temperature for the arc. This does not allow for the fact that in general the source will not be homogeneous and the temperature will vary spatially. On the other hand, the temperature distribution at the walls of the furnaces used in absorption measurements can be measured directly.

The measured equivalent widths are the result of a process of radiative transfer. Emission lines have usually been assumed to be optically thin, but arcs can have cooler gas around the core which can lead to self-absorption of the emission lines.

In the case of absorption lines, it is essential to ensure that the lines are weak enough to be fully on the linear portion of the curve of growth, for one is essentially comparing a line that is weak in a stellar spectrum with the resonance line whose absolute oscillator strength is known. The range in oscillator strength is 10^5 to 10^6 in some cases, and this means that one has to proceed by a series of steps in which the very weak line is compared with a weak line with both on the linear portion of the curve of growth, and then the furnace is adjusted so that the weak line can be compared with a medium strength line, again with both lines on the linear portion of the curve of growth, and so on. Any systematic error at each step is liable to build up to a large error in the final result. The accurate measurement of the equivalent widths of very weak lines requires a high quality spectrograph, for even a small contribution from scattered light or the far wings of the apparatus profile can lead to a large cumulative error.

An alternative to the measurement of equivalent widths is to use the fact that the refractive index of a gas at a wavelength in the vicinity of a line is changed by an amount that is proportional to the oscillator strength. If the absorbing column of gas is placed in one arm of an interferometer and the resulting spectrum is observed, the interference fringes will show displacements on either side of the line wavelength giving a *hook* pattern, from which the oscillator strength of the line can be derived. This method was formerly confined to strong lines, but has now been extended to weaker lines.

Absolute oscillator strengths have been measured using atomic beams, with the resonance line being measured in absorption, and the beam collected for a given time to directly measure the mass flow and hence the density. The degree of excitation is not needed since the line comes from the ground state. Absolute oscillator strengths can also be determined using lifetimes. The lifetime of an excited level t_j defined by $N_j(t) = N_j(0)\exp(-t/t_j)$ is the reciprocal of the transition probability A_{ji} if there is only one downward transition since $dN_j/dt = -N_j t_j = -N_j A_{ji}$ by definition of the transition probability. In general, $t_j = 1/\sum_i A_{ji} A_{ji}$ since all downward transitions will tend to shorten the life of level j. If we know the relative values of the downward transition probabilities, then a measurement of the lifetime of j will allow all of the A_{ji} to be determined.

Lifetime measurements can be made by passing a particle beam through a thin foil which excites the atoms and then observing a line from level j at various distances downstream from the foil. If the beam velocity is known, the decay of intensity with distance can be converted to the decay of intensity with time and so the lifetime t_j can be found. Passage through the foil excites many levels and j may be populated by transitions cascading down from higher levels, thus increasing the apparent lifetime of j. This problem is avoided in the alternative method where the gas is static, and the level j is selectively optically excited by a laser tuned to the wavelength corresponding to the energy difference between j and the ground state. The laser excitation can be pulsed or modulated, and if the emission of a line from level j is monitored, the time delay between excitation and emission gives the lifetime directly.

Use can also be made in some cases of the Hanle effect, where the atom is excited by polarized radiation and observed in a direction perpendicular to the incoming beam but parallel to the plane of polarization. Emission by an atom vibrating parallel to the plane of polarization and perpendicular to the incoming beam should be zero in this direction. However a magnetic field perpendicular to the polarization and the incoming beam will cause the atom (viewed classically for the purposes of this rough argument) to precess at a known rate. The observed intensity will depend on the amount that the atom has precessed between excitation and emission and will hence allow the lifetime to be calculated.

It is interesting to note that up to around 1970, the oscillator strength scale for neutral iron was in error by a factor of about 10 (with the error varying with excitation potential and line strength) largely as a result of systematic errors in relative oscillator strength measurements. The correction of this error essentially increased the abundance of iron in the Universe by an order of magnitude! It is unlikely that errors as large as this remain amongst the commoner elements but errors in individual oscillator strengths could still be as large as 50% in some cases.

Abundance determinations in cool stars present particular problems. The visible spectrum is very crowded with many lines of the metals like iron and molecular bands. This can make it difficult to establish the continuum level against which equivalent widths are measured. The *apparatus profile* of a spectrograph is the output plotted against wavelength when an infinitely narrow line is input, and the full width at half-height measures the resolution of the spectrograph. However apparatus profiles have very weak but very extended wings and these wings overlap in crowded regions producing an apparent lowering of the continuum level and an underestimate of the equivalent widths of weak lines. Spectrum synthesis techniques can include the effect of apparatus profile but the best solution is to work at longer wavelengths, say between 650 and 800 nm, where the spectrum is less crowded.

In cool stars a given element will be present in both atomic and molecular forms, and may have both atomic and molecular lines in the spectrum. As has already been discussed, the interlinked dissociation equations of all relevant molecules have to be solved simultaneously in order to produce the predictions of molecular and atomic number densities (for a given temperature and given abundances) which we need to interpret the observed line strengths. The hydrogen abundance is unlikely to vary between stars, but the abundances of carbon, oxygen and nitrogen in cool giants may well vary, and these elements play an important role in the overall molecular equilibrium. The number of molecules of CN, say, depends not only on the carbon and nitrogen abundances but also on the oxygen abundance. Additional problems may be caused by uncertainties in the values of some dissociation energies, which makes the use of as many indicators as possible desirable.

An example of the sort of procedure that is followed can be found in the work of Lambert and co-workers on the C,N, and O abundances in cool (G and K) giants (see, for example, Lambert and Ries [2]). Spectral features to be used were selected avoiding the blue end of the spectrum where line crowding presents serious problems. Weak permitted lines of CI (659, 711, and 834 nm) and of OI (616, 717, and 926 nm) were rejected because these lines come from very high excitation levels and are therefore very temperature sensitive and liable to non-LTE effects. This left the O I forbidden lines at 630.0 nm and 636.3 nm, the CH electronic system band at around 485 nm, the C_2 Swan electronic band at around 510 nm and bands from the CN red system. The O I, CH and C_2 strengths are fairly insensitive to the nitrogen abundance, but depend differently on the carbon and oxygen abundances so a measurement of an O I line strength corresponds to a curve (for a given temperature) on a plot of oxygen versus carbon abundance, and the intersection of this curve with the curves for CH and C_2 should give the carbon and oxygen abundances. The observed CN strength selects a family of curves on a plot of nitrogen against carbon abundance, each curve corresponding to an oxygen abundance. Using the previous results, the nitrogen abundance can be found. In this particular work a model atmosphere analysis of Fe I and Fe II lines was used to select a model atmosphere with the correct T_{eff} and $\log g$, and also to determine the metallicity. The spectral features used in similar analyses of main sequence stars (Clegg *et al.* [3]) and supergiants (Luck [4]) are slightly different because of differences in line blending problems and line strengths.

The presence of molecular features allows isotopic abundances to be determined. In an atom, the mass that appears in formulae for the energy levels is strictly the reduced mass of nucleus and electron, $m = m_n m_e(m_n + m_e)$ but since even the lightest nucleus (hydrogen) has $m_n = 1840\, m_e$, the reduced mass is very close to the electron mass, and small changes in the nuclear mass between different isotopic species result in very small changes in the energy levels. In the laboratory, the small differences between the wavelengths of different isotopes can be resolved, but line broadening means that this is usually not possible in stars. However, the rotational energy levels of a molecule are due to the rotation of the nuclei, and the rotational energy is inversely proportional to the reduced mass of the nuclei. Similarly, the vibrational energies of the nuclei are inversely proportional to the square root of the reduced mass of the nuclei. Now for $^{12}C^{14}N$ the reduced mass is

$$m(12)m(14)/[m(12) + m(14)] = 6.46 \text{ atomic mass units}$$

while the corresponding reduced mass of $^{13}C^{14}N$ is 6.74 atomic mass units. Clearly the molecular lines of $^{12}C^{14}N$ and $^{13}C^{14}N$ will be well separated. A comparison of the same transition from $^{12}C^{14}N$ and from $^{13}C^{14}N$ should then give the ^{12}C to ^{13}C ratio since the oscillator strengths of the two isotopic species are likely to be nearly the same, as are the degree of excitation and dissociation.

Problems arise because the lines of the more abundant isotopic variant may be saturated. The vibrational–rotational bands of carbon monoxide are the most prominent features in the infrared spectra of cool stars unless the temperature is so low that H_2O is very strong. The fundamental vibrational–rotational bands at around 5000 nm are not accessible to ground-based observations and are strongly saturated. The first overtone bands starting at around 2300 nm lie in an atmospheric window. The individual lines overlap near the bandheads which form very strong features with the $0 \rightarrow 2$, $1 \rightarrow 3$ and $2 \rightarrow 4$ bandheads of $^{12}C^{16}O$ being followed by the $0 \rightarrow 2$ bandhead of $^{13}C^{16}O$. Usually the $^{12}C^{16}O$ lines are saturated so their strengths depend more on Doppler width than on abundance—the bands look stronger in giants than dwarfs because of the greater microturbulence (see Figure 5.8). Hence only the $^{13}C^{16}O$ first overtone lines are useable. The second overtone at around 1600 nm is also in an atmospheric window and the $^{12}C^{16}O$ lines are unsaturated, but the $^{13}C^{16}O$ lines may be so weak as to be unmeasurable. Thus we may have to compare the $^{13}C^{16}O$ first overtone with the $^{12}C^{16}O$ second overtone, with the disadvantage that we are no longer comparing the same transitions with the same oscillator strengths, etc. Determinations of the ^{12}C:^{13}C ratio have used either the CN red electronic system or the CO vibrational–rotational bands.

The cosmic abundance distribution, which is compared with the overall theory of stellar nucleosynthesis in stars of all masses, has been established mainly from the Sun, with solar system values (usually meteoritic) employed for those elements unobservable in the Sun and for isotope ratios. Solar system material is highly fractionated, so that light and heavy elements are found in different places. However the ratios of the abundances of similar elements such as the lanthanides and isotope ratios are likely to be little altered by fractionation, particularly if chondritic meteoritic material which has been little altered in the course of its history is used.

Two important elements whose abundances cannot easily be established in the solar system are helium and neon, neither of which have lines in the photospheric spectrum. Both can be directly collected from the solar wind and can be observed in the solar corona, but it seems likely that the processes that accelerate and transport photospheric material to corona and wind have different efficiencies for different elements and hence that solar wind abundances are not typical of the photosphere. Indeed, helium and neon are not directly observable in any cool stars, in contrast with the situation in hot stars. This is unfortunate as the primeval helium abundance is an important cosmological parameter and since some helium is clearly formed in stars we need to know the variation of helium abundance with epoch. The present-day helium abundance can be found from gaseous nebulae and young stars which of course are hot stars, but old stars are cool stars, which are just those where the measurement of helium in the spectrum is impossible!

Much interest attaches to the measurement of the change of metallicity with age, which is beyond the remit of this book. However it is interesting to see if all the 'metals', the elements heavier than helium, really vary in step. One case where they do not do so exactly is that of oxygen, where in metal-poor stars the deficiency in oxygen relative to solar values is found to be less than that of iron. Both oxygen and iron are produced in supernovae, but most of the iron comes from type I supernovae which probably represent the ends of relatively low mass long-lived stars (many of them in binary systems) whereas most of the oxygen comes from type II supernovae which represent the final stages of massive, short-lived stars. The difference between the behaviour of iron and oxygen abundances might then indicate a difference in the relative frequency of the two types of supernovae with time. The giant stars in globular clusters often show a considerable scatter in the abundances of elements like carbon, nitrogen, oxygen, sodium and aluminium. One might expect that all stars in a globular cluster would have the same age and be formed from material of similar composition. The differences in carbon, nitrogen and oxygen are explicable in terms of the mixing of the products of nucleosynthesis to the surface, as discussed below, but the sodium and aluminium results are hard to understand.

One group of stars that shows very anomalous abundances are the peculiar A stars which are main sequence stars with temperatures between 8000 and 20 000 K (considerably wider than the range of normal A stars) showing unusually strong lines of ionized silicon, chromium, europium, strontium, manganese, and several other elements, the peculiarities depending on the temperature and sub-group of peculiar A stars. These peculiarities are often associated with a strong magnetic field and slow rotation. It is generally accepted that these anomalous line strengths do not indicate a peculiar composition, but are due to radiative diffusion where radiation pressure through the absorption of line radiation selectively pushes the atoms and ions of certain elements high in the atmosphere.

We now turn to stars which show deviations from the cosmic abundance distribution due to nuclear burning within the star. Such abundance changes would be expected to appear, if anywhere, in evolved giants and supergiants, particularly since cool giants and supergiants have deep convective envelopes which might be able to dredge up the products of interior burning to the surface. Red giants and supergiants are often found to show an overabundance of nitrogen and an underabundance of carbon relative to iron compared with solar values in such a way that the total amount of carbon and nitrogen is unchanged from that in the unevolved star. This is precisely what one would expect to be the result of hydrogen burning on the CN cycle where carbon and nitrogen act as catalysts with some carbon being converted to nitrogen but the total amount of carbon and nitrogen remaining unchanged. Red giants and supergiants often show ^{13}C:^{12}C enhanced above the solar value of 1:90, again a predicted result of CN cycle hydrogen burning. In some cases ^{13}C:$^{12}C = 1$:4, which is the equilibrium CN cycle value, and stellar models have some difficulty in obtaining such a high proportion of processed material at the surface.

Main sequence stars have more oxygen than carbon as do most evolved red giants, so the latter show spectra dominated by oxides of the iron group of elements, in particular by titanium monoxide and vanadium monoxide. However a minority of red giants have more carbon than oxygen, which results in dramatic changes in the spectrum which becomes dominated by CH, C_2 and CN. These are called carbon stars. An even smaller minority of red giants are oxygen rich, but only just so, and show features due in particular to zirconium monoxide but also to various rare earths, barium and strontium, which have in common the feature that they are all believed to be made by the process of slow neutron addition, the s process. One noteworthy s process element seen in these red giants is technetium, which has no stable isotopes and so must have been made recently within the star. These are called S stars. The high abundances of s process elements are also found in the cooler carbon stars (sometimes called N stars). It seems likely that N carbon stars and S stars are both on the second ascent of the giant branch (called the asymptotic giant branch). Such stars have a convective envelope, a hydrogen-burning shell and a helium-burning shell around an inert carbon–oxygen core. The helium shell burns in a series of flashes with longer periods of quiescence in between, the flashes leading to a temporary convective region between the two burning shells. It is possible in this way for products of helium burning to reach the surface, partially modified by hydrogen-burning. There are also hotter giant carbon stars of lower luminosity called R stars which are not yet on the asymptotic giant branch, and a different explanation for the large carbon abundance must be looked for in these cases.

Other Stellar Parameters

Apart from abundances and effective temperatures, spectra can also be used to find surface gravities, magnetic fields, and turbulent and rotational velocities, but the last two subjects were discussed in Chapter 3.

For a given effective temperature, the surface gravity determines the gas pressure, which in turn determines the electron pressure. The electron pressure controls the H^- absorption in cool stars and the pressure broadening of lines in hot stars (pressure broadening by neutrals via the Van der Waals interaction only becomes important in stars cool enough to have few ions or electrons). The Balmer jump is mainly dependent on the temperature, but in cooler stars is reduced in size by the increasing importance of H^- absorption, which for a given temperature will increase with the electron pressure. Thus in this temperature range the higher the surface gravity the smaller the Balmer jump. In stars with effective temperatures greater than 10 000 K, the contribution of H^- is negligible and so there is no pressure dependence of the size of the jump, and in stars cooler than the Sun the Balmer jump is small and hard to measure.

The wings of hydrogen lines are formed by pressure broadening by charged particles and hence will increase in importance at high surface gravity. In cooler stars the continuous background is due to H^- which also depends on the electron pressure. Since the observed line wing strength is a function of the ratio of line to continuous absorption coefficient, the dependence on electron pressure and hence on surface gravity tends to cancel out under these conditions. In hotter stars, where H^- is unimportant, the wings contribute much of the equivalent width of hydrogen lines, and hence for a given temperature the equivalent width of Hβ or Hγ can be used as a surface gravity indicator. More detailed work would use the line profile for this purpose.

The broadening of hydrogen lines can also be used to determine the pressure by counting the number of Balmer lines detectable. The energy levels of hydrogen come closer together as the quantum number n increases, for $E = -A/n^2$, where A is a constant. This gives the separation of levels with quantum numbers n and $n-1$ as $\Delta E = A[1/n^2 - 1/(n-1)^2] \sim 2A/n^3$ for large n. Hence if the energy levels are broadened, they will merge at some n, and for higher n run without break through to the continuum so the ionization potential is effectively lowered.

In the stellar spectrum the Balmer lines are closer together as one goes to higher n members of the series at shorter wavelengths, and if the lines are broadened they will merge for some value of n, effectively moving the Balmer jump to longer wavelengths (Figure 5.13). For hydrogen, the broadening is via the linear Stark effect by electrons and ions and is proportional to n^2 and hence is larger for higher n levels. Hence the quantum number of the upper level of the last distinguishable Balmer line is related to the electron density N_e by the Inglis–Teller formula: $\log N_e = 29.0 - 7.5\log(n)$.

The electron pressure can also be determined via Saha's equation for a given temperature from the degree of ionization of an element if lines from two stages of ionization can be observed, and indeed this is the only method for cool stars.

We finally turn to the question of determining stellar magnetic fields, where the Zeeman effect is used. Suppose we are dealing with an energy level with orbital angular momentum quantum number L, spin angular momentum quantum number S, and total angular momentum quantum number J. In a magnetic field of magnetic flux density B (in Tesla) the level specified by J will be split into $2J+1$ sublevels with quantum numbers M_J from $M_J = -J$ to $M_J = +J$, where the angular momentum in the direction of the field is $M_J(h/2\pi)$. The magnetic moment in the field direction is $-g_J\mu_B M_J$ which gives an energy shift of $g_J B\mu_B M_J$, where g_J is the Lande g factor and has the value

$$g_J = 1 + [J(J+1) + S(S+1) - L(L+1)]/\{2J(J+1)\}$$

Transitions called σ transitions obey the rule $\Delta M_J = +1$ or -1. In this case the radiation is polarized perpendicular to the field (and of course to the line of sight) if viewed in a direction that is perpendicular to the field, and is circularly polarized if viewed in a direction parallel to the field, with $\Delta M_J = +1$ and $\Delta M_J = -1$ having opposite senses of circular polarization (Figure 5.14). The other allowed transitions, called π transitions, have $\Delta M_J = 0$ (noting that $M_J = 0$

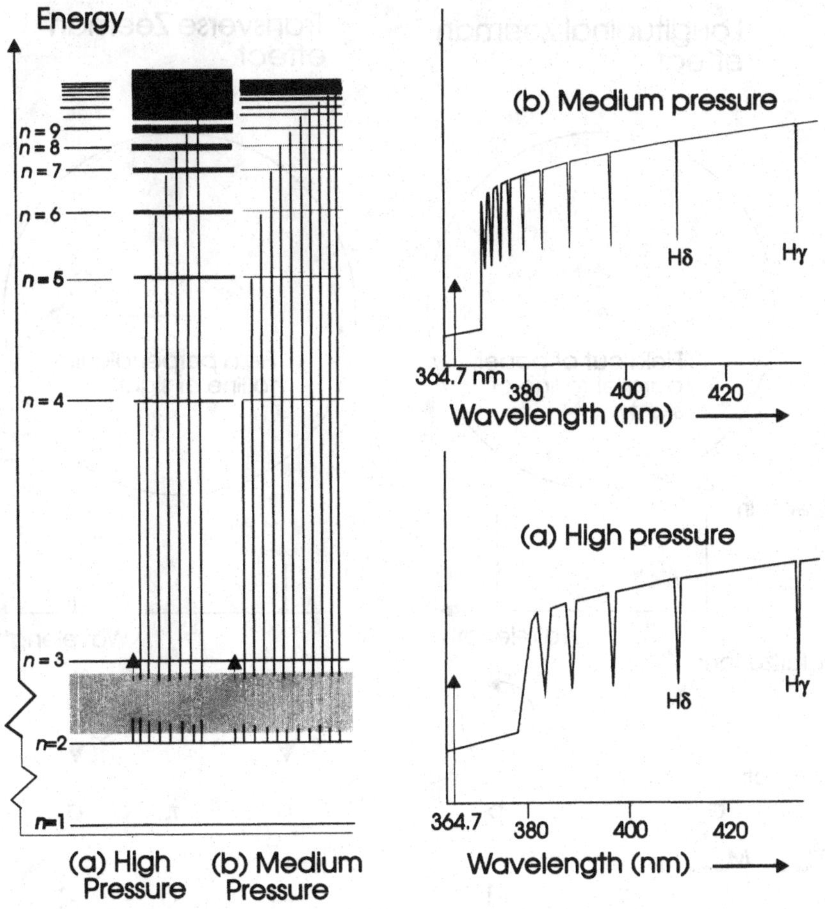

Figure 5.13. Effect of pressure on the number of visible Balmer lines and on the apparent position of the Balmer jump

to $M_J = 0$ is not allowed). In this case the radiation is polarized parallel to the field when viewed in a direction perpendicular to the field, and has zero intensity when viewed in a direction parallel to the field. The quantity $\mu_B = eh/(4\pi m_e)$ is called the Bohr magneton and has the value $9.27*10^{-24}$ J T^{-1}.

If $S = 0$ for the singlet state, $J = L$, $g_J = 1$ and the splitting between adjacent sub-levels is $B\mu_B$ for all levels. Hence a single line transition is split into 3 lines corresponding to $\Delta M_J = +1, 0$, and -1 in order of increasing wavelength. This is called the normal Zeeman effect. However in general S is not zero, and g_J will be different for each level since L will be different, with the result that in the anomalous Zeeman effect, as the more general case is called, $\Delta M_J = +1, 0$ and

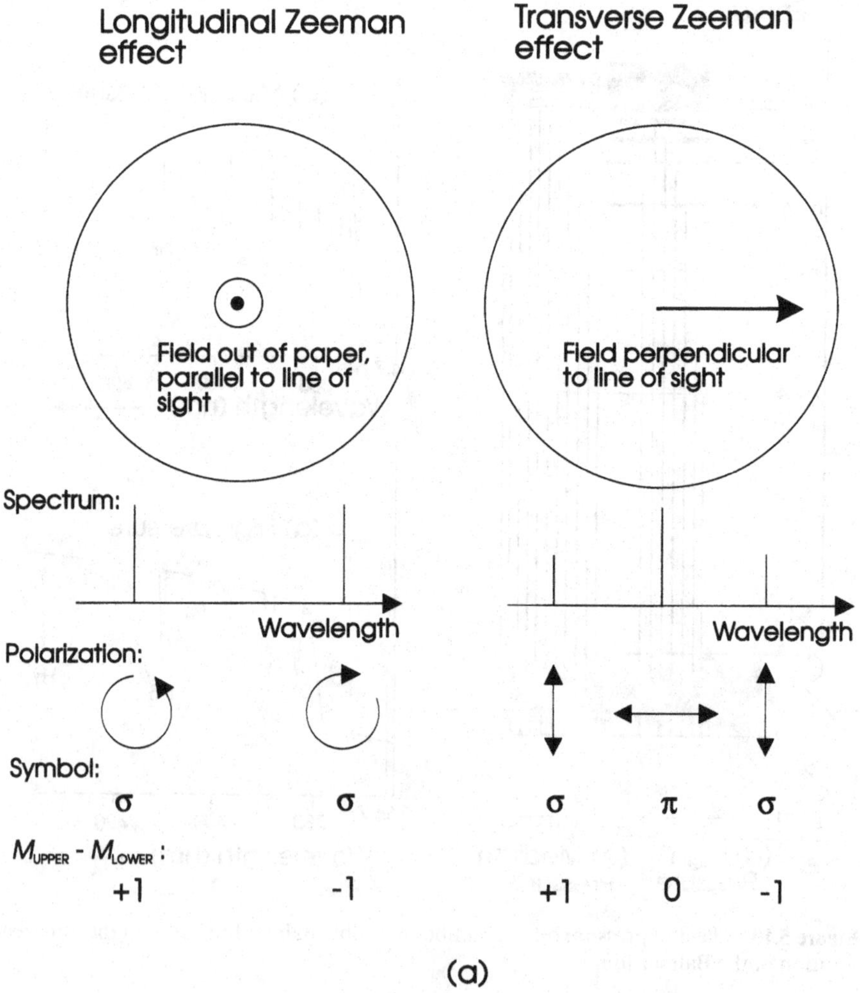

Figure 5.14. (a) Zeeman effect

-1 will each give rise to several lines. If the magnetic field is very large, the orbital and spin angular momenta may couple more strongly to the field than to each other, in which case the shift in energy is given by $B\mu_B M_L + 2B\mu_B M_S$ in the extreme case that the interaction between spin and orbital angular momenta can be neglected altogether. The selection rules are $\Delta M_S = 0$ and the same conditions on M_L and polarization as for M_J, so that the same three-line pattern is obtained as for the normal Zeeman effect. This strong field case is called the Paschen–Back effect. The vast majority of stars, however, correspond to the weak field case.

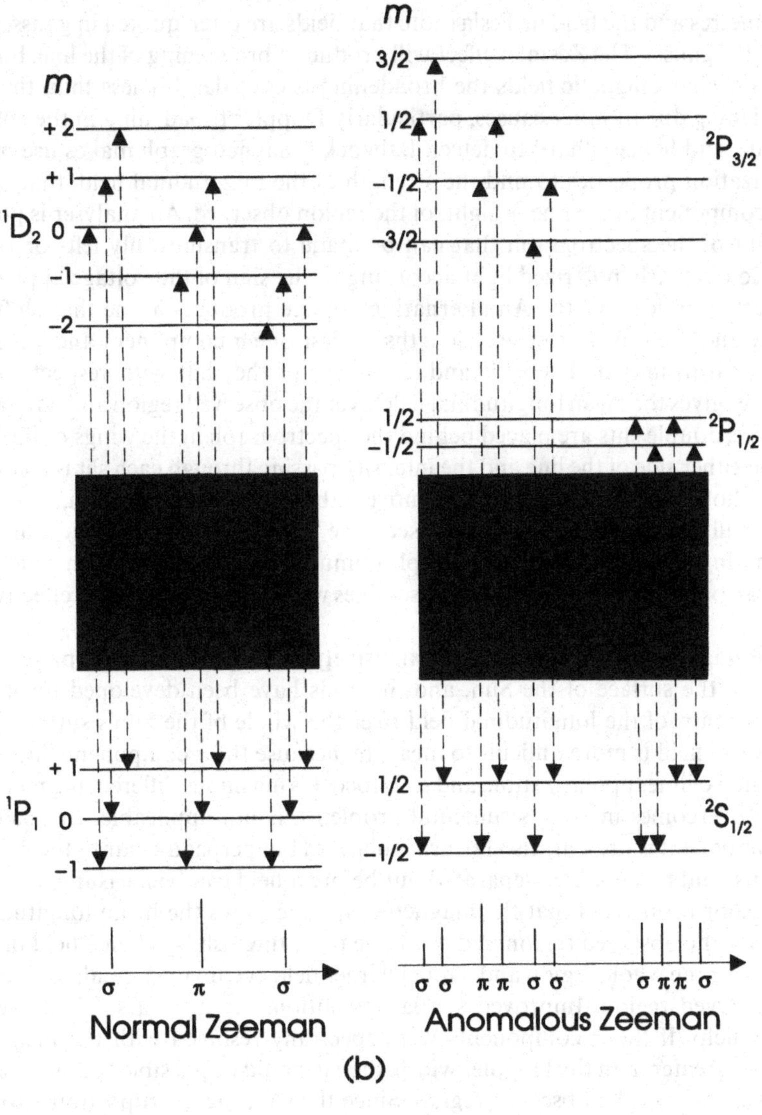

Figure 5.14. (*Contd*)

The wavelength of the σ component is shifted relative to the wavelength in the absence of a magnetic field by

$$\Delta\lambda = -4.67 * 10^{-8}\lambda^2 B[g_J(\text{upper}) + M_J\{g_J(\text{upper}) - g_J(\text{lower})\}]$$

for a transition from M_J (lower) to $M_J + 1$ with the wavelengths measured in nanometres and the field in Tesla (note that fields are often quoted in gauss, with $1\,T = 10^4$ gauss). The Zeeman effect will produce a broadening of the line, but for typical stellar magnetic fields the broadening is considerably less than the line broadening due to other causes, particularly Doppler broadening in the core of the line, and hence is hard to detect. Babcock's magnetograph makes use of the polarization properties to find the strength of the longitudinal field, that is the field component in the line of sight, of the region observed. An analyser is placed in front of the spectrograph that can be made to transmit only left- or right-handed circularly polarized light according to the sign of the voltage applied to an electro-optical crystal. An alternating voltage produces a line that shifts to longer and then to shorter wavelengths as first one σ component and then the other is transmitted. The shift and the phase of the shift with respect to the analyser gives the mean longitudinal field over the observed region and its sign. In practice, double slits are placed behind the spectrograph in the wings of the line, one on either side of the line and the intensity passing through each slit is recorded by a photomultiplier. As the line moves about its mean position, first one photomultiplier and then the other 'sees' the strongest absorption line, and an alternating difference signal from the photomultipliers is received from which the field can be deduced. Obviously one uses lines with the largest possible effective g_J factors.

The magnetograph has been used extensively in studying the small spatial scale fields on the surface of the Sun, and methods have been developed to obtain rapidly maps of the longitudinal field over the whole of the Sun's surface. The transverse field is more difficult to measure, because the σ components have the same sign of linear polarization, and so Babcock's ingenious differencing method, which overcomes many instrumental problems, is not applicable. In this case a π component is present, also linearly polarized but perpendicular to the σ components, and this must be separated out before a field can be measured.

A major problem is that the magnetograph measures the mean longitudinal flux over the observed region and is unable to distinguish between a field of this value over the whole region and a much larger field occupying a small fraction of the observed region. Improved spatial resolution, say from a satellite, would clearly help. If the σ components were spectrally resolved with the magnetic splitting greater than the Doppler width, then it would be possible to measure the strongest field in the observed region. Since the splitting is proportional to the square of the wavelength, and the Doppler width only to the wavelength, the use of infrared lines is highly advantageous. Sunspots are found to have fields of 0.15 to 0.3 T in their central regions (the umbrae), and in pairs of sunspots the polarity of the two spots is usually opposite. Much smaller regions only 100 km in diameter called flux tubes are found to have fields of similar strengths and these flux tubes can combine to make up active regions. However only a small fraction of the Sun's surface is covered by spots and active regions and the average field is only 10^{-4} to 10^{-3} T.

Stars other than the Sun are unresolved by the telescope and if their magnetic fields are like those of the Sun and only have large strengths on a small scale, they would not be detectable by a magnetograph. Regions of opposite polarity giving opposite senses of circular polarization in a longitudinal field measurement would be averaged together to give a very small net circular polarization. Only stars with a strong global magnetic field (with a configuration like that of the Earth), and orientated so the line of sight is not too far from the magnetic axis so that we are looking pole-on, will show longitudinal fields detectable by a magnetograph. A-type main sequence stars rotate fairly rapidly, so that one might expect sharp lined A stars to be nearly pole-on as far as the rotation axis is concerned, and since rotation is involved in generating magnetic fields, it seemed a reasonable guess that such stars might show large longitudinal fields. Indeed it turned out that many of these stars show fields of 10^{-1} T. The present picture is of the class of stars called peculiar A stars already mentioned, with surface temperatures actually running up to 20 000 K, well within the normal B classification and showing spectral peculiarities for a variety of elements including silicon, rare earths, and strontium. It is fairly generally accepted that these peculiarities are due to diffusion rather than anomalies in the internal abundances of these stars. They show dipolar fields from a few times 10^{-2} T to 1 T and it is suspected that the effect of the field on motions that would otherwise spoil diffusive separation may explain the spectral anomalies. The magnetic axis probably does not coincide with the rotation axis, which leads to variations in the observed field as the star rotates, and spectral variations since the distribution of elements on the surface is probably patchy and controlled by the field. Peculiar A stars turn out to be relatively common in the appropriate temperature range, but rotate more slowly than most hot main sequence stars, probably because of magnetic braking.

A second group of stars sometimes showing strong global fields are white dwarfs, as indeed one might expect for collapsed objects. Most white dwarfs do not have strong fields, but about 2% have fields in the range 100 T to 10^5 T. White dwarfs in the cataclysmic binary systems called AM Herculis stars have fields around $3*10^3$ T and are circularly polarized due to cyclotron emission from electrons circling the field lines non-relativistically. A second group of stars called intermediate polars have white dwarfs with field strengths which have been estimated indirectly to be less than 100 T.

Finally we return to cool stars which are not expected to have global fields, so polarization properties cannot be used. Two very similar lines are compared, with one line having a much larger g_J factor than the other. Any difference must be attributed to magnetic broadening, although one must take into account the fact that in regions of high field, 'starspots' may have different spectral properties from the rest of the surface. Fields of a few times 10^{-1} T covering from 10% to 70% of the surface have been found in several main sequence stars with spectral types from G to M. It seems that the area of surface covered by strong fields, rather than the field strength, declines as the star slows down, leading to a decrease in activity with age.

References

[1] H. Jones, A. Longmore, R. Jameson, and M. Mountain, *Monthly Notices Royal Astronomical Society*, **267**, 413, 1994.
[2] D. L. Lambert and L. M. Ries, *Astrophysical Journal*, **217**, 508, 1977.
[3] R. S. Clegg, D. L. Lambert, and J. Tomkin, *Astrophysical Journal*, **250**, 262, 1981.
[4] R. Luck, *Astrophysical Journal*, **219**, 148, 1978.

6 PHOTOIONIZED CLOUDS—GASEOUS NEBULAE AND AGN

Introduction

We find in a wide variety of circumstances warm and bright clouds of gas whose defining characteristic is that a central source emits enough short wavelength photons to photoionize hydrogen in the clouds. Such photons must have wavelengths λ which are less than the Lyman limit of 91.2 nm, which corresponds to the energy of a photon ($E = hc/\lambda$) that can just remove a ground state electron from hydrogen—higher states of hydrogen have too low a population to contribute appreciably to photoionization. The central source may be a star (or several stars) illuminating the surrounding interstellar gas to give an H II *region*, where II refers to ionized hydrogen and I to neutral hydrogen. Only stars of spectral types O and B have high enough surface temperatures to emit appreciable numbers of short wavelength photons and hence to give rise to H II regions. Another stellar example is that of low mass stars shedding their outer envelopes at the ends of their nuclear burning lifetimes to give expanding shells called *planetary nebulae*. Here the central stars have high surface temperatures before cooling off to become white dwarfs, and the ultraviolet photons from the star photoionize the shell. *Active galactic nuclei* (*AGN*) like those found in quasars, Seyfert galaxies and radio galaxies possess compact central sources of uncertain nature (possibly accretion discs around central black holes) which produce copious short wavelength continuum photons which photoionize any surrounding gas clouds out to several hundred parsecs from the central source.

The photoionizing photons not only ionize the gas, but also input energy into the gas in the form of the kinetic energy of the electrons released by ionization, and these contribute to the total heat energy of the nebula. The balance between energy input from photoionization and various cooling radiative processes (recombinations, lines) results in gaseous nebulae having a relatively small range of temperatures from 7000 K to 25 000 K, the higher temperature being found in some AGN clouds. Densities vary a good deal more, of course, but a typical value for a H II region might be $10^9\,m^{-3}$ with considerably lower values in diffuse H II regions and considerably higher values (up to $10^{16}\,m^{-3}$) in some AGN clouds.

The most significant differences between the photoionized nebulae produced by central sources of various kinds are in the states of ionization of the elements other than hydrogen. Helium has an ionization potential greater than that of hydrogen and so to produce a He II zone one requires photons of shorter wavelength (less than 50.4 nm) than those which suffice to ionize hydrogen. The removal of two electrons from helium requires photons of even shorter wavelengths (less than 22.8 nm), and only such photons can produce a He III zone. Now the continuous spectra of stars are fairly close to blackbody distributions, which fall away from peak intensity fairly sharply on the short wavelength side ($I_\nu \propto \nu^3 \exp(-h\nu/kT)$ for a blackbody at temperature T and frequency ν), and hence at wavelengths that are appreciably below that giving peak intensity there will be few photoionizing photons. Thus B stars are too cool to photoionize helium, but O stars have a He II zone, and some central stars of planetary nebulae are hot enough to have a He III zone. Active galactic nuclei, on the other hand, have relatively flat continuous spectra from their central sources with, in typical cases, $I_\nu \propto \nu^{-1}$ or $I_\nu = \text{constant}$, and so produce photons in some numbers at all wavelengths. The result is to produce a very wide range of ionization in the surrounding clouds.

In the case of AGN and very compact H II regions, the gas clouds are not resolved from the central source, and so observationally the spectrum of the gas clouds is seen superimposed on that of the central continuum source. As far as the continuous spectrum of the gas clouds themselves is concerned, gaseous nebulae are usually transparent in the continuum, at least in the visible and down to the Lyman limit at 91.2 nm. Continuum emission is produced by bound–free, free–free and two photon processes, but except at very short and very long wavelengths does not reach the blackbody level appropriate to the local continuum temperature, so that the lines appear as emission lines. The bound-free continuum from the cloud is produced by recombination—the opposite process to photoionization—and appears as a Balmer continuum shortwards of 365 nm due to recombinations to the $n = 2$ level of hydrogen, and as a Lyman continuum shortwards of 91.2 nm due to recombinations to the ground state, dominating the Balmer continuum at these wavelengths. The Lyman continuum may be optically thick. The two-photon process involves transitions from the $n = 2$, $l = 0$ level of hydrogen to the $n = 1$, $l = 0$ ground state. This is forbidden for a single photon (which would give a Lyman α line) because angular momentum could not be conserved, but is allowed if two photons are emitted. As the energy can be divided in any way between the two photons, the spectrum is continuous at all wavelengths above the wavelength of Lyman α, but peaks at about 370 nm. Free–free emission or bremsstrahlung increases in wavelength with the cube of the wavelength and so dominates the continuum at long wavelengths in the infrared and radio regions, and can become optically thick at radio wavelengths. It should be remembered that dust is often present in H II regions and AGN, giving rise to an infrared and millimetre continuum.

The main feature of gaseous nebulae is, however, the emission lines. In any discussion of these lines, the first point to realize is that the LTE relations of Saha and Boltzmann certainly do not apply in the low density conditions of nebulae. Atoms and ions spend most of their time in the ground state, and must be excited by some process before they can produce an emission line. The emission lines are divided into two main categories according to the nature of the excitation process: *recombination lines* and *collisionally excited lines.*

The recombination lines are produced when an ion and an electron recombine to an excited level in the resulting atom (or resulting lower stage of ionization). The excited level decays, often via other excited levels, until the ground state is reached. The set of downward decays from recombinations to all excited levels is called a *cascade*, and produces all the permitted emission lines of the ion or atom concerned (Figure 6.1). Recombinations proceed at a low rate because of the low density of the nebula, and so recombination lines are only detected for the most abundant elements. Hydrogen is the most abundant element of all and in a H II zone the hydrogen is almost completely ionized, so hydrogen recombination lines are bright. The same is true of the helium recombination lines from a He II zone and of the recombination lines of ionized helium from a He III zone. The only other recombination lines often observed are those from carbon which can sometimes be detected at radio wavelengths.

The other process that can put an atom or ion in an excited state from which it can decay radiatively, producing an emission line, is excitation from the ground state by a collision with another particle, nearly always an electron. The average kinetic energy of an electron in a gas with temperature T is $\sim kT$, where k is Boltzmann's constant. We have already seen that the collisional excitation rate for charged particles and an electron number density N_e is proportional to $N_e/T^{1/2}\exp(-E_{ij}/kT)$, where the excitation energy above the ground state is E_{ij}. Hence, to produce appreciable excitation, we require kT to be of the order of E_{ij}. In a nebula, we are normally dealing with ions, which have their energy levels farther apart than neutral atoms, and so have large values of E_{ij}. In many cases, to reach levels above the first excited state requires temperatures approaching 10^6 K, much higher than is normally found in a nebula. However, appreciable excitation of the first excited state above the ground state can often be produced at temperatures of the order of 10^4 K which are typical of nebulae. The resulting emission lines from the first excited levels to the ground states are called *resonance lines.* These lines usually lie in the ultraviolet (the energy difference involved is still quite large). Examples include the C IV doublet at 154.8 nm and 155.0 nm, the

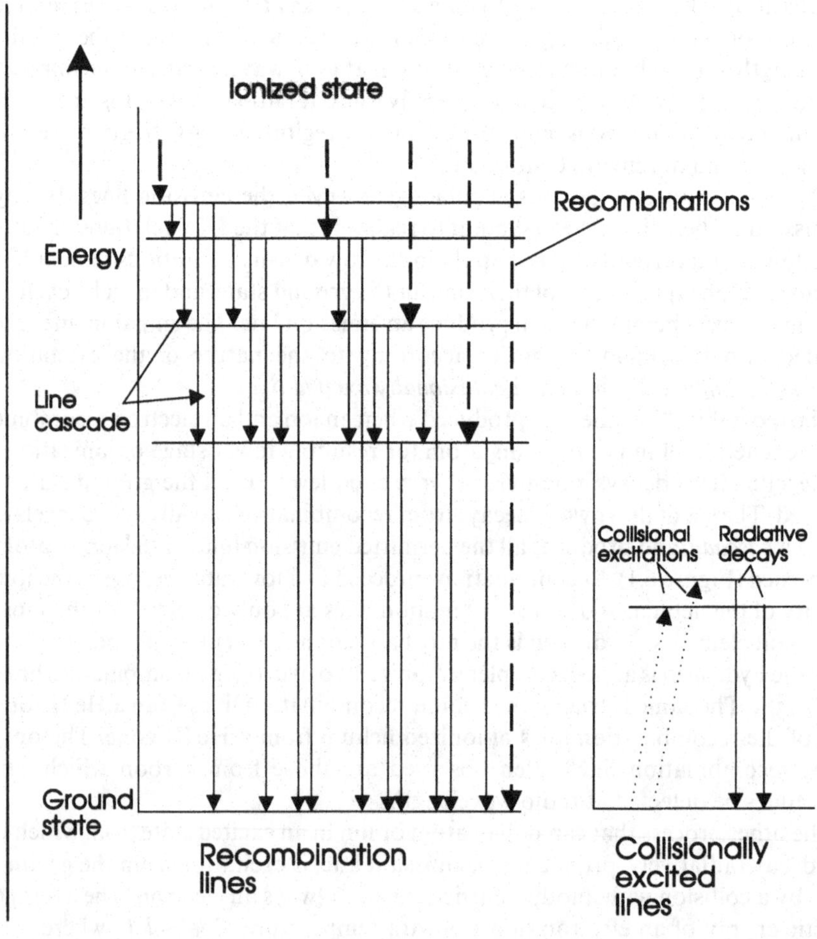

Figure 6.1.

Mg II doublet at 279.8 nm and 280.2 nm, and Lyman α at 121.6 nm, which is both collisionally excited and a recombination line.

Resonance lines are not the only collisionally excited lines observed. Before proceeding, however, it is necessary to be a little more precise about what is meant by first excited level and ground state. It will be recalled from Chapter 2 (p. 48) that the *configuration* of the electrons of an atom is the set of quantum numbers of the individual electrons, so the ground state of an O II ion which has an outer incomplete shell with 3 2p electrons, can be described as having the configuration $1s^2 2s^2 2p^3$, where the superscripts give the numbers of electrons. Only the outer electrons are involved in forming the spectra that we are concerned with here. Permitted radiative transitions involve a change in configuration. The photon emitted carries off angular momentum, and this requires at least one electron to change its angular momentum, giving the selection rule $\Delta l = \pm 1$. It will also be recalled that the electrons in the outermost shell interact to give a total orbital angular momentum with quantum number L, a total spin angular momentum with quantum number S, and a total angular momentum with quantum number J. A given configuration can arrange its angular momenta to give more than one value of L, S and J. Each set of values is called a *term*, which is written ^{2S+1}L. For instance, the ground configuration of O II gives rise to the terms ^{4}S and ^{2}D, with the latter lying higher in energy. Finally spin–orbit interaction splits the terms into fine-structure levels with different values of J, a level being written $^{2S+1}L_J$. The ^{2}D term of O II splits into levels $^2D_{5/2}$ and $^2D_{3/2}$ with slightly different energies, while the ^{4}S term is a singlet as $L = 0$ and $S = 3/2$ can only give $J = 3/2$. The selection rules are $\Delta L = 0, \pm 1, \Delta J = 0, \pm 1$ (with $J = 0$ to $J = 0$ forbidden) and $\Delta S = 0$.

The ground state configuration may give rise to more than one term. If one of the upper terms is excited, a radiative transition may be made to the ground term, but this will be a forbidden transition with a transition probability typically of the order of $0.02\,s^{-1}$, compared with a typical transition probability for a permitted transition of $10^9\,s^{-1}$. In the laboratory such an excited term will be collisionally de-excited long before there is time for a forbidden radiative decay, but under the low density conditions in a nebula where collisional de-excitations are rare, the forbidden line may be seen. Since the terms of the ground state are not very far apart in energy, collisional excitation of the upper terms of the ground state is relatively easy, and the subsequent forbidden downward transitions often lie in the visible. Examples include the $^2D \rightarrow {}^4S$ transition in O II which produces the doublet [O II] 372.6 nm and 372.9 nm . The square brackets are used to denote a forbidden line. Similarly, O III has the ground state configuration $1s^2\,2s^2\,2p^2$ and the two outer p electrons give rise to the terms ^{1}S , ^{1}D and ^{3}P in order of decreasing excitation energy. Fine-structure splitting divides the lowest term into levels $^3P_2$, $^3P_2$ and $^3P_0$ again in order of decreasing excitation energy. The transitions $^1D_2 \rightarrow {}^3P_2$ and $^3D_2 \rightarrow {}^3P_1$ produce the doublet [O III] 500.7 nm, 495.9 nm ($J = 2 \rightarrow J = 0$ is extra forbidden by the selection rule $\Delta J = 0, \pm 1$).

Transitions between fine-structure levels, that is transitions between levels with the same configurations and quantum numbers L and S and only differing in J, also give rise to forbidden lines, although with even smaller transition probabilities of order $10^{-4}\,s^{-1}$. Such transitions are very easily excited since the excitation energies are so small. The energy change is also small, so the wavelengths are long and lie in the infrared or far infrared. Examples are [O III] $^3P_2 \rightarrow {}^3P_1$ at 52 μm wavelength and $^3P_1 \rightarrow {}^3P_0$ at 88 μm.

Finally, there are transitions between terms belonging to different configurations, but having different values of S, thus breaking the selection rule $\Delta S = 0$.

This selection rule is a fairly weak one and we find transition probabilities of the order of $100\,s^{-1}$. Such *intercombination* transitions are called semi-forbidden. An example is the $1s^2\,2s^2\,{}^1S \rightarrow 1s^2\,2s\,2p\,{}^3P$ transition of twice ionized carbon, which gives the line C III] 190.9 nm, where the half bracket is used to designate a semi-forbidden line.

Forbidden lines are never strong enough for radiative transfer effects to be important—in these lines there is emission but no appreciable absorption. Such lines only appear bright because of the large emitting volumes. Recombination lines can be strong enough for absorption to be important and for the line to saturate, and it is then necessary to solve the radiative transfer problem. It should be noted that it is possible to ionize and excite a gas with shockwaves as opposed to ionizing photons, and a shocked nebula can give rise to a spectrum that looks at first sight rather like that from a photoionized nebula. However, when examined in detail, the relationship between ionization and excitation is quite different.

Ionization in Gaseous Nebulae

The interpretation of the spectra of gaseous nebulae requires a knowledge of the state of ionization of the observed elements in the nebula. Indeed in some ways determining the ionization structure in nebulae plays a similar role to the construction of model atmospheres in stellar studies. In this section attention will be largely confined to low density nebulae, leaving on one side the considerably more complicated situation in the high density clouds found in the broad line regions of active galactic nuclei. Collisional processes will be ignored for the moment under the low density conditions considered here. We will first consider the case of a pure hydrogen nebula, obtaining the radius of the region of ionized gas and the degree of ionization, and then go on to consider more briefly the ionization of other elements.

It will be assumed that the photoionized region ends when all the photons capable of photoionizing have been used up. Such a nebula is said to be *photoionization bounded*, which is the usual case. Sometimes, however, as one moves away from the central source, one comes to the end of the gas before all the photons have been used up, in which case the nebula is said to be *density bounded*. As the distance from the central source increases, the flux of photons capable of producing ionization will fall, and so will the degree of ionization. Unionized atoms will absorb photons in being ionized and so will reduce the flux still further. These two effects will determine the extent of the ionized region. Only hydrogen and helium have high enough abundances in most nebulae to control the structure of the ionized region, and only hydrogen will be considered in detail here. The ionization balance for other elements is determined by their ionization potentials, which decide over what wavelength range most photoionizations will take place, and by the spectrum of the central source, which decides at what wavelengths most photons are available.

In detail, the degree of ionization is determined by the balance between photoionizations and recombinations, as in Stromgren's equation, (2.48) of Chapter 2. Writing 'atom' for the lower stage of ionization, and 'ion' for the higher stage of ionization:

$$N_{\text{atom}} \int_{\nu_0}^{\infty} \frac{4\pi J_\nu}{h\nu} a_\nu \, d\nu = \alpha_{\text{rec}} N_{\text{ion}} N_e \tag{6.1}$$

where a_ν is the photoionization cross-section, α_{rec} is the recombination coefficient, J_ν is the mean intensity at frequency ν, and N_e is the electron number density. The threshold frequency for photoionization is $\nu_0 = I/h$ for ionization potential I, and the assumption has been made that the populations of excited levels are so low that only photoionizations from the ground state need to be considered, although α_{rec} represents recombinations to all levels. Often only two stages of ionization of a particular element are significantly populated in any selected region of the nebula, so $N_{\text{ion}} + N_{\text{atom}} = N_{\text{TOT}}$, where N_{TOT} is the total number density of particles of the element concerned and the degree of ionization $z_i = N_{\text{ion}}/N_{\text{TOT}}$ is given by

$$z_i = \frac{\dfrac{N_{\text{ion}}}{N_{\text{atom}}}}{1 + \dfrac{N_{\text{ion}}}{N_{\text{atom}}}} \tag{6.2}$$

For the hydrogenic case, (2.17) gives the photoionization cross-section as $2.8 * 10^{25} Z^4/\nu^3$ for $n = 1$, and (2.21b) gives the recombination coefficient to all levels as $\alpha_{\text{rec}} = 4 * 10^{-17} Z^2/T^{1/2}$, both of these formulae needing to be multiplied by Gaunt factors in more precise work.

The more difficult problem is to calculate the radiation field. Suppose we are dealing with a spherically symmetric situation with a single central source of photoionizing radiation with radius R and a blackbody spectrum characterized by temperature T_{rad}, so the flux from unit area $= \pi B_\nu(T_{\text{rad}})$ and the luminosity per unit frequency interval, L_ν, is given by $4\pi R^2 \pi B_\nu (T_{\text{rad}})$. At a large distance r from the source, the mean radiation field will drop from its value at the surface of the source, $J_\nu = B_\nu$, to $J_\nu = W B_\nu$, where W is the dilution factor given by $W = R^2/4r^2$ for $r \gg R$. The dilution factor means that the photoionization rate will drop with increasing r, whereas in a homogeneous nebula the recombination rate will stay constant, so the degree of ionization will fall outwards.

The radiation field will also decrease with r as photons are absorbed, in a dust-free nebula because they are producing photoionizations. Then $J_\nu = B_\nu(T_{\text{rad}})\exp(-\tau_\nu)$, where τ_ν is the optical depth from the central source to distance r. For the time being it will be assumed that we are dealing with a pure hydrogen nebula. Then

$$\tau_\nu = \int a_\nu N(\mathrm{H}^0)\,\mathrm{d}r = \frac{2.4 \times 10^{25}}{\nu^3}\int N(\mathrm{H}^0)\,\mathrm{d}r$$

$$= 215\frac{N(\mathrm{H}^0)}{\mathrm{N_{TOT}}}\frac{N_{\mathrm{TOT}}}{10^7} \text{ per parsec} \tag{6.3}$$

where (6.3) is evaluated at the threshold frequency for hydrogen, $\nu_0 = 3.28 * 10^{15}\,\mathrm{s}^{-1}$, the Gaunt factor has been allowed for, and $N(\mathrm{H}^0)$ is the number density of neutral hydrogen.

It is also necessary to take into account the fact that recombinations to the ground state produce photons that can themselves photoionize from the ground state, and so give rise to an extra component of the radiation field called the *diffuse* component. When such photons are absorbed, the net result would be the same as if the original recombination had never taken place, were it not for the fact that photoionization and recombination may occur in different parts of the nebula. In the 'on the spot' approximation this spatial separation of the processes is ignored, and it is assumed that the photon is absorbed where it is produced, so that recombinations to the ground state and the effects of the diffuse radiation field cancel each other out and can be ignored to a first approximation. The recombination coefficient α'_{rec} for $n = 2$ and greater is then the appropriate one to use. The inclusion of the temperature dependence of the Gaunt factor then gives for temperatures around 10 000 K:

$$\alpha'_{\mathrm{rec}} = \frac{4.1 \times 10^{-16}}{T^{0.8}} \tag{6.4}$$

At large distances from the central source, the drop in ionization and corresponding increase in $N(\mathrm{H}^0)/N_{\mathrm{TOT}}$ will lead to a much greater increase in optical depth for each parsec travelled outwards (see (6.3)). The photoionizing photons will be rapidly used up, and the ionized region will come to a fairly abrupt end at some radius r_s. Writing $N(\mathrm{H}^0)/N_{\mathrm{TOT}} = 1/2$ at the transition point, (6.3) shows that at the Lyman limit frequency, and a typical N_{TOT} of $10^7\,\mathrm{m}^{-3}$, an optical depth of one is traversed in less than 0.01 parsec confirming that, under typical conditions with a blackbody photoionizing flux the transition from the largely ionized condition to the largely neutral condition takes place in less than 0.01 parsec so that the transition is a sharp one.

Within the radius r_s, all the photoionizing photons are used up and hence in a steady state one must have the condition that the total number of recombinations per second in the whole nebula is equal to the total number of photoionizations, which in turn must equal the total number of photoionizing photons, N_{pi}.

Hence

$$\alpha'_{\rm rec}\langle N({\rm H}^+)N_{\rm e}\rangle \tfrac{4}{3}\pi r_{\rm s}^3 = \int_{\nu_0}^{\infty} \frac{L_\nu}{h\nu}\,{\rm d}\nu = N_{\rm pi}$$

For a pure hydrogen nebula $N({\rm H}^+) = N_{\rm e}$:

$$r_{\rm s}\langle N({\rm H}^+)\rangle^{2/3} = \left[\frac{N_{\rm pi}}{\tfrac{4}{3}\pi\alpha'_{\rm rec}}\right]^{1/3} \tag{6.5}$$

It will be shown shortly that the ionization is high over the whole nebula except near the edge so $N({\rm H}^+) = N_{\rm TOT}$. Hence

$$r_{\rm s} = 2.7 \times 10^{-12}\left(\frac{N_{\rm pi}}{N_{\rm TOT}^2}\right)^{1/3} T_{\rm K}^{0.27}\,\text{parsecs} \tag{6.6}$$

where $T_{\rm K}$ is the kinetic temperature of the nebula which is of the order of 10^4 K, and $r_{\rm s}$ is called the *Stromgren radius.*

If the spectrum of the central source is blackbody with $T_{\rm eff} = T_{\rm rad}$:

$$N_{\rm pi} = L\frac{\displaystyle\int_{\nu_0}^{\infty}\frac{L_\nu}{h\nu}\,{\rm d}\nu}{\displaystyle\int_{0}^{\infty}L_\nu\,{\rm d}\nu} = \frac{L}{kT_{\rm eff}}\frac{\displaystyle\int_{x_0}^{\infty}\frac{x^2}{{\rm e}^x-1}\,{\rm d}x}{\displaystyle\int_{0}^{\infty}\frac{x^3}{{\rm e}^x-1}\,{\rm d}x}$$

$$= \frac{15}{\pi^4}\frac{L}{kT_{\rm eff}}{\rm e}^{-x_0}(x_0^2 + 2x_0 + 2) \quad x_0 \gg 1 \tag{6.7}$$

with

$$x_0 = \frac{h\nu_0}{kT_{\rm eff}} = 158000/T_{\rm eff}, \text{ using} \int_0^{\infty}\frac{x^3}{{\rm e}^x-1}\,{\rm d}x = \frac{\pi^4}{15}$$

However, the ultraviolet continuous spectrum of hot stars deviates strongly from blackbody form because of the Lyman discontinuity, so the ultraviolet flux is much less than that of a blackbody with the given effective temperature and (6.7) overestimates the photoionizing flux by factors of up to 7.

A fit to the spectrum at high frequencies will give a local colour temperature $T_{\rm rad}$ and in terms of this:

$$N_{\rm pi} = 4\pi R^2\pi\int_{\nu_0}^{\infty}\frac{B_\nu(T_{\rm rad})}{h\nu}\,{\rm d}\nu = \frac{8\pi^2k^3}{c^2h^3}R^2T_{\rm rad}^3\int_{x_0}^{\infty}\frac{x^2}{{\rm e}^x-1}\,{\rm d}x$$

$$= 7.91 \times 10^{15}R^2T_{\rm rad}^3{\rm e}^{-x_0}(x_0^2 + 2x_0 + 2) \quad x_0 = \frac{h\nu_0}{kT_{\rm rad}} \gg 1 \tag{6.7a}$$

Model atmosphere predictions (Pannagia [1]) give the following values for the number of photoionizing photons and Stromgren radii of main sequence stars if

$N_{TOT} = 10^7\,m^{-3}$:

O5	$T_{eff} = 47\,000$ K	$N_{pi} = 5*10^{49}\,s^{-1}$	$r_s = 24$ parsecs
O7	$T_{eff} = 38\,500$ K	$N_{pi} = 7*10^{48}\,s^{-1}$	$r_s = 12$ parsecs
O9	$T_{eff} = 34\,500$ K	$N_{pi} = 2*10^{48}\,s^{-1}$	$r_s = 8$ parsecs
B1	$T_{eff} = 22\,600$ K	$N_{pi} = 3*10^{45}\,s^{-1}$	$r_s = 0.9$ parsecs

We now turn to the consideration of the degree of ionization within a H II region. From (6.1) we can express the degree of ionization of a pure hydrogen nebula in terms of a quantity f:

$$f = \frac{1}{N_{TOT}} \frac{N(H^+)^2}{N(H^0)} = \frac{4\pi}{N_{TOT}} \frac{\displaystyle\int_{\nu_0}^{\infty} \frac{a_\nu J_\nu e^{-\tau_\nu}}{h\nu} d\nu}{\alpha'_{rec}} \tag{6.8}$$

where τ_ν is the optical depth at frequency ν from the central source to distance r, as given by (6.3). From equation (6.2):

$$z_i^2 = N(H^+)^2/N_{TOT}^2 = f(1 - z_i) \tag{6.9}$$

so $z_i \sim 1 - 1/f$ if f is large. Let f' be the value of (6.8) if the absorption factor $\exp(-\tau_\nu)$ is omitted from the right-hand side.

First consider the inner part of the nebula. Here τ_ν is small, so let $\tau_\nu = 0$, and $f = f'$. Then

$$\begin{aligned} f &= \frac{1}{N_{TOT}} \frac{4\pi \times 2.5 \times 10^{25}}{4.1 \times 10^{-16}} T_K^{0.8} \frac{R^2}{4r^2} \int_{\nu_0}^{\infty} \frac{2}{c^2 \nu} \frac{1}{e^{h\nu/kT} - 1} d\nu \\ &= \frac{4.3 \times 10^{24} T_K^{0.8} R^2}{N_{TOT} r^2} \int_{x_0}^{\infty} \frac{1}{x(e^x - 1)} dx \end{aligned} \tag{6.10}$$

where the integral has the approximate value of the exponential integral $E_1(x_0)$ if $x_0 \gg 1$. For example, if $T_{rad} = 40\,000$, then $E_1(x_0) \sim 0.004$, so for $N_{TOT} = 10^7\,m^{-3}$, $T_K = 10^4$ K, $R = 10$ solar radii and $r = 1$ parsecs, we find $f = 10^5$.

In the outer part of the nebula the effects of optical depth have to be taken into account. A full solution requires a numerical integration, but an approximate analytical solution can be obtained by evaluating τ_ν at the threshold frequency where it is greatest to give τ_0, and removing the exponential in optical depth from the integral over frequency in (6.8). This will overestimate the effects of absorption but is a reasonable approximation since the greater part of the photoionizations take place at frequencies close to the threshold.

Let $f' = f_0/r^2$ so $f = f_0/r^2 \cdot \exp(-\tau_0)$. Then (6.9) becomes $1 - z_i = z_i^2/f_0 \cdot r^2 \cdot \exp(+\tau_0)$ and from (6.3):

$$\frac{d\tau_0}{dr} = 6.3*10^{-22} N(H^0) = 6.3 \times 10^{-22}(1 - z_i) N_{TOT}$$

$$-\frac{d(e^{-\tau_0})}{dr} = e^{-\tau_0} \frac{d\tau_0}{dr} = \frac{z_i^2 r^2}{f_0} 6.3 \times 10^{-22} N_{TOT}$$

Taking $z_i \sim 1$ and integrating:

$$(1 - e^{-\tau_0}) = \frac{r^3}{3f_0} N_{TOT} 6.3 \times 10^{-22}$$

$$(1 - z_i) \simeq \frac{\dfrac{r^2}{f_0}}{1 - \dfrac{r^3}{3f_0} N_{TOT} 6.3 \times 10^{-22}} \tag{6.11}$$

If dust is present with optical depth $\tau_D = k_D r$, then $f = f_0/r^2 \cdot \exp[-(\tau_0 + \tau_D)]$, so

$$\frac{d(e^{-\tau_0})}{dr} = \frac{z_i^2 r^2}{f_0} 6.3 \times 10^{-22} N_{TOT} e^{+\tau_D}$$

Hence if $z_i \simeq 1$

$$(1 - e^{-\tau_0}) \simeq \frac{N_{TOT}}{f_0} 6.3 \times 10^{-22} \int r^2 e^{k_D r} \, dr$$

which is easily evaluated to give $\exp(\tau_0)$ and hence $\exp(\tau_0 + \tau_D)$ and z_i.

We now turn to the ionization of helium, which will be present with 10–15% of the number density of hydrogen in most nebulae. The threshold frequency for the ionization of helium is $5.94 * 10^{15}$ s^{-1}, and photons at higher frequencies than this threshold can ionize both hydrogen and helium, whereas photons at lower frequencies but above the hydrogen threshold can ionize only hydrogen. The photoionization cross-section of helium at the helium threshold is about 15% higher than that of hydrogen at the hydrogen threshold, but is a further factor $(5.94/3.29)^3$ higher due to the ν^{-3} factor if the comparison is done for both elements at the helium threshold, so that at this and higher frequencies (where both cross-sections scale with the inverse cube of the frequency), the helium absorption coefficient would be about 0.7 times that of hydrogen for typical abundances and the same degree of ionization. Recombination coefficients for the two elements are similar.

If the central source produces very few photons with frequencies greater than $5.94 * 10^{15}$ s^{-1} compared with the number with frequencies greater than $3.29 * 10^{15}$ s^{-1} (as would be the case for a body with the spectrum of a B0 star), then the small flux of high frequency photons will lead to the degree of ionization of helium beginning to fall at a relatively small radius, the optical depth of helium becoming correspondingly large, and the flux of helium photoionizing photons dropping to zero. Lower frequency photons will keep hydrogen highly ionized to much farther out.

On the other hand, if the flux of photons with frequencies greater than $5.94 * 10^{15}$ s^{-1} is comparable with the flux of photons with frequencies greater than $3.29 * 10^{15}$ s^{-1} (as would be the case for a body with the spectrum of an early

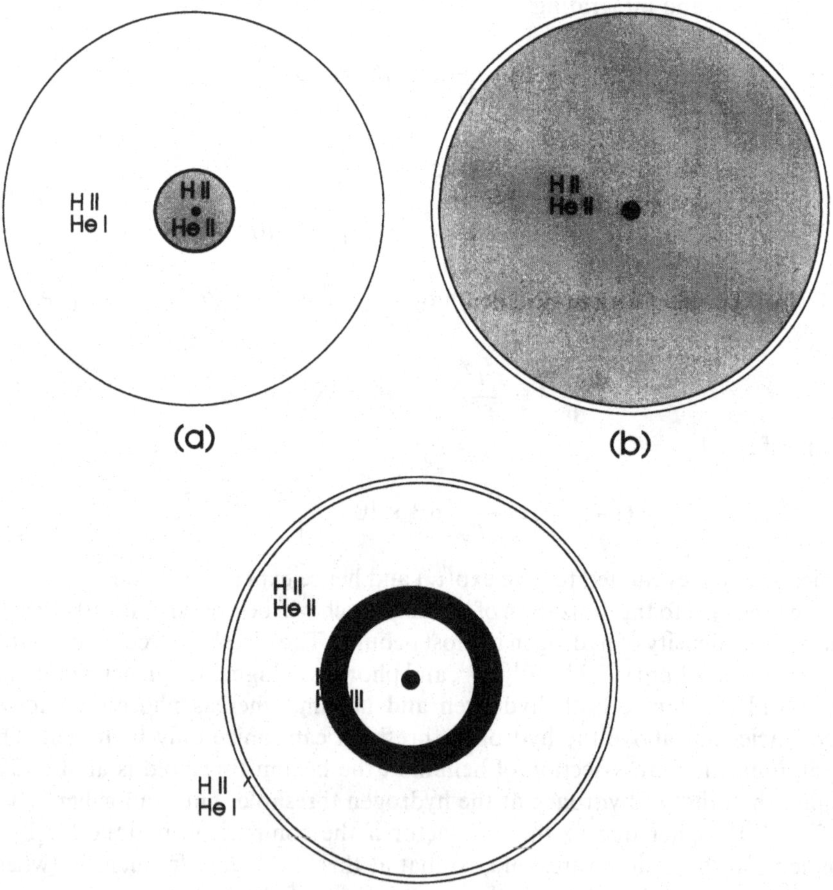

Figure 6.2. (a) B star 30 000 K; (b) O star 40 000 K; (c) planetary nebula 100 000 K

O star), then most photons are capable of ionizing both hydrogen and helium, and hence the ionization zones will end at nearly the same radius, with only a very small shell containing ionized hydrogen but neutral helium (Figure 6.2).

The presence of helium has little effect on the ionization of hydrogen. This is because recombinations to the ground state of helium produce a photon that can ionize hydrogen, and recombinations to excited states of helium cascade down to the ground state, emitting line photons, most of which can ionize hydrogen. Thus most absorptions of a potential hydrogen-ionizing photon by helium are followed by recombination and the production of another photon capable of ionizing hydrogen (see Osterbrock [2] for further discussion of this point).

The relative sizes of H II and He II zones can be approximately estimated by assuming that the presence of hydrogen absorption at frequencies greater than the helium threshold can be neglected in the inner nebula since hydrogen will be highly ionized there. Then

$$\tfrac{4}{3}\pi r_s^3(\text{He})\langle N(\text{He}^+)N_e\rangle\alpha'_{rec}(\text{He}) = N_{pi}[\nu > \nu_0(\text{He})]$$

$$\tfrac{4}{3}r_s^3(\text{H})\langle N(\text{H}^+)N_e\rangle\alpha'_{rec}(\text{H}) = N_{pi}[\nu > \nu_0(\text{H})]$$

In the H II zone, $N_e = N(\text{H}^+)$, but in the He II zone, $N_e = N(\text{H}^+) + N(\text{He}^+)$ so for hydrogen:

$$\langle N(\text{H}^+)N_e\rangle = N(\text{H}^+)^2\left[\frac{r_s(\text{H}) - r_s(\text{He})}{r_s(\text{H})}\right]^3 + N(\text{H}^+)[N(\text{H}^+) + N(\text{He}^+)]\left[\frac{r_s(\text{He})}{r_s(\text{H})}\right]^3$$

If $r_s(\text{H}) \gg r_s(\text{He})$:

$$\left[\frac{r_s(\text{H})}{r_s(\text{He})}\right]^3 \simeq \frac{N_{pi}[\nu > \nu_0(\text{H})]}{N_{pi}[\nu > \nu_0(\text{He})]}\frac{N_{TOT}(\text{He})\alpha'_{rec}(\text{He})}{N_{TOT}(\text{H})\alpha'_{rec}(\text{H})}\left(1 + \frac{N_{TOT}(\text{He})}{N_{TOT}(\text{H})}\right) \quad (6.12)$$

More detailed modelling shows that the hydrogen and helium Stromgren radii are nearly the same if T_{eff} is greater than or equal to 40 000 K, but at lower temperatures of the central source the He II zone is much smaller than the H II zone.

Doubly ionized helium, He III, is produced by ionizing He II which is a single electron hydrogenic ion, so the threshold frequency is 4 times that of hydrogen, $\nu_0(\text{He II}) = 1.32 * 10^{16}\ \text{s}^{-1}$, and similarly He II has a photoionization cross-section at threshold 16 times that of hydrogen. The recombination coefficient to He II is 4 times that to hydrogen. Recombinations of He III produce photons that can photoionize hydrogen and so help to maintain a high degree of ionization of hydrogen in the He III zone. As a result, absorption of radiation with $\nu > 3.29 * 10^{16}\ \text{s}^{-1}$ by hydrogen in the He III zone is negligible, and to a first approximation at $\nu > 1.3 * 10^{16}\ \text{s}^{-1}$ the radiation field falls off purely due to dilution and absorption by once-ionized helium. Thus the Stromgren radius of the He III zone is given by:

$$\tfrac{4}{3}\pi r_s^3(\text{He II})N_{TOT}(\text{He})^2\left[2 + \frac{N(\text{H})}{N(\text{He})}\right]\alpha'_{rec}(\text{He II}) = N_{pi}(\nu > 1.3 \times 10^{16}) \quad (6.13)$$

Models show that the He III region approaches the size of the H II and He II regions for sources with a stellar spectrum corresponding to a temperature greater than 100 000 K. Thus B stars have H II regions with only a small He II region, the hotter O stars have H II and He II regions of nearly the same size, and planetary nebulae (with still hotter central stars in some cases) can have an inner He III region (see Figure 6.2).

Finally, we consider the ionization of the heavier elements. The elements with abundances that are high enough to produce fairly strong lines from the thin gas of nebulae are oxygen, carbon, neon, nitrogen, magnesium, sulphur, silicon and argon, in decreasing order of abundance. The abundances of even carbon and oxygen are usually three orders of magnitude less than that of hydrogen, so where one of these elements is 50% ionized, the geometrical depth corresponding to optical depth one is much greater than for hydrogen. Thus where earlier we estimated the cut-off of hydrogen for a certain gas density would take place over 0.01 parsec, the same calculation for oxygen would give about 10 parsecs. The upshot is that the transitions between ionization zones for these heavier elements are not necessarily sharp.

The main consideration here is the ionization potentials of the various stages of these elements compared with those of hydrogen. An ionization stage for which the ionization potential of the next lower stage of ionization is less than that of hydrogen cannot exist beyond the edge of the H II region, because no photons capable of ionizing it will have escaped being absorbed in the H II region. If the ionization potential concerned is only a little higher than that of hydrogen, then since the photoionization cross-sections and recombination coefficients have the same order of magnitude as those for hydrogen, and hydrogen dominates the absorption at the wavelengths of importance, one would expect the ionization zone to come to an end very close to the edge of the H II region. Similar arguments apply to ionization potentials very close to those of He I and He II (see Table 6.1).

It will be noted that O I is very close in ionization potential to hydrogen so that one would expect the edge of the O II zone to coincide with the edge of the H II region, and similarly one would expect to find the edge of the N II zone not much farther in. On the other hand, C II, Mg II and Si II can exist outside of the H II region. Similarly one might expect the edges of the C III and S III zones to coincide with the edge of the He II region, the edge of the N III zone to be not much farther in, and the edge of the O IV zone to coincide with that of the He III region.

Table 6.1. Ionization potentials in electron-volts (1 eV $= 1.6*10^{-19}$ J)

H I	13.60								
He I	24.59	He II	54.42						
C I	11.26	C II	24.38	C III	47.89	C IV	64.49	C V	392.1
O I	13.62	O II	35.12	O III	54.93	O IV	77.41	O V	113.0
N I	14.53	N II	29.60	N III	47.45	N IV	77.47	N V	97.89
Ne I	21.56	Ne II	40.96	Ne III	63.45	Ne IV	97.11	Ne V	126.2
S I	10.36	S II	23.33	S III	34.83	S IV	47.30	S V	72.68
Ar I	15.76	Ar II	27.63	Ar III	40.74	Ar IV	59.81	Ar V	75.04
Mg I	7.65	Mg II	15.04	Mg III	80.14	Mg IV	109.31	Mg V	141.27
Si I	8.15	Si II	16.35	Si III	33.49	Si IV	45.14	Si V	166.8

Interpolation for other ions is not straightforward except in the most general terms. The photoionization cross-sections are complicated by the fact that we have so far assumed that photoionization proceeded from the ground state to the ground state of the next higher stage of ionization but for several of the heavier elements significant photoionization rates occur to excited levels of the next higher stage of ionization and originate from low lying excited levels of the lower stage of ionization. Each of these processes has a different threshold frequency.

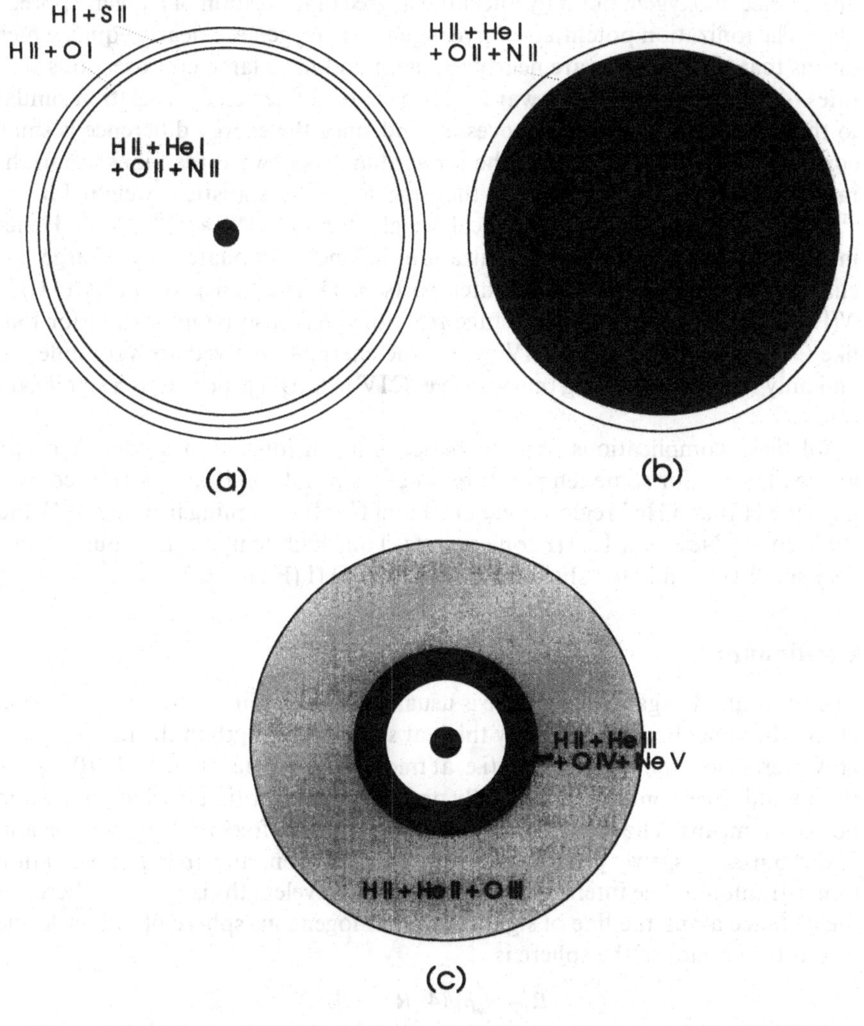

Figure 6.3. (a) B star 30 000 K; (b) O star 40 000 K; (c) planetary nebula 100 000 K

Furthermore, other processes than straightforward photoionization and recombination may influence the ionization balance. Recombination may take place via *dielectronic recombination*, involving two electron excited states and considerably increasing the recombination rates over those calculated assuming single electron processes. The process called *charge exchange* can be important in the ionization rates of some ions, particularly those of oxygen, where reactions like

$$H^+ + O^0 \rightarrow H^0 + O^+$$

can occur. This will not affect the ionization of hydrogen because of the low abundance of oxygen, but may affect the degree of ionization of oxygen appreciably. The ionization potentials of hydrogen and oxygen are nearly equal, which means that the reactions are nearly resonant and have large cross-sections. The rates of the forward and backward reactions are connected by a relation similar to those for other collisional processes, and since the energy difference is small compared with kT, the rates of the forward and backward reactions are in the ratio of the statistical weights of the products. The statistical weight for O I (^{3}P)$*H^+ = 9*1$, while the statistical weight for O II (^{4}S)$*$ $H^0 = 4*2$. Hence in full equilibrium with the ionization balance dominated by charge exchange (which happens in the outer parts of H II regions), N(O I)/N(O II) = $N(H^0)/N(H^+)*9/8$. Charge exchange reactions may also be important for ions like C IV, N IV, O IV and Ne IV, where the energies involved are very different and only the reaction going one way (e.g. C IV with H^0) proceeds at a significant rate.

All these complications need to be fed into an ionization model. A rough approximation for some temperature ranges is to take O II and N II as coming from the H II and He I regions, and O III and Ne III as coming from the H II and He II zones. Ne V is a He III zone species. Thus high temperature sources with very small H II and He I shells have very little O II (Figure 6.3).

Continuum

The continuum in gaseous nebulae is usually optically thin in the ultraviolet and visible, but may become optically thick at some wavelength in the radio region. However, below the Lyman limit (i.e. at frequencies greater than $3.29*10^{15}\,s^{-1}$), the 'bound–free' continuum is usually optically thick, for that is what 'ionization bounded' means. The processes involved are 'bound–free' recombinations and photoionizations, two photon emission by hydrogen, and free–free radiation (bremstrahlung). The intensity at optically thin wavelengths is $\int j_\nu \rho\, dz$, where z is the distance along the line of sight. For a homogeneous sphere of radius R, the flux at the surface of the sphere is

$$(4/3\pi R^3 4\pi j_\nu \rho)/(4\pi R^2) = 4\pi/3 j_\nu \rho R$$

At optically thick long wavelengths, $I = S_\nu = B_\nu(T_e)$ for the kinetic temperature

T_e, since the dominant process at these wavelengths, free–free, is a collisional one and therefore in LTE. This gives a flux at the surface of a homogeneous cloud of πB_ν. For intermediate cases and for a line of sight which at closest approach passes a distance p from the centre of the cloud:

$$I(p) = \int_0^\infty S(z) e^{-\tau(z)} \mathrm{d}\tau(z) \quad \text{(from (1.6))}$$
$$= S(1 - e^{-\tau_p}) \tag{6.20}$$

where the second line follows if the source function is constant and the total optical depth along the line of sight is τ_p. The flux received from a spherical cloud at distance d is given by (1.13):

$$F = \frac{1}{\mathrm{d}^2} \int_0^\infty I(p) 2\pi p \mathrm{d}p$$

The source function and the absorption coefficient will in general be functions of the distance r from the central source, and from Figure 6.4 we see that $r^2 = p^2 + z^2$ if $z = 0$ is taken at the point of closest approach to the centre, and $\mathrm{d}z = r\mathrm{d}r/\sqrt{(r^2 - p^2)}$. Hence:

$$\tau = \int \frac{\kappa\rho(r) r}{\sqrt{r^2 - p^2}} \mathrm{d}r$$
$$\tau_p = 2 \int_p^\infty \frac{\kappa\rho(r) r}{\sqrt{r^2 - p^2}} \mathrm{d}r \tag{6.21}$$

which can be used directly to find $I(p)$ if S is constant.

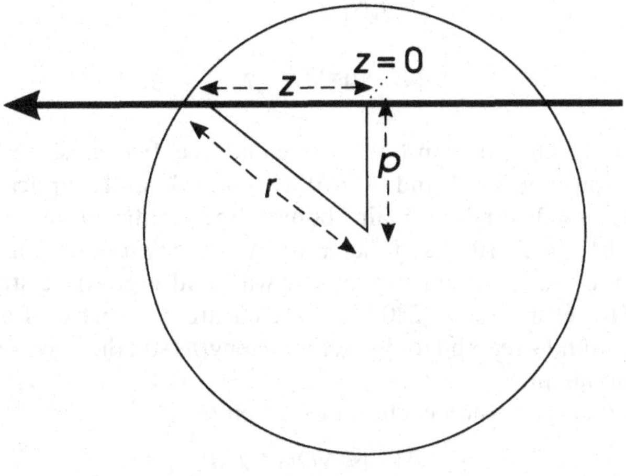

Figure 6.4.

The continuum in the visible and ultraviolet regions will be considered first. The emission coefficient for recombinations can be obtained from equation (2.20) for the recombination coefficient by taking the emission coefficient over a particular frequency range, $d\alpha_{rec}(\nu)/d\nu$, and assuming the frequency is high enough for the stimulated emission correction to be ignored:

$$\frac{d\alpha_{rec}}{d\nu} = 3.26 \times 10^{-12} \frac{Z^4}{n^3} \frac{1}{T^{3/2}} e^{\chi_i/kT} \frac{e^{-h\nu/kT}}{\nu}$$

where the recombinations are to the lower level with principal quantum number n, and the ionization energy is χ_i from that level. Hence

$$j_\nu \rho = \frac{1}{4\pi} \frac{d\alpha_{rec}}{d\nu} h\nu N_e N_{ion}$$

$$= 1.7 \times 10^{-46} \frac{Z^4}{n^3} \frac{N_e N_{ion}}{T^{3/2}} e^{\chi_i/kT} e^{-h\nu/kT} \quad (6.22)$$

which only applies for $\nu > \chi_i(n)/h$. Here N_e is the electron number density and N_{ion} is the number density of ions of net charge Z. At the limiting ν_0 for each n:

$$j_\nu \rho = 1.7 * 10^{-46} Z^4/n^3 N_e N_{ion}/T^{3/2}$$

with higher levels making much smaller contributions.

The free–free emission coefficient at optical wavelengths can be obtained ((2.22a)) from the absorption cross-section a_ν ((2.22)), for since the process is collisional and LTE holds:

$$j_\nu \rho = B_\nu \kappa_\nu \rho = B_\nu a_\nu (1 - e^{-h\nu/(kT)}) N_{ion}$$

$$= \frac{2h\nu^3}{c^2} a_\nu N_{ion} e^{-h\nu/(kT)}$$

$$= 5.44 \times 10^{-52} N_{ion} N_e \frac{Z^2}{T^{1/2}} e^{-h\nu/kT} \quad (6.23)$$

The main contributor to both bound–free and free–free emission in nebulae is hydrogen, so putting $Z = 1$ and $T = 10^4$ K, and taking the optical wavelength range where $n = 3$ for the main bound–free contribution, we find that $\exp[\chi_i/kT]/(n^3 T) \sim 2 * 10^{-5}$ and therefore conclude that bound–free dominates free–free in the visible, a conclusion that will hold even more strongly in the ultraviolet. However for $\lambda > 2280$ nm in the infrared, $n = 5$, free–free emission is greater than bound–free, and for longer wavelengths still the free–free emission is completely dominant.

The two-photon emission coefficient is given by

$$j_\nu \rho = 1/(4\pi) N(2s) A(2s) P_\nu 2h\nu$$

where $N(2s)$ is the population of the 2s level of hydrogen (the 2p level with the same energy can decay directly to the ground state as a Lyman α photon), $A(2s)$ is the transition probability for a two-photon decay to the ground state and has the value $8.23\,s^{-1}$, and P_ν is the probability that photon X, say, emitted in the two photon process, has frequency ν. The two photons X and Y must have frequencies ν_1 and ν_2 satisfying $\nu_1 + \nu_2 = \nu$ (Lyman α). P_ν is symmetrical about frequency $\nu = 0.5\nu$ (Lyman α) $= 1.23*10^{15}\,s^{-1}$, where it reaches a maximum of $5.27*10^{-16}$. If we take P_ν as the probability that $\nu_1 = \nu$ and hence that $\nu_2 = 2.47*10^{15} - \nu$, then there is an equal probability that $\nu_1 = 2.47*10^{15} - \nu$ and so $\nu_2 = \nu$. In other words, the probability that one or other photon has a frequency ν is $2P_\nu$.

Recombinations populate 2s directly and via downward cascades from higher levels that pass through 2s, and the rate of populating 2s is $0.3\,\alpha'_{rec}\,N_eN_{ion}$, where the recombination coefficient is for recombinations to all levels except $n = 1$, and the factor 0.3 approximately allows for the fact that not all these recombinations will eventually pass through 2s. Then

$$0.3\alpha'_{rec}N_eN_{ion} \sim N(2s)A(2s)$$

$$j_\nu\rho \sim 2h\nu/(4\pi)0.3\alpha'_{rec}N_eN_{ion}P_\nu$$

$$\sim 6.3*10^{-54}N_eN_{ion}\nu P_\nu \text{ at } P_\nu \text{ at } T = 10^4\,\text{K} \tag{6.24}$$

Just above the Balmer limit at $\nu = 8.22*10^{14}\,s^{-1}$, P_ν has the value $5.01*10^{-16}$, and it can be seen that the two-photon emission coefficient ($\sim 3*10^{-54}N_eN_{ion}\,J\,m^{-3}$) exceeds the typical bound–free emission coefficient at this wavelength, and does so up to a wavelength of about 480 nm. This will reduce the size of the Balmer jump in emission. However, at high densities ($> 10^{10}\,m^{-3}$), collisional transfers between 2s and 2p take place, and since the two-photon transition probability is many orders of magnitude less than that of Lyman α, the population of 2s will reach the ground state via Lyman α and the two photon process will become unimportant.

At long wavelengths, free–free emission is dominant. Full allowance now has to be made for stimulated emission, and also for the fact that the Gaunt factor G_{ff} is appreciably different from one. The absorption coefficient is

$$\kappa_\nu\rho \simeq a_\nu(1 - e^{-h\nu/kT})G_{ff}N_{ion}$$

$$\simeq a_\nu \frac{h\nu}{kT} N_{ion}G_{ff}, \quad \text{if } h\nu \ll kT$$

$$\simeq 1.77 \times 10^{-12}\frac{N_eN_{ion}}{\nu^2T^{3/2}}G_{ff} \tag{6.25}$$

where for a typical temperature of 10^4 K, the error in using the low frequency approximation becomes less than 10% for frequencies less than $3*10^{13}\,s^{-1}$, that is for wavelengths greater than $10\,\mu m$. The Gaunt factor is given by $G_{ff} = 0.5513\ln(T^{3/2}/\nu) + 9.75$ for hydrogen. Writing this in power law form

$G_{ff} \sim G_0 T^a \nu^b$, we find that since $dG_{ff}/d\nu = bG_{ff}/\nu = -0.5513/\nu$ and $dG_{ff}/dT = aG_{ff}/T = 0.5513 * 1.5/T$, and noting that at $T = 10^4$ K and $\nu = 10^9\ s^{-1}$ $G_{ff} = 5.95$, then $b = -0.093$ and $a = 0.139$. Thus in the vicinity of $T = 10^9$ K and $\nu = 10^9\ s^{-1}$ one can write

$$G_{ff} \sim 11.9 T^{0.15}/\nu^{0.1}$$

so

$$\kappa_\nu \rho \simeq \frac{2.1 \times 10^{-11}}{\nu^{2.1}} \frac{N_e N_{ion}}{T^{1.35}} \tag{6.26}$$

Hence a cloud of geometrical thickness R will become optically thick ($\tau_\nu = 1$) when

$$\nu = \frac{[N_e N(H^+) R]^{0.476} 8.27 \times 10^{-6}}{T^{0.643}}$$

$$\simeq 8 * 10^5 \left[\left(\frac{N_e}{10^6}\right)\left(\frac{N(H^+)}{10^6}\right)\left(\frac{R}{pc}\right)\right]^{0.48} \quad \text{at } T = 10^4\ \text{K} \tag{6.27}$$

The long wavelength free–free continuum from a point on the surface of the cloud can be described by the intensity (surface brightness) or equivalently by the brightness temperature. For a homogeneous cloud, $I_\nu = S_\nu(1 - \exp[-\tau_\nu])$. If $\tau_\nu \gg 1$, $I_\nu = S_\nu = B_\nu$ since free–free is a collisional process and so the source function takes on its LTE value. Hence

$$I_\nu = 2kT\nu^2/c^2, \quad \text{since } h\nu \ll kT \tag{6.28a}$$

The brightness temperature T_B, given by $B_\nu(T_B) = I_\nu$, for the optically thick case is then

$$T_B = T \tag{6.28b}$$

On the other hand if the cloud is optically thin with $\tau_\nu \ll 1$, $I_\nu = S_\nu \tau_\nu = B_\nu \tau_\nu$ so

$$I_\nu = \frac{2kT\nu^2}{c^2} \frac{2.1 \times 10^{-11} N_e N(H^+) R}{\nu^{2.1} T^{1.35}}$$

$$= \frac{6.4 \times 10^{-51}}{\nu^{0.1} T^{0.35}} N_e N(H^+) R \tag{6.29a}$$

and

$$T_B = \frac{2.1 \times 10^{-11}}{\nu^{2.1} T^{0.35}} N_e N(\mathrm{H}^+) R \tag{6.29b}$$

where it is worth noting that the intensity is almost independent of frequency.

Finally, at short wavelengths, the free–free continuum should certainly be optically thin and we can take the Gaunt factor as being close to 1, but can no longer assume that $h\nu \ll kT$. Then

$$I_\nu = B_\nu \tau_\nu = \frac{2h\nu^3}{c^2} \frac{1}{e^{h\nu/kT} - 1} a_\nu (1 - e^{-h\nu/kT}) N_e N(H^+) R$$

$$\simeq \frac{5.4 \times 10^{-52}}{T^{0.5}} N_e N(H^+) R e^{-h\nu/kT} \tag{6.30}$$

A typical free–free continuum is plotted in Figure 6.5.

The flux received from the whole cloud can also be measured. If the cloud is at distance d and observations are made at a frequency at which the cloud is optically thin, then

$$F_\nu = \frac{L_\nu}{4\pi d^2} = \frac{1}{4\pi d^2} \int 4\pi j_\nu \rho \, dV$$

$$= \frac{1}{d^2} \int \kappa_\nu \rho B_\nu \, dV$$

$$= \frac{6.4 \times 10^{-51}}{d^2 \nu^{0.1} T^{0.35}} \int N_e N(H^+) \, dV \tag{6.31}$$

using (6.26) or (6.29a) with $\kappa_\nu \rho = \tau_\nu / R$. It should be noted that radio astronomers often call flux 'flux density' and measure it in units of *Janskys*, where one Jansky $= 10^{-26}$ W m^{-2}. If the cloud is uniform and for simplicity we assume that helium is mainly neutral, $N_e = N(H^+) =$ constant, (6.31) can be inverted to give

$$N_e^2 = 1.6 * 10^{50} d^2 \nu^{0.1} T^{0.35} F_\nu / V \tag{6.31a}$$

where V is the volume of the cloud. If the cloud becomes optically thick at some observable frequency so a turnover between forms (6.28a) and (6.29a) is seen in the intensity spectrum, then observations of the brightness temperature at lower frequencies give the kinetic temperature via (6.28b), although measurement of an intensity or brightness temperature requires that the source is resolved. The turnover frequency where the cloud becomes optically thick, which can be observed in flux or intensity measurements, gives $N_e^2 R$ from (6.27). If spherical geometry is assumed, $V = 4/3\pi R^3$, and (6.31a) and $N_e^2 R$ from (6.27) then give N_e and R. Sometimes R can be found directly from the distance and the angular size. The mass for a pure hydrogen cloud is then $M = N_e m_H V$, which is easily corrected for the presence of helium so that 10% neutral helium gives the result $M = N_e V(m_H + 4m_H/9) = 13/9 m_H$. If the helium is ionized it will contribute both to the free–free emission and to the electron density, but allowance for this is straightforward provided one knows the relative extent of the He II, He III and H II zones.

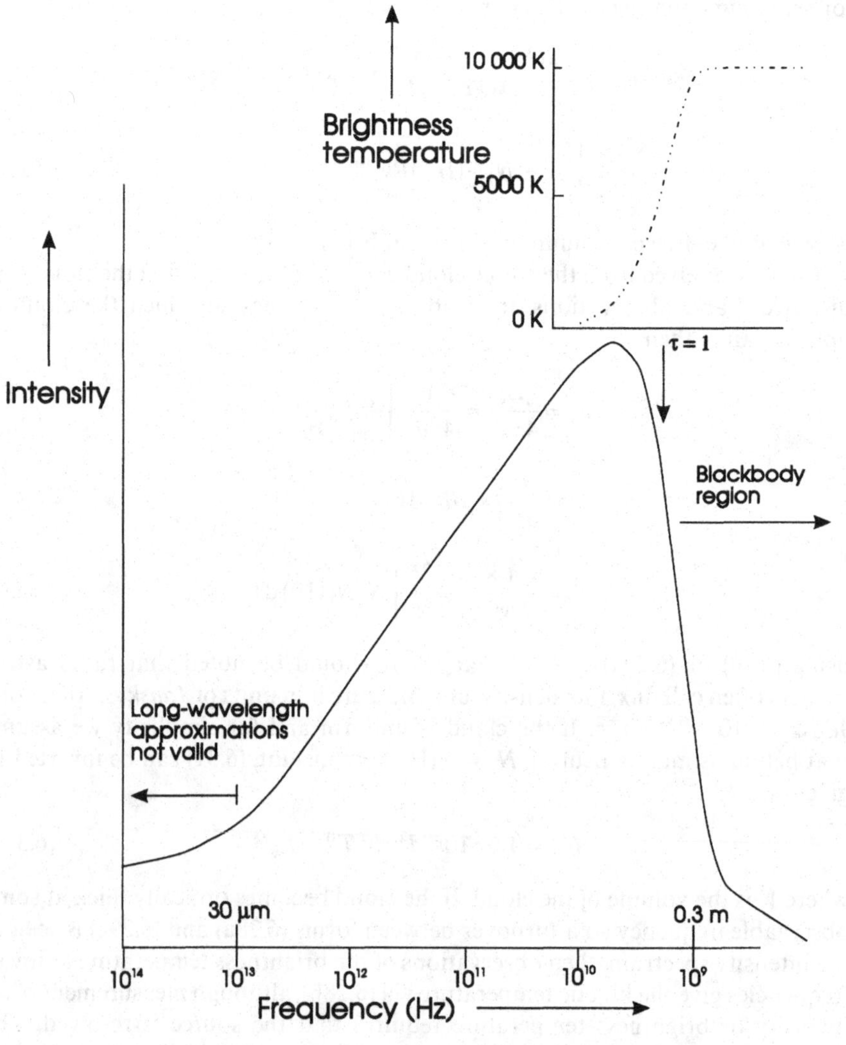

Figure 6.5. Free–free continuum spectrum of nebula: $N_e^2 R = 3{*}10^{35}\,\mathrm{m}^{-5}$

If the nebula is photoionization bounded, the entire flux capable of photoionizing hydrogen is used up in photoionizations, and the number of photoionizations per second equals the number of recombinations per second, where as usual recombinations to the ground state are ignored. Hence, in unit time:

$$\text{number of photons emitted with } \nu < \nu_0 = \text{number of recombinations in whole nebula} = 6.6{*}10^{34} d^2 \nu^{0.1} T^{-0.45} F_\nu\,\mathrm{s}^{-1} \qquad (6.32)$$

from (6.4) and (6.31), and where the photoionizing flux may come from more than one star. Relation (6.32) can be useful because it is by no means always obvious which stars are ionizing a particular nebula, nor what their spectral types are, particularly if the nebula is obscured by dust.

Major problems arise in determining the densities and masses of ionized hydrogen clouds because such clouds are not in general homogeneous. The emission is proportional to the square of the density so the denser parts of the cloud contribute disproportionately to the intensity received from the cloud, which leads to an overestimate of the average density. Only in a very few cases is information available on the density distribution in the cloud. The same problem arises in interpreting line strengths, which are again usually proportional to the square of the density.

Recombination Lines

In this section we first consider predictions of the relative strengths of recombination lines in the optical and the use of such lines in determining the volume of ionized gas and the abundances of hydrogen and helium. We then discuss the interpretation of radio wavelength recombination lines.

Consider a level j in the atom or ion that is the product of recombination. As was noted in the preceding section on ionization, recombinations take place at a rate per unit volume and unit time given by $\alpha_{\text{rec}}(j)N_{\text{e}}N_{\text{ion}}$, where N_{e} and N_{ion} are the number densities of electrons and recombining ions. The recombination coefficient α_{rec} is a weak function of temperature and is roughly proportional to $T^{-1/2}$, which follows from the fact that this is a collisional process between charged particles with a zero threshold energy. Since this is a collisional process it also follows that LTE rates should hold. For hydrogen or once ionized helium, $N_{\text{e}} = N_{\text{ion}}$. Level j will decay to lower levels i emitting recombination lines, and higher levels k will decay to j emitting other recombination lines, the transition probabilities being A_{ji} and A_{kj}, respectively (Figure 6.6). The statistical equilibrium balance is then

$$\alpha_{\text{rec}}(j)N_{\text{e}}N_{\text{ion}} + \sum_{k(k>j)} N_k A_{kj} = N_j \sum_{i(i<j)} A_{ji}, \quad j > 1 \tag{6.33}$$

where it has been assumed that photoionizations only take place from the ground state, and that there are no line absorptions. Starting from a high value of j, one can solve for N_j/N_{ion}, working one's way downwards through the set of equations (6.33). The values of N_j thus obtained can then be used to predict recombination line intensities:

$$I_\nu(ji) = \int j_\nu(ji)\rho\,\mathrm{d}x \quad \text{(along the line of sight)}$$

$$= \int A_{ji} N_j \frac{h\nu}{4\pi}\mathrm{d}x \tag{6.34}$$

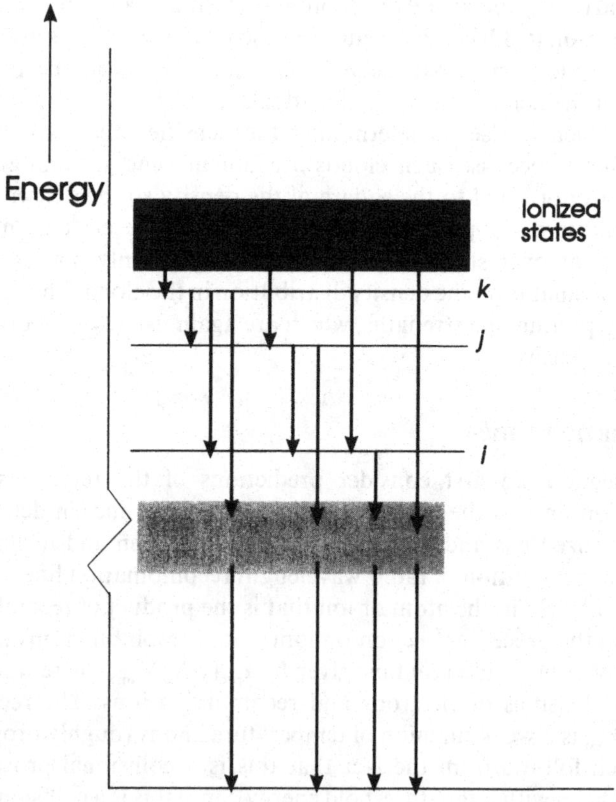

Figure 6.6.

Writing $\alpha''_{rec}(j) = \alpha_{rec}(j)T^{1/2}$ and $N'_j = N_j T^{1/2}/(N_e N_{ion})$, (6.33) becomes

$$\alpha''_{rec}(j) + \sum_{k(k>j)} N'_k A_{kj} = N'_j \sum_{i(i<j)} A_{ji} \tag{6.35}$$

where, since $\alpha''_{rec}(j)$ is nearly independent of N_e and T, so must be the solutions N'_j, and hence so must be the relative values of N_j and the line intensities.

This is not quite true, for the temperature dependence of the recombination coefficient is not exactly proportional to $T^{-1/2}$ and depends slightly on the level being considered, thus introducing a slight temperature dependence into the relative recombination line intensities. The levels of hydrogen are degenerate, each being composed of $n-1$ sublevels of different orbital angular momentum quantum number but the same energy. Radiative single photon transitions are only allowed for $\Delta l = 1$, so 3d can decay to 2p but not to 2s. However, collisions, particularly those with hydrogen ions, can cause transitions between degenerate levels, tending to drive the relative populations of the degenerate sub-levels to their equilibrium values, which are proportional to their statistical weights, $2l + 1$. The inclusion of collisions introduces an additional weak dependence on temperature, together with a weak dependence on density.

It is also necessary to examine more closely the assumption that the recombination lines are optically thin, and hence that absorption in these lines can be ignored. Usually, at least the Lyman series of hydrogen lines, with $n = 1$ as the lower level, is optically thick. This is certainly true if the Lyman continuum is optically thick, which is the same as the statement that the nebula is ionization bounded. We shall assume for the moment that the Balmer continuum and Balmer lines are optically thin.

Under these circumstances, every Lyman continuum photon from the central source that is intercepted by the nebula will be absorbed. The ions thus formed will recombine. If the recombination goes to the ground state ($n = 1$), a Lyman continuum photon will be produced, reborn only to be absorbed again, etc. If the recombination goes to the $n = 2$ excited level, a Balmer continuum photon will be produced and will escape, and the atom will then decay from $n = 2$ to $n = 1$, producing a Lyman α photon. The Lyman α photon will be scattered many times since the line is optically thick and the density is too low for collisional de-excitation, and will eventually escape from the edge of the nebula. If the recombination goes to the $n = 3$ level, a Paschen continuum photon will be produced, which will escape. There are two possibilities for the de-excitation of the $n = 3$ level. It can decay to $n = 2$, giving an Hα Balmer photon, which will escape, the $n = 2$ level then decaying to $n = 1$, producing Lyman α which will be scattered many times before escaping from the edge of the nebula as before. Alternatively, $n = 3$ can decay directly to $n = 1$, producing a Lyman β photon. The Lyman β line is optically thick, and the Lyman β photon will be scattered many times. However, at each scattering, there is a chance that $n = 3$ will decay via Hα and Lyman α, so that eventually all the Lyman β photons will be converted to Hα and Lyman α, and the end product of all recombinations to $n = 3$ is always this combination. Similar arguments hold for higher levels. We conclude that *every Lyman continuum photon eventually gives rise to a Lyman α and a Balmer line photon.*

It also follows that the cascade essentially ends at $n = 2$, for $n = 2$ to $n = 1$ emission is always followed by $n = 1$ to $n = 2$ absorption, and equation (6.33) can therefore be modified to take account of optical thickness in the Lyman lines by starting the sum over i at $i = 2$ instead of $i = 1$. This is called 'case B' hydrogen recombination. In the discussion that follows, we shall assume that the Balmer lines themselves are also optically thin, although this may well not be true in some AGN, and the consequences of this will be discussed later.

The results of recombination calculations for hydrogen line emission coefficients are as follows at $T = 10\,000$ K:

$$j(H\alpha)/j(H\beta) = 2.87$$
$$j(H\gamma)/j(H\beta) = 0.466$$
$$j(\mathrm{Ly}\alpha)/j(\mathrm{H}\beta) = 34 \text{ if Ly } \alpha \text{ is optically thick}$$

A change in temperature to 20 000 K only alters $j(\mathrm{H}\alpha)/j(\mathrm{H}\beta)$ to 2.69, and the effect of a doubling in electron density is smaller still. Thus the ratio of line intensities in a recombination cascade is essentially fixed irrespective of temperature and density. The ratio of Balmer line intensities,

$$I(\mathrm{H}\alpha):I(\mathrm{H}\beta):I(\mathrm{H}\gamma)$$

is called the *Balmer decrement*. Observationally, the Balmer decrement can differ from the theoretical value owing to the presence of dust. Dust absorbs blue light more than red light, so $I(\mathrm{H}\alpha):I(\mathrm{H}\beta)$ is increased by dust. Suppose the dust optical depth at the wavelength of Hβ, 486.1 nm, is $\tau_{\mathrm{D}}(\mathrm{H}\beta)$. The dust absorption coefficient is approximately proportional to the reciprocal of the wavelength, or if the normal observed reddening law is used, more exactly $\tau_{\mathrm{D}}(\mathrm{H}\alpha) = 0.65\ \tau_{\mathrm{D}}(\mathrm{H}\beta)$. Hence if the dust cloud lies between us and the gaseous nebula:

$$\frac{I(\mathrm{H}\alpha)}{I(\mathrm{H}\beta)} = \left[\frac{I(\mathrm{H}\alpha)}{I(\mathrm{H}\beta)}\right]_0 \frac{\mathrm{e}^{-\tau_{\mathrm{D}}(H\alpha)}}{\mathrm{e}^{-\tau_{\mathrm{D}}(H\beta)}}$$

where the subscript zero refers to the unreddened value. Hence

$$\frac{I(\mathrm{H}\alpha)}{I(\mathrm{H}\beta)} = \left[\frac{I(\mathrm{H}\alpha)}{I(\mathrm{H}\beta)}\right]_0 \mathrm{e}^{0.35\tau_{\mathrm{D}}(H\beta)} \tag{6.36}$$

However it is quite likely that the dust may be mixed with the gas of the nebula, and in this case the dust affects the line formation process instead of just acting as an external absorber. Suppose the line itself is optically thin with uniform line emission coefficient per unit volume $j_\nu\rho$, the uniform dust absorption coefficient per unit volume is $\kappa_{\mathrm{D}}\rho$, and the thickness of the cloud is R. Then:

$$I(\mathrm{Balmer}) = \int_0^R j_\nu \rho\, \mathrm{e}^{-t_{\mathrm{D}}} \mathrm{d}x$$
$$= \frac{j\rho}{\kappa_{\mathrm{D}}\rho} \int_0^{\tau_{\mathrm{D}}} \mathrm{e}^{-t_{\mathrm{D}}} \mathrm{d}t_{\mathrm{D}} \quad \text{where} \quad \mathrm{d}t_{\mathrm{D}} = \kappa_{\mathrm{D}}\rho\, \mathrm{d}x$$
$$= \frac{j\kappa_{\mathrm{D}} R}{\tau_{\mathrm{D}}}(1 - \mathrm{e}^{-\tau_{\mathrm{D}}})$$

$$\frac{I(\mathrm{H}\alpha)}{I(\mathrm{H}\beta)} = \frac{j(\mathrm{H}\alpha)\rho R\, \tau_{\mathrm{D}}(\mathrm{H}\beta)(1 - \mathrm{e}^{-\tau_{\mathrm{D}}(\mathrm{H}\alpha)})}{j(\mathrm{H}\beta)\rho R\, \tau_{\mathrm{D}}(\mathrm{H}\alpha)(1 - \mathrm{e}^{-\tau_{\mathrm{D}}(\mathrm{H}\beta)})}$$
$$= \left[\frac{I(\mathrm{H}\alpha)}{I(\mathrm{H}\beta)}\right]_0 \frac{1}{0.65} \frac{(1 - \mathrm{e}^{-0.65\tau_{\mathrm{D}}(\mathrm{H}\beta)})}{(1 - \mathrm{e}^{-\tau_{\mathrm{D}}(\mathrm{H}\beta)})} \tag{6.37}$$

Equation (6.37) predicts a different Balmer decrement from (6.36) for a given dust optical depth. By using two line ratios (say Hα:Hβ and Hβ:Hγ), it is in principle possible to distinguish between internal dust (mixed with the gas) and external dust.

Recombination lines can also be used to estimate the volume of the ionized cloud, to find the relative abundances of hydrogen and helium, and to determine something about the slope of the photoionizing continuum from the central source. All these determinations assume that the nebula is ionization bounded—that is, that going outwards from the central source one reaches a point where all the ionizing photons have been used up before one has run out of gas to photoionize. The assumption that a nebula is ionization bounded can be observationally confirmed if spectral lines from species existing on the edge of an I II region are seen, such as those of O I or S II.

The volume of ionized gas can be determined as follows. The total number of recombinations equals the total number of ionizing photons. We know that a fraction f_β of all Balmer recombination photons are Hβ photons (from the Balmer decrement). Hence if the volume of gas is V:

$$\alpha_{\text{rec}} N_e N_{\text{ion}} V = (\text{number of H}\beta \text{ photons})/\,f_\beta$$

Now $N_e \sim N_{\text{ion}}$ since the gas is mainly hydrogen and ionized, so

$$V = (\text{number of H}\beta \text{ photons})/(\alpha'_{\text{rec}} f_\beta N_e^2)$$

The flux $F(\text{H}\beta)$ of Hβ photons equals the Hβ luminosity, $L(\text{H}\beta)$, divided by $4\pi d^2$, where d is the distance of the cloud. $L(\text{H}\beta)$ is the number of Hβ photons produced in the whole cloud per second multiplied by the energy of an Hβ photon, $h\nu(\text{H}\beta) = hc/\lambda(\text{H}\beta)$. Hence

$$V = \frac{F(\text{H}\beta) 4\pi d^2 \lambda(\text{H}\beta)}{hc\alpha'_{\text{rec}} f_\beta N_e^2} \tag{6.38}$$

The electron density can be estimated from the forbidden line intensities as we shall see later. Hβ is the line most often used for volume estimates of this sort, but in principle any other Balmer line could be employed.

Xanstra's method uses recombination lines to estimate the slope of the photoionizing continuum. The short wavelength part of this continuum often cannot be observed directly. The number of photoionizing photons emitted per second is the monochromatic continuum luminosity L_ν divided by the photon energy and integrated from the threshold frequency ν_0 (the Lyman jump for hydrogen) to infinity. Each such photoionizing photon gives rise to a Balmer line photon, and hence on average to f_β Hβ photons. Hence

$$\frac{L(\mathrm{H}\beta)}{h\nu_\beta}=f_\beta\int_{\nu_0}^{\infty}\frac{L_\nu}{h\nu}\mathrm{d}\nu$$

$$\frac{L(\text{visible})}{\int_{\nu_0}^{\infty}\frac{L_\nu}{h\nu}\mathrm{d}\nu}=\frac{L(\text{visible})}{\frac{L(\mathrm{H}\beta)}{h\nu_\beta}}\frac{\frac{L(\mathrm{H}\beta)}{h\nu_\beta}}{\int_{\nu_0}^{\infty}\frac{L_\nu}{h\nu}\mathrm{d}\nu}$$

$$=\frac{F(\text{visible})}{F(\mathrm{H}\beta)}h\nu_\beta f_\beta \tag{6.39}$$

where L (visible) and F (visible) are the continuum outputs in some visible waveband, say the B or the V band, and the ratio of luminosities on the right-hand side has been replaced by a ratio of fluxes, which can be observed. The left-hand side of the equation represents the ratio of continuum in a visible band to that at photoionizing (and often inaccessible) wavelengths, less than 91.2 nm for hydrogen, for example.

Xanstra's method can also be applied to helium. The recombination lines of He II come from the He III zone. He II is a single electron ion that has a spectrum very like that of hydrogen, except that the energy levels for the same quantum number n have 4 times the energy of those of hydrogen because the square of the nuclear charge appears in the expression for energy. Alternate lines of He II therefore coincide with those of hydrogen, e.g. $n=4$ to $n=2$ of He II coincides with $n=2$ to $n=1$ of hydrogen, but the other lines of He II like $n=4$ to $n=3$, 468.6 nm, do not coincide and hence can be used observationally. From (6.39):

$$\frac{L(\text{visible})}{\int_{\nu_1}^{\infty}\frac{L_\nu}{h\nu}\mathrm{d}\nu}=\frac{\text{visible flux}}{468.6\ \text{nm flux}}h\nu_{468.6}f_{468.6} \tag{6.40}$$

where $f_{468.6}$ is the fraction of the He II cascade that goes through 468.6 nm and ν_1 is the threshold for double ionization of helium (corresponding to a wavelength of 304 nm). One is therefore effectively measuring the continuum at wavelengths less than 304 nm, and the observed visible, 486.1 nm and 468.6 nm fluxes therefore give three widely separated points in the continuum.

The recombination lines can also be used to obtain the helium to hydrogen abundance ratio in a gaseous nebula. Firstly, one can obtain the ratio of once ionized helium to once ionized hydrogen by using the known fraction f_β of all hydrogen recombinations in an H II region passing through Hβ, and the similarly calculated fraction $f_{587.6}$ of all helium recombinations in the He II zone passing through a particular He I line, usually 587.6 nm. Then

$$\frac{\text{flux(He I 587.6)}}{\text{flux(H I 486.1)}}=\frac{f_{587.6}h\nu_{587.6}}{f_{486.1}h\nu_{486.1}}\frac{\alpha_{\text{rec}}(\text{He})}{\alpha_{\text{rec}}(\text{H})}\frac{N(\text{He})}{N(\text{H})}\frac{V(\text{He II})}{V(\text{H II})} \tag{6.41}$$

For O stars, planetary nebulae and AGN, the H II zone and He II zone outer edges are coincident. For O stars there is little He III so $V(\text{He II}) = V(\text{H II})$. For the planetary nebulae and AGN we have to write $V(\text{He II}) + V(\text{He III}) = V(\text{H II})$. The relative volumes of He II and He III can be found from the relative intensities of the two cascades:

$$\frac{\text{flux(He II 468.6)}}{\text{flux(He I 587.6)}} = \frac{f_{468.6}}{f_{587.6}} \frac{\alpha_{\text{rec}}(\text{He II})}{\alpha_{\text{rec}}(\text{He I})} \frac{h\nu_{468.6}}{h\nu_{587.6}} \frac{V(\text{He III})}{V(\text{He II})} \tag{6.42}$$

giving $V(\text{He III})/V(\text{He II})$ with $V(\text{H II}) = V(\text{He II})[1 + V(\text{He III}/)V(\text{He II})]$.

Finally, for B stars, there will, of course be no He III, but the helium ionized zone will be substantially smaller than the hydrogen ionized zone. However, we can use the fact that O III comes from the He II zone, whereas O II comes from the H II + He I zone. The relative strengths of [O III] and [O II] lines enable $V(\text{O III})/V(\text{O II})$ and hence $V(\text{He II})/V(\text{He I})$ to be estimated, and so using $V(\text{H II}) = V(\text{He I}) + V(\text{He II}) = V(\text{He II})[1 + V(\text{He I})/V(\text{He II})]$ one can find $N(\text{He})/N(\text{H})$.

We now turn to *radio recombination lines* between levels with high values of the quantum number $n(\sim 100)$. The transition probabilities are small, as are the populations of the levels involved, so the energy emitted in one of these lines is much less than that from a visible recombination transition between small values of n. However the radio recombination lines, seen as emission lines on top of free–free continuum, are relatively easily detected, and have the important advantage of being unaffected by interstellar dust.

The energy of level n in a single electron atom or ion with atomic number Z is given by

$$E_n = -\frac{2\pi^2 m e^4 Z^2}{h^2 n^2 (4\pi\varepsilon_0)^2} = -\frac{2.179 \times 10^{-18} Z^2}{n^2} J$$

where m is the mass of an electron to a first approximation. However, m is strictly the reduced mass of the electron–nucleus system:

$$m = \frac{m_e m(\text{nucleus})}{m_e + m(\text{nucleus})}$$

$$= m_e\left(1 - \frac{m_e}{m(\text{nucleus})}\right)$$

taking the first term of a power expansion with $m_e/m(\text{nucleus}) \ll 1$.
Thus

$$m = m_e\left(1 - \frac{5.486 \times 10^{-4}}{A}\right)$$

where A is the mass number of the nucleus. The frequency of a transition between upper level n_j and lower level n_i is $\nu = E_{ij}/h$ with $E_{ji} = E_j - E_i$, which becomes

$$\nu = 3.289842 \times 10^{15} Z^2 \left[1 - \frac{5.486 \times 10^{-4}}{A}\right]\left[\frac{1}{n_i^2} - \frac{1}{n_j^2}\right] \tag{6.43}$$

At large n:

$$\nu \simeq 6.580 \times 10^{15} Z^2 \left[1 - \frac{5.486 \times 10^{-4}}{A}\right]\frac{\Delta n}{n_i^3} \tag{6.43a}$$

and

$$E_{ji} \simeq 4.358 \times 10^{-18} Z^2 \frac{\Delta n}{n_i^3} \tag{6.44}$$

where $n_j = n_i + \Delta n$ and $\Delta n \ll n$, and in (6.44) the mass correction has been ignored. These formulae should apply to all single electron ions when the appropriate Z and A are inserted. In the particular case of He II, transitions between even values of n coincide with transitions of hydrogen if the small mass correction factor is ignored. Lines are labelled with a number and a Greek letter where the number refers to n_i and the Greek letter to Δn with $\Delta n = 1$ as α, $\Delta n = 2$ as β and so on, so the hydrogen transition from $n_j = 101$ to $n_i = 100$ is called H100α and according to (6.43) has a frequency of 6.478 GHz. Around a particular frequency one will see transitions with different values of Δn, but those of larger Δn will also have larger n_i. In many electron ions and atoms, an electron with large n will orbit at a much larger radius than the other Y electrons, and these Y electrons will be very effective in shielding the nucleus, so the outer large n electron will move in the field of a nucleus with charge $Z - Y$. Thus neutral atoms with $Y = Z - 1$ will have lines at the same frequencies as lines of the same n and Δn of hydrogen, except that the larger mass of the nucleus will shift the lines of heavier elements to slightly higher frequencies. Lines of helium (often), carbon (sometimes) and sulphur (occasionally) have been detected as satellites to hydrogen lines in this way.

Even at quantum numbers around 20, frequencies are very close to hydrogenic values for all elements, and lines from neutral elements will nearly coincide for the same n. Lines from once ionized elements will nearly coincide for the same n with each other but of course will be at quite different frequencies from those of neutral elements because of the Z^2 factor, and so on for higher stages of ionization. Lines of carbon, oxygen and nitrogen ions have been detected in the infrared originating in the thick winds of Wolf–Rayet stars. Of course by the time that n has come down to values of the order of 10 the shielding of the outer electron is becoming incomplete so the effective charge on the nucleus is larger as are the binding energies. It then becomes necessary to consider electron–electron interactions.

It has so far been assumed that recombination cascades can be calculated considering only radiative transitions, with the effect of collisions being solely to transfer atoms between sub-levels of the same n but different l, which introduces only small corrections to the radiative level populations. However, collisions can play a significant role in populating the high n levels, so it is necessary to consider how the relative collisional and radiative rates can be calculated.

The transition probability in the hydrogenic case is given by (2.13a)

$$A_{ji} = \frac{1.575 \times 10^{10} Z^4}{\left[\frac{1}{n_i^2} - \frac{1}{n_j^2}\right]} \frac{1}{n_i^3 n_j^5}$$

For large n so $\Delta n \ll n$ we have

$$A_{ji} = \frac{1.575 \times 10^{10} Z^4}{2 n_j^5 \Delta n} \tag{6.45}$$

This must be multiplied by a Gaunt factor in accurate work. Clearly transitions with $\Delta n = 1$ are the strongest of those with small Δn. For $n \sim 50$ to 100, the total downward radiative transition probability to all lower levels is about one order of magnitude greater than the value given by (6.45) for $\Delta n = 1$ alone. However, collisions between levels with different values of n are of increasing importance as n increases, because the 'size' of an atom grows as n^2 so that one would expect the collisional cross-section to grow with something like the fifth power of n. Collisional excitation rates include a factor $\exp(-E_{ij}/kT)$ so for a typical value for kT of 10^{-19} J at $T = 10^4$ K, $E_{ij} \ll kT$ for high n transitions and the exponential factor will be of order 1. Since A_{ji} is inversely proportional to the fourth power of n, $C_{ji}/A_{ji} \propto n^9$ and collisional excitation must dominate at sufficiently high n.

Similar considerations apply to collisional ionization (see Chapter 2), where $\exp(-I_n/kT)$ is also of the order of 1 since the ionization energy from high n levels is much less than kT. Hence we must consider the possibility of collisional ionization. Collisional excitation and ionization will force high levels to have LTE populations.

There are a number of different formulae for collisional excitation rates. If one takes (2.29) (there is some evidence that this equation may be applicable exceptionally to neutral hydrogen), with f converted to A_{ji} using (2.10), and if (6.44) is substituted for E_{ij}, one obtains:

$$C_{ji} \simeq \frac{4 \times 10^{-22} n^9}{(\Delta n)^3} N_e \frac{A_{ji}}{T^{1/2}} \tag{6.46}$$

where the constant is multiplied by 0.27 to give Sareph's strong coupling formula. Starting from (2.30a) one obtains:

$$C_{ji} \simeq \frac{4 \times 10^{-26} n^{11}}{(\Delta n)^{3.68}} N_e A_{ji} T^{0.18} \tag{6.47}$$

while Goldberg uses $C_{ji} = 8.4 * 10^{-14} n^4 \ln_e(0.047n) N_e$ for $\Delta n = 1$. These various formulae give rather different values, but because of the steep dependence on n, we find that for $T = 10^4$ K, $N_e = 10^{10}\,\mathrm{m}^{-3}$ they all predict for $\Delta n = 1$ that

$C_{ji} \sim A_{ji}$ for n in the range 20 to 30, and that $C_{ij} = \sum A_{ji}$ for n in the range 30 to 40. Thus for this electron density, the relative values of the level populations begin to approach those given by Boltzmann's equation for $n > 40$. At higher values of n, collisional ionization will bring the bound level populations into a Saha relation with the population of the ionized state.

It is customary, therefore, for large n to use LTE (Boltzmann and Saha) populations with departure coefficients $b_n = N_n(\text{actual})/N_n(\text{Boltzmann})$ to express deviations from LTE, with $b = 1$ assumed for the ionized state. Calculation of the b_n requires taking into consideration for each level of all the collisional transitions into and out of the level, of all radiative transitions downward from the level, and of all radiative transitions leading into the level from above. For example the calculations of Brocklehurst [3] show that at $N_e = 10^8\,\mathrm{m}^{-3}$, b_n reaches 0.9 for $n \sim 105$, and at $N_e = 10^{10}\,\mathrm{m}^{-3}$, b_n reaches 0.9 for $n \sim 55$, where $T = 10^4$ K is assumed in both cases, although the temperature dependence is very weak. On the low n side of the transition from collisional to radiative dominance, in the $N_e = 10^8\,\mathrm{m}^{-3}$ case b_n reaches 0.7 for $n \sim 65$ and for lower n collisions can be neglected (Figure 6.7).

We now use these considerations to find the strengths of radio recombination lines. In what follows 'NSE' means without correction for stimulated emission, and ϕ_ν is the normalized line profile. First consider the LTE case, denoted by an asterix:

$$\begin{aligned}\kappa_\nu^*\rho &= \kappa_\nu^*(\mathrm{NSE})\rho[1 - \mathrm{e}^{-h\nu/kT}]\\ &= N_i^* B_{ij} h\nu\phi_\nu[1 - \mathrm{e}^{-h\nu/kT}]\\ &\simeq \frac{h\nu}{kT} N_i^* \frac{g_j}{g_i}\frac{c^2}{8\pi h\nu^3} A_{ji} h\nu\phi_\nu, \quad h\nu \ll kT\end{aligned}$$

Using Saha and Boltzmann with $I_n = I - E_n$, $g_n = 2n^2$, $U(\mathrm{H}^0) = 2$, $U(\mathrm{H}^+) = 1$:

$$\begin{aligned}N_i^* &= \frac{N_i}{N(\mathrm{H}^0)}\frac{N(\mathrm{H}^0)}{N(\mathrm{H}^+)} N(\mathrm{H}^+)\\ &= \frac{g_n}{U(\mathrm{H}^0)}\mathrm{e}^{-E_n/kT}\frac{U(\mathrm{H}^0)}{2U(\mathrm{H}^+)}\left[\frac{h^2}{2\pi mkT}\right]^{3/2}\mathrm{e}^{I/kT} N_e N(\mathrm{H}^+)\\ &= 4.14 \times 10^{-22} n_i^2 \mathrm{e}^{I_n/kT}\frac{N_e N(\mathrm{H}^+)}{T^{3/2}}\end{aligned}$$

Substituting the approximations for ν and A_{ji} valid at large n, and noting that $\exp(I_n/kT) \sim 1$:

$$\kappa_\nu^*\rho = 8.5 \times 10^{-23}\frac{Z^2}{(\Delta n)^2}\phi_\nu\frac{N_e N(\mathrm{H}^+)}{T^{5/2}}\,\mathrm{m}^{-1} \tag{6.51}$$

$$= 4.8 \times 10^{-23}\frac{Z^2}{(\Delta n)^2}\frac{\mathrm{e}^{-(\Delta\nu/\Delta\nu_D)^2}}{\Delta\nu_D}\frac{N_e N(\mathrm{H}^+)}{T^{5/2}}\,\mathrm{m}^{-1} \tag{6.51a}$$

for a Doppler profile with Doppler width $\Delta\nu_D$.

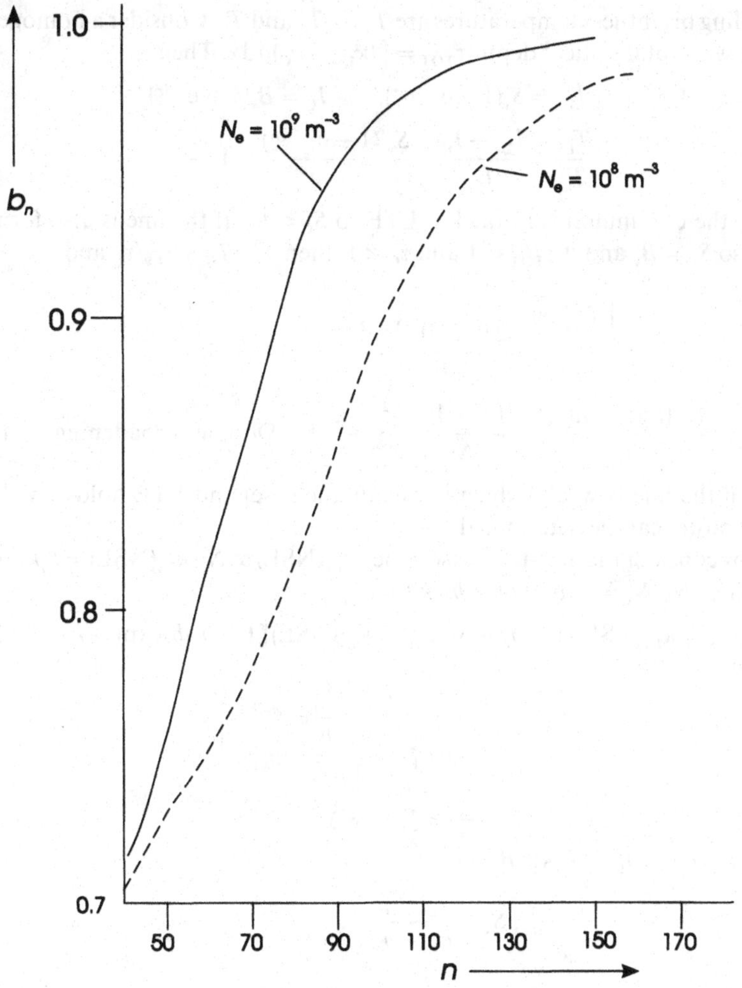

Figure 6.7. Departure coefficients from LTE for hydrogen levels at $T = 10^4$ K (after Brocklehurst)

Using (6.26) for the continuum free–free absorption coefficient:

$$\frac{\kappa^*_{L\nu}}{\kappa_C} = 4 \times 10^{-12} \frac{Z^2}{(\Delta n)^2} \frac{\nu^{2.1}\phi_\nu}{T^{1.15}} \tag{6.52}$$

where the constant is altered to $2.9 * 10^{-12}$ if allowance is made for helium continuous absorption. Suppose the intensity at a point in the line profile is $I_{L\nu}$ and the intensity of the neighbouring continuum is I_C, and suppose the corresponding brightness temperatures are $T_{L\nu} + T_C$ and T_C. Consider a homogeneous cloud with total optical depth $\tau_{TOT} = \int(\kappa_{L\nu} + \kappa_C)\rho\,dx$. Then

$$I_{L\nu} = S_\nu(1 - e^{-\tau_{TOT}}), \qquad I_C = B_\nu(1 - e^{-\tau_C})$$

$$\frac{T_{L\nu}}{T_C} = \frac{I_{L\nu} - I_C}{I_C} = \frac{S_\nu}{B_\nu}\frac{(1 - e^{-\tau_{TOT}})}{(1 - e^{-\tau_C})} - 1 \tag{6.53}$$

where the continuum is formed in LTE so $S_C = B_\nu$. If the line is also formed in LTE so $S_\nu = B_\nu$ and if $\tau_{TOT} \ll 1$ and $\tau_C \ll 1$, then $T_{L\nu}/T_C \sim \tau_{L\nu}/\tau_C$ and

$$\frac{\int T_{L\nu}\,d\nu}{T_C} \simeq \frac{2.9 \times 10^{-12}\nu^{2.1}}{T^{1.15}} \quad \text{for } \Delta n = 1 \tag{6.54a}$$

$$\frac{T_{L\nu}(\text{line centre})}{T_C} \simeq \frac{1.6 \times 10^{-12}\nu^{2.1}}{\Delta\nu_D T^{1.15}} \quad \text{for Doppler broadening} \tag{6.54b}$$

Thus if the line is weak (which is usually the case) and LTE holds, the kinetic temperature can be determined.

Now consider the non-LTE case. Since $\kappa_{L\nu}(\text{NSE}) \propto N_i$, $\kappa_{L\nu}(\text{NSE}) = b_i\kappa^*_{L\nu}(\text{NSE})$ and since $N_j/N_i = g_j/g_i \exp(-h\nu/kT)b_j/b_i$,

$$\kappa_{L\nu} = \kappa_{L\nu}(\text{NSE})(1 - g_i/g_j N_j/N_i) = \kappa_{L\nu}(\text{NSE})[1 - b_j/b_i \exp(-h\nu/kT)]$$

Hence

$$\kappa_{L\nu} = b_i\kappa^*_{L\nu}\frac{\left[1 - \frac{b_j}{b_i}e^{-h\nu/kT}\right]}{1 - e^{-h\nu/kT}}$$

$$= b_i\kappa^*_{L\nu}\beta \quad \text{say} \tag{6.55}$$

Since $j_{L\nu} \propto N_j$, $j_{L\nu} = b_j\kappa^*_{L\nu}B_\nu$:

$$S_\nu = \frac{j_{L\nu} + j_C}{\kappa_{L\nu} + \kappa_C}$$

$$= B_\nu\frac{1 + b_j\frac{\kappa^*_{L\nu}}{\kappa_C}}{1 + b_i\frac{\kappa^*_{L\nu}}{\kappa_C}\beta} \tag{6.56}$$

Now

$$\tau_{TOT} = \tau_C + \tau_{L\nu} = \tau_C(1 + b_i\beta\kappa^*_{L\nu}/\kappa_C) \tag{6.57}$$

(6.53) gives

$$\frac{T_{L\nu}}{T_C} = \frac{\left(1 + b_j\frac{\kappa^*_{L\nu}}{\kappa_C}\right)(1 - e^{-\tau_{TOT}})}{\left(1 + b_i\frac{\kappa^*_{L\nu}}{\kappa_C}\beta\right)(1 - e^{-\tau_C})} - 1$$

For $\tau_T \ll 1$ and $\tau_C \ll 1$, the exponentials can be expanded in a power series. To first order, κ_{TOT}/κ_C cancels in numerator and denominator, so the expansion must be taken to second order:

$$\frac{T_{L\nu}}{T_C} = \frac{S_\nu}{B_\nu}\frac{\tau_{TOT}}{\tau_C}\frac{\left(1-\frac{\tau_{TOT}}{2}\right)}{\left(1-\frac{\tau_C}{2}\right)} - 1 \tag{6.58}$$

Substituting from (6.56) and (6.57) and rearranging:

$$\frac{T_{L\nu}}{T_C} = \frac{\frac{\kappa^*_{L\nu}}{\kappa_C} b_j\left[1-\frac{\tau_C}{2}\left\{1+\frac{b_i}{b_j}\beta\left(1+b_j\frac{\kappa^*_{L\nu}}{\kappa_C}\right)\right\}\right]}{1-\frac{\tau_C}{2}}$$

If $\tau_C \ll 1$ and $\kappa^*_{L\nu}/\kappa_C \ll 1$, $|\beta| \ll 1$:

$$\frac{T_{L\nu}}{T_C} \simeq \frac{T^*_{L\nu}}{T_C} b_j[1-\tfrac{1}{2}\tau_C b_i \beta] \tag{6.59}$$

The definition of β, (6.55), shows that

$$\begin{aligned}\beta &= \frac{\left[1-\frac{b_j}{b_i}e^{-h\nu/kT}\right]}{[1-e^{-h\nu/kT}]}\\ &\simeq \frac{b_j}{b_i}\left[1-\frac{kT}{h\nu}\frac{b_j-b_i}{b_j}\right], \quad \text{if } h\nu \ll kT\\ &\simeq \frac{b_j}{b_i}\left[1-\frac{kT}{h\nu}\frac{d\ln_e(b_n)}{dn}\Delta n\right]\end{aligned} \tag{6.60}$$

It will be seen from (6.59) that non-LTE effects came into play in two different ways. Firstly the line is weakened because b_j is less than 1. Secondly, with $kT \gg h\nu$ in the radio region, if b_n changes fairly rapidly with n, $\beta \sim -b_j/b_i \Delta n\, kT/(h\nu) d\ln(b)/dn$ will be large and negative, thus strengthening the line, an effect due to stimulated emission and thus constituting a kind of maser. For small n, $b_n < 1$ and constant so the line is simply weaker than LTE would predict. For large n, $b_n = 1$, so the line has its LTE strength. At intermediate quantum numbers b_n can change rapidly enough with n to more than compensate for the fact that $b_n < 1$ and hence give a stronger line than an LTE calculation would predict. However, the question of departures from LTE remains somewhat controversial. Nebulae are very inhomogeneous, and regions of greatly varying density may contribute along a given line of sight. The dense regions will give an LTE contribution, but this may be amplified by maser effects in low density regions. On the other hand,

Stark broadening of lines will take place in the denser regions, giving extended wings which may not be picked up observationally, leading to an underestimate of the integrated brightness temperature of the line.

Collisionally Excited Lines

Consider an optically thin transition from upper level j to lower level l. The intensity of the line is given by

$$I_\nu = \int j_\nu \rho \, \mathrm{d}x \tag{6.61}$$

$$= \int \frac{A_{jl}}{4\pi} N_j h\nu \, \mathrm{d}x \tag{6.61}$$

In the simplest cases, the upper level of the transition involved, j, is populated only by collisional excitation from the ground state, i. Then

$$N_j \sum_n A_{jn} + N_j \sum_n C_{jn} = N_i C_{ij}$$

where $n < j$. It has been assumed that the processes populating j by recombinations from the ionized state and by radiative decays from higher levels can be neglected. The optical thinness of the line itself means that radiative transitions from l to j can be ignored. Then

$$I_\nu = \frac{h\nu}{4\pi} A_{jl} \int \sum_n \frac{N_i C_{ij}}{(A_{jn} + C_{jn})} \mathrm{d}x \tag{6.62}$$

If the density is too low for collisional de-excitations to be significant compared with radiative decays, and if the observed transition is the main radiative decay from level j (which will usually be the case if l is the ground state), (6.62) reduces to

$$I_\nu = \frac{h\nu}{4\pi} \int N_i C_{ij} \mathrm{d}x \tag{6.62a}$$

N_i in these equations is essentially the total number density of the ion concerned, since the populations of all excited levels are likely to be small.

The collisional excitation rate for forbidden lines (which are always optically thin) is given by (2.28)

$$C_{ij} = \frac{8.6 \times 10^{-12}}{T^{1/2}} N_e \frac{\Omega(i,j)}{g_i} \mathrm{e}^{-E_{ij}/kT}$$

so that if (6.62a) holds:

$$I_\nu = 6.8 \times 10^{-13} E_{lj} \frac{\Omega(i,j)}{g_i} \int \frac{N_i N_e}{T^{1/2}} \mathrm{e}^{-E_{ij}/kT} \mathrm{d}x \tag{6.63}$$

If the line originates in a homogeneous volume of gas with volume V and at distance d, then the flux received is $(4\pi j_\nu \rho V)/(4\pi d^2)$, or

$$\text{flux received} = \frac{6.8 \times 10^{-13}}{d^2} E_{ij} \frac{\Omega(i,j)}{g_i} V N_e^2 \frac{N_{\text{ion}}}{N_H} \frac{N_H}{N_e} e^{-E_{ij}/kT} \tag{6.64}$$

where $N_H/N_e \sim 0.9$ for a helium abundance 10% that of hydrogen and singly ionized hydrogen and it has been assumed that $N_{\text{ion}} \sim N_i$. In general, of course, N_e, T, and N_{ion}/N_H will vary over the nebula.

It is worth noting that if one of the levels has fine structure (sub-levels with different values of J), then

$$\Omega(i,jJ) = \frac{(2J+1)}{(2L+1)(2S+1)} \Omega(i,j)$$

where the sub-level has the quantum number J and statistical weight $g_J = 2J + 1$, and the whole level has quantum numbers L and S and statistical weight $= (2L+1)(2S+1)$.

The ratio of lines belonging to a particular ion can be used as a diagnostic of conditions in the nebula. Ideally, we would like to find line ratios that are sensitive to just one parameter, say temperature.

Take, for example, the case of ions with outer electron configuration np^3, which have ^4S as their ground state, $^2D_{3/2}$ and $^2D_{5/2}$ as the sub-levels of the first excited term, and $^2P_{1/2}$ and $^2P_{3/2}$ as the sub-levels of their second excited term. The ^2P levels can decay radiatively directly to the ground state or to the ^2D sub-levels, all the transitions involved being forbidden and therefore optically thin. The initial levels of the emission lines are the same in both cases, so the ratio of the strength of the ^2P to ^2D lines to that of the ^2P to ^4S lines should be proportional to the ratio of the quantities (transition probability * photon energy) for the two sets of lines. However, the wavelengths of the ^2P to ^2D lines are much longer than those of the ^2P to ^4S transitions, and hence are much less affected by the presence of dust. Since the ratio of the strengths of the two sets of lines can be predicted for the dust-free case and is independent of the means of excitation and hence of density and temperature, observations of this ratio enables the amount of dust extinction to be calculated.

This approach is usually applied to the case of S II, which has the $^2P_{3/2}$ level higher than the $^2P_{1/2}$ level, and the $^2D_{5/2}$ level higher than the $^2D_{3/2}$ level (Figure 6.8).

The longer wavelength transitions are:

$^2P \rightarrow {}^2D$	$J = 3/2 \rightarrow 5/2$	1032.05 nm	$A_{ji} = 0.18\,s^{-1}$
	$J = 3/2 \rightarrow 3/2$	1028.67 nm	$A_{ji} = 0.13\,s^{-1}$
	$J = 1/2 \rightarrow 5/2$	1037.05 nm	$A_{ji} = 0.08\,s^{-1}$
	$J = 1/2 \rightarrow 3/2$	1033.64 nm	$A_{ji} = 0.16\,s^{-1}$

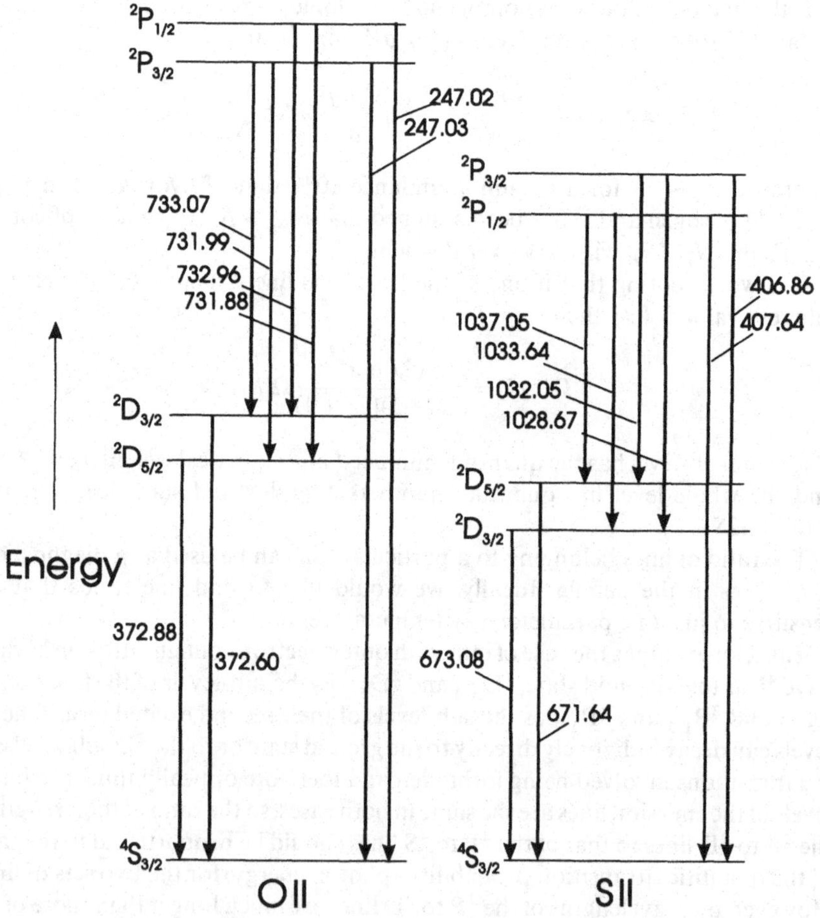

Figure 6.8. p^3 configuration; wavelengths in nm

and the shorter wavelength transitions are:

$$
\begin{array}{llll}
{}^2\mathrm{P} \rightarrow {}^4\mathrm{S} & J = 3/2 \rightarrow 3/2 & 406.86\,\mathrm{nm} & A_{ji} = 0.22\,\mathrm{s}^{-1} \\
 & J = 1/2 \rightarrow 3/2 & 407.64\,\mathrm{nm} & A_{ji} = 0.09\,\mathrm{s}^{-1}
\end{array}
$$

Hence

$$
\frac{I({}^2\mathrm{P} \rightarrow {}^2\mathrm{D})}{I({}^2\mathrm{P} \rightarrow {}^4\mathrm{S})} = \frac{\sum g_j A_{ji}(P \rightarrow D) h\nu(P \rightarrow D)}{\sum g_j A_{ji}(\mathrm{P} \rightarrow \mathrm{S}) h\nu(\mathrm{P} \rightarrow \mathrm{S})}
$$

$$
\simeq 0.394 \times \frac{1.720}{1.062} \simeq 0.64
$$

where the sub-levels are weighted by their statistical weights $2J + 1$ to represent their relative populations under conditions where collisions drive them to the Boltzmann values with $\Delta E \ll kT$. Then

$$\left[\frac{I(^2\mathrm{P}\rightarrow{}^2\mathrm{D})}{I(^2\mathrm{P}\rightarrow{}^4\mathrm{S})}\right]_{\mathrm{observed}} = 0.638\,\frac{e^{-\tau_{\mathrm{D}}(1032\,\mathrm{nm})}}{e^{-\tau_{\mathrm{D}}(407\mathrm{nm})}} \tag{6.65}$$

where the dust optical depth is τ_{D}. If the wavelength dependence of the dust optical depth is known (and very approximately it varies inversely with the wavelength), the dust optical depth at any wavelength can be calculated. The lines at 1032 nm are observed with difficulty because of terrestrial absorption at these wavelengths. Similar lines are found for the other ions with the same configuration (O II, Ne IV, Ar IV, etc.), but the short wavelength pairs in these cases are at wavelengths that cannot be observed from the Earth's surface.

The remaining forbidden transitions for the np^3 configuration, namely the $^2\mathrm{D}\rightarrow{}^4\mathrm{S}$ doublet, can be used to determine the electron density through the collisional rates, which are proportional to N_e. In this case the ratio is

$$\frac{I(^2\mathrm{D}_{5/2}\rightarrow{}^4\mathrm{S})}{I(^2\mathrm{D}_{3/2}\rightarrow{}^4\mathrm{S})} = \frac{j(^2\mathrm{D}_{5/2}\rightarrow{}^4\mathrm{S})}{j(^2\mathrm{D}_{3/2}\rightarrow{}^4\mathrm{S})}$$

$$= \frac{N(^2\mathrm{D}_{5/2})A(^2\mathrm{D}_{5/2}\rightarrow{}^4\mathrm{S})h\nu(^2\mathrm{D}_{5/2}\rightarrow{}^4\mathrm{S})}{N(^2\mathrm{D}_{3/2})A(^2\mathrm{D}_{3/2}\rightarrow{}^4\mathrm{S})h\nu(^2\mathrm{D}_{3/2}\rightarrow{}^4\mathrm{S})} \tag{6.66}$$

where the last term, the ratio of the frequencies, can be taken to be 1.0 since the wavelengths are close. For the same reason dust extinction will have no effect on the line ratio.

In what follows, the Einstein probabilities will be written $A_{5/2}$ and $A_{3/2}$, and the multiplicities will be omitted. The radiative transitions are forbidden and therefore so weak that radiative excitation of the $^2\mathrm{D}$ levels is negligible, and the population of these levels by radiative decay from the higher $^2\mathrm{P}$ levels is of no importance since the populations of all excited levels are small. Thus the $^2\mathrm{D}$ levels are mainly populated by collisional excitation from the ground state and depopulated by radiative de-excitation and collisional de-excitation. It is the competition between the latter two processes that leads to the dependence on density. Collisions leading to the transfers between $^2\mathrm{D}_{5/2}$ and $^2\mathrm{D}_{3/2}$ must also be taken into account—these will tend to drive the relative populations of the two levels to their Boltzmann values, and will be designated $C_{5/2,3/2}$ and $C_{3/2,5/2}$.

The equations of statistical equilibrium can be written:

$$N(^4\mathrm{S})C(^4\mathrm{S}\rightarrow{}^2\mathrm{D}_{5/2}) + N(^2\mathrm{D}_{3/2})C_{3/2,5/2} = N(^2\mathrm{D}_{5/2})[A_{5/2} + C(^2\mathrm{D}_{5/2}\rightarrow{}^4\mathrm{S}) + C_{5/2,3/2}]$$

$$N(^4\mathrm{S})C(^4\mathrm{S}\rightarrow{}^2\mathrm{D}_{3/2}) + N(^2\mathrm{D}_{5/2})C_{5/2,3/2} = N(^2\mathrm{D}_{3/2})[A_{3/2} + C(^2\mathrm{D}_{3/2}\rightarrow{}^4\mathrm{S}) + C_{3/2,5/2}]$$

Hence:

$$\frac{N(\mathrm{D}_{5/2})}{N(\mathrm{D}_{3/2})}=\frac{C(\mathrm{S}\rightarrow\mathrm{D}_{5/2})}{C(\mathrm{S}\rightarrow\mathrm{D}_{3/2})}\frac{\left[A_{3/2}+C(\mathrm{D}_{3/2}\rightarrow\mathrm{S})+C_{3/2,5/2}\left\{1+\dfrac{C(\mathrm{S}\rightarrow\mathrm{D}_{3/2})}{C(\mathrm{S}\rightarrow\mathrm{D}_{5/2})}\right\}\right]}{\left[A_{5/2}+C(\mathrm{D}_{5/2}\rightarrow\mathrm{S})+C_{5/2,3/2}\left\{1+\dfrac{C(\mathrm{S}\rightarrow\mathrm{D}_{5/2})}{C(\mathrm{S}\rightarrow\mathrm{D}_{3/2})}\right\}\right]}$$

Now $C(i,j)=8.6*10^{-12}\Omega(i,j)g_i\exp(-E_{ij}/kT)/T^{1/2}\,N_e=r*\Omega(i,j)/g_i\exp(-E_{ij}/kT)$, where $r=8.6*10^{-12}N_e/T^{1/2}$, but in the ratio of collisional excitation rates the factor $8.6*10^{-12}N_e(T^{1/2}g_i)$ cancels between numerator and denominator, and the ratio of the exponentials is close to one since $\Delta E(5/2,3/2)\ll kT$. For transitions to a particular level of a term:

$$\Omega(i,jJ)=\Omega(i,j)g_J/g_j,\ \text{so}\ C(\mathrm{S}\rightarrow\mathrm{D}_{5/2})/C(\mathrm{S}\rightarrow\mathrm{D}_{3/2})=g_{5/2}/g_{3/2}.$$

The ratio of populations then becomes

$$\frac{N(\mathrm{D}_{5/2})}{N(\mathrm{D}_{3/2})}=\frac{g_{5/2}}{g_{3/2}}\frac{\left[A_{3/2}+C(\mathrm{D}_{3/2}\rightarrow\mathrm{S})+C_{3/2,5/2}\left\{1+\dfrac{g_{3/2}}{g_{5/2}}\right\}\right]}{\left[A_{5/2}+C(\mathrm{D}_{5/2}\rightarrow\mathrm{S})+C_{5/2,3/2}\left\{1+\dfrac{g_{5/2}}{g_{3/2}}\right\}\right]}$$

and the ratio of intensities becomes

$$\frac{I(\mathrm{D}_{5/2}\rightarrow\mathrm{S})}{I(\mathrm{D}_{3/2}\rightarrow\mathrm{S})}=\frac{g_{5/2}}{g_{3/2}}\frac{\left[1+\dfrac{C(\mathrm{D}_{3/2}\rightarrow\mathrm{S})+C_{3/2,5/2}\left\{1+\dfrac{g_{3/2}}{g_{5/2}}\right\}}{A_{3/2}}\right]}{\left[1+\dfrac{C(\mathrm{D}_{5/2}\rightarrow\mathrm{S})+C_{5/2,3/2}\left\{1+\dfrac{g_{5/2}}{g_{3/2}}\right\}}{A_{5/2}}\right]}$$

which, substituting for C, becomes

$$\frac{I(\mathrm{D}_{5/2}\rightarrow\mathrm{S})}{I(\mathrm{D}_{3/2}\rightarrow\mathrm{S})}=1.5\frac{\left[1+\dfrac{r}{A_{3/2}}\left\{\dfrac{\Omega(3/2,\mathrm{S})}{4}+\dfrac{\Omega(3/2,5/2)}{4}\times 1.67\right\}\right]}{\left[1+\dfrac{r}{A_{5/2}}\left\{\Omega(5/2,\mathrm{S})/6+\dfrac{\Omega(3/2,5/2)}{6}\times 2.5\right\}\right]}$$

$$=1.5\frac{\left[1+\dfrac{r}{A_{3/2}}\{0.1\Omega(\mathrm{D},\mathrm{S})+0.417\Omega(3/2,5/2)\}\right]}{\left[1+\dfrac{r}{A_{5/2}}\{0.1\Omega(\mathrm{D},\mathrm{S})+0.417\Omega(3/2,5/2)\}\right]} \qquad (6.67)$$

If the density is high, then collisional excitation will be dominant, the populations will be determined by collisional processes and so will be given by Boltzmann's equations for $D_{5/2}$ and $D_{3/2}$. The ratio of exponentials is again close to 1, so $N(\mathrm{D}_{5/2})/N(\mathrm{D}_{3/2}) = g_{5/2}/g_{3/2} = 1.5$. Hence

$$\frac{I(\mathrm{D}_{5/2}\rightarrow\mathrm{S})}{I(\mathrm{D}_{3/2}\rightarrow\mathrm{S})} = 1.5\frac{A(\mathrm{D}_{5/2}\rightarrow\mathrm{S})}{A(\mathrm{D}_{3/2}\rightarrow\mathrm{S})} \tag{6.68}$$

which otherwise follows from (6.67) in the limit of large N_e.

If the electron density is low, then de-excitation is radiative only. Then

$$\frac{N(\mathrm{D}_{5/2})}{N(\mathrm{D}_{3/2})} = \frac{C(\mathrm{S}\rightarrow\mathrm{D}_{5/2})\,A_{3/2}}{C(\mathrm{S}\rightarrow\mathrm{D}_{3/2})\,A_{5/2}}$$

$$\frac{I(\mathrm{D}_{5/2}\rightarrow\mathrm{S})}{I(\mathrm{D}_{3/2}\rightarrow\mathrm{S})} = \frac{g_{5/2}}{g_{3/2}} = 1.5 \tag{6.69}$$

which again otherwise follows from (6.67), in this case in the limit of low N_e.

The changeover from (6.68) to (6.69) occurs at around the *critical density* $N_e = N_c$, which is defined as the density at which $C_{ji} = A_{ji}$, and which will have somewhat different values for $D_{5/2}$ and $D_{3/2}$. More precisely (6.67), which also takes into account collisional transitions between $\mathrm{D}_{5/2}$ and $\mathrm{D}_{3/2}$, can be used to find the exact value of the electron density which gives a line ratio half way between the limiting values, and to predict the value of the line ratio for intermediate electron densities.

Take first the case of S II, for the D→S transition of which:

$$J = 5/2 \rightarrow 3/2 \quad 671.64\,\mathrm{nm} \quad A_{ji} = 2.6*10^{-4}\,\mathrm{s}^{-1}$$

$$J = 3/2 \rightarrow 3/2 \quad 673.08\,\mathrm{nm} \quad A_{ji} = 8.8*10^{-4}\,\mathrm{s}^{-1}$$

$$\Omega(\mathrm{S},\mathrm{D}) = 6.98 \quad \Omega(\mathrm{D}_{3/2},\mathrm{D}_{5/2}) = 7.59$$

The low density limit predicts that $I(671.6)/I(673.1) = 1.5$, while the high density limit predicts that the same intensity ratio is 0.44. Critical densities for the two transitions are $4.3*10^9\,\mathrm{m}^{-3}$ and $1.4*10^{10}\,\mathrm{m}^{-3}$, and use of (6.67) suggests the line ratio is half way between the limiting values at a density of about $7*10^8\,\mathrm{m}^{-3}$.

In the case of O II (where the order of the energy levels $\mathrm{D}_{5/2}$ and $\mathrm{D}_{3/2}$ is inverted compared with S II, see Figure 6.8), we have for the D→S transition:

$$J = 5/2 \rightarrow 3/2 \quad 372.88\,\mathrm{nm} \quad A_{ji} = 3.6*10^{-5}\,\mathrm{s}^{-1}$$

$$J = 3/2 \rightarrow 3/2 \quad 372.60\,\mathrm{nm} \quad A_{ji} = 1.8*10^{-4}\,\mathrm{s}^{-1}$$

$$\Omega(\mathrm{S},\mathrm{D}) = 1.34 \quad \Omega(\mathrm{D}_{3/2,3/2}) = 1.17$$

The low density limit predicts that $I(372.9)/I(372.6) = 1.5$, while the high density limit predicts that the same intensity ratio is 0.3. (6.67) gives a density dependence very similar to that of S II.

The oxygen lines are only 160 km s^{-1} apart in terms of Doppler shift so the lines may not be resolved in some sources. They did however play an important role in early work on nebulae because they are readily detected on blue photographic plates. The S II lines are better separated and present no problems to modern detectors. The temperature dependence of both line ratios (due to decays from the higher P term to the D term) is very slight. Both line ratios will give a good indication of density in the range from 10^8 m^{-3} to a few times 10^9 m^{-3}, but outside this range will only indicate that the density is less than $10^8 m^{-3}$ or greater than a few times 10^9 m^{-3}

One problem with using the O II and S II line ratios as density indicators is that the lines originate in relatively low ionization regions of the nebula, and if the nebula is inhomogeneous, the density values obtained may not be applicable to the higher ionization zones. It is possible to sample the latter using the Ar IV doublet (configuration $2p^3$) 471.13 nm from $^2D_{5/2}$ and 474.02 nm from $^2D_{3/2}$, although 471.13 nm is blended with He I 471.3 nm and an estimate of the helium contribution has to be made using unblended He I lines. If ultraviolet spectra are available, the Ne IV ($2p^3$ configuration) doublet 242.45 nm from $^2D_{5/2}$ and 242.18 nm from $^2D_{3/2}$ can be used.

At higher densities, use may be made of line ratios from other configurations. Calcium III has a ground state configuration of $2s^2$ which gives rise to the single term ^1S, but the configuration 2s2p gives rise to the first excited term ^3P as well as to the term ^1P (Figure 6.9). Transitions from the ^3P to the ground state ^1S are semi-forbidden since ΔS is not zero, but while $^3P_0 \rightarrow {}^1S_0$ is completely forbidden since it requires $J=0 \rightarrow J=0$, and $^3P_2 \rightarrow {}^1S_0$ 190.7 nm is strongly forbidden ($A_{ji} = 5.2*10^{-3}\,s^{-1}$) since $\Delta J = 2$, the transition $^3P_1 \rightarrow {}^1S_0$ is only forbidden by the rule on S, and hence has the typical semi-forbidden value of $A_{ji} = 95.9\,s^{-1}$ (note that there is an allowed configuration change in all these transitions). The ratio of the intensities of 190.7 nm to 190.9 nm can be calculated in a similar way to that used for O II and S II line ratios with excitation as always being collisional but with competition between radiation and collisions for de-excitation. The high density limit is

$$\frac{I(190.7)}{I(190.9)} = \frac{g_2}{g_1} = \frac{5}{3}$$

and the low density limit is about $9*10^{-1}$, the intensity ratio reaching about half its maximum value at an electron density of around $4*10^{10}$ m^{-3} ($\Omega(^3P, {}^1S) \sim 1.0$). Thus a higher density range can be probed than was possible with O II and S II.

We now consider line ratios that can be used to estimate kinetic temperatures, where we use the fact that the relative collisional excitation rates of levels with considerably different excitation potentials are strongly dependent on temperature.

Consider the configuration np^3, which has ^3P as the ground state, the term ^1D as the first excited state and the ^1S term as the second excited state (Figure. 6.10).

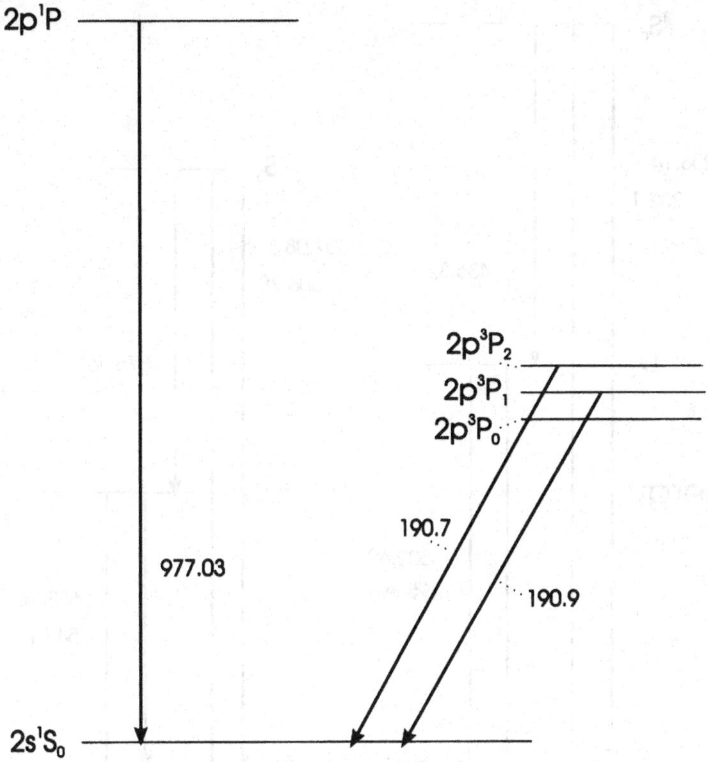

Figure 6.9. C III

The most conveniently observed transitions (they lie in the visible for O III and N II) are $^1S_0$ to $^1D_2$ and the doublet $^1D_2$ to $^3P_2$ and $^3P_1$. These three transitions are all forbidden because there is no change in configuration, singlet to doublet transitions are additionally forbidden because they break the weak rule $\Delta S = 0$, and $^1D_2 \rightarrow {}^3P_0$ is much weaker than its two companion transitions because $\Delta J = 2$ and is neglected in what follows.

Now

$$\frac{I(^1S \rightarrow {}^1D)}{I(^1D \rightarrow {}^3P_2) + I(^1D \rightarrow {}^3P_1)} = \frac{N(^1S)A(^1S \rightarrow {}^1D)h\nu(^1S \rightarrow {}^1D)}{N(^1D)A(^1D \rightarrow {}^1P)h\nu(^1D \rightarrow {}^3P)}$$

Suppose the electron density is low so collisional de-excitation from ^1D and ^1S can be neglected, as can collisional excitation from ^1D to ^1S. Suppose also that the population of ^1D by radiative decays from ^1S is also unimportant. The

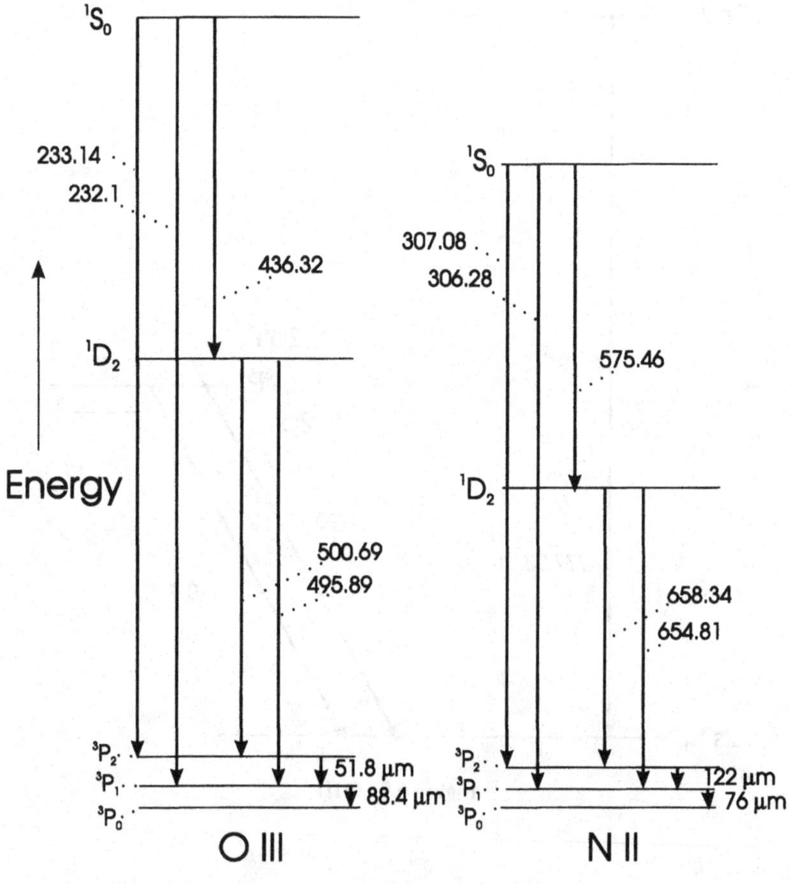

Figure 6.10. p^2 configuration; wavelengths in nm except for IR transitions

equations of statistical equilibrium (in simplified notation) give

$$N(\mathrm{D})A(\mathrm{D}\rightarrow\mathrm{P}) = N(\mathrm{P})C(\mathrm{P}\rightarrow\mathrm{D})$$

$$N(\mathrm{S})[A(\mathrm{S}\rightarrow\mathrm{P}) + A(\mathrm{S}\rightarrow\mathrm{D})] = N(\mathrm{P})C(\mathrm{P}\rightarrow\mathrm{S})$$

$$\frac{N(\mathrm{S})}{N(\mathrm{D})} = \frac{C(\mathrm{P}\rightarrow\mathrm{S})}{C(\mathrm{P}\rightarrow\mathrm{D})}\frac{A(\mathrm{D}\rightarrow\mathrm{P})}{A(\mathrm{S}\rightarrow\mathrm{P}) + A(\mathrm{S}\rightarrow\mathrm{D})}$$

$$\frac{I(\mathrm{S}\rightarrow\mathrm{D})}{I(\mathrm{D}\rightarrow\mathrm{P})} = \frac{C(\mathrm{P}\rightarrow\mathrm{S})}{C(\mathrm{P}\rightarrow\mathrm{D})}\frac{\nu(\mathrm{S}\rightarrow\mathrm{D})}{\nu(\mathrm{D}\rightarrow\mathrm{P})}\frac{A(\mathrm{S}\rightarrow\mathrm{D})}{A(\mathrm{S}\rightarrow\mathrm{P}) + A(\mathrm{S}\rightarrow\mathrm{D})}$$

Now

$$C \propto 8.6 \times 10^{-12} \frac{\Omega \, e^{-E/kT}}{g \;\; T^{1/2}}$$

so

$$\frac{I(\mathrm{S}\rightarrow\mathrm{D})}{I(\mathrm{D}\rightarrow\mathrm{P})} = \frac{\Omega(\mathrm{P}\rightarrow\mathrm{S})}{\Omega(\mathrm{P}\rightarrow\mathrm{D})} e^{-\Delta E/kT} \frac{\nu(\mathrm{S}\rightarrow\mathrm{D})}{\nu(\mathrm{D}\rightarrow\mathrm{P})} \frac{A(\mathrm{S}\rightarrow\mathrm{D})}{A(\mathrm{S}\rightarrow\mathrm{P}) + A(\mathrm{S}\rightarrow\mathrm{D})} \tag{6.70}$$

with $\Delta E = E(\mathrm{S}) - E(\mathrm{D})$. Thus measurement of the line ratio will provide a value for the kinetic temperature.

In general, collisional de-excitation will be important as well as collisional excitation from ^{1}D to ^{1}S. Assuming that ^{3}P can be treated as a single level:

$$N(\mathrm{P})C(\mathrm{P}\rightarrow\mathrm{D}) + N(\mathrm{S})[C(\mathrm{S}\rightarrow\mathrm{D}) + A_{\mathrm{SD}}] = N(\mathrm{D})[C(\mathrm{D}\rightarrow\mathrm{S}) + C(\mathrm{D}\rightarrow\mathrm{P}) + A_{\mathrm{DP}}]$$

$$N(\mathrm{P})C(\mathrm{P}\rightarrow\mathrm{S}) + N(\mathrm{D})C(\mathrm{D}\rightarrow\mathrm{S}) = N(\mathrm{S})[C(\mathrm{S}\rightarrow\mathrm{D}) + C(\mathrm{S}\rightarrow\mathrm{P}) + A_{\mathrm{SD}} + A_{\mathrm{SP}}]$$

Hence

$$\frac{N(\mathrm{S})}{N(\mathrm{D})} = \frac{\left[\{C(\mathrm{D}\rightarrow\mathrm{S}) + C(\mathrm{D}\rightarrow\mathrm{P}) + A_{\mathrm{DP}}\} \times \dfrac{C(\mathrm{P}\rightarrow\mathrm{S})}{C(\mathrm{P}\rightarrow\mathrm{D})} + C(\mathrm{D}\rightarrow\mathrm{S})\right]}{\left[C(\mathrm{S}\rightarrow\mathrm{D}) + C(\mathrm{S}\rightarrow\mathrm{P}) + A_{\mathrm{SD}} + A_{\mathrm{SP}} + \dfrac{\{C(\mathrm{S}\rightarrow\mathrm{D}) + A_{\mathrm{SD}}\} \times C(\mathrm{P}\rightarrow\mathrm{S})}{C(\mathrm{P}\rightarrow\mathrm{D})}\right]}$$

$$= \frac{\Omega_{\mathrm{PS}} e^{-\Delta E/kT}\left[A_{\mathrm{DP}} + \dfrac{r}{5}\left\{\Omega_{\mathrm{PD}} + \Omega_{\mathrm{DS}} e^{-\Delta E/kT} + \dfrac{\Omega_{\mathrm{DS}}\Omega_{\mathrm{PD}}}{\Omega_{\mathrm{PS}}}\right\}\right]}{\Omega_{\mathrm{PD}}\left[A_{\mathrm{SP}} + A_{\mathrm{SD}} + A_{\mathrm{SD}} \times \dfrac{\Omega_{\mathrm{PS}}}{\Omega_{\mathrm{PD}}} e^{-\Delta E/kT} + r\left\{\Omega_{\mathrm{PS}} + \Omega_{\mathrm{DS}} + \Omega_{\mathrm{DS}} \dfrac{\Omega_{\mathrm{PS}}}{\Omega_{\mathrm{PD}}} e^{-\Delta E/kT}\right\}\right]}$$

where $C(\mathrm{P}\rightarrow\mathrm{S})/C(\mathrm{P}\rightarrow\mathrm{D})$ has been removed from the square brackets and the formulae for collisional excitation and de-excitation substituted with $g_{\mathrm{D}} = 5$ and $g_{\mathrm{s}} = 1$.

Hence

$$\frac{I(\mathrm{S}\rightarrow\mathrm{D})}{I(\mathrm{D}\rightarrow\mathrm{P})} = I_{\mathrm{R}} \frac{\left[1 + \dfrac{r}{5 \times A_{\mathrm{DP}}}\left\{\Omega_{\mathrm{PD}} + \Omega_{\mathrm{DS}} e^{-\Delta E/kT} + \Omega_{\mathrm{DS}} \dfrac{\Omega_{\mathrm{PD}}}{\Omega_{\mathrm{PS}}}\right\}\right]}{\left[1 + \dfrac{A_{\mathrm{SD}}}{A_{\mathrm{PS}} + A_{\mathrm{SD}}} \dfrac{\Omega_{\mathrm{PS}}}{\Omega_{\mathrm{PD}}} e^{-\Delta E/kT} + \dfrac{r}{A_{\mathrm{SP}} + A_{\mathrm{SD}}}\left\{\Omega_{\mathrm{PS}} + \Omega_{\mathrm{DS}} + \Omega_{\mathrm{DS}} \dfrac{\Omega_{\mathrm{PS}}}{\Omega_{\mathrm{PD}}} e^{-\Delta E/kT}\right\}\right]}$$

with

$$I_{\mathrm{R}} = \frac{\Omega_{\mathrm{PS}}}{\Omega_{\mathrm{PD}}} \frac{\nu_{\mathrm{SD}}}{\nu_{\mathrm{DP}}} e^{-\Delta E/kT} \frac{A_{\mathrm{SD}}}{A_{\mathrm{SP}} + A_{\mathrm{SD}}} \tag{6.71a}$$

If the terms in $\exp(-\Delta E/kT)$ inside the square brackets can be neglected, we finally find

$$\frac{I(\mathrm{S}\rightarrow\mathrm{D})}{I(\mathrm{D}\rightarrow\mathrm{P})}=\frac{\Omega_{\mathrm{PS}}}{\Omega_{\mathrm{PD}}}\frac{\nu_{\mathrm{SD}}}{\nu_{\mathrm{DP}}}\mathrm{e}^{-\Delta E/kT}\frac{A_{\mathrm{SD}}}{A_{\mathrm{SP}}+A_{\mathrm{SD}}}\frac{\left[1+\dfrac{r}{5A_{\mathrm{DP}}}\Omega_{\mathrm{PD}}\left(1+\dfrac{\Omega_{\mathrm{DS}}}{\Omega_{\mathrm{PS}}}\right)\right]}{\left[1+r\dfrac{(\Omega_{\mathrm{PS}}+\Omega_{\mathrm{DS}})}{A_{\mathrm{SP}}+A_{\mathrm{SD}}}\right]} \tag{6.71b}$$

Thus at high electron densities, an electron density dependence (via r) appears in the ratio. Clearly, if N_e is known, say from O II or S II measurements, then the temperature can be found from (6.71).

The two main applications of (6.71) are to O III and N II. For O III we have:

$^1S_1\rightarrow{}^1D_2$	436.32 nm	$A_{SD}=1.8\,\mathrm{s}^{-1}$		$\Omega_{DS}=0.62$
$^1D_2\rightarrow{}^3P_2$	500.69 nm	$A=2.0*10^{-2}\,\mathrm{s}^{-1}$	$A_{DP}=2.67*10^{-2}\,\mathrm{s}^{-1}$	$\Omega_{PD}=2.17$
$^1D_2\rightarrow{}^3P_1$	495.89 nm	$A=6.7*10^{-3}\,\mathrm{s}^{-1}$		
$^1S_1\rightarrow{}^3P_2$	233.14 nm	$A=7.8*10^{-9}\,\mathrm{s}^{-1}$	$A_{SP}=2.21*10^{-1}\,\mathrm{s}^{-1}$	$\Omega_{PS}=0.28$
$^1S_1\rightarrow{}^3P_1$	232.10 nm	$A=2.2*10^{-1}\,\mathrm{s}^{-1}$		

For N II we have:

$^1S_1\rightarrow{}^1D_2$	575.46 nm	$A_{SD}=1.1\,\mathrm{s}^{-1}$		$\Omega_{DS}=0.41$
$^1D_2\rightarrow{}^3P_2$	658.34 nm	$A=3.0*10^{-3}\,\mathrm{s}^{-1}$	$A_{DP}=4*10^{-3}\,\mathrm{s}^{-1}$	$\Omega_{PD}=2.68$
$^1D_2\rightarrow{}^3P_1$	654.81 nm	$A=1.0*10^{-3}\,\mathrm{s}^{-1}$		
$^1S_1\rightarrow{}^3P_2$	307.08 nm	$A=1.5*10^{-4}\,\mathrm{s}^{-1}$	$A_{SP}=3.42*10^{-2}\,\mathrm{s}^{-1}$	$\Omega_{PS}=0.35$
$^1S_1\rightarrow{}^3P_1$	306.28 nm	$A=3.4*10^{-2}\,\mathrm{s}^{-1}$		

Hence using (6.71b) for the O III data, we obtain:

$$\frac{I(436.3)}{I(500.7)+I(495.9)}=0.1316\mathrm{e}^{-32990/T}\frac{\left[1+4.49\times10^{-10}\dfrac{N_e}{T^{1/2}}\right]}{\left[1+3.83\times10^{-12}\dfrac{N_e}{T^{1/2}}\right]} \tag{6.72}$$

where the square bracket in the denominator can be neglected except at very high electron densities. At $T=10^4$ K, the line ratio is doubled compared with the value ignoring collisional de-excitation when $N_e=2.2*10^{11}\,\mathrm{m}^{-3}$.

Using (6.71b) for the N II data we obtain:

$$\frac{I(565.5)}{I(658.3)+I(654.8)}=0.1447\mathrm{e}^{-25010/T}\frac{\left[1+2.50\times10^{-9}\dfrac{N_e}{T^{1/2}}\right]}{\left[1+5.78\times10^{-12}\dfrac{N_e}{T^{1/2}}\right]} \tag{6.73}$$

At $T=10^4$ K, the line ratio is doubled compared with the value obtained ignoring collisional de-excitation when $N_e=4*10^{10}\,\mathrm{m}^{-3}$. Note however that at $T=10^4$ K the ratio for O III has a value of $4.9*10^{-3}$ and for N II of $1.2*10^{-2}$.

Thus the 436.3 nm line of O III in particular is very weak and hard to detect unless the temperature is in the upper part of the range of nebular temperatures.

Of course, in general all forbidden line ratios are functions of both T and N_e, and the observed value of any such ratio allows a curve to be selected on a T versus N_e diagram. A second observed line ratio allows another curve to be selected, and in the absence of errors the intersection of the two curves fixes both T and N_e. The examples given so far (O II and S II on the one hand and O III and N II on the other) provide curves that are nearly horizontal and nearly vertical, and which therefore intersect at right angles.

Advances in technology not only enable line ratios in the ultraviolet to be used, but also extend the observable range far into the infrared. The fine structure levels of the ^3P ground state of a p^2 configuration give rise to far-infrared forbidden transitions that can be observed from above the atmosphere. For instance the O III has transitions $^3P_2 \rightarrow {}^3P_1$ at 51 800 nm (more commonly quoted as 52 (μm)) with a transition probability of $9.8*10^{-5}\,s^{-1}$ and $^3P_1 \rightarrow {}^3P_0$ at 88 400 nm (88 μm) with a transition probability of $2.6*10^{-5}\,s^{-1}$ (the transition $^3P_2 \rightarrow {}^3P_0$ has $\Delta J = 2$ and so has a very small transition probability). The excitation energies of these levels are much less than kT and so collisional processes are dominant. The equations of statistical equilibrium are:

$$N_0[C(0\rightarrow 1) + C(O\rightarrow 2)] = N_1[A_{10} + C(1\rightarrow 0)] + N_2C(2\rightarrow 0)$$

$$N_1[C(1\rightarrow 2) + C(1\rightarrow 0) + A_{10}] = N_0C(0\rightarrow 1) + N_2[C(2\rightarrow 1) + A_{21}]$$

$$N_2[C(2\rightarrow 0) + C(2\rightarrow 1) + A_{21}] = N_0C(0\rightarrow 2) + N_1C(1\rightarrow 2)$$

Hence:

$$\frac{N_2}{N_1} = \frac{C(0\rightarrow 2)A_{10}}{C(0\rightarrow 1)A_{21}} \frac{\left[1 + \frac{1}{A_{10}}\left\{C(1\rightarrow 2) + C(1\rightarrow 0) + C(1\rightarrow 2) \times \frac{C(0\rightarrow 1)}{C(0\rightarrow 2)}\right\}\right]}{\left[1 + \frac{C(0\rightarrow 2)}{C(0\rightarrow 1)} + \frac{1}{A_{21}}\left\{C(2\rightarrow 0) + C(2\rightarrow 1) + C(0\rightarrow 2) \times \frac{C(2\rightarrow 1)}{C(0\rightarrow 1)}\right\}\right]}$$

If we multiply the numerator and denominator by $A_{ji}h\nu$, write out the expressions for the collisional rates and put $\exp(-\Delta/EkT) \sim 1$ in the ratios of collisional rates inside the square brackets, we obtain:

$$\frac{I_{21}}{I_{10}} = \frac{88400\,\Omega_{02}}{51800\,\Omega_{01}} e^{-\Delta E/kT} \frac{\left[1 + \frac{r}{3A_{10}}\left\{\Omega_{01} + \Omega_{12} + \Omega_{01} \times \frac{\Omega_{12}}{\Omega_{02}}\right\}\right]}{\left[1.5 + \frac{r}{5A_{21}}\left\{\Omega_{02} + \Omega_{12} + \Omega_{02} \times \frac{\Omega_{12}}{\Omega_{01}}\right\}\right]}$$

$$\simeq 0.55e^{-278/T}\left[1 + 4.7 \times 10^{-7}\frac{N_e}{T^{1/2}}\right]$$

where $\Omega_{01} = 0.54$, $\Omega_{02} = 0.27$ and $\Omega_{12} = 1.29$ have been substituted. Thus the ratio $I(52\,\mu\text{m})/I(88\,\mu\text{m})$ is strongly dependent on electron density but only very weakly dependent on temperature, and so is an excellent determinant of electron density for densities up to a few times $10^9\,\text{m}^{-3}$.

We may also want to determine abundances from forbidden lines. The strength of a line is given by (6.62):

$$I = h\nu \frac{A_{ji}}{4\pi} \int \frac{N_j}{N_{\text{TOT}}} N_{\text{TOT}}\,\text{d}x$$

where N_{TOT} is the number density of the ion concerned. Suppose for simplicity that i is the ground state, and that only two levels are involved so $N_i C_{ij} = N_j(C_{ji} + A_{ji})$ with $N_i \sim N_{\text{TOT}}$. Then

$$I = \frac{8.6 \times 10^{-12}}{4\pi} \frac{h\nu\Omega_{ij}}{g_i} \int \frac{e^{-E_{ij}/kT} N_e N_{\text{TOT}}}{T^{1/2}\left(1 + \dfrac{8.6 \times 10^{-12} N_e \Omega_{ij}}{T^{1/2} A_{ji} g_j}\right)} \text{d}x$$

Hence if $T(x)$ and $N_e(x)$ are known, then $\int N_{\text{TOT}}\text{d}x$ can be found, and if $\int N_{\text{H}}\text{d}x$ is also known from recombination lines or optically thin free–free continuum, then

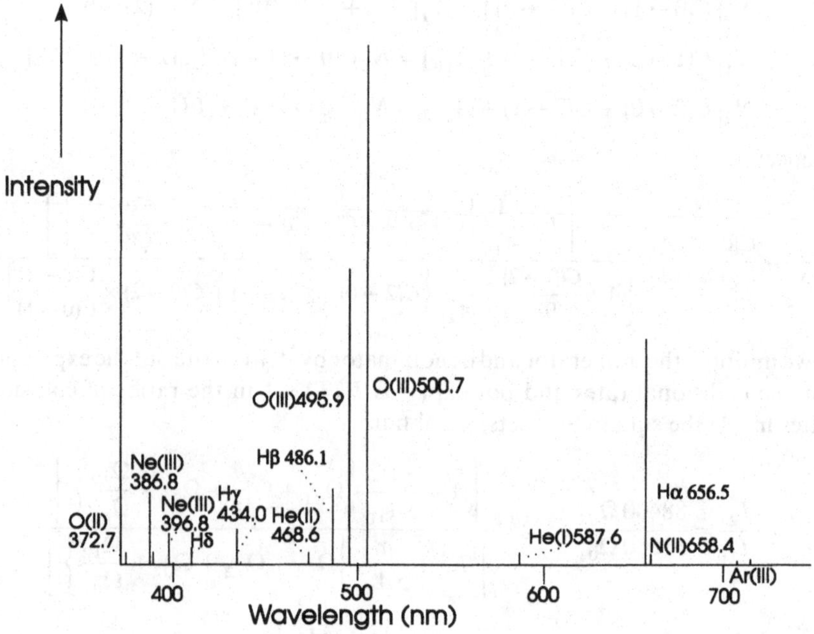

(a)

Figure 6.11. (a) Planetary nebula spectrum: fairly high excitation. (b) Planetary nebula spectrum: very low excitation

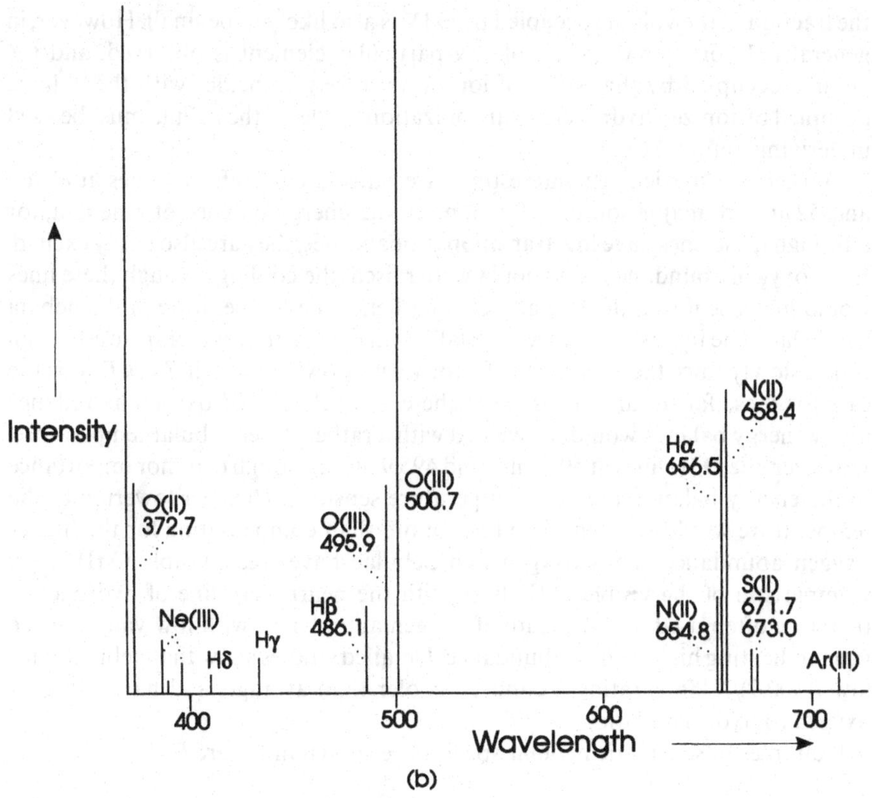

Figure 6.11. (*Contd*)

the abundance of the ion relative to hydrogen can be found. Indeed if T and N_e are constant throughout the nebula, the factors in T and N_e can be taken outside the integral sign and the intensity of the forbidden line compared directly with that from optically thin hydrogen lines or continuum to give the ratio of the column densities of ionized hydrogen and the ion concerned. However, most nebulae are not homogeneous and temperature variations can cause large variations in the collisional excitation rate for the ion through the factor $\exp(-E_{ij}/kT)$. This is a major problem in making abundance estimates. It must also be remembered that the equation above gives only the abundance of the ion being observed. In some cases forbidden lines from more than one stage of ionization are observed and the ionic column densities can be added to give the total abundance of the element. A case in point is oxygen, where both the O II and O III lines are observed as indeed can O I, although only a small fraction of the H II region is likely to have oxygen in neutral form. For stellar ionized nebulae, the fraction of the volume occupied by O IV is also likely to be small. However, in general only one ionic species of any particular element is observed, and the volume occupied by that stage of ionization rarely coincides with the volume occupied by ionized hydrogen, so an ionization model of the nebula must be used in deriving abundances.

Oxygen is a particularly interesting case, since the far infrared lines at 88 μm and 52 μm are major sources of cooling in the energy balance of a nebula, for although these lines have low transition probabilities, they are also easily excited. If the oxygen abundance of a nebula were raised, the cooling through these lines would increase if nothing else altered, and hence the temperature of the nebula would fall. The intensity of these lines falls with decreasing temperature although rather slowly since the exponential factor varies slowly with T if $E \ll kT$, as is the case for these far infrared lines, so in the case of the raised oxygen abundance a new energy balance would be reached with a rather lower nebular temperature. However, the O III lines at 500.7 nm and 495.9 nm, although of minor importance in the energy balance, are very temperature sensitive. Hence a lowering of the temperature would weaken these lines, more than compensating for the higher oxygen abundance. Thus oxygen rich nebulae have weak visible O III lines! Comparison of the visible O III lines with the nearby Hβ line of hydrogen is therefore often used as a measure of oxygen abundance, with low values of the ratio indicating high oxygen abundance. Detailed studies show that nebulae with low $I(500.7)/I(486.1)$ ratios are indeed cooler than average, and have a higher oxygen to hydrogen ratio.

Examples of schematic nebular spectra are shown in Figure 6.11.

Resonance Fluorescence

There is a third way of producing emission lines besides collisional excitation and recombination. The presence of strong emission lines in the nebula means that if another transition coincides in wavelength with one of these emission lines, then absorption will take place, raising the ion to which the transition belongs to an excited level. The ion will normally radiatively decay back to the ground state, returning the photon initially absorbed to the radiation field. However, if the excited level can decay to the ground state via an intermediate level or levels, then a line or lines at longer wavelengths than the absorbing transition will be produced in the course of decay back to the ground state. The excited level may lie too high to be populated appreciably by collisions, and the strengths of the longer wavelength lines may be much greater than predicted by recombination theory. Such unexpectedly strong lines are called *resonance fluorescence* lines. It will be noted that the excited level will normally decay directly back to the ground state, but it may be that conditions are such that the strong original line is scattered many times, thus providing many opportunities for absorption by the fluorescing line and hence of degradation into longer wavelength photons. It is also worth

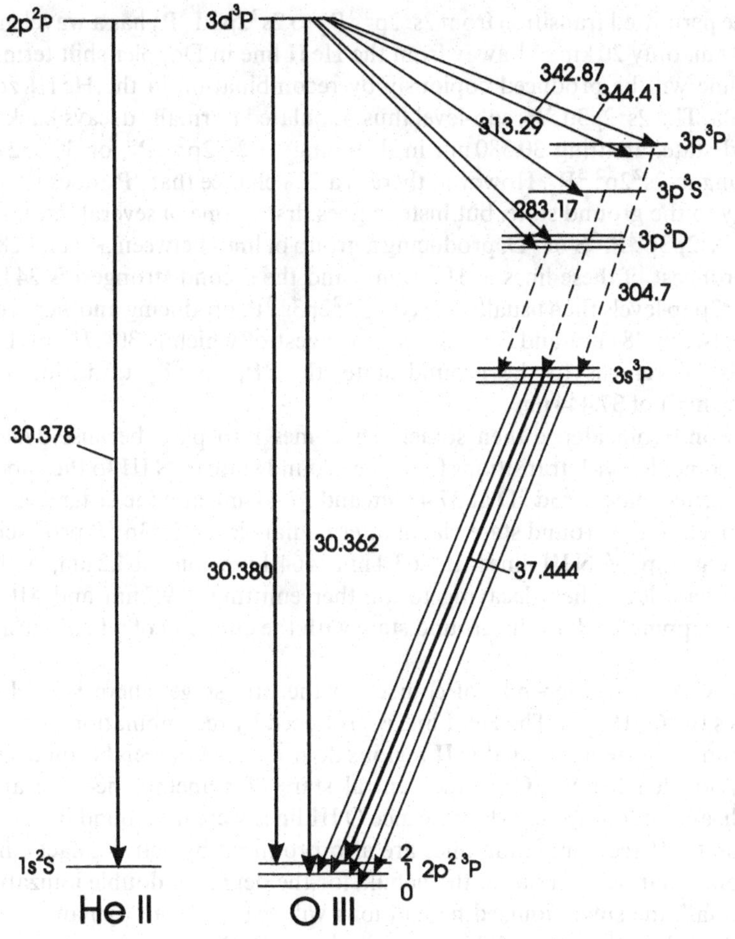

Figure 6.12. Resonance fluorescence; not all levels or lines shown

noting that the wavelength coincidence does not have to be exact because of the finite Doppler width of the lines, and this is particularly true in active galactic nuclei where the velocities involved are large.

The classic example of resonance fluorescence is the production of a group of O III permitted lines between 280 μm and 380 nm first elucidated by Bowen (Figure 6.12). The He II Lyman α line connecting the ground state and the first excited state of ionized helium has a wavelength of 30.378 nm (one-quarter that of hydrogen Lyman α because of the Z^2 factor). The O III ground state is $2s^22p^2\ ^3P$ with sub-levels with $J = 0, 1$ and 2. There is an excited state of O III, $2s^22p3d\ ^3P$, and the permitted transition from $2s^22p^2\ ^3P_2$ to $2s^22p3d\ ^3P_2$ has a wavelength of 30.380 nm, only 20 km s^{-1} away from the He II line in Doppler shift terms. The He II line will be produced copiously by recombination in the He III zone of a nebula. The $2s^22p3d\ ^3P_2$ sub-level thus populated normally decays back to the ground state, emitting 30.380 nm in dropping to $2s^22p^2\ ^3P_2$, or 30.362 nm in dropping to $2s^22p^2\ ^3P_1$. However, there is a 2% chance that $^3P_2$ does not decay directly to the ground state, but instead goes first to one of several intermediate levels, $2s^22p3p\ ^3P$, ^3S or ^3D, producing a group of lines between 344 and 281 nm. The strongest of these lines is 313.29 nm, and the second strongest is 344.4 nm. The $2s^22p3p$ levels then usually decay to $2s^22p3s\ ^3P$, producing another group of lines between 381 nm and 302 nm, the strongest of which is 304.71 nm. Finally $2s^22p3s\ ^3P$ returns to the ground state, the $^3P_1$ to $^3P_2$ transition having a wavelength of 37.444 nm.

A second coincidence then sometimes comes into play, because 37.444 nm nearly coincides with transitions from the ground state of N III to the sub-levels of the excited state $2s^23d\ ^2D$ at 37.43 nm and 37.44 nm, and the latter sometimes decay back to the ground state via an intermediate level, $2s^23p\ ^2P$ producing en route a group of N III lines at 463.4 nm, 464.1 nm and 464.2 nm, with this intermediate level then decaying to another emitting 409.7 nm and 410.3 nm, before dropping back to the ground state with the emission of a final ultraviolet line.

This whole cycle depends, of course, on the first stage where He II Lyman α excites the O III ions. The He II line is produced by recombinations in a He III zone, and most stellar excited H II regions do not have appreciable quantities of doubly ionized helium. Only the central stars of planetary nebulae are hot enough, and indeed Bowen fluorescence O III lines were first found in planetary nebulae. He II recombination lines are also produced by active galactic nuclei. Once one is far enough out in the nebula for the degree of double ionization to start to fall, the singly ionized helium ions will scatter He II Lyman α and the optical depth in this line will become very large. In the same zone there will be appreciable quantities of O III. The Lyman α photons in their multiple scatterings will eventually be caught by an O III if they are not lost in photoionizing neutral hydrogen or helium (their frequency is, of course, high enough for both of these continuum processes, although only a fraction of hydrogen and helium is likely to be neutral). Once caught by O III, 30.38 nm is usually re-emitted as the O III ion returns to the ground state, or if the return is to the $^3P_1$ sub-level of the ground state, 30.36 nm is produced, with the Bowen decay happening only occasionally. However the O III ultraviolet lines are liable to be reabsorbed by O III, thus leading through multiple scattering to further opportunities for the Bowen decay to take place. Only those Lyman a photons that escape from the nebula or are changed into 30.36 nm or 30.38 nm photons that escape will not end up producing the Bowen lines. The situation is complicated in the case of planetary nebulae by the geometrical situation with the expanding shell. However, for both planetary nebulae and AGN, something approaching one-half of the He II Lyman α photons are eventually converted into Bowen lines.

A second important case of resonance fluorescence is that involving the Lyman β line of hydrogen at 102.577 nm and the transition between the ground state of O I, $2s^22p^4\ ^3P$, and the excited state $2s^22p^33d\ ^3D$ with a wavelength of 102.572 nm. The excited state decays first to $2s^22p^33p\ ^3P$, emitting a near infrared photon (1028.6 nm or 1028.7 nm), then drops to $2s^22p^33s\ ^3S$, emitting 844.6 nm, and finally to the ground state, emitting the 130.2 nm line. The 844.6 nm line is observed in some active galactic nuclei as a resonance fluorescence line, and 130.2 nm can be seen in redshifted objects.

Photodissociation Regions

Stars are born from molecular clouds. Newborn massive stars photoionize the gas around them to produce H II regions, but their ultraviolet radiation can also dissociate molecules, in particular the most common species, H_2. Now the dissociation energy of H_2 is 4.48 eV as compared with the ionization energy of hydrogen of 13.6 eV, so that stellar photons with wavelengths between the Lyman limit at 91.2 nm and 277 nm will escape from the H II region but can dissociate molecular hydrogen. Thus farther out than the region of ionized hydrogen will lie a zone of neutral gas in which hydrogen is partially or fully dissociated into hydrogen atoms, called a *photodissociation region* or PDR for short. The ultraviolet input will heat the region either via the hydrogen molecules or via interstellar grains. Farther out still before the cold molecular cloud proper is reached, there may be a shock front which can heat hydrogen molecules to quite high temperatures but only temporarily for the molecules will cool once they have passed through the front. There appears to be no rigorous definition of a PDR, but in general terms one speaks of the volume of gas in which hydrogen is neutral but in which molecular abundances are affected by far ultraviolet radiation. H_2 is by far the most important molecule in molecular clouds, and the discussion which follows is largely confined to H_2 and its spectrum, but CO is also of importance. Although CO is more tightly bound than H_2 (dissociation energy 11.09 eV, corresponding to a wavelength of 111.8 nm), detailed investigation of the radiative dissociation mechanisms show that molecular CO can only exist farther into the cloud and away from the far ultraviolet source than the transition point between atomic hydrogen and H_2. Carbon has a lower ionization potential than hydrogen, and so will be in the form of C II over much of the PDR, with a relatively thin layer of C I as one moves farther into the molecular cloud before becoming associated as CO. An easily dissociated molecule like O_2 only forms in appreciable quantities even farther into the cloud. Thus some workers include within the definition of a PDR essentially the whole region from the shock front to the H II/H I interface (Figure 6.13).

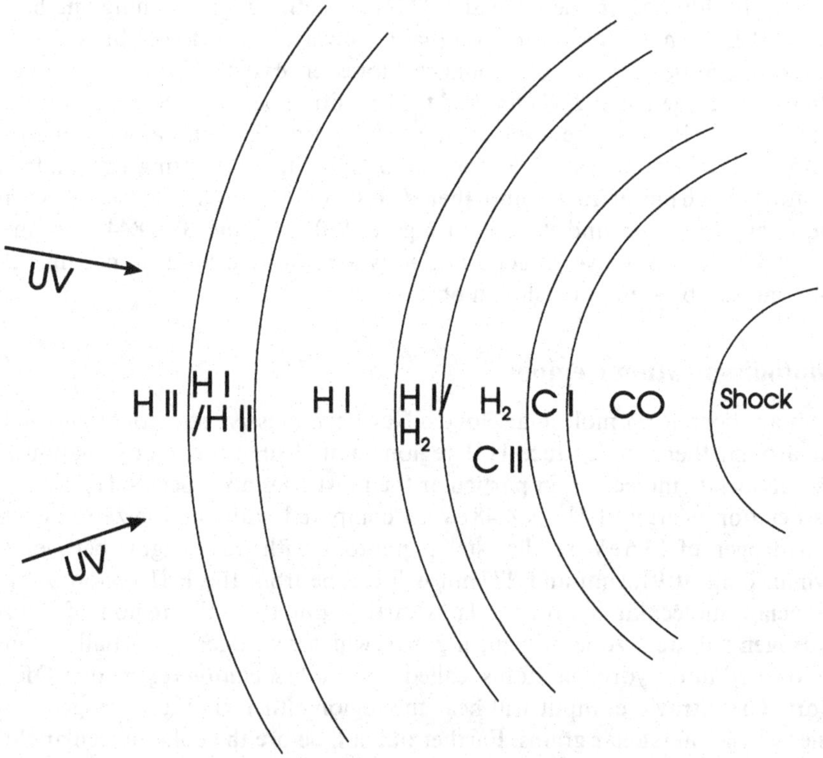

Figure 6.13. Photodissociation region (schematic)

Returning to the question of the degree of H_2 association, we find that in fact hydrogen molecules are mainly dissociated by spontaneous radiative dissociation. To understand this process, we need to examine the energy level structure of molecular hydrogen a little more closely. Hydrogen molecules in a molecular cloud will be found largely in the ground electronic state, X $^1\Sigma_g$, where X (and A,B,C...) are labels for the electronic states, with Σ indicating that the component of electronic orbital angular momentum along the internuclear axis, $\Lambda\, h/(2\pi)$, has $\Lambda = 0$ and the superscript referring to the multiplicity $2S + 1$ for spin angular momentum $Sh/(2\pi)$. The subscript g (even) refers to the fact that the molecular wave function of this state remains unchanged on reflection through the centre of the molecule, as opposed to changing sign when it would be designated u for odd (g and u being used only for homonuclear molecules which of course must have a centre of symmetry on the internuclear axis). The ground electronic state has 15 bound vibrational levels specified by quantum number v with corresponding energy $E = hcG(v)$ with

$$G(v) = \omega_e(v + 1/2) - \omega_e x_e(v + 1/2)^2 + \omega_e y_e(v + 1/2)^3 \qquad (6.75)$$

using the conventional terminology for molecular constants ω_e, etc. If excited to a vibrational quantum number above 15 in the ground electronic state, the molecule is unbound, and the energy to achieve this is the dissociation energy of the hydrogen molecule quoted above. If the gas was very dense and collisional processes were dominant, then when the average thermal energy approached this value, hydrogen would become dissociated, as indeed happens in stellar atmospheres. It must be remembered that temperatures of several thousand degrees are required to produce appreciable collisional excitation of the unbound vibrational levels.

However, in the low density and cooler conditions of interstellar clouds, radiative processes may be dominant. Now the hydrogen molecule has no dipole moment since it is homonuclear, and hence pure rotational and vibrational–rotational dipole transitions are forbidden and can only take place through weak and slow quadrupole transitions. Hence the radiative excitation rate to these vibrational levels is small. Electronic transitions from the ground state are permitted, provided the multiplicity does not change, and the even/odd symmetry does change. There are excited bound electronic states at excitation energies of around 11.4 eV ($1.8*10^{-18}$ J), B $^1\Sigma_u$, and of around 12.2 eV ($2*10^{-18}$ J), C $^1\Pi_u$, which are connected to the ground state by the permitted Lyman ($\sim$ 110 nm) and Werner ($\sim$ 101 nm) systems, respectively, with transitions to excited vibrational levels of these states being, of course, possible at shorter wavelengths. Radiative transitions to the neighbouring C $^3\Pi$ and E $^1\Sigma_g$ states are forbidden by the selection rules just mentioned. There are higher electronic levels, but absorption by neutral atomic hydrogen at wavelengths shortwards of 91.2 nm prevents radiative excitation of these, as it prevents excitation of high vibrational levels of B $^1\Sigma_u$ and C $^1\Pi_u$. Thus radiative excitation is essentially confined to the region between 91.2 nm and 110.8 nm (Figure 6.14).

Radiative excitation of B $^1\Sigma_u$ and C $^1\Pi_u$ will be followed by radiative decay to the electronic ground state. In 10% of cases, this will be to a vibrational level above $v = 15$, and so will result in an unbound molecule which will spontaneously dissociate. This is the process of spontaneous radiative dissociation which dominates in the case of molecular hydrogen. The direct formation of H_2 in the singlet ground state from two hydrogen atoms with the emission of a photon to carry off the energy released (radiative association) is forbidden under interstellar conditions. The hydrogen molecule is therefore formed by the attachment of hydrogen atoms to grains, the atoms then associating to form a molecule with the energy released transferred to the grain. Let the number density of hydrogen atoms be n_1 and the number density of hydrogen molecules be n_2. The rate of formation of hydrogen molecules per unit volume will then be proportional to the number of hydrogen atoms hitting a grain per second $\sim \sigma_{gr} v_{av} n_1$, where σ_{gr} is the

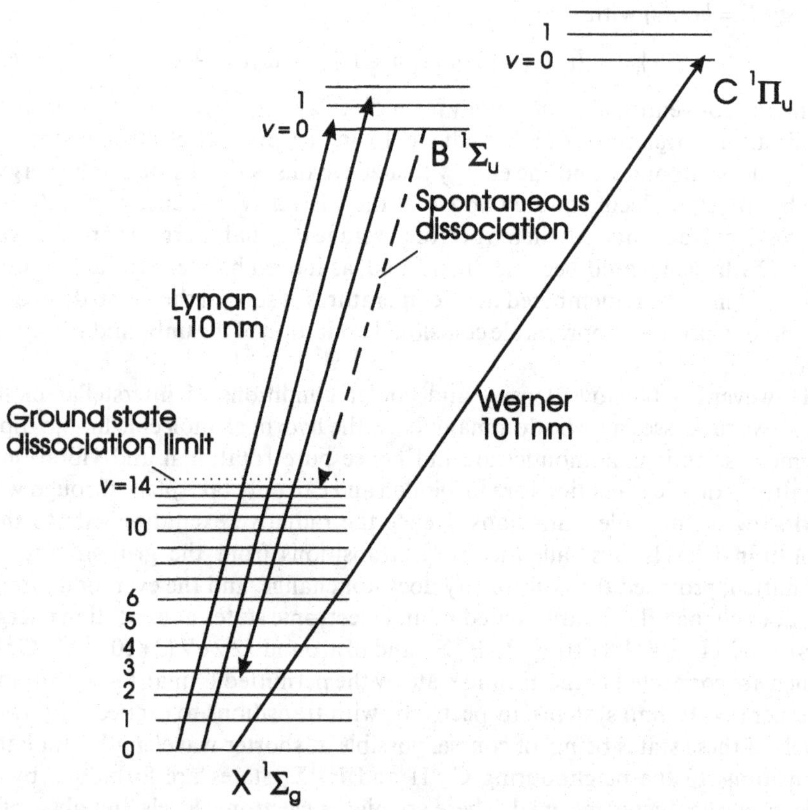

Figure 6.14. Electronic transitions of H_2

cross-section of the grain, πa^2, and to the number of grains per unit volume, $n_d = (n_1 + 2n_2)m_H/m_{gr}r_{dg}$, where m_{gr} is the mass of a grain, $4/3\pi a^3\pi_{gr}$, and r_{dg} is the dust to gas mass ratio. Hence we can write the production rate of molecular hydrogen per unit volume, P, as $n_1\ (n_1 + 2n_2)m_H/m_{gr}r_{dg}\sigma_{gr}v_{av}\eta$, where η is the efficiency of conversion of hydrogen atoms to hydrogen molecules on the grain and is close to 1. Thus P can be written: $P = R\eta n_1(n_1 + 2n_2)$, where $R \propto v \propto T^{1/2}$ and has a value of about $3 * 10^{-24}\ T^{1/2}\ m^3\ s^{-1}$.

The destruction rate of hydrogen molecules is $F_{UV}\sigma_{PD}n_2\xi$, where F_{UV} is the ultraviolet flux at the wavelength of the Lyman and Werner transitions to which the molecule is exposed, σ_{PD} is the total cross-section for electronic excitation in the same transitions and ξ is the fraction of excited molecules that spontaneously dissociate and is about 0.1. However the ultraviolet flux at the edge of the region, F_0, is reduced at a particular point in the region by dust absorption, and by absorption in the electronic transitions themselves, the latter process being known as self-shielding. Writing σ_{PD} as $\int \sigma_\nu(\mathrm{PD})d\nu = \Sigma_k \int \sigma_\nu(k)d\nu$, where the sum is over individual lines k, we find

$$\text{destruction rate } D = F_0\xi n_2 \sum_k \int \sigma_\nu(k)e^{-\tau_\nu}d\nu \tag{6.76}$$

where the optical depth to the dissociating radiation can be written as the sum of continuum (due to dust) and line (due to self-shielding) parts:

$$\tau_\nu = \tau_C + \tau_{\text{line}}$$
$$= \int \sigma_{dc}n_d dz + \sum_k \sigma_\nu(k)n_2 dz$$

where σ_{dc} is the dust absorption cross-section.

If σ_p is the dust cross-section per proton so $\sigma_p = \sigma_{dc}m_H/m_{gr}r_{dg} \sim 10^{-25}\,\mathrm{m}^{-2}$ and N_1 and N_2 are the column densities of atomic and molecular hydrogen measured from the irradiated side of the cloud:

$$\tau_\nu = \sigma_p(N_1 + 2N_2) + \sum_k \sigma_\nu(k)N_2 \tag{6.77}$$

The line cross-section can then be written in terms of the oscillator strength for a Lorentzian profile with Lorentz width Γ_k:

$$\sigma_\nu(k) = \frac{\pi e^2}{4\pi\varepsilon_0 m_e c} f_k \frac{\Gamma_k}{4\pi^2} \frac{1}{(\Delta\nu)^2 + \left(\dfrac{\Gamma_k}{4\pi}\right)^2}$$
$$\simeq \frac{\pi e^2}{4\pi\varepsilon_0 m_e c} f_k \frac{\Gamma_k}{4\pi^2(\Delta\nu)^2}, \quad \text{for } \Delta\nu \gg \frac{\Gamma}{4\pi}$$
$$= \frac{\sigma_k\Gamma_k}{4\pi^2(\Delta\nu)^2} \quad \text{say} \tag{6.78}$$

The destruction rate then becomes

$$D = F_0\xi n_2 e^{-\sigma_p(N_1+2N_2)} \sum_k \int_0^\infty \frac{\sigma_k\Gamma_k}{4\pi^2(\Delta\nu)^2} e^{-\sigma_k\Gamma_k N_2/4\pi^2(\Delta\nu)^2} d\nu$$
$$= F_0\xi n_2 e^{-\sigma_p(N_1+2N_2)} \sum_k \frac{\sqrt{\sigma_k\Gamma_k N_2}}{2\pi N_2} 2\int_0^\infty e^{-u_k^2}du_k, \quad \text{with} \quad u_k^2 = \frac{\sigma_k\Gamma_k N_2}{4\pi^2(\Delta\nu)^2}$$
$$= F_0\xi n_2 e^{-\sigma_p(N_1+2N_2)} \frac{1}{2\sqrt{\pi N_2}} \sum_k \sqrt{\sigma_k\Gamma_k} \tag{6.79}$$

Equating the destruction and formation rates:

$$R\eta n_1(n_1+2n_2)=F_0\xi e^{-\sigma_p(N_1+N_2)}\frac{n_2}{2\sqrt{\pi N_2}}\sum_k\sqrt{\sigma_k\Gamma_k}$$

Hence

$$\frac{dN_1}{dN_2}=\frac{n_1\,dz}{n_2\,dz}=\frac{F_0\xi\sum_k\sqrt{\sigma_k\Gamma_k}}{\eta R\,(n_1+2n_2)}\frac{e^{-\sigma_p(N_1+2N_2)}}{2\sqrt{\pi}\sqrt{N_2}}$$

$$\int_0^{N_1}e^{\sigma_p N_1'}\,dN_1'=\frac{F_0\xi}{\eta R}\frac{1}{2\sqrt{\pi}}\frac{\sum_k\sqrt{\sigma_k\Gamma_k}}{(n_1+2n_2)}\int_0^{N_2}\frac{e^{-2\sigma_p N_2'}}{\sqrt{N_2'}}\,dN_2'$$

$$\frac{(e^{\sigma_p N_1}-1)}{\sigma_p}=\frac{F_0\xi}{\eta R}\frac{1}{2\sqrt{\pi}}\frac{\sum_k\sqrt{\sigma_k\Gamma_k}}{(n_1+2n_2)}\sqrt{\frac{2}{\sigma_p}}\int_0^{\sqrt{2\sigma_p N_2}}e^{-q^2}\,dq \tag{6.80}$$

where the total density which is proportional to n_1+2n_2, has been assumed to be constant.

Equation (6.80) can be solved to obtain a relation between N_1 and N_2. However the total thickness of the hydrogen dissociation region can be obtained by letting N_2 become very large so that we are taking a point beyond the distance at which the hydrogen becomes totally molecular, when N_1 is the column density of hydrogen atoms through the whole of the region in which hydrogen is partially dissociated. Then

$$\begin{aligned}N_1&=\frac{1}{\sigma_p}\ln_e\left(\frac{F_0\xi}{\eta R}\frac{\sum_k\sqrt{\sigma_k\Gamma_k}}{(n_1+2n_2)}\frac{1}{2\sqrt{2}}\sqrt{\sigma_p}+1\right)\\&=\frac{1}{\sigma_p}\ln_e(1+\alpha D_0)\end{aligned} \tag{6.81}$$

where

$$\alpha=\frac{G_0}{(n_1+2n_2)},\qquad D_0=\frac{\xi}{\eta R}\frac{F_{is}}{2\sqrt{2}}\sum_k\sqrt{\sigma_k\Gamma_k\sigma_p}$$

and

$$G_0=\frac{\text{far UV flux irradiating region}}{F_{is}}$$

F_{is} is 4π times the interstellar mean intensity in the vicinity of the Sun in the appropriate wavelength range, which has a value of about $6*10^{-22}\ \mathrm{Wm^{-22}\,Hz^{-1}}$ so $G_0=F_0/F_{is}$. If we crudely model the neutral gas region as having a thickness t where the hydrogen is almost entirely atomic at constant density so $n_p=n_1$ and then a thickness 2δ where the atomic hydrogen density falls linearly to zero, then $N_1 = (t+\delta)n_p$ and for a given n_p we can find the distance into the cloud at which the hydrogen is half dissociated, although such a model has a varying pressure, and more realistically we need to calculate the temperature, density and pressure consistently. It is customary to measure distance into the region in terms of visual dust extinction $A_V = 2.5\log(\exp(-\tau_D)) = 1.086\tau_D = 1.086\sigma_p$ (visual)$n_p(t+\delta) \sim 0.5\sigma_p N_1$ since the visual dust extinction cross-section is about half that in the far ultraviolet.The ratio α is then the main parameter characterizing PDRs with D_0 a constant except for the possibility that σ_p may vary if grain sizes are unusual. Typically, α has values of the order of $10^{-8}\,\mathrm{m}^3$ although of course there is a wide range.

Jura [4] argues that since $\Gamma_k = \Sigma A_{kl}$ for natural line broadening (collisional broadening will of course be negligible under low density conditions) and roughly $\Sigma A_{kl} \sim 6*10^{10} f_k$, where f_k is the oscillator strength for the $R(0)$ line of the given vibrational band of the electronic transition with $\sigma_k = 2.66*10^{-6} f_k\,\mathrm{m}^2$, then $\sqrt{(\sigma_k\Gamma_k)} = 1.5*10^8\,\sigma_k$ and the destruction rate is

$$D = F_0\xi n_2 e^{-\tau_c} \frac{1}{2\sqrt{\pi}\sqrt{N_2}} 1.5\times 10^8 \sum_k \sigma_k$$

Since $D_{us} = F_0\xi n_2 \Sigma\sigma_k \sim 5*10^{-11} G_0 n_2$ is the unshielded destruction rate per unit volume, D can be written $D = D_{us}\beta/\sqrt{N_2}\exp(-\tau_c)$ with $\beta/\sqrt{N_2}$ representing the effect of self-shielding with a value for β of $4.2*10^7\,\mathrm{m}^{-1}$ and D_0 in (6.81) becomes $\xi/(\eta R) F_{is}\sqrt{\sigma_p}\sqrt{(\pi/2)}\,\beta\Sigma\sigma_k \sim 10^{-7}\,\mathrm{m}^{-3}$.

If α is small so $\alpha D_0 \ll 1$, (6.81) becomes:

$$N_1 \sim 1/\sigma_p \alpha D_0 \tag{6.81a}$$

while if α is large so $\alpha D_0 \gg 1$,

$$N_1 \sim 1/\sigma_p \ln_e(\alpha D_0) \tag{6.81b}$$

so for large ultraviolet fluxes or low total densities, the self-shielding means that the thickness of the dissociated region increases relatively slowly with the ultraviolet flux. These thicknesses can be turned into visual dust absorptions A_V, and it is then found that dissociation comes to an end for A_V about 1.

Carbon monoxide is destroyed by the process of predissociation where a radiative electronic transition is made to an excited bound electronic state which is linked to another excited electronic state which is unbound (Figure 6.15). Experiments in the late 1980s increased estimates of the dissociation rate of CO by a factor of about 40. Since a line absorption process is involved, self-shielding can become important in the same way as for hydrogen molecules, with the added complication that since the lines involved are in the same fairly narrow region of the far ultraviolet spectrum as the H_2 dissociating lines, it quite often happens that a CO line is shielded by an H_2 line. The upshot of detailed calculations is that

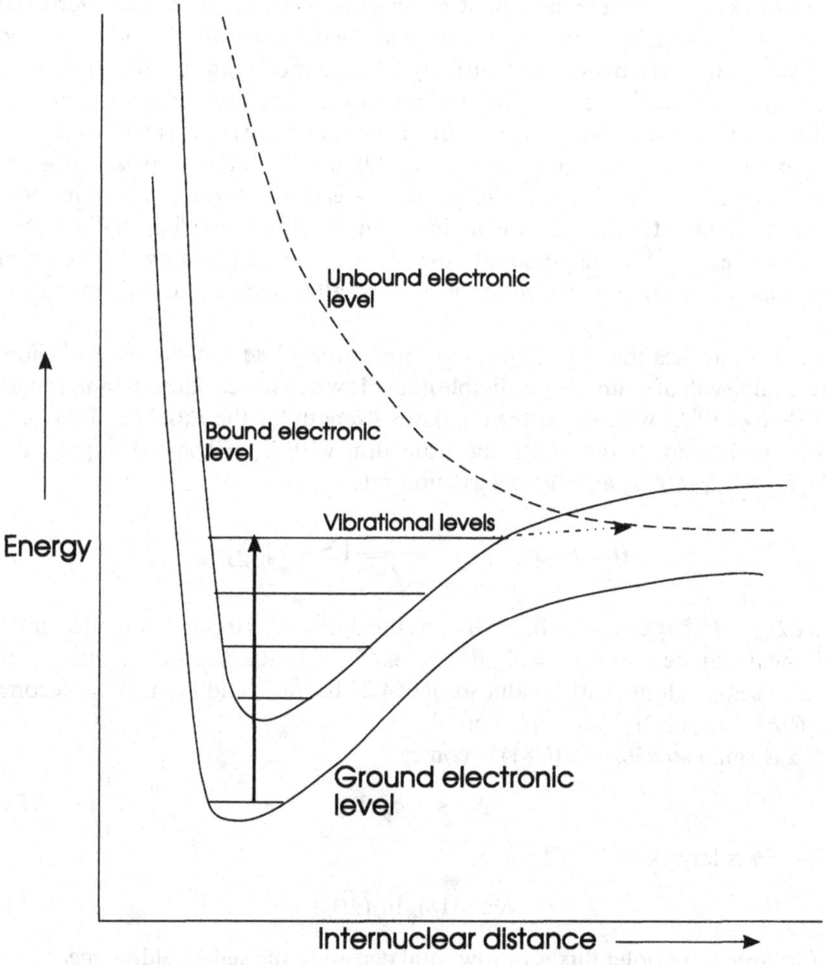

Figure 6.15. Predissociation; bound levels have minimum energy

CO becomes dissociated considerably deeper into the cloud than H_2 at visual absorptions of about 2 to 4.

We now consider the molecular hydrogen spectrum produced by a PDR. The ultraviolet excitations which can return to highly excited vibrational levels of the ground state leading to dissociation can also return to bound vibrational states of the ground electronic state, and in fact do so 90% of the time (Figure 6.16). These bound vibrational levels of the ground state then cascade downwards to $v = 0$ and $J = 0$ or $J = 1$ via radiative vibrational-rotational transitions and pure rotational transitions although the latter have lower transition probabilities and

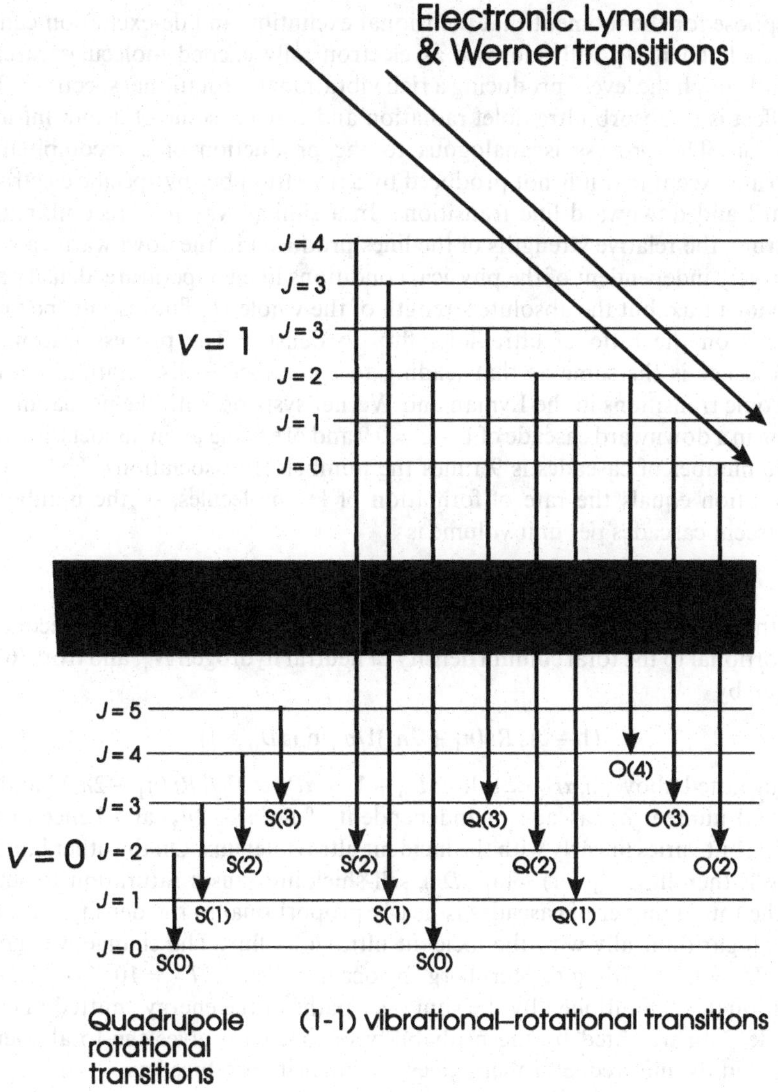

Figure 6.16.

are only of importance within the $v = 0$ vibrational state. The quadrupole vibrational–rotational transitions are slow, with transition probabilities of the order of 10^{-6} to 10^{-7} s^{-1}. Thus collisional de-excitation may compete with radiative de-excitation, and at least the lower levels can be thermalized at a critical density that is relatively modest ($\sim 10^{12} m^{-3}$)—the concept of the critical density is discussed in Chapter 8.

Suppose for the moment that collisional excitation and de-excitation can be neglected. Then H_2 will fluoresce as electronically excited molecules cascade down through the levels, producing a rich vibrational—rotational spectrum. The net effect is to absorb ultraviolet radiation and convert some of it into infrared radiation. The process is analogous to the production of a recombination spectrum except that it is not produced by association but by specific electronic upward and downward line transitions. In a similar way to a recombination spectrum, the relative strengths of the lines produced in the downward cascade are largely independent of the physical conditions like temperature, density and ultraviolet flux, but the absolute strength of the whole H_2 fluorescent spectrum depends on the ratio of ultraviolet flux to density. The process leading to fluorescence is the same as that leading to spontaneous dissociation, namely electronic transitions in the Lyman and Werner systems, with the probability of ending in a downward cascade of $1 - \xi = 0.9$ and of ending up in dissociation of ξ, so the number of cascades is 9 times the number of dissociations. The rate of dissociation equals the rate of formation of H_2 molecules, so the number of fluorescent cascades per unit volume is

$$(1 - \xi)/\xi R\eta n_1(n_1 + 2n_2)$$

and the total number of fluorescent cascades from the PDR per second is proportional to the total column density of neutral hydrogen N_1, and from (6.81) is given by

$$(1 - \xi)\xi R\eta(n_1 + 2n_2)1/\sigma_p \ln_e(\alpha D_0 + 1) \tag{6.82}$$

As was noted above, if $\alpha D_0 \ll 1$, $\ln_e(\alpha D_0 + 1) \sim \alpha D_0 \propto F_0/[R\eta(n_1 + 2n_2)]$ and so the total number of cascades is independent of $R\eta(n_1 + 2n_2)$ and hence of the density, but varies linearly with the incident ultraviolet flux. On the other hand, if $\alpha D_0 \gg 1$, then $\ln_e(\alpha D_0 + 1) \sim \ln_e(\alpha D_0)$, self-shielding causes saturation to set in and the total number of cascades is nearly proportional to the density but only varies logarithmically with the incident ultraviolet flux. The changeover point with $\alpha D_0 = 1$, according to Sternberg [5] occurs when $F_0/F_{is} = 10^{-8}(n_1 + 2n_2)$. Equation (6.82) multiplied by the sum over paths of the energy emitted in each cascade path weighted by the probability of that path gives the total energy emitted in the infrared. Sternberg gives the intensity produced as

$$I = 3*10^{-17}(n_1 + 2n_2)\ln_e[9*10^7 F_0/F_{is}\, 1/(n_1 + 2n_2) + 1] \quad \mathrm{W\,m^{-2}\,sr^{-1}} \tag{6.83}$$

We now need to consider how this energy is divided between the various lines of the cascade. The first step is to calculate the entry rate into the various levels of the ground state first via excitation from the ground state to an excited electronic state which will depend on the ultraviolet radiation field and then via radiative line decays from the excited electronic state to the various vibrational and rotational states of the ground electronic state, which will depend only on the transition probabilities in the appropriate lines. In a cold molecular gas only the

$J = 0$ and $J = 1$ levels of the $v = 0$ vibrational level of the electronic ground state will be appreciably populated, and the rotational quantum number can only change by one in an electronic transition, so that the return transitions to the ground electronic state will only produce appreciable populations in the $J = 0, 1, 2$ and 3 rotational levels. On the other hand all the vibrational levels will receive appreciable populations, even $v = 14$ having something of the order of 10% of the population of $v = 0$. The subsequent downward cascade is now calculated working from the uppermost level $v = 14$ and the highest J value considered and moving downwards, for each level considering the input from the excited electronic states and from higher levels of the ground electronic states and the output by quadrupole transitions to lower levels, thus obtaining the population of each level.

The quadrupole transitions obey the selection rules $\Delta J = \pm 2, 0$ where $\Delta J = 0$ gives the Q branch ($J = 0$ to $J = 0$ forbidden so Q(1) is the first line of this branch with the convention of labelling being $Q(J_L)$), $J_U - J_L = 2$ gives the S branch starting with S(0) and $J_U - J_L = -2$ giving the O branch with lowest member O(2). The downward transition probabilities are proportional to the fifth power of the frequency, a matrix element depending mainly on the vibrational quantum numbers involved and a rotational line strength factor S_L that can be written in terms of J_U as $S_L(J) = J(J-1)/[(2J-1)(2J+1)$ for the S branch, $S_L(J) = (J+1)(J+2)/[(2J+1)(2J+3)]$ for the O branch, and $S_L(J) = 2J(J+1)/[(2J-1)(2J+3)]$ for the Q branch, but usually the v^5 factor makes the S branch transitions from a given upper level the strongest. The lifetimes of most levels lie between 10^6 and 10^5 s, except those of the ground vibrational state which can only drop to the lowest rotational levels ($J = 0$ or $J = 1$) by pure rotational quadrupole transitions giving lifetimes ranging downwards as $1/J^5$ from the $3 * 10^{10}$ s for $J = 2$. The upshot is that high vibrational levels can have relatively large populations compared with thermal equilibrium, but only the lowest rotational levels of each vibrational state have appreciable populations.

The level populations then enable the relative line strengths to be estimated, since the line emission is optically thin so the line strengths are proportional to the emission coefficient which in turn is equal to the product of the level population and the transition probability divided by 4π. The cascade populations lead us to expect quite strong lines from fairly high vibrational levels but only from the first few rotational levels. The line wavelengths can be calculated from the standard equations:

vibrational wavenumber

$$G(v) = 1/\lambda = 440121\,(v+1/2) - 12134(v+1/2)^2 + 81.29(v+1/2)^3\,\mathrm{m}^{-1}$$

rotational wavenumber

$$F(J) = [6085.30 - 306.22(v+1/2) + 5.77(v+1/2)^2 + 0.51(v+1/2)^3)]J(J+1) - [4.666 - 0.18(v+1/2)]J^2(J+1)^2\,\mathrm{m}^{-1}$$

$$1/\lambda = G(v_U) - G(v_L) + F(J_U) - F(J_L)$$

with $J_U - J_L = 2$ for the S branch, 0 for the Q branch and -2 for the O branch. The same formula will give the pure rotational lines for 0–0 where effectively there is only an S branch, although for high J values the power series need extra terms. Thus for the 1–0 vibrational transition we have the Q branch starting at Q(1) at 2.407 μm, Q(2) at 2.412 μm and Q(3) at 2.424 μm, while the S branch starting with S(0) at 2.223 μm moves more rapidly to shorter wavelengths with S(1) at 2.122 μm, S(2) at 2.034 μm and so on, while the O branch starting at O(2) at 2.626 μm moves rapidly to longer wavelengths with O(3) at 2.802 μm and so on. The 2–1 vibrational band is at somewhat longer wavelengths with a large degree of overlap, with Q(1) at 2.561 μm, S(0) at 2.364 μm and O(2) at 2.798 μm, and so on. Finally it is worth noting that the pure rotational lines start with S(0) at 28 μm, but are very widely spaced so that S(17) is near 3.5 μm. The large degree of wavelength overlap between lines from different excitations makes comparisons more straightforward from the point of view of variations of the atmosphere and instrumental sensitivity with wavelength.

Predictions of fluorescent cascades (see Sternberg [5]) suggest that the strongest lines include 1–0 S(1), S(2) and S(3), Q(1) and Q(3) and O(3), but 2–1 S(1), Q(1) and Q(3), and O(3) are not much weaker, as is the case for the Q(1) and S(1) lines of 2–0, 3–1, 4–2 and 5–3. The level populations derived from the strengths of these lines will give a high vibrational excitation temperature but a low rotational excitation temperature. The nuclei of H_2 are identical, so the nuclear triplet states with parallel nuclear spins must combine with odd J rotational states if the overall wave function is to be antisymmetric (*ortho*-hydrogen) and the nuclear singlet states must combine with even J rotational states (*para*-hydrogen). Radiative transitions never link *ortho*- and *para*-hydrogen, which therefore behave as separate species, with ortho levels having statistical weight $3(2J+1)$ and para levels having statistical weight $(2J+1)$. Observationally, fluorescent lines suggest an ortho to para ratio rather less than 3, in the range 1 to 2.

Now because of the low transition probabilities, collisional de-excitation can compete with radiative de-excitation at relatively low densities of around $10^{10}\,\mathrm{m}^{-3}$, and at even lower densities in the case of the rotational levels of the ground vibrational state. The critical density for this to occur for level j is given by $C_{ji} \sim \sigma_{ji} v_{av} n_{crit} = A_{ji}$, where only one downward transition to level i has been included but strictly all downward transitions should be considered. The population of level j is given by

$$N_j = (\text{entry rate from higher levels, } P_j)/(A_{ji} + C_{ji})$$

so that if $A_{ji} \gg C_{ji}$, line strength is proportional to $N_j A_{ji} \propto P_j$, whereas if $C_{ji} \gg A_{ji}$, the line strength is proportional to $P_j A_{ji}/C_{ji} \propto P_j\, n_{crit}/n$ and since P_j will increase with n, the line strengths will tend to a constant value as the density is increased further.

At high densities the level populations will approach a Boltzmann distribution. In PDRs, the temperature is determined by the balance between heating by the ultraviolet flux photo-ejecting electrons from grains, which will tend to be concentrated near the illuminated side of the region, and cooling by forbidden fine-structure emission lines which are collisionally excited. At high densities collisional de-excitation of hydrogen molecules that have been excited by ultraviolet radiation will lead to heating of the gas, and high densities will also tend to suppress the forbidden line atomic radiation so that temperatures can reach 1000 K. The collisional excitation will then populate a wider range of rotational levels in the ground and first excited vibrational levels, but at these temperatures will not populate appreciably the higher vibrational levels. Thus the strongest lines are all 1–0 transitions, covering a wide range in J. The vibrational and rotational excitation temperatures are the same and the ratio of ortho- to *para* hydrogen is 3:1. Even a relatively small fraction of dense hot gas will dominate the emission. The ratio of S(1) 2–1 to S(1) 1–0 intensities (the lines are at 2.25 μm and 2.12 μm respectively) are often used to discriminate between fluorescent and collisional excitation, with a predicted ratio of about 0.5 for the fluorescent case, and less than 0.1 for the collisional case, although the use of more than this one line ratio is highly desirable, particularly as in many cases the excitation is by shocks.

The discussion of PDRs has concentrated on the H_2 spectrum, but of course there are many other line transitions from the PDR. Low lying fine-structure atomic levels can be collisionally excited and de-excited radiatively. Examples including the C II line at 158 μm ($^2P_{3/2}$–$^2P_{1/2}$) from the large portion of the PDR in which carbon is singly ionized and the O I lines (oxygen is neutral throughout) at 63 μm ($^3P_1$–$^3P_2$) and 145 μm ($^3P_2$–$^3P_3$). Recombination lines of C I are also seen, whereas from the transition region to associated CO one finds fine structure lines of C I at 370 μm and 609 μm and high J rotational lines of hot CO.

References

[1] N. Pannagia, *JA, Astronomical Journal*, **78**, 929, 1972.
[2] D. E. Osterbrock, *The Astrophysics of Gaseous Nebulae*, University Science Books, 1989.
[3] M. Brocklehurst, *Monthly Notices Royal Astronomical Society*, **148**, 417, 1971.
[4] M. Jura, *Astrophysical Journal*, **191**, 375, 1974.
[5] A. Sternberg, in T. Hartquist (ed.), *Molecular Astrophysics*. Cambridge University Press, 1990.

7 PHOTOIONIZED CLOUDS—AGN

Introduction and Classification

Active galaxies are galaxies where very large luminosities are produced in very small volumes near the centre of the galaxy. These regions, called active galactic nuclei or AGN for short, have 'non-thermal' continuous spectra and, where they display emission line spectra (the majority of cases), have lines with large widths, presumably indicating large velocities. Most AGN also exhibit variability.

Making these defining characteristics more precise is difficult. Firstly, there is the 'luminous but compact' characteristic. Most of the objects on lists of AGN have nuclei with blue absolute magnitudes of -20 (equivalent typically to a total luminosity of 10^{46} W) or greater, in some cases very much greater, with most of the luminosity coming from a region less than a parsec in radius. However nuclei showing other AGN characteristics such as broad lines but having low luminosity are known but are naturally hard to detect. Some of the emission line luminosity (the narrow lines) comes from a rather larger region typically with a radius of tens of parsecs, but luminous examples extending over several kiloparsecs are known. Secondly, there is the 'non-thermal' characteristic. A thermal spectrum resembles a blackbody spectrum, with the flux per unit frequency interval rising rapidly to a peak with increasing wavelength and then falling away more slowly at longer wavelengths, the wavelength giving the peak output falling with increasing temperature. The spectrum of a normal galaxy covers a broader range in wavelength because it represents the sum of the spectra of individual stars with a variety of temperatures. A typical AGN continuous spectrum, on the other hand, varies with frequency or wavelength fairly slowly right through the ultraviolet and visible, and AGN have large X-ray and far infrared fluxes. However, 'non-thermal' has to be put in inverted commas because slowly varying with frequency continua can also be produced by objects with large ranges of temperature contributing significantly to the total emission, such as some dust clouds and accretion discs, so 'non-stellar' would be a safer description. The third characteristic is the presence of high velocities. Normal spiral galaxies have rotation velocities between 150 and 300 $\mathrm{km\,s^{-1}}$ and normal elliptical galaxies have velocity dispersions in the same range. The spectrum of a normal galaxy is the sum over many stars and the lines are therefore Doppler broadened by amounts corresponding to these velocities. The AGN have line widths ranging from 400 up to several thousand $\mathrm{km\,s^{-1}}$ if interpreted as Doppler broadening. No other explanation of the broadening seems tenable in many cases.

AGN are to be distinguished from galaxies with luminous central regions and emission lines produced by a rapid rate of star formation. Many galaxies have some star formation and weak emission lines from the nucleus, but the *starburst galaxies* have very large star formation rates and may as a result have high luminosity, which in some cases makes them among the most luminous galaxies known. Starburst galaxies have extensive H II regions with accompanying emission lines, but much of the ultraviolet radiation from hot newly born stars is re-radiated by dust at long wavelengths so that the bulk of the luminosity of a starburst galaxy may appear in the far infrared. The temperatures of the dust grains cover a wide range so the infrared spectrum can be fairly flat. The enhanced star formation typically occurs within a radius of one to two kiloparsecs, much larger than the region from which most of the luminosity of an AGN comes, and starburst galaxies do not have the broad lines and variability of an AGN. Many AGN do have dust heated by the central source and hence can have infrared spectra resembling starburst galaxies. On the other hand, although starburst galaxies have some radio and X-ray emission from supernova remnants, AGN in general have much higher X-ray luminosity, and some AGN are very powerful radio emitters. It will be seen that although the distinction between AGN and starburst galaxy is often easily made, it can sometimes be difficult to decide which is the dominant process in a particular galaxy.

We now return to AGN proper, which of course now constitute an enormous area of research. For our present purposes, the most important aspect of AGN is the large continuum fluxes from the visible to the X-ray region, which are capable of photoionizing any gas clouds in the vicinity of the nucleus to produce a very wide range of ionizations. Most of this chapter will be devoted to exploring the consequences in the form of the line emission from such photoionized nebulae. First, however, to introduce the terminology, we will make a brief tour of the AGN zoo. As we list the various types of AGN, it must be remembered that the classification of AGN is naturally biased towards the means of discovery—bright nucleus, nucleus with broad emission, X-ray source, radio source, variable object—rather than to the essential nature of the object, which we as yet hardly understand.

Seyfert galaxies are spiral galaxies with bright nuclei producing broad emission lines. The nuclei are ultraviolet bright, producers of strong X-ray fluxes and are variable, but Seyferts are never strong radio sources. *N-type galaxies* are galaxies with bright nuclei that dominate the whole galaxy, and are often strong radio sources. *Radio galaxies* are strong radio sources that are always elliptical galaxies, usually with bright nuclei showing broad emission lines, variability, and ultraviolet bright continua. The category 'N type' is largely of historical interest and non-stellar (resolved) galaxies with AGN are classified either as Seyferts or as radio galaxies. *Quasi-stellar objects* or QSOs are unresolved (apparently stellar) objects with luminosity as least as great as that of a very bright galaxy (usually much greater than that of the brightest normal galaxies), whose spectra show a non-thermal continuum with broad emission lines, and are also characterized by large X-ray fluxes and variability. *Quasars* are QSOs that are also powerful radio sources (although often quasars and QSOs are lumped together and called indifferently 'quasars' and 'QSOs', with the phrases 'radio-loud' and 'radio quiet' being used to distinguish radio power). *BL Lac objects* are objects with apparently stellar images with non-thermal continua that are strong radio sources and show strong variability, but have no line emission. The radiation from BL Lac objects is polarized, and non-thermal continuum, radio emission, strong variability and the polarization are shared with a sub-group of quasars sometimes called *optically violent variables* (OVV), which differ from BL Lacs mainly in the presence of strong emission lines. OVV and BL Lac objects are sometimes grouped together as *blazars*. It should be noted that weak lines have been seen in some BL Lac objects, which enable a redshift to be obtained and confirm that they are indeed very luminous objects like the quasars.

Quasars, QSOs and BL Lac objects are more luminous than Seyferts and radio galaxies, the dividing line in the optical coming at a blue magnitude of -23. The 'stellar' group—Quasars, QSOs and BL Lac objects—also have much lower space densities by a factor of about 100 than the 'galaxies'—Seyferts and radio galaxies. The 'stellar' group can therefore be seen to much larger distances, but because of their rarity very nearby examples are unlikely to be found. At the distances of most observed quasars and QSOs, an underlying galaxy would be quite faint, and the very luminous nucleus with its finite sized image when observed through the Earth's atmosphere would usually swamp the light from such a galaxy. Thus if 'stellar' AGN are the nuclei of galaxies in the same way that Seyfert nuclei are the nuclei of spiral galaxies, one would not expect to see the underlying galaxy in most cases. In fact faint traces of an underlying galaxy of uncertain type have now been found for a number of QSOs and BL Lac objects. It would then seem plausible that the 'stellar' group of AGN (QSOs, etc.) represents at least partially a brighter continuation of the phenomenon represented at lower luminosity by Seyfert galaxies and radio galaxies. This is born out by the fact that variability can make an AGN appear to change type so there is at least one example of an object classified as a QSO appearing when less bright as a Seyfert galaxy, and another of a BL Lac object (as seen at one time) appearing as more like a Seyfert with broad lines on another occasion. In both 'stellar' and 'galaxy' groups, the radio loud objects are in a minority by a factor of about 100. In the matter of observational bias, it is also worth noting that observations in the visible of QSOs and quasars are mainly of distant, redshifted objects, so that it is the ultraviolet rest frame that is moved into the wavelength region observed, whereas most Seyferts are observed at low redshift. Thus, unless ultraviolet or infrared observations are available, rather different rest frame spectral regions are observed in QSOs/quasars and in Seyferts/radio galaxies.

The emission lines fall into two different categories: the 'broad' lines with full widths at half maximum of the order of 1000 to 5000 $km\,s^{-1}$ and the 'narrow' lines with full widths at half maximum of the order of 500 $km\,s^{-1}$, although the term 'narrow' is something of a misnomer since these lines are about twice the width of those found in a normal galaxy. Forbidden transitions are only found as narrow lines whereas permitted lines may have both broad and narrow components. Now a wide range of ionizations are represented in the broad line spectrum and since O VI is seen as a broad line as well as lower stages of ionization of oxygen, O III is certainly present in part of the broad line region. The fact that [O III] forbidden line transitions are seen only as 'narrow' lines then suggests that broad forbidden lines are suppressed by collisional de-excitation. Lines that are collisionally excited and then decay radiatively have intensities that are proportional to the density squared as do recombination lines, but if a line is collisionally excited but the excited level normally decays by collisional de-excitation and only occasionally by radiative de-excitation (so the level population tends towards its Boltzmann equilibrium value) then the intensity of the line will only be proportional to the density. Hence at high densities, forbidden lines are weak compared with both permitted collisionally excited lines and recombination lines. It must be concluded that the 'narrow' lines in AGN come from a lower density region than the 'broad' lines. A quick glance at the observational data suggests that the narrow lines in a particular object all have much the same width in velocity units, and similarly the broad lines in a particular object have much the same width, although more detailed inspection shows considerable variation from line to line, particularly for the broad lines. This leads to a division in models of line-forming regions in AGN into *broad line regions* (BLR) and *narrow line regions* (NLR), the first having higher velocities and densities than the second.

Some Seyferts show lines from both BLR and NLR and are called *Seyfert 1* galaxies. Others show only lines from NLR (even the permitted lines only have a narrow component) and are called *Seyfert 2* galaxies. Intermediate cases are classified as Seyfert 1.5, 1.8 and 1.9 galaxies. Similarly, some radio galaxies have both broad and narrow lines and are termed *broad line radio galaxies* (BLRG), whereas others have only narrow lines and are called *narrow line radio galaxies* (NLRG). Seyfert 1 galaxies are more luminous on average than Seyfert 2 galaxies, although both types cover a range of luminosity. Interestingly, examples are known where variability has been accompanied by a change of spectrum from Seyfert 1 to Seyfert 2 . No cases of quasars or QSOs having only narrow lines are known.

Continuous Spectra

We now turn from classification to a brief consideration of the continuous spectrum of AGN. Continuous spectra are plotted as flux per unit frequency interval, F_ν, versus frequency, ν, or if a wide wavelength range is involved, versus log ν. In the latter case, the plot is very commonly drawn with log νF_ν as the abscissa, since log νF_ν is approximately proportional to the energy emitted in a given range of log ν and hence the areas under the curve in such a plot give directly the relative amounts of energy in various frequency ranges. 'Non-thermal' continua are often characterised by a spectral index by fitting a power law $F_\nu \propto \nu^{-\alpha}$ to the observed spectrum.

A detailed consideration of synchrotron radiation lies outside the remit of this book, but a brief summary must be given to allow the arguments about the nature of AGN continua to be appreciated. Synchrotron radiation is the radiation produced by relativistic electrons spiralling around a magnetic field, and has the property that if the distribution of electron energies has a power law form, then the resulting synchrotron flux per unit frequency interval also has a power law form. Any realistic acceleration mechanism will produce fewer high energy electrons than low energy electrons, and unless the dependence is a very weak one, the resulting synchrotron spectrum will have a positive spectral index, with larger fluxes at low frequencies. Synchrotron absorption will also become important at low frequencies and at some frequency the plasma will become optically thick. As absorption and emission come into balance, the spectrum will turn over and at lower frequencies the flux will decrease with decreasing frequency according to $F_\nu \propto \nu^{2.5}$. If the source is made of several components of different thicknesses and hence of different turnover frequencies, and if the components are unresolved, the observed spectrum will be a composite of the individual synchrotron spectra and may look fairly flat. Synchrotron radiation is about 70% linearly polarized with the plane of polarization determined by the magnetic field direction, independently of frequency if the source has a single component and the magnetic field is in the same direction over the whole source. A less uniform field direction will reduce the degree of polarization, and the presence of several components with different thicknesses and different field directions could result in changes in the plane of plane of polarization with frequency.

Radio galaxies were originally divided into *flat spectrum* sources, usually compact, and *steep spectrum* sources, usually extended, where the dividing line depended on whether the spectral index of the radio spectrum was greater (steep) or less (flat) than 0.4. Many radio galaxies turn out when mapped to radiate in two *radio lobes* on opposite sides of the optical galaxy but often lying well outside the optical extent of the galaxy, hundreds of kiloparsecs apart. The lobes have a steep radio spectrum that is polarized and extends to the visible and there is no doubt that the synchrotron emission is the mechanism involved, and equally no doubt that the lobes must be being continually resupplied with energy, presumably from the central galaxy. The most powerful (in radio terms) radio galaxies have *hot spots* near the outer edges of their radio lobes. There is also a compact source, sometimes relatively weak and commonly with a flat spectrum, that coincides with the optical galaxy. Closer examination shows linear features called *jets* extending from the core towards the lobes, with not more than two jets observed. The intensities of the jets can be very different, and in some cases only one jet is seen.

Radio galaxy morphology is used to divide radio loud AGN into those in which the lobes dominate the radio luminosity or *lobe-dominated* sources, and those in which the core is more important or *core-dominated* sources. BL Lac objects and OVV quasars are all core dominated, while some quasars are lobe dominated and some are core dominated, and the majority of radio galaxies are lobe dominated. A division is also made amongst the powerful radio galaxies between the most powerful radio galaxies with radio powers greater than $5*10^{25}\,\mathrm{W\,Hz^{-1}}$ at a frequency of 1.4 GHz, called FR II (where FR stands for Faranoff–Riley) and which have hot spots at the outer edges of the lobes, and slightly less powerful radio galaxies where the brightest points in the lobes are well in from the edges, which are termed FR I. The cores and jets in FR II sources are more luminous than those in FR I sources, but since the lobes are much more luminous in FR II sources, the jets and cores represent a smaller fraction of the total radio luminosity in these galaxies. The FR II sources are associated with giant elliptical galaxies, whereas FR I sources are associated with the even more luminous D and cD elliptical galaxies. The lobe-dominated quasars have an FR II radio morphology.

After this brief excursion into the radio properties of AGN, we start by considering the continuous spectra of BL Lac objects. These are core-dominated radio sources with continuous spectra that rise smoothly with increasing frequency from a fairly flat radio spectrum to a maximum (in νF_ν terms), which is most commonly in the infrared but may be at higher frequencies, and then declines at shorter wavelengths. Observations of variability provide an upper limit to the source size, since a source cannot vary coherently in a shorter time scale than the light travel time across the object. The large radio luminosity and the small sizes combine to give very high radio brightness temperatures of up to 10^{12} K which cannot have a thermal origin. This, together with the polarization and the radio spectrum, strongly suggests that synchrotron radiation provides the radio luminosity. The smooth variation of the continuous spectrum into the visible and the polarization also found at optical wavelengths then further suggest a synchrotron origin for the continuous whole spectrum, with the possible exception of the X-ray region in some objects where the flux lies above an extrapolation of the spectrum from lower frequencies (Figure 7.1).

High resolution observations of the very luminous cores show miniature luminous jets on milli-arcsecond scales (corresponding to lengths of parsecs),

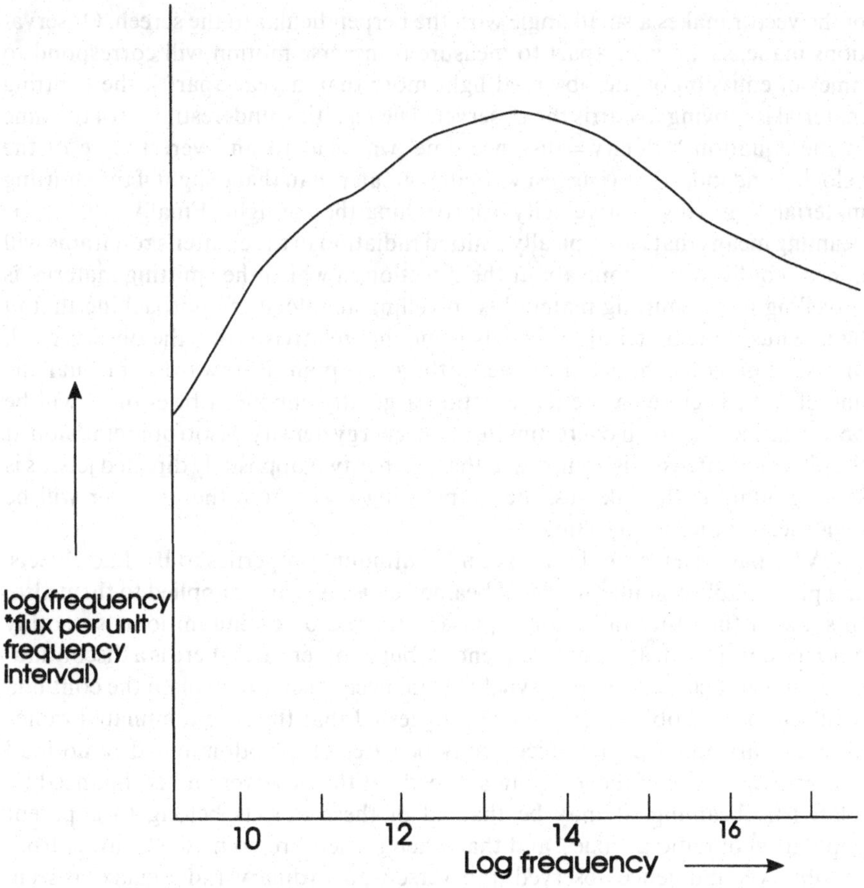

Figure 7.1. Continuous spectrum of blazar; note units

usually extending on one side of the nucleus only, and pointing towards outer jets with an arcsecond (kiloparsec) size. Knots in the mini-jet often show outward movements (displacements on a radio map over a period of time) which apparently imply transverse velocities greater than the velocity of light (*superluminal velocities*). Bends through large angles are sometimes seen in the mini-jets. The high brightness temperatures imply radiation energy densities at which inverse Compton scattering to higher frequencies should lead to severe energy losses to the scattering electrons and to very large X-ray fluxes, which are not observed. All these phenomena can be explained if the material in the jets is in bulk motion at velocities close to the velocity of light and if the jet is making a small angle with the line of sight. A modest change in direction of a vector in three dimensions can appear as a large angular deflection when projected onto a screen if the direction of the vector makes a small angle with the perpendicular to the screen. Observations made, say, a year apart to measure transverse motion will correspond to times of emission of the observed light more than a year apart if the emitting material is moving towards the observer. The resulting underestimate of the time in the equation 'velocity = distance/time' will lead to an overestimate of the velocity, and indeed can suggest velocities greater than that of light if the emitting material is moving at a velocity approaching that of light. Finally, relativistic beaming means that isotropically emitted radiation in the emitter's rest frame will appear confined to a cone about the direction in which the emitting material is travelling if the emitting material is travelling at velocities approaching that of light. Thus if the material in the jets is moving relativistically, the observer will only see emission if the beam and hence the jet are pointing towards him, but if the line of sight is close to the jet direction a greatly enhanced brightness will be observed, leading to an overestimate of the energy density if isotropic emission at the observed intensity is assumed. If the core has two oppositely directed jets, as is seen in many radio galaxies, the jet pointing away from the observer will be invisible, or at least very faint.

OVV quasars have similar radio and continuum properties to BL Lac objects, and presumably a similar model of beamed emission can be applied to them also. In some of the other radio-loud quasars, the radio continuum joins smoothly onto the continuum at higher frequencies, but in other cases there is a discontinuity, so it is not clear whether a synchrotron mechanism can explain the continua of all radio-loud objects. It has been suggested that the core-dominated radio-loud quasars and BL Lac objects may be cases of lobe-dominated radio-loud quasars where one of the jets points directly at the observer and is enhanced by relativistic beaming. It may be that all of these objects belong to a parent population of radio galaxies, and that when the jets are pointed well away from the observer and hence observed transversely an 'ordinary' radio galaxy is seen. Radio lobes radiate isotropically and so should be seen when a jet is pointing towards the observer as a halo around the core. The observed luminosities of such haloes are broadly consistent with the hypothesis of a single population with varying orientation, as is the fact that radio galaxies have weaker cores and lobes which are farther apart than is the case with radio-loud quasars.

The continua of radio-quiet AGN (Seyferts, QSOs) are more complicated. Of course the fact that these objects are called radio-quiet does not mean that radio emission is absent, and indeed many of these objects have radio luminosities greater than those of normal spirals, but only that the radio power is less than 10^{23} WHz^{-1} at a frequency of 1 GHz, whereas radio-loud objects have greater powers than this. The relatively weak radio spectra of these objects are probably also due to synchrotron radiation, but the strong far-infrared continuous spectrum does not join on smoothly to the radio spectrum, with a rapid increase in flux as one goes from a frequency of $3*10^{11}$ Hz to $3*10^{12}$ Hz (i.e. from a wavelength of 1 mm to a wavelength of 100 μm). In a plot of log νF_ν versus log ν, the spectrum peaks between $3*10^{12}$ Hz and $3*10^{13}$ Hz (between 100 and 10 μm), then drops a little to a local minimum around $3*10^{14}$ Hz. Between $3*10^{14}$ Hz and $3*10^{15}$ Hz (1 μm to 100 nm in the ultraviolet) there is an appreciable rise, which can partially be attributed to the Balmer continuum for wavelengths less than 375 nm, and partly to many Fe II lines blended together between 350 nm and 180 nm, the two features together being referred to as the 300 nm bump. The underlying continuum also appears to rise, a feature called the blue bump. Between $3*10^{15}$ Hz and $3*10^{16}$ Hz (the latter frequency corresponding to a photon energy of 0.3 keV in the soft X-ray region) there lies a region that is at present unobserved, but in some objects when the soft X-ray region is approached from higher frequencies there is a rise, suggesting that the peak of the blue bump may lie in the unobserved wavelength range. The value of νF_ν shortwards of the blue bump is substantially lower than the value at longer wavelengths than that of the blue bump. At hard X-ray wavelengths, the spectrum seems to follow a power law with spectral index 0.7. The continua of these objects are clearly complex, with more than one mechanism involved, and there is a lot of variation from object to object (Figure 7.2).

The origins of the continuous spectra in these AGN are controversial. The steep drop in flux between the sub-millimetre and radio regions seems to occur in many objects at about the same frequency and has a spectral index less than -2.5, so that it cannot be attributed to self-absorption in a synchrotron spectrum. However, an optically thin thermal spectrum at a frequency less than that of the blackbody peak has a flux proportional to $\kappa_\nu B_\nu \propto 2kT\nu^{-2}$, and interstellar dust has an absorption coefficient that falls as one goes to lower frequencies beyond $2*10^{-11}$ Hz (see Chapter 8). Dust resembling that found in our galaxy with temperatures in the range 50 to 1000 K (blackbody peaks from $3*10^{12}$Hz to $6*10^{13}$ Hz) will give a broad millimetre to far infrared flux distribution with a low frequency drop such as is observed in AGN. The same sort of spectral component is observed in star burst galaxies where there is little doubt about the dust origin of the emission. The temperature distribution could be produced by dust lying at distances ranging from parsecs to kiloparsecs away from the central source. Dust cannot exist at temperatures much in excess of 2000 K, so it is to be expected that there will be little dust emission at frequencies higher than $1.5*10^{14}$ Hz, which could be part of the reason for the 1 μm local minimum in the continuous spectrum.

In support of the dust hypothesis, it has been noted that some, although by no means all, Seyferts show a spectral feature characteristic of dust, and that the infrared emission from some Seyferts is extended on a kiloparsec scale. The lack of far-infrared variability in radio-quiet objects also suggests an extended region as the source of the far-infrared emission, although the evidence is limited and some radio-loud AGN show rapid infrared variations. It seems likely that at least the sub-millimetre and far-infrared continua in many AGN should be attributed to radiation from dust.

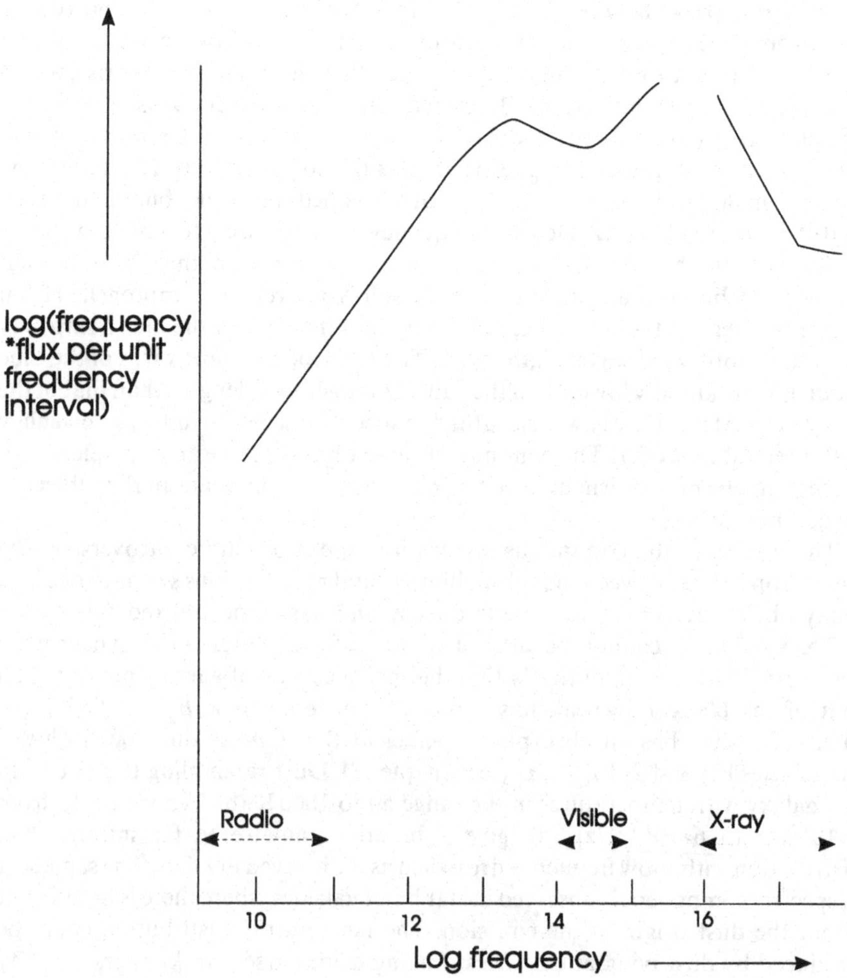

Figure 7.2. Continuous spectrum of radio-quiet AGN

It has been suggested that the blue bump is due to an accretion disc around a massive black hole. Variability of the optical light from AGN has been observed on periods down to minutes, and of the X-ray flux on even shorter periods, with the most rapid X-ray variations implying source sizes less than 10^{11} m. Most workers in the field would agree that only accretion onto a massive black hole could produce the very large luminosities from very small volumes that seem to be required of the ultimate power sources of AGN. Material with appreciable angular momentum, such as one would expect galactic material to have, will accrete gravitationally via a rotating accretion disc. A thin, opaque disc has a temperature that falls with distance from the central source of gravitational attraction and will radiate with an overall flux distribution that is proportional to $\nu^{1/3}$. The accretion disc around a black hole will not extend indefinitely near the centre of the system, for there is a stable orbit of minimum radius under these circumstances that must mark the innermost possible extent of the disc and the corresponding highest disc temperature will be around 10^6 K under conditions at the centre of an AGN. An accretion disc will then give a spectrum not unlike that of a blue bump, flattening off around the (inaccessible) frequency of 10^{16} Hz. At the moment neither theory nor observations are good enough to make a precise test of the accretion disc hypothesis for the blue bump. The hard X-ray power law continuum would seem to have a non-thermal origin, possibly Compton scattering of free–free radiation from the disc by electrons in the disc atmosphere.

We now turn to discuss the photoionization that would be produced by the central source continuum in surrounding clouds. As we have seen, AGN have central sources and sometimes have continua that can be approximated over not too large a wavelength range by flat spectra: $F_\nu \propto \nu^{-a}$ with a usually between 1 and 0. Thus photoionizing photons of all frequencies are present in appreciable quantities and a very wide range of ionizations will be present in the nebula, in some cases ranging from S II to Fe XIV. Suppose the flux incident upon a cloud is represented by $F_\nu = (\nu/\nu_0)^{-a} h F_H$, where ν_0 is the hydrogen threshold frequency. The number of photoionizing photons is

$$\int_{\nu_0}^{\infty} \frac{F_\nu}{h\nu}\,d\nu = F_H \nu_0^a \int_{\nu_0}^{\infty} \frac{1}{\nu^{a+1}}\,d\nu = \frac{F_H}{a}$$

so F_H is of the order of the number of photoionizing photons. The number of photoionizations per unit volume is then:

$$N(H^0)F_H h 6 \times 10^{-22} \int_{\nu_0}^{\infty} \left(\frac{\nu}{\nu_0}\right)^{-(3+a)} \frac{d\nu}{h\nu} = N(\mathrm{H}^0)F_H \frac{6.0 * 10^{-22}}{3+a}$$

where the photoionization cross-section has been written as $6 * 10^{-22}(\nu/\nu_0)^{-3}$ m^2.

The recombination rate at $T = 10^4$ K is approximately $2.6 * 10^{-19} N_e N(\mathrm{H}^+)$, so

$$\frac{N(\mathrm{H}^+)}{N(\mathrm{H}^0)} \simeq \frac{2 \times 10^{-3}}{3+a} \frac{F_H}{N_e} \tag{7.1}$$

The ratio of the flux of photoionizing photons to the number of electrons per unit volume is called the *ionization parameter* (sometimes defined as above but multiplied by a factor $1/c$ to make it dimensionless) and is a useful parameter for describing the ionization of AGN nebulae.

It is important to recognize that there are copious soft X-ray very high frequency photons produced by the central source in AGN (although, of course, the power law spectrum must eventually have a high frequency cut-off—

otherwise the total flux would be infinite). Absorption cross-sections go as ν^{-3} and so, except for the contributions from very highly ionized species of heavy ions, absorption coefficients at very high frequencies are small. This makes the transitions between ionized zones lacking in sharpness, and can produce an extended region of partially ionized hydrogen. High frequency photons may eject an inner shell electron from an ion or atom, and the resulting ion may then fill the gap in its inner shell with an outer electron, ejecting another outer electron to remove the energy in a radiationless transition called an *Auger transition*. Such a process will link stages of ionization differing in the number of electrons by two, and will also produce very energetic electrons which will lose their energy in heating up the gas. A few soft X-ray photons may penetrate beyond the H II region proper and produce occasional ionizations, particularly of helium. These events will not be sufficiently frequent to give an He II region, but will eject electrons with a lot of energy, since the X-ray photons have far more energy than is needed just to ionize helium. This will heat up the gas, so hot gas may be present outside the H II region proper, where it may excite ions like S II.

Emission Lines

We now turn to the interpretation of the narrow line region spectrum. There appear to be no systematic differences between the line spectra of Seyfert 2 galaxies and NLRG radio galaxies, and systematic differences between the spectra of Seyfert 2 and NLRG galaxies on the one hand, and of the narrow line components of Seyfert 1 galaxies, BLRG radio galaxies and QSOs on the other, are small. At first glance the average NLR spectrum looks not unlike that of an H II region or a planetary nebula, except for the presence of lines from high stages of ionization such as [Ne V] (sometimes seen in planetary nebulae), [Fe VII], and [Fe X], and the presence of fairly strong lines from low stages of ionization such as [O I], [N I], and [S II].

The methods of analysis described in the previous chapter can be applied here. The [O III] I(495.9 + 500.7)/I(436.3) intensity ratio and the corresponding [N II] ratio can be used to obtain temperatures if the density is not too high, and the [S II] I(671.6)/I(673.1) intensity ratio can be used to obtain an electron density (after allowing for a modest dependence on temperature). Typically, temperatures in the range 10^4 to $2*10^4$ K, and densities in the range $3*10^8$ to $5*10^4\ \mathrm{m}^{-3}$, are found. The ionization could be produced either by photoionization or by shock waves, but in the latter case ionization should be correlated with temperature, so for instance oxygen twice ionized by shocks must be at a temperature of 50 000 K, which is considerably higher than that found in most NLR. It must be concluded that the NLR are photoionized.

The Balmer line decrement in NLR is steeper than predicted by recombination theory, implying reddening by dust, a result confirmed in a few cases by comparison of near-infrared and violet [S II] lines. The ratio $I(\mathrm{H}\alpha)/I(\mathrm{H}\beta)$ gives a slightly discrepant dust extinction from that deduced from other line pairs, but this is readily explained by the fact that the dust-free $I(H\alpha)/I(H\beta)$ ratio is slightly increased above its pure recombination value by a contribution to Hα from collisional excitation in partly neutral gas. The extinction due to dust at any wavelength can then be calculated and line strengths appropriately corrected if it is assumed that the properties of the dust are the same as those in our galaxy.

The fraction of recombinations passing through Hβ is known, so that if a nebula is photoionization limited we can calculate the number of Hβ photons produced by a given ionizing continuum. This ionizing continuum is not directly observed in low redshift AGN, but if we assume a power law continuum we can extrapolate from the visible in an inverse use of Xanstra's method, and predict the ratio of the flux in Hβ to that in the neighbouring continuum. In fact, there is strong evidence that much of the continuum radiation escapes from the central source without being absorbed by the surrounding clouds, so the prediction is an upper limit. The observational results show that there is more than sufficient ionizing continuum flux available to produce the observed strength of Hβ. The slow variation with frequency of the continuous spectrum means that there are many high energy photons available to produce high stages of ionization. As has already been remarked, such an ionizing spectrum also produces less strong stratification of ionization than is found with blackbody sources, and there will be regions in which hydrogen is partially ionized and partially neutral with considerable numbers of neutral oxygen and nitrogen atoms, and S II ions. Analyses of line strengths using detailed ionization models suggest abundances not very different from solar.

We can also use the known fraction of recombinations passing through Hβ and the observed Hβ flux to estimate the total number of recombinations and hence, given the density, the emitting volume (6.38). Strictly N_e^2 in that equation should be replaced by $N_e N(H^+)$ but in a fully ionized gas $N_e = N(H) + 2N(He)$ with $N(H^+) = N(H)$ and for normal abundances $N(He) = 0.1N(H)$. The volume and the electron density then give the mass

$$\begin{aligned} M &= [N(H) + 4N(He)]m_H V = 1.4N(H)m_H V \\ &= \frac{1.4}{1.2} N_e m_H V \end{aligned} \tag{7.2}$$

$$V = 1.2 \frac{F(H\beta)4\pi d^2}{h\nu_\beta f_\beta \alpha_{rec} N_e^2} \tag{7.3}$$

Note that the distance d to the nebula is required to turn the observed flux F into a luminosity. Total masses of 10^6 solar masses and volumes of $3 * 10^4\ pc^3$ are typically obtained.

Direct observations of some nearby Seyfert 2 galaxies have given radii of the overall NLR of several hundred parsecs, and there is some evidence that the radii of NLR may be considerably larger for QSOs. It follows that only a small fraction of the NLR region, less than 1%, actually contains gas of any significant density. A thin shell would satisfy this requirement, but an arrangement of the gas in a number of individually fairly optically thick clouds in a largely empty 100 parsec or so radius sphere seems more likely, i.e. the *filling factor* of the NLR is small.

The narrow line profiles often have extended wings, the extension being greater to short wavelengths (the blue wing). The mean full width at half maximum (FWHM) in velocity units and the degree of asymmetry vary from galaxy to galaxy, but the variation between different lines from the same galaxy is much less. These smaller variations within the spectrum of a particular Seyfert 1 galaxy show a tendency for lines from the more highly ionized species to have larger FWHM in velocity units, whereas in Seyfert 2 galaxies there is a tendency for lines which are easily collisionally de-excited to have smaller FWHM. Now collisional de-excitation takes place if the electron density is greater than a critical value which is different for each line, and has the effect of weakening that line in comparison with collisionally excited but radiatively de-excited lines and with recombination lines. Hence only collisionally excited lines with large critical densities will be seen in the spectrum of a high density gas. In an inhomogeneous environment with varying density the greatest contribution to a line's strength is likely to come from regions with densities just below the critical density, since the strength of collisionally excited lines will increase with density up to the critical density.

The correlation in some objects of FWHM with critical density implies that the denser clouds are faster moving. The correlation in other objects of the FWHM with ionization implies that in these objects the more highly ionized clouds are faster moving. Now the degree of ionization depends on the ionization parameter (the ratio between the ionizing flux and electron density). The flux falls with the inverse square of distance from the central source, so if the density within clouds does not depend on distance from the centre, the ionization will be greatest near the centre, and in these objects the largest velocities will be found nearest the centre. If the largest velocities are also found nearest the centre in clouds where there is a correlation between FWHM and critical density, then for these objects the clouds with the largest internal densities are found nearest the centre. A model which would be consistent with all the observations would have velocities falling as one moves farther from the centre, and the average density of a cloud also falling as one moves farther from the centre, but at different rates in different objects. An AGN with a slow fall-off in average cloud density would be dominated by a falling ionization parameter with distance from the centre, whereas a system in which the average density fell as fast as the inverse square of the distance would have an almost constant ionization parameter with distance from the centre, and observations would be dominated by the critical density factor. However, it must be remembered that the correlations are only found in some AGN and the model is certainly not unique. The asymmetry in the line profiles can be explained if the line radiation from the gas clouds moving directly away from the observer is blocked by obscuring dusty material. Presumably gas on the far side of the nucleus from the observer will be more obscured by any symmetric distribution of obscuring material, provided the dust is not wholly outside the gas. A component of outflow in the NLR velocity field would then give a blue asymmetry.

We now turn to the broad line region, which is much less well understood than the narrow line region. In the visible it is dominated by the Balmer lines of H I and the lines of He II, and in some AGN the lines of Fe II are prominent. In the ultraviolet the lines of Mg II at 279.8 nm are strong and C III] at 190.9 nm is often present, while at shorter wavelengths we find the Lyman lines and a variety of collisionally excited lines, those of C IV at 154.9 nm being particularly conspicuous, with other strong lines including Si IV at 140.0 nm, N V at 124.0 nm, and O VI at 103.5nm.

No broad line [O III] lines have been detected, and since oxygen is certainly present in the broad line region (O VI lines, for instance) and since the range of ionization seen in broad lines goes from low ionization stages like Mg II to high ionization stages (O VI again), O III must certainly be present. The O III forbidden lines must therefore be suppressed by comparison with other lines by collisional de-excitation, which sets in at a density of about $10^{12}\,m^{-3}$. The complete suppression of the forbidden lines needs a density of $10^{14}\,m^{-3}$, and this must represent a lower limit to the BLR density. The semi-forbidden line of C III] at 190.9 nm is seen in most AGN however. This line can also be collisionally suppressed, but because the radiative transition probability is much higher than for a true forbidden line, the critical density is also higher at about $10^{16}\,m^{-3}$. Other weaker semi-forbidden lines sometimes observed such as N III] 175.0 nm, N IV] 148.6 nm and O III] 166.3 nm confirm this upper limit. The presence of these semi-forbidden lines means that a considerable fraction of the BLR must have a density less than $10^{16}\,m^{-3}$, and so a suitable representative density for BLR is about $10^{15}\,m^{-3}$.

The determination of the temperature in BLR is more difficult. The absence of forbidden lines means that the cooling effect of these lines is missing from the energy balance in comparison with the NLR and so if the BLR are also photoionized, we might expect them to be a little hotter than the NLR. The fact that iron is present in singly ionized form means that the temperature is less than 35 000 K, at which iron would become collisionally twice ionized at the densities of BLR. The C III line at 97.7 nm and the C III line at 190.9 nm are collisionally excited with the former having a much larger excitation energy, so the ratio of line strengths gives a measure of the temperature (provided that collisional de-excitation is not beginning to affect the 190.9 nm line strength). In fact 97.7 nm is usually weak or absent, so an upper limit of 25 000 K for the temperature of the BLR in a few QSOs can be obtained. A temperature of 20 000 K would seem reasonable for the BLR of most AGN.

The strengths of the broad lines and in particular of the hydrogen broad lines correlates with the visible continuum strength over a very wide range of AGN luminosity. This suggests that BLR are photoionized, for in that case the Balmer hydrogen lines would be produced mainly by recombination, and a correlation of line luminosity with the luminosity of the photoionizing continuum would be expected with, for power law continua, the visible luminosity correlating with the luminosity of the photoionizing continuum. Broad lines vary on time scales down to weeks or even days, which gives an upper limit to the size of the region containing the BLR clouds. The variation of the broad lines is correlated with the variation of the continuum, following after a delay of the order of a week. This is again consistent with photoionization since the recombination time of the high density gas in BLR is short so the BLR will respond rapidly to changes in the photoionizing flux. Thus the delay gives fairly directly the light travel time from the central continuum source to the BLR. The radius of the BLR zone is about 0.1 parsec for an AGN of luminosity 10^{46} W, perhaps scaling with the square root of the luminosity. The photoionizing luminosity with the distance from central source to the average BLR cloud gives the photoionizing flux at the cloud. The flux combined with the density enables the ionization parameter to be calculated. This parameter is the main input into ionization models, and line ratios can be calculated as a function of ionization parameter and compared with observations, with in particular the ratio of O VI 103.5 nm to C IV 154.9 nm being very sensitive to the ionization parameter in the range of conditions appropriate for AGN BLR. A typical value for the parameter of $3*10^7\ \mathrm{m\,s^{-1}}$ is found in AGN BLR.

Further information about the structure of BLR comes from the observation of strong broad lines of Fe II and Mg II. Mg II and Fe II have ionization potentials similar to that of H I, so these lines should come from a region in which hydrogen is mainly neutral. Photoionization models with power law continua have large zones in which hydrogen is partially neutral but which are heated and kept partially ionized by soft X-ray photons, *if* the cloud involved is sufficiently optically thick. The Mg II and Fe II lines therefore show that BLR clouds should be optically thick to the Lyman continuum. However, spectra of high redshift AGN show no signs of a Lyman discontinuity, which would normally mean that the optical depth in the Lyman continuum is less than 0.1. These observations can be reconciled if the BLR clouds are optically thick and therefore ionization bounded but only block a small fraction of the continuum source so most continuum light reaches us without passing through BLR gas. If the Balmer lines are mainly produced by recombination, broad component Balmer line luminosities together with the electron density give the total volume and mass of the broad line gas by the same method that was employed in the case of the NLR. Typical values are $3*10^6\ \mathrm{pc}^3$ and 50 solar masses for Seyfert 1 galaxies and rather higher values for QSOs. Comparing the volume with the radius of the region from which the broad lines come (estimated from delay times or ionization parameters), it is clear that the filling factor is small, perhaps around 0.01. The line profiles are often quite smooth, suggesting that the number of contributing clouds with differing Doppler shifts is large. Thus, as for the NLR, we must picture the region producing the broad lines as containing a large number of optically thick clouds, but most of the region as being empty or filled with a very low density high temperature gas.

The broad line Balmer line ratios have caused considerable argument since they do not accord with recombination theory and the observed ratio of Lyman α to Hβ, with a typical value of 10, is very discordant with the normal recombination theory value of 23. In fact the recombination prediction for the high density conditions found in BLR approaches 34, because Lyman is a 2p to 1s transition only, 2s to 1s being forbidden, but at high densities collisions can transfer atoms from 2s to 2p whence they can decay radiatively through Lyman α. The Lyman α to Hβ ratio is, of course, difficult to observe in a particular object, and early estimates were made by patching together spectra of high redshift AGN showing Lyman α and spectra of low redshift AGN showing Hβ. Ultraviolet and infrared observations make it possible to cover a wide wavelength range in a single object. $I(\mathrm{H}\alpha)/I(\mathrm{H}\beta)$ is typically about twice the value predicted by recombination theory in AGN, and IH(β)/I(Hγ) is larger than predicted by recombination theory, but by a smaller factor. Applying a simple galactic dust reddening law to recombination predictions does not produce the observed results for any value of the extinction because the absorption needed to fit the observed Lyman α/Hβ ratio does not reproduce the Balmer line ratios, and indeed the H(α)/H(β) ratio is not consistent with the H(β)/H(γ) ratio for any value of external dust extinction.

It should be noted that at high densities collisional excitation may contribute to hydrogen line intensities. The difference between AGN and H II regions in this respect is that the former have large volumes where hydrogen is partly neutral but the temperature is high so H^0 is available to be excited and there are electrons around with the energy to produce the excitation, whereas the latter only have high temperatures in the fully ionized region. The excitation is of neutral atoms by electrons, and hence the cross-section is zero at the threshold energy but rises rapidly at higher collisional energies. The collisional excitation rate is proportional to $N_e N(\mathrm{H}^0)$, whereas the recombination rate leading to Lyman α is proportional to $N_e N(\mathrm{H}^+)$ with the result that the ratio of collisional excitation leading to Lyman α to recombination leading to Lyman α is proportional to $N(\mathrm{H}^0)/N(\mathrm{H}^+)$. The threshold is high, so the collisional excitation rate is very dependent on temperature. Equality of collisional excitation leading to Lyman α to recombination leading to Lyman α occurs at a temperature of about 10 000 K for $N(\mathrm{H}^0) = N(\mathrm{H}^+)$ and 12 500 K for $N(\mathrm{H}^0) = 0.1N(\mathrm{H}^+)$. Levels of hydrogen with $n > 2$ have higher thresholds and smaller excitation cross-sections. Thus collisions make a small contribution to Hα and a very small contribution to Hβ. Predicting the overall line strengths requires a model of the photoionized region with varying temperature (collisionally excited Lyman α is an important coolant) and varying degree of ionization. However, clearly the result of including the effects of collisional excitation is to increase the Lyman α flux relative to that of $H\beta$.

It is therefore necessary to consider possible radiative transfer effects on the hydrogen line strengths. The continuum absorption coefficient at the threshold for Lyman continuum absorption is $\kappa_C = 6.3*10^{-22} N(H^0)$. The line absorption coefficient at the centre of Doppler broadened Lyman α is

$$\kappa_L = \frac{\pi e^2}{4\pi\varepsilon_0 mc} f \frac{1}{\sqrt{\pi}\Delta\nu_D} N(H^0)$$

where $\Delta\nu_D$ is the Doppler width and it has been assumed that essentially all the hydrogen is in the ground state. Using $f = 0.4$ and taking the Doppler width appropriate to 12 500 K (the much larger observed line width is assumed to be due to macroscopic motions), $\kappa_L = 5.3*10^{-18} N(H^0)$. Suppose that the Lyman continuum optical depth of the region where hydrogen is partly neutral and the Mg II and Fe II lines originate is about 100, a figure derived from more detailed models, although we have already seen that it must be considerably greater than 1. The Lyman α line centre optical depth must then be κ_L/κ_C 100, i.e. of the order of one million, and the average Lyman α photon formed in the first place by recombination or collisional excitation will be scattered a very large number of times before escaping. The photon will zigzag through the partly neutral region in a random walk, and hence will traverse a longer path than the direct distance from the point of creation of the photon to the edge of the nebula. This effect will also increase the population of the excited level $n = 2$.

The scattering of resonance line photons is usually treated using the escape probability method developed in Chapter 4 (equations (4.30)–(4.36) and (4.40)–(4.45). An order of magnitude estimate of the effects of resonance line scattering can be derived by considering a Lyman α photon emitted at the centre of a spherical cloud of radius R and line centre optical radius τ_{LC}. The distance in frequency from the line centre is x in units of Doppler width, so the line profile is $\phi_x = 1/\sqrt{\pi}\exp(-x^2)$ and the optical depth is $\tau_x = \tau_{LC}\exp(-x^2)$.

Equation (4.36), for instance, gives the one-sided escape probability from a slab for a Doppler line profile as approximately:

$$\frac{1}{2\sqrt{\pi\tau_{LC}}}\frac{1}{\sqrt{\ln_e \tau_{LC}}} \simeq \frac{1}{6\tau_{LC}}, \quad \text{if } 10^4 < \tau_{LC} < 10^6 \tag{7.4}$$

where it will be recalled from Chapter 4 that photons only escape if redistributed into the optically thin wings of the line with $x > x_1$, and that while the photon remains in the core of the line it does not travel far from its point of origin, so the distance to the surface remains close to R. The escape probability is the reciprocal of the *average number of scatterings to escape*, N_{esc}, strictly plus one, but the 'one' can be ignored for the very large optical depths involved in this case, so the average number of scatterings before escape is about $6\tau_{LC}$. Between each scattering a line photon travels an average optical distance of about one (by definition) and since $\tau_{LC} = \kappa_{LC}\rho R$, this corresponds to a geometrical distance of R/τ_{LC}. Of course the photon will not stay in the line centre, but will move around in frequency at each scattering with an average line absorption coefficient of $\kappa_{LC}/\sqrt{\pi}\int \exp(-2x^2)dx = \kappa_{LC}/\sqrt{2}$ until it reaches the line wings and escapes. Hence the average distance travelled per scattering $\sim \sqrt{2}R/\tau_{LC}$ and

total path travelled $\simeq$ (number of scatterings) $\times$ (average distance travelled)

$$\simeq \frac{6\tau_{LC}\sqrt{2}R}{\tau_{LC}} = 9R$$

where a final escape distance from the wings of about R has been included.

These simple estimates are for a photon emitted at the centre of a spherical volume. N_{esc} and the path length will, of course, be less for an average photon which may be generated anywhere in the volume. The geometrical situation will also be somewhat different in an AGN as a BLR is made up of many clouds, each highly ionized on the side facing the central source and neutral on the opposite side, so Lyman α photons mainly escape towards the central source through the highly ionized, transparent side of the cloud. A more serious inaccuracy arises because we have assumed complete redistribution in the scattering process. For very strong lines, x_1 occurs at a frequency in the line wings where the broadening is Lorentzian and the scattering coherent. Coherent scattering in the wings at a frequency at which the line is still optically thick (but much less so than in the core proper) can take a photon quite large distances from the point of origin. For Lyman α and a slab of half-geometrical thickness R and half-optical thickness at the line centre τ_{LC}, $N_{esc} \sim s_1\tau_{LC} \sim 3\tau_{LC}$ and the path length travelled $= s_2R \sim 6R$, where the numerical values of s_1 and s_2 are very approximate (and controversial).

With the values just given, a Lyman α photon travels six times farther before escaping than an Hβ photon, provided that Hβ is still optically thin. Galactic dust has an extinction coefficient 2.5 times larger at the wavelength of Lyman α than at the wavelength of Hβ, so that if such dust were mixed in some way with the partly neutral region of a BLR, it would have a 15 times greater effect on Lyman α than on Hβ.

We now turn to the question of the effects of optical depth on the Balmer line ratios. Resonance line scattering will increase the probability of finding a hydrogen atom in the $n = 2$ level and hence increase the optical depths in the Balmer lines. Every photoionization is balanced by a recombination, and every recombination produces a Lyman α photon. Each time a Lyman α photon is scattered, the $n = 2$ level of some hydrogen atom is populated for a time given by the reciprocal of the transition probability of Lyman α. Hence the column density of hydrogen atoms in the $n=2$ level is given by:

$$\text{(number flux of photoionizing photons)} \times \frac{\text{(number of scatterings)}}{\text{(transition probabilty of Lyman } \alpha)}$$

since all photoionizing photons are absorbed somewhere in the column.

The optical depth in the centre of a Balmer line is then:

$$\tau(B) = \text{(number flux of photoionizing photons)} \frac{s_1 \tau_{\mathrm{LC}}}{A(\text{Lyman } \alpha)} \frac{\pi e^2}{4\pi\varepsilon_0 mc} \frac{f}{\sqrt{\pi}\Delta\nu_{\mathrm{D}}}$$

Hence

$$\tau(\mathrm{H}\alpha) = \text{(number flux of photoionizing photons)}\, \tau_{\mathrm{LC}} s_1\, 7.4 \times 10^{-26}$$

Typical BLR have a flux of photoionizing photons of $2.5*10^{21}$ photons m^{-2} s^{-1}, which, with $s_1 = 3$ gives $\tau(\mathrm{H}\alpha) = 6*10^{-4}\,\tau_{\mathrm{LC}}$. In fact collisional excitation will also generate Lyman α photons, increasing the estimate of $\tau(\mathrm{H}\alpha)$ by a factor of up to 5. Thus if $\tau(\text{Lyman }\alpha) \sim 10^6$, the H$\alpha$ optical depth will exceed 100. The Hβ optical depth is 0.2 of the Hα optical depth and $\tau(\mathrm{H}\gamma) = 0.60\,\tau(\mathrm{H}\alpha)$, remembering that all these lines have the same lower level as Hα, so Hβ and possibly Hγ may also be affected by optical thickness.

Consider first the fate of an Hα photon when the Hα line is very optically thick. It will be scattered, exciting the $n = 3$ level, which will decay either by emitting Hα again or by emitting Lyman β. Lyman β has a very large optical depth and so will be scattered many times, on each occasion having a good chance of being converted to Hα and Lyman α. On the other hand an Hβ photon, if the Hβ line is optically thick, will be absorbed, populating $n = 4$. This level will then decay either by emitting Hβ again or by emitting Paschen α or by emitting Lyman γ, which will be scattered many times until it is converted into Paschen α or Hβ. If Hβ is produced it will be absorbed again, but if Paschen α is produced it will escape if this line is optically thin. Hence the net effect is to convert Hβ into Hα and Paschen α, and hence to increase $I(\mathrm{H}\alpha)/I(\mathrm{H}\beta)$ over the standard recombination value. Indeed, for some values of $\tau(\mathrm{H}\alpha)$, the increase can be by a factor of greater than 2. The optical depth of Hγ is less than that of Hβ, so it will be affected less by optical depth effects and so $I(\mathrm{H}\beta)/I(\mathrm{H}\gamma)$ will be less than the recombination value. If the Lyman α optical depth is very large, Hα scattering will increase the population of $n = 3$ sufficiently for Paschen α to become optically thick. Then Paschen α excitation of $n = 4$ can lead to Hβ emission and $I(\mathrm{H}\alpha)/I(\mathrm{H}\beta)$ will decrease back towards the recombination value. It seems possible that optical depth effects can explain at least some of the anomalous hydrogen line ratios, but equally make it difficult to use hydrogen line strengths to detect dust obscuration.

Indeed, the amount of dust obscuration in BLR is a controversial question. The observed intensities of lines in the optical region in general are large relative to Lyman α compared with model predictions, whereas the observed intensities of lines in the ultraviolet relative to Lyman α are in much better agreement with the models, which could be due to dust extinction dimming all UV lines, including Lyman α. More precise indications can be obtained from the ratio of the intensity of the He II $n = 4$ to $n = 3$ line at 468.6 nm to the intensity of the He II $n = 3$ to $n = 2$ line at 164.0 nm. These two lines are successive stages in the He II recombination cascade, and as the first is certainly optically thin and the second is probably only slightly affected by optical depth considerations, the ratio can be reliably predicted to be about 1:10. Values much less than 1:10 for the ratio must be due to dust affecting the 164.0 nm line more strongly than the 468.6 nm line. A similar use can be made of the fluorescence lines of O I previously mentioned, which are excited by absorption of Lyman β. The subsequent decay to the ground state goes through 844.6 nm and 130.2 nm, so these lines have an almost fixed intensity ratio. Any large observed deviation from this ratio is likely to be due to the effect of dust reducing the intensity of the 130.2 nm line, the large ratio of the wavelengths of the two lines making the differential effect of dust particularly large, although also making accurate observations difficult. The He II and O I line ratios in the few objects in which they have been accurately measured indicate some reddening to be present.

The most dramatic effect of dust occurs in Seyfert 2 galaxies. A small degree of linear polarization has been found in some Seyfert 2 galaxies. The polarization is wavelength independent, which suggests it must be due to electron scattering as opposed to dust scattering. The plane of polarization is perpendicular to the radio axis of the galaxies. However, when the spectrum of the galaxy is taken in polarized light, broad lines show up—indeed the polarized spectrum looks very like that of a Seyfert 1 galaxy. The broad lines are presumably present in spectra taken in unpolarized light but are too weak to be detected. The broad lines and the continuum from Seyfert 2 galaxies are quite strongly polarized, but the narrow lines are unpolarized. The strong broad line polarization confirms that it is electron scattering that is producing the polarization, for a synchrotron process would give rise to a polarized continuum, but would not produce lines. The continua of Seyfert 2 nuclei are much weaker than those of Seyfert 1 nuclei, and in some cases are difficult to disentangle from the spectrum of the underlying galaxy, but as far as can be told have similar spectral distributions to those of Seyfert 1 galaxies. In some cases the Seyfert 2 photoionizing flux, as deduced from extrapolating the observed continuum, seems too weak to produce the ionization seen in the NLR. Finally it may be noted that in the intermediate Seyfert 1.8 and 1.9 types, the Hα to Hβ ratios are very large.

The generally accepted explanation of all these observations is that Seyfert 2 galaxies have a Seyfert 1 central source and BLR, but have obscuring material in the line of sight which prevents us seeing the BLR and central source directly. The obscuring material might be in the form of a torus 1 parsec or so from the centre and so farther out than the BLR but much farther in than the NLR. Free electrons above and below the torus, perhaps from ionized regions on the inner surface of the torus, scatter some of the emission from the central source and BLR back into the line of sight, at the same time polarizing the scattered radiation. Most of the NLR clouds in their much larger volume will not have their line of sight to us obscured by the torus, nor will the ionizing radiation from the central source be blocked by the torus. If this model is correct, Seyfert 2 galaxies are Seyfert 1 galaxies viewed from an angle at which the torus blocks the direct line of sight to the inner regions. Seyfert 1.8 and 1.9 galaxies are cases where the angle of viewing is such that the BLR are partially obscured and heavily reddened. It is not clear whether this model explains the differences between Seyfert 1 and 2 galaxies in all cases, nor whether something similar is happening in NLRG and BLRG.

The broad line profiles show great variety, but there is no general trend in favour of a blue asymmetry such as was found for the narrow lines, and hence no general evidence for a radial outflow. The velocities are consistent with the rotational velocities expected at distances of the order of 0.1 parsec from a 10^8 solar mass black hole such as is often postulated as the ultimate power source for AGN. The Fe II lines often have smaller widths in velocity units than the hydrogen lines, which in turn are narrower than the He I lines, with the He II lines widest of all. Rotational velocities should increase as one move inwards as should the degree of ionization in the BLR clouds. However the results just mentioned are only rough trends, and the dynamics of BLR are still not understood.

8 THE SPECTRUM OF THE COLD INTERSTELLAR MEDIUM

The interstellar medium in our part of our galaxy is composed of a mixture of clouds in which hydrogen is atomic and clouds in the interior of which hydrogen is molecular, together with a hotter and less dense intercloud medium. An atomic cloud has typically a temperature of 100 K and a number density of $2.5*10^7\ m^{-3}$ with a radius of a few parsecs. A molecular cloud typically has a lower temperature of 15 K, a higher density of $2*10^8\ m^{-3}$ and a radius of 20 pc . Most of the intercloud medium has a temperature of around 8000 K and a number density of $2*10^5\ m^{-3}$, the product of the temperature and the number density being of the same order in the two types of cloud and the intercloud medium so that pressure equilibrium is maintained. About half of the intercloud medium is ionized so it is a low density HII region. Finally there is a very hot minor component of the intercloud medium with a temperature of 10^6 K and a number density of $10^{-9}\ m^{-3}$. The clouds, however, are closely confined to the galactic plane in a thin disc, whereas the warm intercloud gas is probably in the form of a thick disc, and the hot intercloud gas extends thousands of parsecs above and below the plane and may be better described as a halo.

The subject of this chapter is the cold clouds. Gas in the low density medium will be far from full thermodynamic equilibrium and so Boltzmann's equation and Saha's equation are unlikely to hold. In the cold gas and a weak radiation field there are no processes capable of exciting atoms very far, and so only the ground state and levels very close to the ground state are likely to be populated. However, for small energy changes, radiative excitation by the microwave background may be important. When dealing with stellar atmospheres, the temperature gradient in the outer layers of the star was all important, but in an interstellar cloud there is no obvious major energy source, and so it is often possible as a first approximation to assume that the cloud has constant temperature. It must be remembered that this is an approximation, since a radiating optically thick cloud almost certainly will have some temperature variations.

An important aspect of cold clouds is the presence of dust. The dust makes up only a small fraction of the mass of the interstellar medium, but has important effects. Firstly, dust removes ultraviolet radiation and to a lesser extent visible radiation, partly by scattering and partly by absorption and reradiation in the infrared. Virtually all astronomical observations in the visible and ultraviolet wavelength bands have to be corrected for dust extinction. The reddening produced by dust is often used as a readily detectable sign of the presence of interstellar medium. In some cases the long wavelength reradiation from dust grains represents the greater part of the luminosity of the system concerned, an example being starburst galaxies. Secondly, dust enables the formation of interstellar molecules. Molecules are dissociated by ultraviolet radiation unless they are shielded by dust, and the formation of molecules is sometimes assisted by the presence of a solid surface.

The edges of dense clouds are subject to molecular dissociation by ultraviolet radiation. These photodissociation regions were dealt with in Chapter 6 because of the analogy between photodissociation and photoionization. Molecular lines sometimes show greatly enhanced intensities due to maser action. The kinetic temperature of the gas producing the lines is low and in some respects, therefore, astronomical masers belong in this chapter, but the brightness temperatures of these lines are often extremely high and consideration of the maser phenomenon is left to a separate chapter, Chapter 9.

Atoms in Interstellar Clouds

In this section we are mainly concerned with the observation of atomic hydrogen in cold clouds, which means that we are leaving the denser clouds in which hydrogen is entirely molecular to the next section. First we consider briefly optical observations of atoms other than hydrogen.

Interstellar absorption lines are lines produced by atoms in the interstellar medium and seen in absorption against the spectrum of a background star. In other words, interstellar absorption lines are observed in stellar spectra but have no connection with the star. They were discovered in the spectra of spectroscopic binaries, which have two sets of spectral lines from the two (usually unresolved) stars with the two sets shifting in wavelength relative to each other as the stars move in their orbits around the centre of mass and their Doppler shifts change (Figure 8.1). Thus there will be a phase in the orbital motion when one star is moving directly away from the observer so that its spectrum is redshifted and the other star is moving directly towards the observer so that its spectrum is blueshifted, with the net result that the two sets of lines are well separated. There will be another phase in the orbital motion when both stars are moving across the

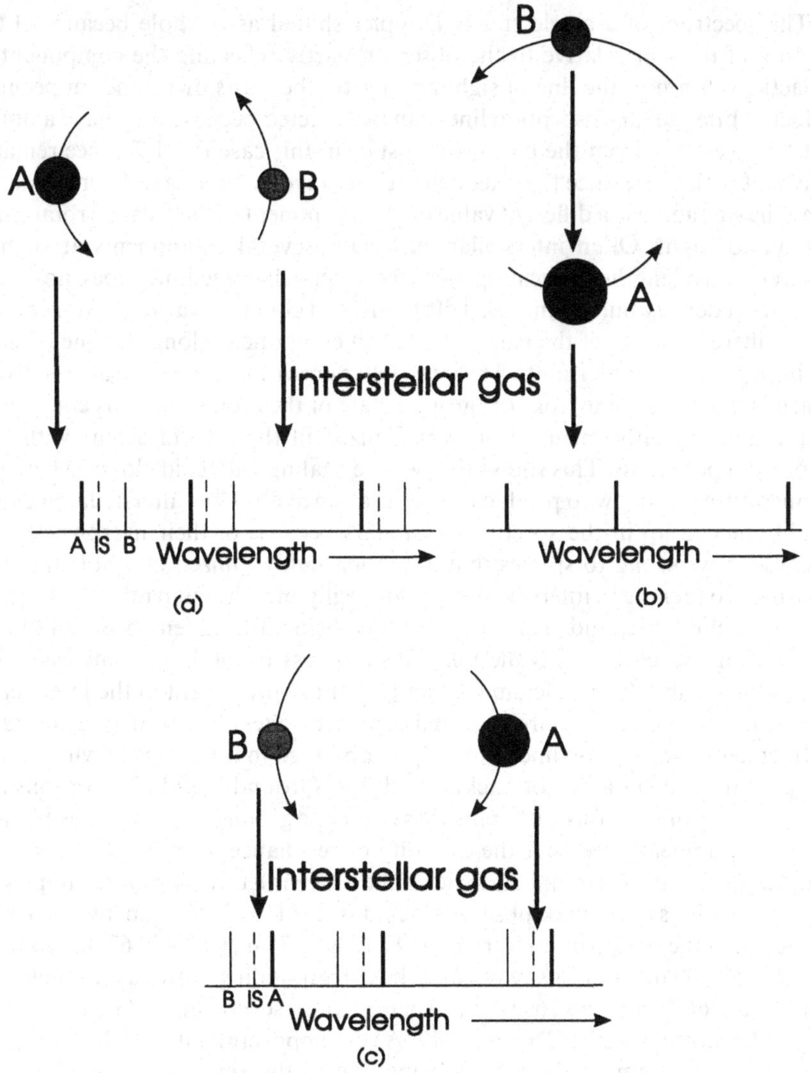

Figure 8.1. Spectroscopic binary with interstellar lines (IS)

line of sight in opposite directions, and the transverse motions give no Doppler shift so the two sets of lines coincide. In the case of the particular spectroscopic binaries in which interstellar lines were discovered, the spectrum showed calcium lines that did not move with either of the sets of stellar lines but stayed at a constant wavelength and hence must originate in material between us and the star.

The spectrum of a single star is Doppler shifted as a whole because of the motion of the star relative to the observer, partly reflecting the component of galactic rotation in the line of sight and partly the star's own random peculiar velocity. Interstellar absorption lines can be detected because they have a different Doppler shift from the background star (in this case the difference remains constant with time) since the interstellar cloud involved is at a different distance than the star and has a different value of the component of the galactic rotation in the line of sight. Often interstellar lines have several components at slightly different wavelengths, indicating that the interstellar medium does not have a constant density but is composed of clouds, and clouds at various distances will have different values of the galactic rotation component along the line of sight giving different Doppler shifts. The interstellar lines thus discovered are relatively sharp and always come from the ground state of the atoms and ions concerned. The atoms are either neutral or once ionized in the case of atoms with low ionization potentials. This shows that we are dealing with cold clouds where the temperature is too low to produce any excitation at all. Thus interstellar lines are readily picked up in the spectra of hot stars because of their narrowness and because they belong to species that are more highly ionized in a hot star. The thermal broadening in interstellar clouds is negligible, although microturbulence may be quite large, and there will be no collisional broadening or rotational broadening, which explains the relative sharpness of the lines compared with stellar lines. Lines from calcium Ca I and Ca II are prominent in the interstellar spectrum but calcium is entirely ionized to higher stages of ionization in hot stars.

Interstellar absorption lines can only be observed in directions in which there happens to be a suitable hot background star. Ground-based observations are confined to atoms or ions with transitions from the ground state in the visible and it must be remembered that the majority of resonance lines and other strong transitions from the ground state lie in the ultraviolet. Important examples of those transitions that can be observed include Ca II H at 396.85 nm (blended with hydrogen in the background star), Ca II K at 393.37 nm, Ca I 422.67 nm, sodium Na D at 589.00 nm and 589.59 nm and the corresponding but weaker potassium doublet at 766.49 nm and 769.90 nm. There are also several lines of Fe I and Ti II, and of the molecules CH, CN, and CH^+. A very important but weak line is that of lithium at 670.8 nm, the weakness being due to the very small abundance of lithium. Ultraviolet observations have greatly extended the list of elements, although these require a background star that is bright in the ultraviolet (it must be remembered that interstellar gas is associated with interstellar dust and that dust removes ultraviolet radiation!). Examples include C I and C II, N I, O I, Mg II, Al II, Si I, S I and S II, and Fe I and Fe II.

The interpretation of the interstellar absorption line spectrum is, in some ways, more straightforward than that of stellar spectra. The interstellar medium does not emit so one essentially has a curve of growth with $R_c = 1.0$. The Doppler broadening is largely turbulent, and the Lorentzian broadening in the wings is entirely natural since collisional broadening is negligible at the low densities involved. Very few interstellar lines are strong enough to be on the damping portion of the curve of growth, which leaves the microturbulent velocity in the Doppler width as the only free parameter apart from the column density of absorbing atoms. The population of the lower levels of the transitions is the total population of the atom or ion in question since collisional excitation is negligible for atoms at the temperature of the interstellar medium.

The degree of ionization must, however, be estimated to obtain the total abundance of the element under consideration. One might naively imagine that the degree of ionization would be negligibly small at the low temperatures of the interstellar medium but this quantity is determined by the balance between photoionizations by the diluted starlight radiation field and recombinations. Radiation from stars with wavelength less than the Lyman limit at 91.2 nm will be strongly absorbed by neutral hydrogen and indeed little or none will emerge from ionization bounded H II regions. On the other hand, longer wavelength radiation will largely escape and is available to ionize atoms or ions with ionization potentials less than that of hydrogen. The stars producing the radiation field at a typical point in the interstellar medium will be parsecs away and so the field will be fairly dilute, but the density and recombination rate are also low. Using the mean radiation field in our neighbourhood, one finds that sodium is mainly Na II although no lines can be observed in the visible from Na II but only from neutral Na I. Calcium lines from Ca I and Ca II are observed, but since the ionization potential of Ca II is less than that of hydrogen, one would expect that a considerable fraction of calcium is in the form of unobservable Ca III. It is worth noting that if a large fraction of sodium, say, is in the form of Na II, then the recombination rate will be proportional to N(Na II) N_e which will be nearly equal to N(Na)N_e, and the photoionization rate to N(Na I), so N(Na I) will be approximately proportional to the electron density. Na I lines will preferentially originate in the denser parts of the cloud, and the same is true for other atoms or ions where the greater part of the element is in the next higher stage of ionization.

The equivalent widths of weak lines, if spectral resolution and signal noise are good enough to allow them to be measured accurately, enable the column density (the integrated number density along the line of sight through the cloud) of the atom or ion concerned to be determined at once using 4.5a with $R_C = 1.0$. For stronger lines saturation effects start to become important and the shape of the curve of growth or the microturbulence velocity must be known. One approach has been to use the relative strengths of the two Ca II lines H and K, for these lines come from the same lower level and have oscillator strengths in the ratio 1:2, so an observed ratio of the equivalent widths that is less than 2 indicates saturation. A high resolution spectrum of a weak line enables the Doppler width to be measured directly from the profile. The large number of lines available from ultraviolet observations enables curves of growth to be drawn for a few atoms and ions. It appears that the form of the curve of growth can differ for those lines that belong to a minority stage of ionization, which are of course the lines that prefer the denser parts of the cloud. It should be noted that if two clouds have very similar Doppler shifts they will not be resolved spectroscopically, so the different curves may represent the effects of the super-position of sub-clouds with different densities and turbulent motions.

By far the most important probe of the atomic parts of the interstellar medium is the hydrogen line at 21.1 cm or 1420 MHz. This line is the transition between the hyperfine structure levels of the ground state of hydrogen. In the upper of these levels, the proton and electron have parallel spins to give a total angular momentum quantum number of 1, while in the lower level the proton and electron spins are antiparallel to give a total angular momentum quantum number of 0. The electron has a magnetic moment of $9.284*10^{-24}$ JT^{-1} and the proton has a magnetic moment of $1.411*10^{-26}$ JT^{-1} with the opposite sign for the relationship between magnetic moment and spin to that of the electron. The interaction of the two magnetic moments gives rise to a difference in energy of

$$\frac{\mu_0}{4\pi}\frac{32\pi}{3}\mu_e\mu_n\frac{1}{\pi a_0^3} = 9.43*10^{-25}\,\text{J}$$

where μ_0 is the permeability of a vacuum, μ_e is the magnetic moment of an electron and μ_n is the magnetic moment of the proton, with a_0 = Bohr radius. The frequency corresponding to this energy difference is 1420.4 MHz, the wavelength is 0.21106 m and the transition probability is $2.85*10^{-15}\,s^{-1}$. The energy difference between the two levels is much less than kT even at a temperature of 10 K and we can expect collisional excitation and de-excitation to drive the populations to their equilibrium values. The exponential factor in Boltzmann's equation is 1.0 here so

$$N(\text{upper})/N(\text{lower}) = g(\text{upper})/g(\text{lower}) = (2*1+1)/(2*0+1) = 3$$

This particular line is of importance because it can be produced in emission, even by the cold interstellar medium, by reason of its very low excitation energy. For emission lines, we are no longer confined to lines of sight that happen to have a suitable background star. The transition probability is small, for the line is forbidden, but hydrogen atoms comprise 90% of the atoms in the interstellar medium, most of them are neutral and about half are not combined into molecules, and the excitation is extremely high. Furthermore at this wavelength interstellar dust has no effect. Hydrogen can therefore be detected by the 21 cm line all over our galaxy and indeed in external spiral galaxies out to 100 Mpc and more.

The analysis of the 21 cm line is particularly straightforward because of the simple excitation conditions with only the two hyperfine levels populated. As we have already seen, the line absorption coefficient can be written in terms of the

Einstein coefficients and allowing for stimulated emission:

$$\kappa_\nu = B_{ij} h\nu N_i \phi_\nu [1 - e^{-h\nu/kT_{ex}}]$$

$$= A_{ji} \frac{c^2}{8\pi h\nu^3} \frac{g_j}{g_i} h\nu N_i \phi_\nu [1 - e^{-h\nu/kT_{ex}}]$$

$$= A_{ji} \frac{g_j}{g_i} \frac{c^2 h}{8\pi k} \frac{\phi_\nu}{\nu} \frac{N_i}{T_{ex}}, \quad \text{if } \frac{h\nu}{kT_{ex}} \ll 1 \tag{8.1}$$

where the transition is from upper level j to lower level i at frequency ν with lower level population N_i, ϕ_ν is the line profile and T_{ex} is the excitation temperature. The condition $h\nu/kT_{ex} \ll 1$ always holds for the 21 cm line.

In radio astronomy, the intensity is normally replaced by the brightness temperature T_B, and for a homogeneous source we have already shown that $T_B = T_{ex}(1 - \exp[-\tau_\nu])$ for optical depth τ_ν. If the source is optically thin at this frequency so $\tau_\nu \ll 1$, $T_B = T_{ex}\tau_\nu$ with $\tau_\nu = \kappa_\nu \rho R$ for a cloud of thickness R along the line of sight. Hence the brightness temperature integrated over the line profile in the optically thin case is:

$$\int T_B d\nu = T_{ex} \rho R \int \kappa_\nu d\nu$$

$$= T_{ex} \rho R A_{ji} \frac{g_j}{g_i} \frac{c^2 h}{8\pi k} \frac{N_i}{T_{ex}\nu} \int \phi_\nu d\nu$$

$$= \frac{c^2 h}{8\pi k} \frac{A_{ji} g_j}{g_i} \frac{1}{\nu} \frac{N_i}{N_{TOT}} (N_{TOT} \rho R) \tag{8.2}$$

where $N_{TOT}\rho R$ is the column density of the species concerned.

In the case of the 21 cm line of hydrogen, we have $N_j/N_i = 3/1$, so $N_i/N_{TOT} = 1/4$ since higher levels in hydrogen are not populated at all. Substituting we find:

$$\text{column density} = \frac{8\pi k}{c^2 h} \frac{1}{3A_{ji}} \frac{\nu}{1/4} \int T_B d\nu$$

$$= 3.9 * 10^{18} \int T_B d\nu \quad m^{-2} \tag{8.3}$$

If the line is Doppler broadened, for the optically thin case $T_B(\Delta\nu) = T_B(0) \exp(-[\Delta\nu/\Delta\nu_D]^2)$, where $T_B(0)$ is the line centre brightness temperature, $\Delta\nu$ is the distance in frequency from the centre of the line and $\Delta\nu_D$ is the Doppler width. Now $\int \exp(-[\Delta\nu/\Delta\nu_D]^2) d\nu = \sqrt{\pi}\Delta\nu_D$, so

$$\int T_B d\nu = T_B(0)\sqrt{\pi}\Delta\nu_D \tag{8.4}$$

Sometimes observational results are summarized in the form of $T_B(0)$ and the line width (full width at half height = $1.66\Delta\nu_D$) with the latter given in velocity units of kilometres per second, the Doppler width in kilometres per second being equal to $\Delta\nu_D$ (in frequency units)/ν * (speed of light in kilometres per second). Radio telescopes (except for long base line interferometers) have a poor resolution. If the telescope measurement is taken as an intensity, but the beam of the telescope is greater than the source, then the true brightness temperature will be underestimated.

One can also estimate the total flux received from a cloud as opposed to the surface brightness of part of a cloud. The flux is all that is measurable if the cloud is not spatially resolved by the telescope. Suppose the cloud is spherical and of radius R, is optically thin and homogeneous, and is at a distance d. Then the emission coefficient equals the absorption coefficient multiplied by the source function, and the total luminosity of the cloud is given by 4π times the emission coefficient per unit volume times the volume of the cloud. The flux received is the luminosity/$(4\pi d^2)$, and the source function $B_\nu(T_{ex}) = (2kT_{ex}/c^2)\nu^2$. Using equation (8.1) for the absorption coefficient per unit mass multiplied by the density ρ (N_{TOT} is the total number of hydrogens per unit volume) we obtain:

$$F_\nu = \frac{1}{3}\frac{R^3}{d^2} h A_{ji} \frac{g_j}{g_i} \nu\phi_\nu \frac{N_i}{N_{TOT}} N_{TOT}\rho \qquad (8.5)$$

If the cloud is resolved and is optically thick, then $T_B = T_{ex}$. If optical thickness happens at any frequency in the line profile it will happen at the line centre, so where optical thickness is suspected, $T_B(0)$ gives an estimate of the excitation temperature. For hydrogen and a wavelength of 21 cm, collisions will dominate the factors determining the excitation temperature and the excitation temperature should approximate to the kinetic temperature. The big difficulty is in knowing whether a given observation is looking at an optically thick or an optically thin transition. Equation (8.1) shows that the optical depth at the line centre for a Doppler broadened line is

$$\tau_0 = \frac{1.46 * 10^{-19}}{T_{ex}\Delta\nu_D} N_{TOT}\rho R \qquad (8.6)$$

which for $T_{ex} = 100\,\mathrm{K}$ and for a microturbulence of $3\,\mathrm{km\,s^{-1}}$ in the Doppler width gives a line centre optical depth of about $10^{-25} N_{TOT}\rho R$.

An isolated cloud in which the turbulence velocities truly have a Gaussian distribution has a Gaussian profile if weak becoming more flat-topped when saturated, but in the galaxy clouds often have overlapping (in velocity) profiles so use of the line profile is not easy. The distribution of hydrogen in our galaxy is in the form of a thin (about 100 pc thick) disc, with the gas rotating with the stars at about $225\,\mathrm{km\,s^{-1}}$ around the centre with little variation in velocity until one comes within 2 kpc of the centre. Observations perpendicular to the plane are unlikely to involve large optical depths, and the same is true of external galaxies unless we observe them close to edge on. The situation when observing in the plane of our galaxy is more complicated. One is talking here about looking through, say, 10 kpc of hydrogen but the different radial velocities in the line of sight of the clouds at different distances means that they only partially overlap in line centre frequency. The effective R for a homogeneous medium is of the order of $\Delta R = \Delta\nu_D/(\nu/c^* dV/dr)$, where dV/dr is the gradient of the line of sight velocity with distance along the line of sight. Now observations directly towards the galactic centre will have all rotational motion transverse to the line of sight giving no Doppler shift, so that the optical depth will be the sum of the depths of all the clouds to the centre and hence will be large, whereas at galactic longitude $l = 45°$ (measured from the direction to the centre), dV/dr will be quite large and the optical depth much smaller. One might therefore use observations towards the centre in the hope that they were optically thick to determine the gas temperature. The largest brightness temperatures found in this direction suggest an excitation temperature of 135 K, although there could be a contribution from 1000 K optically thin gas outside the main clouds. One could then use observations in directions like $l = 45°$ to find column densities at various distances and hence volume densities, on the assumption that the gas in these directions is optically thin.

It is also possible to observe the 21 cm line in absorption against some bright background continuous source of radiation at these wavelengths (free–free or synchrotron). Suppose the background source has a brightness temperature $T_B(BS)$ at 21 cm, and is observed through a homogeneous foreground cloud with excitation temperature T_{ex} and optical depth τ. Then

$$T_B(1) = T_B(BS)e^{-\tau} + T_{ex}(1 - e^{-\tau}) \tag{8.7}$$

that is, we observe at the line wavelength the background source absorbed by the cloud plus the emission from the cloud. If the background source is smaller than the foreground cloud, the telescope can be pointed so that the background source is missed but the foreground source is included within the beam. Then

$$T_B(2) = T_{ex}(1 - e^{-\tau}) \tag{8.8}$$

as only the emission from the cloud is observed.

Finally the telescope can be pointed in the direction of the background source tuned off the line wavelength. If the background source has a continuous spectrum slowly varying with wavelength, then a slight change in wavelength will not alter the brightness temperature of the background source appreciably and so

$$T_B(3) = T_{BS} \tag{8.9}$$

Then $[T_B(1) - T_B(2)]/T_B(3) = \exp(-\tau)$ so the optical depth can be determined. Substitution of τ in (8.8) then enables the excitation temperature to be found. The observations are actually performed slightly differently with the receiver being chopped on and off the line wavelength and the difference signal being measured.

Thus, pointing at the background source:

$$\Delta T_B(1) = T_B(1) - T_B(3) = (T_{ex} - T_{BS})(1 - e^{-\tau}) \tag{8.7a}$$

and pointing away from the background source but through the foreground cloud:

$$\Delta T_B(2) = T_B(2) = T_{ex}(1 - e^{-\tau}) \tag{8.8a}$$

so

$$\Delta T_B(2) - \Delta T_B(1) = T_{BS}(1 - e^{-\tau}) \tag{8.10}$$

which with a separate measurement of the background source brightness temperature (8.9) gives the optical depth, and substituting in (8.8a) again enables the excitation temperature to be calculated.

One problem that can arise is that the background source is smaller than the beam size, in which case T_{BS} in (8.7a) and (8.10) has to be multiplied by the ratio of the background source solid angle to the beam solid angle. A major difficulty is that the assumption has to be made that the foreground cloud has the same optical depth and excitation temperature at the two different positions for the observations represented by (8.7a) and (8.8a), and in view of the inhomogeneity of the neutral hydrogen distribution this is unlikely to be the case, at least for the optical depth. This difficulty can be avoided in a few special cases where the background source varies with time in a known manner, the main examples being pulsars. The beam is not moved, but observations are taken at two times when the background source has brightness temperatures T'_{BS} and T''_{BS}. Then at the first time:

$$\Delta T'_B(1) = (T_{ex} - T'_{BS})(1 - e^{-\tau})$$

and at the second time:

$$\Delta T''_B(1) = (T_{ex} - T''_{BS})(1 - e^{-\tau})$$

so

$$\Delta T'_B(1) - \Delta T''_B(1) = [T''_{BS} - T'_{BS}](1 - e^{-\tau})$$

thus enabling the optical depth to be measured at a single point, and by substitution the excitation temperature to be found.

Molecules in the Cold Interstellar Medium

In this section we shall concentrate on the interpretation of molecular lines in the radio and millimetre regions of the spectrum, although as we have already remarked a few molecular absorption lines are seen as interstellar lines in the visible spectrum, and other lines are seen from hotter parts of the interstellar medium. Much of what has already been said about the 21 cm line of hydrogen applies to molecules as well, and as with the 21 cm line, molecular lines are observed both in emission and, if a suitable background is available, in absorption. However, there are a number of important differences. The rotational levels of molecules are usually farther apart than the hyperfine structure sub-levels of hydrogen but are much closer together than the excited levels of an atom in most cases. It is therefore possible for quite a number of rotational levels of a molecule to be populated as compared with the hydrogen where only the two hyperfine sub-levels need to be considered. On the other hand the larger excitation energies often involved (compared with the excitation energy of the upper hyperfine level in hydrogen) mean that the level populations are no longer roughly proportional to the statistical weight but that the temperature-dependent exponential factor must be taken into account. Furthermore, we can no longer assume that the levels are populated collisionally and that populations have their LTE values, partly because collisions with the larger threshold energies needed are not so frequent, and partly because radiative processes are available. Energy levels separated by amounts corresponding to millimetre transitions can be populated by the ubiquitous microwave background radiation which peaks in the sub-millimetre region. Optically thick lines can also alter the level populations, which means we may have to solve the equations of radiative transfer. In the case of the more opaque clouds, temperature gradients may be present. Care must be exercised at the shorter wavelengths that long wavelength approximations are still valid. Observationally, one always observes millimetre emission lines superimposed on a microwave background continuum.

Molecular spectra were briefly discussed in the section on Photodissociation Regions in Chapter 6. We now consider in a little more detail the rotational spectrum. In the cold interstellar medium, we will be dealing with the ground electronic state and the lowest vibrational level with vibrational quantum number zero. For simplicity we shall initially consider the case of molecules which are linear (all the atoms are in a line) with ground electronic states with total electronic angular momenta of zero. Diatomic molecules are, of course, necessarily linear.

Classically, a rigid rotator with a single axis has an energy of rotation $E = 1/2I\omega^2$, where ω is the angular velocity of rotation and I is the moment of inertia about the axis of rotation, with $I = \mu r^2$ for reduced mass μ. Now the angular momentum of such a rotator is $I\omega$, so

$$E_{\rm rot} = \frac{1}{2I}(\text{angular momentum})^2$$

In quantum mechanics:

$$(\text{angular momentum})^2 = J(J+1)\left(\frac{h}{2\pi}\right)^2$$

where h is Planck's constant and $J = 0, 1, 2, 3\ldots$ is the angular momentum quantum number. Hence:

$$E_{\text{rot}} = J(J+1)\left(\frac{h}{2\pi}\right)^2 \frac{1}{2I} \tag{8.11}$$

which is also the result obtained by a more rigorous derivation.

It is usual to write

$$E_{\text{rot}}(J) = hBJ(J+1) \tag{8.11a}$$

which defines the rotational constant B so

$$B = h/(8\pi^2 I) \tag{8.12}$$

although sometimes $E = hcBJ(J+1)$ or $E = BJ(J+1)$ are used with appropriate changes in (8.12). The selection rule for transitions is $\Delta J = +1$ or -1 which gives just one series of lines since $J = 0$ to $J = 1$ is the same line as $J = 1$ to $J = 0$. Thus if J'' is the rotational quantum number of the lower level of a transition (whether in absorption or emission) then the frequency of the transition is

$$\nu = \Delta E/h = B[(J''+1)(J''+2) - J''(J''+1)] = 2B(J''+1) \tag{8.13}$$

So far we have assumed that the nuclei behave as a rigid rotator with fixed separation (in a diatomic molecule) r_e, given by the minimum in a plot of potential energy of the nuclei against separation. The lowest order approximation assumes that a displacement $r - r_e$ from the equilibrium position will produce a restoring force proportional to the displacement, $F = -k(r - r_e)$, so that the potential energy in the neighbourhood of r_e is $V = (1/2)k(r - r_e)^2$. Classically, the nuclei will oscillate about separation r_e as a harmonic oscillator with frequency $\nu_{\text{osc}} = 1/(2\pi)\sqrt{(k/\mu)}$. Quantum mechanically, the harmonic oscillator has a series of energy levels:

$$E_{\text{vib}}(v) = h\nu_{\text{osc}}(v + 1/2) \tag{8.14}$$

where $v = 0,1,2,3...$ is the vibrational quantum number. It should be noted that the potential cannot be of a pure quadratic form, for it must rise very steeply at small separations and tend to zero at large separations, and an exact expression for the vibrational energy will include anharmonic terms. Now if the molecule is rotating, there will be a centrifugal force which will increase the equilibrium separation of the nuclei and moment of inertia and effectively decrease B so the separation of neighbouring energy levels and line frequencies will become less than indicated by (8.11a) and (8.13) at large values of J. In the cold interstellar medium we are only concerned with low values of J for which the correction involved is negligible, but it is convenient to complete our discussion of rotational energy levels here for future reference.

The centrifugal force is $\mu\omega^2 r = \mu$ (angular momentum$/I)^2 r = J(J+1)h^2/(4\pi^2)1/(\mu r^3)$ for reduced mass of the molecule μ, so equilibrium in the rotating molecule will be reached when

$$k(r-r_e)=\frac{J(J+1)\left(\dfrac{h}{2\pi}\right)^2}{\mu r^3}$$

$$=\frac{J(J+1)\left(\dfrac{h}{2\pi}\right)^2}{\mu r_e^3\left(1+\dfrac{r-r_e}{r_e}\right)^3}$$

At the new equilibrium position,

$$\text{total energy} = \text{kinetic energy} + \text{potential energy}$$

$$=\frac{J(J+1)\left(\dfrac{h}{2\pi}\right)^2}{2\mu r^2}-\tfrac{1}{2}k(r-r_e)^2$$

$$=\frac{J(J+1)\left(\dfrac{h}{2\pi}\right)^2}{2\mu r_e^2\left(1+\dfrac{r-r_e}{r_e}\right)^2}-\tfrac{1}{2}k(r-r_e)^2$$

$$\simeq hBJ(J+1)-\frac{J^2(J+1)^2h^4}{32\pi^4\mu^2r_e^6k}$$

noting that $(r-r_e)/r_e \ll 1$ so $1+(r-r_e)/r_e \sim 1$, and substituting for $r-r_e$ in the potential energy term.

It is customary to write

$$E = hBJ(J+1) - hDJ^2(J+1)^2 + \cdots \tag{8.15}$$

where

$$D = 16\pi^2B^3\mu/k = 4B^3/\nu_{osc}^2 \tag{8.16}$$

so

$$\nu = 2B(J''+1) - 4D(J''+1)^3 \tag{8.17}$$

The frequency ν_{osc} is usually quoted in the form of the wavenumber $\omega_e = \nu_{osc}/c$ and if D and B are both also divided by c to be in wavenumber form, $D' = D/c$ and $B' = B/c$, then $D' = 4B'^3/\omega_e^2$. Equation (8.15) may be taken as the first two terms of a power series, with the constants B, D, etc to be determined empirically, as ω_e is. Finally it should be noted that high values of the vibrational quantum number v mean that the nuclei spend time at separations considerably different from r_e, and the deviation from quadratic form at such $r-r_e$ has a steep slope inhibiting small r and a shallow slope allowing large r. The result is that the average value of

r will be greater than r_e for large v, decreasing the value of B. Thus in general B and D will be functions of v, and we write

$$B_v = B_e - \alpha(v + 1/2) + \cdots$$

and sometimes

$$D_v = D_e - \beta(v + 1/2) \tag{8.18}$$

Again, we are here only concerned with $v = 0$, but it is convenient to complete our discussion of rotational levels to include higher vibrational states, especially since B is normally tabulated as B_e and α, with $B_0 = B_e - 1/2\alpha$.

The transition probability for a line is given by (2.11):

$$A_{ji} = \frac{64\pi^4}{3hc^3 4\pi\varepsilon_0} \nu^3 |\boldsymbol{d}|^2$$

where

$$\boldsymbol{d} = e \int \Phi_j \boldsymbol{r} \Phi_i \, \mathrm{d}V$$

represents the dipole moment. The rigid rotator wave functions can be factorized into radial and angular parts where only the angular parts change in a transition in which only the rotational quantum number changes. Hence the integral can be factorized into a term depending on J which has the value zero unless $\Delta J = +1$ or -1, and a term depending only on radial distributions which can be written with the product of initial wave functions since the product of initial angular wavefunctions is 1. So

$$d^2 = f(J)(e \textstyle\int \phi_i r \phi_i \mathrm{d}V)^2$$
$$= \frac{(J+1)}{(2J+3)} d_p^2$$

where d_p is the *permanent dipole moment*. Then

$$A_{ji} = 1.046 * 10^{21} \nu^3 d_p^2 \frac{(J+1)}{(2J+3)} \tag{8.19}$$

with J the quantum number of the lower level.

The prediction of the emission coefficient or the optical depth requires that we also know the statistical weight of the level with rotational quantum number J and the partition function. If the nuclei have no spin, the statistical weight of each level is simply $2J + 1$. If Boltzmann's equation holds or if at least a single excitation temperature can be used to describe the populations of all relevant levels, then the partition function is given by

$$U = \sum_j (2J+1) \exp\left[-\frac{hBJ(J+1)}{kT_{ex}}\right]$$

$$\simeq \int_0^\infty (2J+1)\exp\left[-\frac{hBJ(J+1)}{kT_{ex}}\right] \mathrm{d}J$$

$$\simeq \int_0^\infty \exp\left[-\frac{hBJ'}{kT_{ex}}\right] \mathrm{d}J', \quad \text{with } J' = J^2 + J$$

$$\simeq \frac{kT_{ex}}{hB} \tag{8.20}$$

A rather better approximation to the partition function is obtained by noting that an integral can be approximated by a sum using Simpson's rule that for a sum in which the terms are at unit intervals (as here) the integral is approximately represented by 2/3 the first term plus 8/3 the second plus 4/3 the third term and so on. The integrand in (8.20) varies smoothly with J falling to zero at large J so the fact that equal weights are used in the sum to be approximated makes little error, but the first term in the sum is 1.0 and not 2/3, so a better approximation is

$$U = (kT_{ex})/(hB) + 1/3 \tag{8.20a}$$

Carbon monoxide is a particularly simple case with no electronic angular momentum in the ground state and no nuclear spin. The equilibrium separation is $1.128*10^{-10}$ m, a typical value for diatomic molecules, and $\mu = 6.859\, m_H$ so $B \sim 5.76*10^{10}\,\mathrm{s}^{-1}$ (equivalent to $190\,\mathrm{m}^{-1}$ in wavenumber units). The vibrational spectrum has $\omega_e = 217\,021\,\mathrm{m}^{-1}$ so $D \sim 1.8*10^5\,\mathrm{s}^{-1}$ ($6.0*10^{-4}\,\mathrm{m}^{-1}$ in wavenumber form). Exact values in wavenumber units are $B_e = 193.13\,\mathrm{m}^{-1}$, $\alpha_e = 1.748\,\mathrm{m}^{-1}$ so $B_0 = 192.23\,\mathrm{m}^{-1}$ and $D = 6.43*10^{-4}\,\mathrm{m}^{-1}$. Hence $\nu(J=1 \text{ to } J=0)$ is 115.26 GHz (a wavelength of 2.60 mm). The permanent dipole moment of CO is unusually small at 0.111 Debyes or $3.71*10^{-31}$ CM (unit charge at $0.02r_e$ since one Debye $= 3.336*10^{-30}$ CM). Hence the transition probability is $7*10^{-8}\,\mathrm{s}^{-1}$ for the 1–0 transition.

For hydrides like CH the mean molecular weight is about $0.9\,\mathrm{m_H}$ while r_e is similar to the value for CO, so the rotational constant B is larger and the wavelengths of the corresponding transitions lie in the sub-millimetre range rather than at millimetre wavelengths. The permanent dipole moments of hydrides and indeed of most polar molecules is of the order of a few Debyes. Polyatomic linear molecules tend to have fairly large moments of inertia and hence to have low J lines at long wavelengths. Molecules that have a centre of symmetry, that is a point (the centre of mass) the inversion relative to which of all the coordinates of the nuclei leaves the system indistinguishable from its initial state, have no permanent dipole moment. Examples include homonuclear molecules like H_2 and N_2 and also linear molecules like C_2H_2 where the atoms form a line H—C≡C—H with the centre of mass in the middle. A molecule with two nuclei of the same element but with different isotopic forms has a permanent

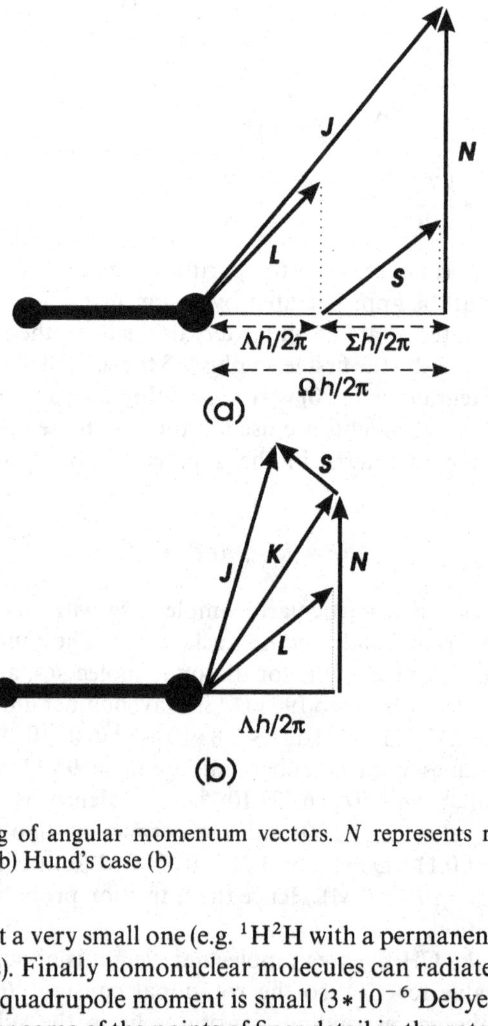

Figure 8.2. Coupling of angular momentum vectors. *N* represents rotation of nuclei. (a) Hund's case (a), (b) Hund's case (b)

dipole moment but a very small one (e.g. $^1H^2H$ with a permanent dipole moment of $6*10^{-4}$ Debyes). Finally homonuclear molecules can radiate via quadrupole radiation, but the quadrupole moment is small ($3*10^{-6}$ Debyes for H_2).

We now consider some of the points of finer detail in the rotational spectra of diatomic molecules. First we consider the effects of the angular momentum of the electronic system on the rotational energy levels (Figure 8.2). For diatomic molecules and linear molecules, electrons move in a field which is symmetric about the axis of the system, and it is the projection of the total orbital angular momentum (***L***) on this axis, designated $\Lambda(h/2\pi)$, which influences the energy of the system. Λ is analogous to M_L in an atom and can take values from 0 up to L. States with $\Lambda = 0$ are called Σ states (compared with S states in an atom), states with $\Lambda = 1$ are called Π states, and states with $\Lambda = 2$ are called Δ states. The energy differences between states of a given L but different Δ are large and of the order of the energy differences between atomic levels. It should be noted that the projection of $\boldsymbol{L}$ onto the axis can have positive or negative sign for a given Λ if $\Lambda > 0$, but these two projections to a first approximation have the same energy so each Λ state ($\Lambda > 0$) is doubly degenerate. The electrons also have spin angular momentum $\boldsymbol{S}$ and again it is often the projection onto the nuclear axis that is important. This is rather confusingly called Σ (compared with M_S in the atomic case) and has $2S + 1$ values. If $\Lambda > 0$, there will be a magnetic field in the axis direction produced by the electronic motion, and this will interact with the electronic magnetic moment associated with Σ to give a splitting of a given Λ state into a multiplet with $2S + 1$ levels since in this case the sign of Σ alters the sign of the energy change. Since the magnetic field is proportional to the component of the orbital angular momentum along the axis, energy $\propto$ constant $\Lambda\Sigma$. The magnitude of the total electronic angular momentum component along the axis quantum number is called $\Omega = |\Lambda + \Sigma|$ and a particular electronic state is written ${}^{2S+1}\Lambda_\Omega$.

If the electronic motion is strongly coupled to the axis, we have what is called Hund's case (a), and the total electronic angular momentum $\boldsymbol{\Omega}$ along the axis and the nuclear rotation $\boldsymbol{N}$ perpendicular to the axis couple to give the total angular momentum $\boldsymbol{J}$. Then the total angular momentum quantum number $J = \Omega, \Omega + 1, \Omega + 2$, etc. and values of J less than Ω are not allowed. Classically, the kinetic energy of a body with three principle axis, say x, y, and z, is

$$\begin{aligned} E &= \tfrac{1}{2}I_x\omega_x^2 + \tfrac{1}{2}I_y\omega_y^2 + \tfrac{1}{2}I_z\omega_z^2 \\ &= \frac{(\text{angular Momentum about } x)^2}{2I_x} + \frac{(\text{angular Momentum about } y)^2}{2I_y} \\ &\quad + \frac{(\text{angular Momentum about } z)^2}{2I_z} \end{aligned} \tag{8.21}$$

Here we may take the z axis as the molecular axis so (angular momentum about $z)^2 = \Omega^2(h/2\pi)^2$, and $I_y = I_x$ with

$$\begin{aligned} &(\text{angular momentum about } x)^2 + (\text{angular momentum about } y)^2 \\ &\quad = \boldsymbol{N}^2 = \boldsymbol{J}^2 - \boldsymbol{\Omega}^2 = (J(J+1) - \Omega^2)\,[h/(2\pi)]^2 \end{aligned}$$

Thus

$$E = hBJ(J+1) + h(A - B)\Omega^2 \tag{8.22}$$

where B has the same meaning as before and $A = h/(8\pi^2 I_z)$. Apart from an additive constant and the fact that the minimum value of J is Ω and not zero this is no different from (8.11).

If the nuclear spin is very weakly coupled to the molecular axis, then $\boldsymbol{\Lambda}$ and $\boldsymbol{N}$ form a resultant angular momentum normally designated $\boldsymbol{K}$ where the associated quantum number K can take values Λ, $\Lambda + 1$, etc. and $\boldsymbol{K}$ and $\boldsymbol{S}$ then form the total angular momentum $\boldsymbol{J}$ so $J = K + S, K + S - 1, \ldots, |K - S|$. There is a slight splitting in the energies of the levels with the same value of K but different J. Then

$$\text{energy} = hBK(K+1) + \text{constant} * K \quad \text{for } J = K + 1/2, \text{ if } S = 1/2$$

$$\text{energy} = hBK(K+1) - \text{constant} * (K+1) \quad \text{for } J = K - 1/2, \text{ if } S = 1/2$$

where the constant is very small compared with hB. This is called Hund's case (b) and always applies if $\Lambda = 0$, and also in some light molecules including CH.

Levels with a given value of Λ and J (ignoring electron spin for the moment) are doubly degenerate because Λ, the projection of $\boldsymbol{L}$ on the axis of the molecule, can for a given magnitude have either a positive or negative sign. Quantum mechanically we have two states, labelled Ψ^+ and Ψ^- or c and d, which represent symmetric and anti-symmetric combinations of the two projections. Interaction between the orbital angular momentum $\boldsymbol{L}$ and the rotation of the nuclei $\boldsymbol{N}$ leads to an energy difference between the two states since $\boldsymbol{L} \cdot \boldsymbol{N}$ does depend on the sign as well as the magnitude of $\boldsymbol{L}$. This splitting of energy levels is called Λ *doubling*. It increase as $J(J+1)$ or $K(K+1)$ as one goes to higher rotational quantum numbers. The doubling decreases as one goes to higher values of Λ but of course is zero for $\Lambda = 0$, so that in practice it is mainly significant for Π states. Transitions can take place between the two Λ doubled levels of a given J state since the parity is different, and the very small splitting means that these transitions will be at long wavelengths. The most important case for astronomical purposes is the transition between the c and d states of the ground state ($^2\Pi_{3/2}$, $J = 3/2$) of OH at a wavelength of 18 cm.

Finally in this consideration of points of detail we come to the hyperfine structure, the effects of nuclear spin I. The nuclear spin adds vectorially to the total angular momentum without nuclear spin to give the final total angular momentum F, with quantum number F with $2I+1$ values (if $J > I$) from $J + I$ to $|J - I|$. A very weak splitting is produced by magnetic effects, but more relevantly the hyperfine levels have different energies if the nuclear electrical charge distribution possesses an appreciable quadrupole moment of the electrical charge distribution which will interact with the electrical field produced by the electrons of the molecule. The nuclear quadrupole moment is given in terms of the nuclear charge density distribution ρ and charge on a proton e by

$$Q = \frac{1}{e} \int \rho(3z^2 - r^2)\mathrm{d}x\mathrm{d}y\mathrm{d}z$$

where the z coordinate is taken along the spin axis of the nucleus. Assuming cylindrical symmetry, Q expresses the departure from spherical symmetry of the nucleus. Let ϕ represent the electric field potential of the electrons at some point and let $q = \mathrm{d}^2\phi/\mathrm{d}z_\mathrm{m}^2$ be the second derivative of this potential in a direction parallel to the axis of a linear molecule so q is a fixed quantity for a given molecule. We also define $C = F(F+1) - J(J+1) - I(I+1)$ which represents $\boldsymbol{I} \cdot \boldsymbol{J}$, and hence the angle between the nuclear spin axis and the molecular angular momentum. Then the quadrupole energy is

$$E_Q = -eqQ\frac{\frac{3}{4}C(C+1) - I(I-1) - J(J+1)}{2I(2I-1)(2J-1)(2J+3)} \tag{8.23}$$

There is also a smaller magnetic interaction between the magnetic field produced by electrons outside a closed shell and the magnetic moment of the nucleus. This is proportional to $\boldsymbol{I}\cdot\boldsymbol{J}$ which gives

$$E_M = \frac{a}{2}(F(F+1) - J(J+1) - I(I+1) \tag{8.24}$$

where a is proportional to the magnetic moment of the nucleus and to $1/J(J+1)$ for the electron or electrons in the simplest case with Hundt's case (a) coupling.

Now the proton and neutron both have spin 1/2, orbital angular momenta must have integral quantum numbers, and the proton spins in a nucleus are paired as are the neutron spins. Hence a nucleus with even numbers of both protons and neutrons (even mass number A and atomic number Z) will have zero total spin in the ground state, while a nucleus with odd A and hence with either odd proton number and even neutron number or even proton number and odd neutron number will have one nucleon unpaired and half integral spin and a nucleus with even mass number but odd atomic number will have both proton number and neutron number odd and will have integral spin. All nuclei with $I > 1/2$ have quadrupole moments. Transitions between hyperfine levels are subject to the selection rule $\Delta F = -1, 0, +1$, with the strongest transitions having $\Delta F = \Delta J$. ^{16}O has zero spin but ^{1}H has $I = 1/2$ and a quadrupole moment so OH levels are hyperfine split into 2. Thus the ground state of OH with $J = 3/2$ has hyperfine levels with $F = 2$ and 1, and the Λ doubled c and d levels thus form four sub-levels (the hyperfine splitting is much less than the Λ doubling), the transitions between these levels making up a quadruplet with $\Delta J = 0$ so the strongest transitions are those from $F = 2$(c) to $F = 2$(d) and from $F = 1$(c) to $F = 1$(d).

Nuclear spin also has an important effect on the spectra of molecules containing two or more identical nuclei. Exchanging such nuclei leaves the molecule and hence the square of the wavefunction (strictly the product with the complex conjugate) unchanged, so the total wavefunction must be either symmetric (unchanged) or antisymmetric (changed only in sign) under such an operation. If the identical nuclei have half-integral spin they are fermions and the overall wavefunction must be anti-symmetric, whereas if the identical nuclei have integral spin, they are bosons and the total wavefunction must be symmetric. The total wavefunction is made up of the electronic wavefunction, the nuclear rotational and vibrational wavefunctions and the nuclear spin function. The vibrational wavefunction is symmetric since it involves only on the magnitude of the internuclear distance, and the rotational wavefunction is symmetric for even values of J and anti-symmetric for odd values of J.

Suppose that we are dealing with two identical nuclei, and suppose they both have spin $I = 0$, so the total wavefunction must be symmetric and the nuclear spin wavefunction must be symmetric. If the electronic wave function is symmetric then so must be the rotational wave function, and only even values of J are allowed, while if the electronic wavefunction is anti-symmetric, then so must be the rotational wavefunction and only odd values of J are allowed, with ground state $J = 1$. If there are two identical nuclei and both have spin $I = 1/2$, the total nuclear spin quantum number T can be 1 (spins parallel) to which correspond three combinations of individual nuclear spin wavefunctions that are symmetric, or the total nuclear spin quantum number can be 0, to which corresponds a single combination of individual nuclear spin wavefunctions which is anti-symmetric. The symmetric case has statistical weight $2T + 1 = 3$ and the anti-symmetric case has statistical weight 1. The overall wavefunction must be anti-symmetric (fermions) so if the electronic wavefunction is symmetric (as it is in the ground state of $^1H_2$) then J must be even for the $T = 0$ case and J must be odd for the $T = 1$ case. It follows that the odd J levels of H_2 (ortho-hydrogen) have statistical weights of $3\,(2J + 1)$ and the even J levels of H_2 (para-hydrogen) have statistical weights proportional to $(2J + 1)$, and the populations of the rotational levels and the intensities of lines from them will show a 3:1 alternation (leaving aside for the moment the effects of the exponential factor in Boltzmann's equation). Similarly for $I = 1$ there will be a 2:1 alternation in the populations of alternate levels.

We now turn to consider the case of polyatomic molecules. Diatomic molecules have a single vibrational mode of vibration about the mean internuclear distance which is called a stretching mode, but in polyatomic molecules both *stretching* and *bending* modes are possible. In general for a molecule with N atoms, there are $3N - 6$ normal vibrational modes ($3N$ degrees of freedom minus 3 for translational motion of the whole molecule and 3 for rotational motion, all in three dimensions), and the vibrational state of the molecule can be expressed in terms of a superimposition of these modes. A linear molecule has $3N - 5$ normal vibrational modes since one of the rotations about the axis possesses no energy. The vibrational frequency of a classical oscillator is proportional to the square root of the force constant over the mass, where the force constant depends on the shape of the potential energy curve about its minimum value. Clearly the larger the reduced mass, the lower the frequency, and in general terms the stronger the binding the higher the frequency, with stretching vibrations having higher frequencies than bending vibrations. Each bond, say O—H, has its characteristic frequencies which are often only slightly influenced by the rest of the molecular environment. However in this chapter we are mainly interested in rotational transitions. Linear molecules have been included here with diatomic molecules having similar rotational spectra in the vibrational ground state.

A number of important polyatomic molecules are *symmetric tops*. These have the same moment of inertia about two principal axes. Equivalently one can say that molecules with an axis with three (or greater)-fold symmetry, that is an axis about which a rotation through 120° (for three-fold) leaves the molecule unchanged, are symmetric tops. A well known example is ammonia, NH_3. The moments of inertia about the three principal axes are given the symbols I_a, I_b, and I_c, with by convention $I_c \geqslant I_b \geqslant I_a$. A linear molecule has $I_a = 0$. A symmetric top may either have the two larger moments of inertia equal, so $I_c = I_b > I_a$, in which case it is said to be *prolate*, or it may have the two smaller moments of inertia equal, so $I_c > I_b = I_a$, in which case it is said to be *oblate*. Consider first the prolate case. Classically, the rotation energy $= B'_a j_a^2 + B'_b j_b^2 + B' j_c^2 = B'_a j_a^2 + B'_b(j_b^2 + j_c^2)$, where j_a, j_b and j_c are the angular momenta about the appropriate axes, and $B'_a = 1/(2I_a)$, etc. The total angular momentum j is given by $j^2 = j_a^2 + j_b^2 + j_c^2$ so the rotational energy is $B'_b j^2 + (B'_a - B'_b) j_a^2$. Quantum mechanically we take $\boldsymbol{J}$ and $\boldsymbol{J}_a$ as the total angular momentum operator and the operator for the component of the angular momentum about the a axis, and J and K as the associated quantum numbers. The (total angular momentum)^2 is $J(J+1)(h/[2\pi])^2$ and the component of the angular momentum is $K(h/[2\pi])$. Then the rotational energy is given by

$$E(J, K) = hBJ(J+1) + h(A-B)K^2 \tag{8.25}$$

where

$$B = \frac{B'_b}{h}\left(\frac{h}{2\pi}\right) = \frac{h}{8\pi^2 I_b}$$

$$A = \frac{B'_a}{h}\left(\frac{h}{2\pi}\right) = \frac{h}{8\pi^2 I_a}$$

A is always greater than B and K must be less than or equal to J. If $K > 0$, each level will have a double degeneracy, which can be imagined as being due to the fact that a given projection of the angular momentum on the axis can still correspond to clockwise or anticlockwise rotation.

Selection rules are $\Delta J = 1$ or -1, as before, with the additional requirement that $\Delta K = 0$ since the dipole lies along the symmetry axis. Thus for a rigid rotor the line frequencies are given by the same formula as for the linear case, $\nu = 2B(J+1)$ for lower level J, with no dependence on K.

However for a non-rigid rotor with centrifugal distortion:

$$\begin{aligned} E(J, K) = hBJ(J+1) + h(A-B)K^2 - hD_J J^2(J+1)^2 \\ - hD_{JK} J(J+1)K^2 - hD_K K^4 \end{aligned} \tag{8.26a}$$

$$\nu = 2(B - D_{JK}K^2)(J+1) - 4D_J(J+1)^3 \tag{8.26b}$$

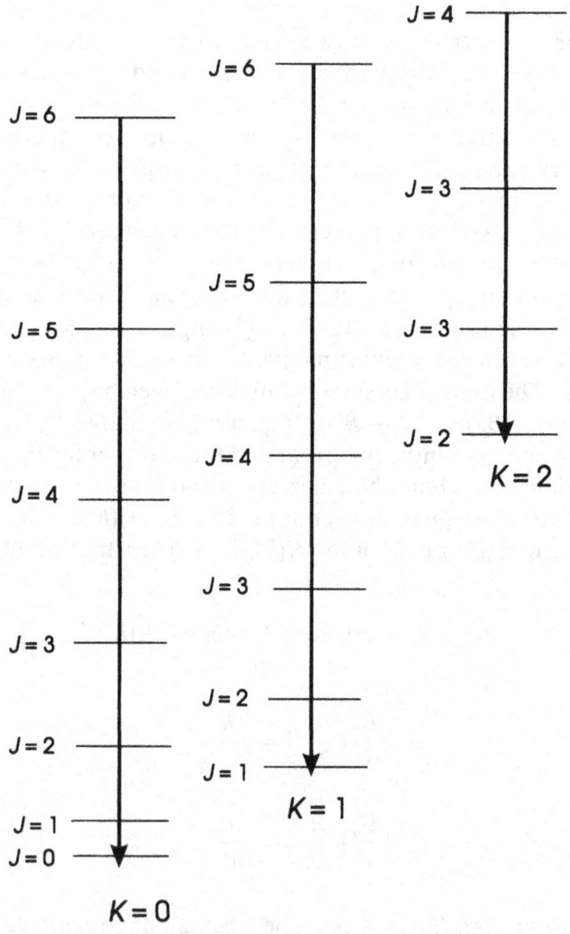

Figure 8.3. Rotational levels of symmetric top (prolate case—for the oblate case, levels of the same J but different K have nearly the same energy)

so that for a transition from a level with a given value of J there are $J + 1$ components with values of K from 0 to J.

In the case of an oblate symmetric top the same argument can be used with $C = h/(8\pi I_C)$ replacing A so $E = hBJ(J+1) + h(C - B)K^2$ for the rigid case. Thus symmetric top molecules have rotational energy schemes with separate ladders (of J values) for different values of K. Each ladder has a lowest J value equal to K, and there are no transitions between ladders. For a given value of J, the energy goes up with increasing K for the prolate case since $A > B$, and goes down very slightly with increasing K for the oblate case since $C < B$ (Figure 8.3).

The transition probability for the symmetric top is given by

$$A_{J+1,J} = \frac{64\pi^4}{3(4\pi\varepsilon_0)hc^3}\nu^3\mu_p^2\frac{(J+1)^2 - K^2}{(J+1)(2J+3)}$$

where the factor $\{(J+1)^2 - K^2)\}/\{(J+1)(2J+3)\}$ replaces $(J+1)/(2J+3)$ in the corresponding formula for a linear molecule.

The partition function is:

$$Q_{\mathrm{rot}} = \sum_{J=0}^{\infty}(2J+1)e^{-hBJ(J+1)/kT}\sum_{K=0}^{J}g_K e^{-(A-B)K^2(h/kT)} \tag{8.27}$$

$$\simeq e^{Bh/4kT}\sqrt{\left[\frac{\pi}{B^2A}\left(\frac{kT}{h}\right)^3\right]}\left[1+\frac{1}{12}\left(1-\frac{B}{A}\right)\frac{Bh}{kT}+\cdots\right]$$

$$\simeq\sqrt{\left[\frac{\pi}{B^2A}\left(\frac{kT}{h}\right)^3\right]},\quad \text{if } \frac{Bh}{kT}\ll 1 \tag{8.27a}$$

where $g_K = 2$ except for $K = 0$ when $g_K = 1$.

It is also possible to have a spherical rotor like CH_4, where $I_A = I_B = I_C$. Such molecules have the same energy levels as a linear molecule, but have no permanent dipole moment to first order although centrifugal distortion can produce a small dipole moment of the order of 10^{-5} to 10^{-6} Debyes. It follows that rotational transitions of such molecules are very weak.

Molecules with all three moments of inertia different are called asymmetric rotors. Often the moments of inertia about two axis are close in value, in which case one can treat the system to first order as the corresponding symmetric rotor with its levels split to remove the remaining degeneracy. The levels of an asymmetric rotor are often labelled

$$JK_{K_{-1}K_1}$$

where K_{-1} refers to the K value that a prolate symmetric top would have and K_1 refers to the K value that an oblate symmetric top would have. If the molecule is close to the prolate case, then the levels will be mainly separated for a given J value by the value of K_{-1} with a slight splitting according to the value of K_1, and vice versa for the oblate case.

Vibrational transitions normally occur at infrared wavelengths, and except when observed in absorption against a bright infrared background source are not of interest here. However, for molecules with a pyramidal shape like NH_3, *inversion* transitions can occur at long wavelengths. The nitrogen atom can vibrate with respect to the plane defined by the three hydrogen atoms either 'above' or 'below' the plane in the potential well created by the atoms (Figure 8.4). The energy of the vibration should be independent of whether the nitrogen is 'above' or 'below', so there should be two degenerate states. However, although classically there is an impassable barrier between the two states for the ground

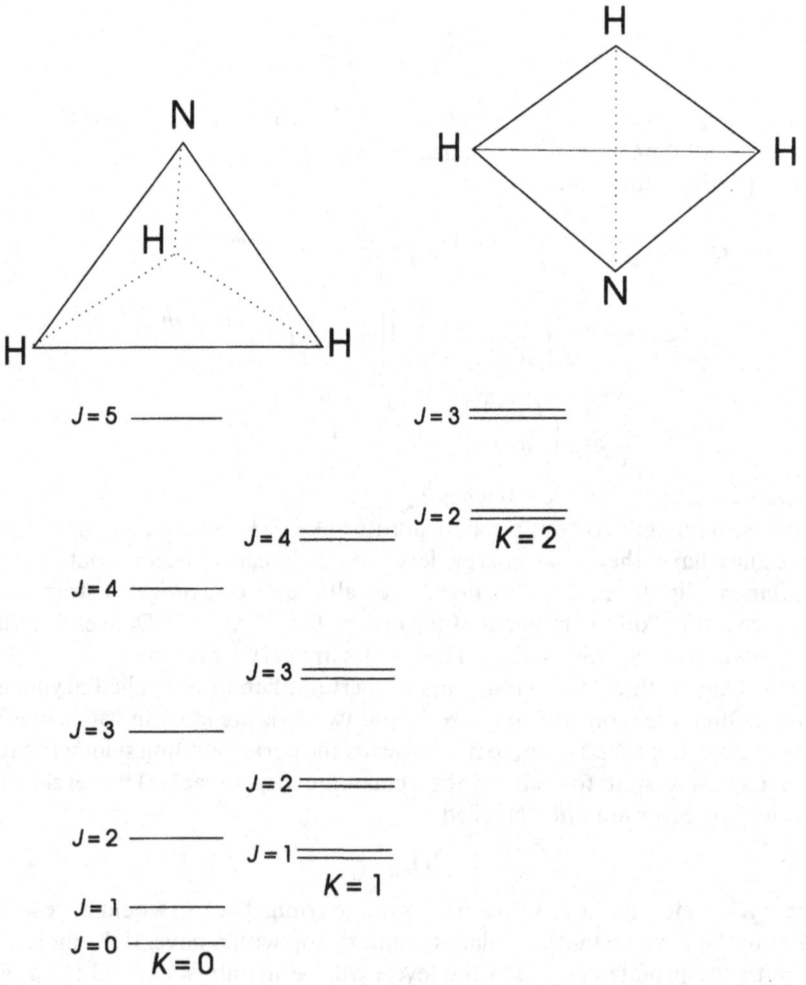

Figure 8.4. NH_3 showing inversion doubling of all levels except those of $K = 0$

vibrational state (the potential energy becomes much greater than the minimum value when the nitrogen passes through the plane) making transitions impossible, quantum mechanically tunnelling can happen, mixing the 'above' and 'below' states. As usual, this results in a symmetrical combination of the 'above' and 'below' wave functions, and an anti-symmetrical combination. The two combinations have different energies, the symmetric combination having the lower energy. Transitions between the two levels thus produced have long wavelengths, in the case of NH_3 in the lowest ($v = 0$) vibrational state at 1.25 cm. Such a transition can clearly be excited in the low temperatures of the interstellar medium, and the 1.25 cm line is seen in some astronomical sources.

Interpreting Molecular Line Strengths

We now turn to a more detailed consideration of the determination of temperatures and column densities from molecular lines, with particular reference to the most abundant molecule after hydrogen, CO. Some aspects of the problem are similar to those already discussed in the case of the 21 cm line of hydrogen, for instance the question of whether the observed line is optically thin or optically thick, and the possibility that the cloud does not fill the telescope beam (resolution for a given telescope aperture is much better at millimetre wavelengths, but telescope diameters are smaller so the gain in angular resolution is of the order of 15). However, as has already been pointed out, there are also additional complications in the presence of the universal microwave background which not only affects observations but also excitation, and the fact that the long wavelength approximation is not necessarily valid (at a wavelength of 2.64 mm, $h\nu/kT$ is 0.55 at $T = 10\,\mathrm{K}$). In the case of the 21 cm line, only the two hyperfine levels are populated according to their statistical weights, but many rotational levels in a molecule may be populated and even for a line with the ground state as lower level, the partition function introduces an extra factor proportional to the temperature. Furthermore, LTE does not necessarily hold, the excitation temperature may differ from the kinetic temperature, and the temperature characterizing the partition function may be different from the excitation temperature of the line. In compensation, observations of more than one line of a particular molecule may be available.

The brightness temperature of a transition from an optically thin homogeneous cloud of geometrical depth R (if the long wavelength approximation for stimulated emission is valid) is given by equation (8.2) with N_i/N_{TOT} given by the Boltzmann equation characterized by the excitation temperature T_{ex} and the partition function U approximated by (8.20). The transition is from an upper level with rotational quantum number $J + 1$ to a lower level with quantum number J so the frequency of the transition is $2B(J + 1)$, $g_j = 2J + 3$ and $g_i = 2J + 1$. The nuclear spin and the electronic angular momentum are assumed to be zero. Then:

$$\int T_{\mathrm{B}}\,\mathrm{d}\nu = \frac{c^2 h}{8\pi k}\frac{A_{ji}g_j}{g_i}\frac{1}{\nu}\frac{g_i}{U}\exp\left[-\frac{E}{kT_{\mathrm{ex}}}\right](N_{\mathrm{TOT}}\rho R)$$

$$= \frac{c^2 h}{8\pi k}\frac{A_{ji}(2J+3)}{(2J+1)}\frac{1}{2B(J+1)}\frac{hB}{kT_{\mathrm{ex}}}(2J+1)\exp\left[-\frac{hBJ(J+1)}{kT_{\mathrm{ex}}}\right](N_{\mathrm{TOT}}\rho R)$$

$$= \frac{c^2 h^2}{16\pi k^2}\frac{1}{T_{\mathrm{ex}}}A_{ji}\frac{(2J+3)}{(J+1)}\exp\left[-\frac{hBJ(J+1)}{kT_{\mathrm{ex}}}\right](N_{\mathrm{TOT}}\rho R) \qquad (8.28)$$

where cases other than linear molecules can be dealt with by substituting the appropriate quantities for the statistical weights, partition function and excitation energy in the first equation.

Using (8.19) for A_{ji} we obtain:

$$\int T_{\rm B}\,{\rm d}\nu = \frac{4\pi^3 h}{3ck^2(4\pi\varepsilon_0)}\frac{\nu^3}{T_{\rm ex}} d_p^2 \exp\left[-\frac{hBJ(J+1)}{kT_{\rm ex}}\right](N_{\rm TOT}\rho R) \tag{8.29}$$

$$= \frac{32\pi^3 h}{3ck^2(4\pi\varepsilon_0)}\frac{B^3(J+1)^3}{T_{\rm ex}} d_{\rm p}^2 \exp\left[-\frac{hBJ(J+1)}{kT_{\rm ex}}\right](N_{\rm TOT}\rho R) \tag{8.29a}$$

Consider first the question of optical thickness in the case of CO. In the solar system, the abundance of ^{12}C is about 90 times that of ^{13}C. If the same abundance ratio holds in interstellar clouds, then the fact that the ratio of the intensity of the $^{12}C^{16}O$ line to that of the $^{13}C^{16}O$ line is often much less than 90 suggests that the $^{12}C^{16}O$ lines are saturated and the $^{13}C^{16}O$ lines are fairly optically thin, because the optical depths will be proportional to the abundance of the carbon isotope involved but other factors will be nearly the same. If the $^{12}C^{16}O$ line is saturated, then the brightness temperature at the line centre will give the excitation temperature, and one might expect the excitation temperature to be the same for $^{13}C^{16}O$. If the $^{13}C^{16}O$ line is really weak, then equations (8.29) and (8.29a) will give the column density of $^{13}C^{16}O$ molecules, which only needs to be multiplied by the abundance ratio of ^{12}C to ^{13}C to give the total column density of CO. More generally if one assumes the long wavelength approximation still holds and ignores for the moment the effect of background radiation, then since $T_{\rm B} = T_{\rm ex}(1 - {\rm e}^{-\tau})$ and $T_{\rm ex}$ is the same for both lines:

$$\frac{T_{\rm B}(^{12}{\rm C}^{16}{\rm O})}{T_{\rm B}(^{13}{\rm C}^{16}{\rm O})} = \frac{(1-{\rm e}^{-\tau(^{12}{\rm C}^{16}{\rm O})})}{(1-{\rm e}^{-\tau(^{13}{\rm C}^{16}{\rm O})})} \tag{8.30}$$

If an isotopic abundance ratio is assumed so $N(^{12}{\rm C})/N(^{13}{\rm C}) = R$, say, then $\tau(^{12}{\rm C}^{16}{\rm O}) = R\tau(^{13}{\rm C}^{16}{\rm O})$, and (8.30) can be solved for $\tau_\nu(^{13}{\rm C}^{16}{\rm O})$ at each frequency in the line. Note that the fraction of the beam filled by the cloud should not affect the determination of the optical depth in (8.30). Equations (8.28) and (8.29) express the column density in terms of the integral of the brightness temperature over frequency, but since those expressions are derived for the optically thin case, we can write $\tau = T_{\rm B}/T_{\rm ex}$. It is more customary to write the integral as one over velocity V, where ${\rm d}\nu = \nu/c\,{\rm d}V$, thus giving

$$\int \tau_\nu\,{\rm d}V = \frac{c^3 h}{8\pi k}\frac{A_{ji}g_j}{\nu^2 T_{\rm ex} U}\exp^{-E_i/(kT_{\rm ex})} N_{\rm TOT}\rho R \tag{8.31}$$

$$= \frac{16\pi^3 h}{3k^2 4\pi\varepsilon_0}\frac{B^2(J+1)^2}{T_{\rm ex}^2} d_{\rm p}^2 N_{\rm TOT}\rho R \exp^{-hBJ(J+1)/kT_{\rm ex}} \tag{8.32}$$

where (8.32) is for the case of a CO transition with lower level J. These relations hold irrespective of the degree of saturation.

How can we show that the isotope ratio has the solar system value? In a few cases, it is possible to measure the lines of both $^{13}C^{16}O$ and $^{12}C^{18}O$, both of which are weak since both contain a rare isotope. Hence $T_B(^{13}C^{16}O)/T_B(^{12}C^{18}O) = N(^{13}C)/N(^{12}C) \cdot N(^{16}O)/N(^{18}O)$, and one can show that the observations are consistent with the solar system isotope ratios of both carbon and oxygen. Secondly, the highest velocity clouds from Orion show $T_B(^{13}C^{16}O)/T_B(^{12}C^{16}O)$ approaching 80, and this would be consistent with optical thinness even for the more abundant isotope, together with a solar system isotope ratio.

Before proceeding further, we should note that the long wavelength approximation may no longer be a good one for, for example CO at low temperatures. Without this approximation, (8.31) becomes

$$\int \tau_v \mathrm{d}V = \frac{c^2}{8\pi v^3} \frac{A_{ji} g_j}{U} \exp[-E_i/kT_{\mathrm{ex}}] N_{\mathrm{TOT}} \rho R (1 - \mathrm{e}^{-hv/kT_{\mathrm{ex}}}) \tag{8.31a}$$

Equation (8.32) for linear molecules (which includes all diatomic molecules), with the better approximation for the partition function, becomes

$$\int \tau_v \mathrm{d}V = \frac{8\pi^3 B d_{\mathrm{p}}^2}{3k4\pi\varepsilon_0} \frac{N_{\mathrm{TOT}} \rho R (J+1) \mathrm{e}^{-hBJ(J+1)/kT_{\mathrm{ex}}}}{T_{\mathrm{ex}} + \dfrac{hB}{3k}} (1 - \mathrm{e}^{-hv/kT_{\mathrm{ex}}}) \tag{8.32a}$$

For $^{12}C^{16}O$, $(hB)/(3k) = 0.93$ and for $^{13}C^{16}O$, $hB/(3k) = 0.89$. One can invert (8.35a) for $^{12}C^{16}O$ and $^{13}C^{16}O$ to give the column densities:

$$N_{\mathrm{TOT}} \rho R(^{12}C^{16}O) = 2.39 \times 10^{15} \frac{\mathrm{e}^{hB_{12}J(J+1)/kT_{\mathrm{ex}}}}{J+1} \frac{T_{\mathrm{ex}} + 0.93}{1 - \mathrm{e}^{-hv_{12}/hT_{\mathrm{ex}}}} \int \tau \mathrm{d}V \mathrm{m}^{-2} \tag{8.33}$$

$$N_{\mathrm{TOT}} \rho R(^{13}C^{16}O) = 2.49 \times 10^{15} \frac{\mathrm{e}^{hB_{13}J(J+1)/kT_{\mathrm{ex}}}}{J+1} \frac{T_{\mathrm{ex}} + 0.89}{1 - \mathrm{e}^{-hv_{13}/hT_{\mathrm{ex}}}} \int \tau \mathrm{d}V \mathrm{m}^{-2} \tag{8.33a}$$

where the velocity is in metres per second.

One can also compare two transitions of the same isotopic form, say transitions with lower levels l and m. Then at any particular point in the line profile (any particular velocity), we have

$$\frac{\tau_V(m \text{ to } m+1)}{\tau_V(l \text{ to } l+1)} = \frac{(J_m+1)\,\mathrm{e}^{-E_m/kT_{\mathrm{ex}}}}{(J_l+1)\,\mathrm{e}^{-E_l/kT_{\mathrm{ex}}}} \frac{1 - \mathrm{e}^{-hv_m/kT_{\mathrm{ex}}}}{1 - \mathrm{e}^{-hv_l/kT_{\mathrm{ex}}}}$$

where v_l and v_m are the frequencies of the transitions from l to $l+1$ and from m to $m+1$. If $m = l+1$, then $E_m - E_l = hv_l$ and

$$\frac{\tau_V(l+1 \text{ to } l+2)}{\tau_V(l \text{ to } l+1)} = \frac{J_l+2}{J_l+1} \frac{1 - \mathrm{e}^{-hv_m/kT_{\mathrm{ex}}}}{\mathrm{e}^{hv_l/kT_{\mathrm{ex}}} - 1} \tag{8.34}$$

where it has been assumed that the same excitation temperature characterizes both transitions. Thus the optical depths in $^{13}C^{16}O$ 2–1 and 1–0 transitions can be used to find the excitation temperature if this is less than 30 K. For higher temperatures a comparison of 4–3 and 3–2 gives a more accurate result. If the optical depths in these lines are not very small, one can use the ratio of the $^{13}C^{16}O$ to $^{12}C^{16}O$ lines, as in (8.28), to find the $^{13}C^{16}O$ optical depth. If the excitation temperature of a particular transition can be estimated, then the ratio of the brightness temperature for an observation that is believed to be of an optically thick transition (say the centre of a $^{12}C^{16}O$ line) to the excitation temperature will give the fraction of the beam that is filled by the cloud, for the brightness temperature of an optically thick cloud that fills the beam should be equal to the excitation temperature.

We now consider some of the finer details of the interpretation of diatomic molecule spectra. First we turn to the effect of background radiation on the brightness temperature T_{BCK}. In the long wavelength approximation:

$$T_B = T_{ex}(1 - e^{-\tau}) + T_{BCK}e^{-\tau} \quad \text{or} \quad T_{OBS} = T_B - T_{BCK} = (T_{ex} - T_{BCK})(1 - e^{-\tau})$$

since observationally one normally chops between cloud and a cloud-free region where there is only background radiation to pick up a difference signal. However, one should also notice that the brightness temperature is often defined in terms of the Rayleigh–Jeans approximation to a blackbody when it is called a radiation temperature so $T'_B = c^2/(2k\nu^2)I_\nu$. If we take both background and source function to be given by Planck functions with no approximation:

$$I_\nu = \frac{2h\nu^3}{c^2}\frac{1}{e^{h\nu/kT_{BCK}} - 1}e^{-\tau_\nu} + \frac{2h\nu^3}{c^2}\frac{1}{e^{h\nu/kT_{ex}} - 1}(1 - e^{-\tau_\nu})$$

so

$$T'_{OBS} = T'_B - T'_{BCK} = \frac{h\nu}{k}\left[\frac{1}{e^{h\nu/kT_{ex}} - 1} - \frac{1}{e^{h\nu/kT_{BCK}} - 1}\right](1 - e^{-\tau_\nu}) \qquad (8.35)$$

$$T'_{OBS}(1\text{–}0) = 5.532\left(\frac{1}{[e^{5.532/T_{ex}} - 1]} - 0.1479\right)(1 - e^{-\tau_\nu}) \quad \text{for } ^{12}C^{16}O \qquad (8.36)$$

$$= 5.289\left(\frac{1}{[e^{5.289/T_{ex}} - 1]} - 0.1642\right)(1 - e^{-\tau_\nu}) \quad \text{for } ^{13}C^{16}O \qquad (8.36a)$$

where the background radiation temperature has been taken to be 2.7 K.

If the cloud fills only a fraction f of the beam, then the brightness temperature of the cloud radiation will be reduced by a fraction f, and a fraction f of the background radiation will be exponentially absorbed and a fraction $1 - f$ of the background radiation will be unaffected. The net effect will to be to reduce T_{OBS} by a factor f.

If $T_{ex} \gg T_{BCK}$ so the second factor on the right-hand side of (8.36) and (8.36a) can be ignored, then

$$f = \frac{kT'_{OBS}}{h\nu}\frac{(e^{h\nu/kT_{ex}} - 1)}{(1 - e^{-\tau_\nu})} \qquad (8.37)$$

The mean column density over a partly filled beam will be given by (8.33) times f, and substituting for f we obtain for the $^{12}C^{16}O$ $J=1$ to $J=0$ transition:

$$N_{TOT}\rho R = 4.3 \times 10^{14} \frac{(T_{ex}+0.93)}{e^{-5.53/T_{ex}}} \int \frac{T'_{OBS}\tau_V}{(1-e^{-\tau_V})} dV \tag{8.38}$$

So far it has been implicitly assumed that the excitation temperature is considerably greater than the background temperature. However at millimetre wavelengths, the cosmic microwave background may be the main cause of excitation, which will drive T_{ex} towards T_{BCK}. If $T_{ex}=T_{BCK}$, then in the long wavelength approximation, $T_{OBS}=T_B-T_{BCK}=(T_{ex}-T_{BCK})(1-e^{-\tau})=0$. In other words, if the excitation temperature is reduced to the temperature of the microwave background, then as much energy is absorbed from the microwave background as is emitted, and no emission line is seen. Excitation can of course be by collisions as well as by radiation, with the former tending to drive the excitation temperature towards the kinetic temperature of the gas. Hence the detection of an emission line from a cloud depends on the ratio of the collisional rate to the radiative rate, and hence on the density since collisional rates are proportional to the density. There will be a critical density for the appearance of a particular emission line.

Consider a two-level molecule (levels i and j) excited by radiation with mean intensity J_ν and blackbody temperature T_R and by collisions characterized by kinetic temperature T_K. From equation (2.33) we have

$$S_\nu = \frac{2h\nu^3}{c^2}\frac{1}{e^{h\nu/kT_{ex}}-1} = \frac{\int \phi_\nu J(T_R)_\nu + \varepsilon B_\nu(T_K)}{1+\varepsilon}$$

with

$$\varepsilon = \frac{C_{ji}}{A_{ji}}(1-e^{-h\nu/kT_K})$$

If J_ν varies slowly with frequency, then $\int \phi_\nu J_\nu d\nu = J_\nu \int \phi_\nu d\nu = J_\nu(T_R)$ and in the long wavelength approximation $S_\nu=(2k\nu^2)/c^2\,T_{ex}$, $J_\nu=(2k\nu^2)/c^2\,T_R$ and $B_\nu=(2k\nu^2)/c^2\,T_K$, so

$$T_{ex} = \frac{T_R+\varepsilon T_K}{1+\varepsilon} \tag{8.32}$$

with

$$\varepsilon = \frac{C_{ji}}{A_{ji}}\frac{h\nu}{kT_K}$$

Hence

$$T_{ex} = \frac{T_R+t_0}{1+t_0/T_K} \tag{8.39}$$

with

$$t_0 = \frac{h\nu}{k}\frac{C_{ji}}{A_{ji}} = T_0\frac{C_{ji}}{A_{ji}} \simeq 5.5\frac{C_{ji}}{A_{ji}} \quad \text{for } 2.64\,\text{mm}$$

Dropping the long wavelength approximation, we have more exactly:

$$\frac{1}{e^{h\nu/kT_{ex}}-1} = \frac{1}{1+\varepsilon}\frac{1}{[e^{h\nu/kT_R}-1]} + \frac{\varepsilon}{1+\varepsilon}\frac{1}{[e^{h\nu/kT_K}-1]}$$

or

$$\frac{T_0}{T_{ex}} = \frac{T_0}{T_K} + \ln_e\left(\frac{e^{T_0/T_R} + \frac{C_{ji}}{A_{ji}}[e^{T_0/T_R}-1]}{e^{T_0/T_K} + \frac{C_{ji}}{A_{ji}}[e^{T_0/T_R}-1]}\right) \tag{8.40}$$

Both of these expressions give $T_{ex} = T_K$ when $C_{ji}/A_{ji} \gg 1$, and $T_{ex} = T_R$ when $C_{ji}/A_{ji} \ll 1$.

When $T_{ex} = T_K$ the line is said to be thermalized, being called sub-thermal for smaller values of the excitation temperature. The dividing line occurs when $\varepsilon = 1$, which gives an excitation temperature half-way between the radiation temperature characterizing the radiation and the kinetic temperature. This happens at a density such that

$$C_{ji}/A_{ji} * \{1 - \exp(-h\nu/kT_K)\} = 1 \tag{8.41}$$

or in the long wavelength approximation when $C_{ji}/A_{ji} = kT_K/h\nu$, with $C_{ji} = \langle\sigma(v)v\rangle n$ for collisional cross-section σ at velocity v averaged over the Maxwellian velocity distribution and $n =$ number density of colliding particles.

The critical density is defined as the density of colliding particles when the collisional de-excitation rate equals the radiative de-excitation rate (spontaneous plus induced by the radiation field), so

$$C_{ji}(n_{crit})/A_{ji} * \{1 - \exp[-T_0/T_{rad}]\} = 1 \tag{8.41a}$$

or, since the factor in curly brackets represents typically a correction of 15%or so, simply as the density of colliding particles at which

$$C_{ji}/A_{ji} = 1 \tag{8.41b}$$

This will give an excitation temperature a few degrees above the microwave background temperature and hence a detectable line, the density at which the excitation temperature comes half-way between the temperature of the background and the kinetic temperature being roughly an order of magnitude higher since typical kinetic temperatures might be 30 to 100 K whereas the background temperature is 2.7 K.

Collisions by charged particles have the largest cross-sections for excitation, but it is unlikely that the inner regions of dense molecular clouds have a high degree of ionization. The most abundant molecules are H_2 since hydrogen is the most abundant element and is largely associated into molecules in molecular clouds, and to a first approximation we can assume that the critical density is that of hydrogen molecules. Hence the critical density is the total density of the gas. Charged particle collisional excitation involves a change produced by an electrical field as does a radiative transition so it is not surprising that cross-sections for collisional excitation are proportional to the Einstein radiative transition probabilities A_{ji} if the radiative transition is permitted. For such transitions C_{ji}/A_{ji} is independent of A_{ji}, and so all transitions should have much the same critical density. On the other hand, the cross-section for collisional excitation by neutral particles at typical interstellar temperatures does not vary much from molecule to molecule and has a typical value of $10^{-19}\,\mathrm{m}^2$. Now $v = \sqrt{[8kT/(\pi m_H)\{1/A + 1/2\}]}$, where A is the molecular weight of the molecule so for $A \sim 28$, $v = 5.8*10^2\,(T/30)^{1/2}\,\mathrm{m\,s^{-1}}$. Thus the critical density ($C_{ji}/A_{ji} = 1$) at $T = 30\,\mathrm{K}$ is roughly $A_{ji}/(6*10^{-17})\,\mathrm{m}^{-3}$. This gives a density of about $10^9\,\mathrm{m}^{-3}$ for the $J = 1$ to 0 transition of CO which has a very low transition probability but around $4*10^{11}\,\mathrm{m}^{-3}$ and $10^{12}\,\mathrm{m}^{-3}$ for the 1–0 transitions of HCN and CS respectively. This allows us to put lower limits on the density of clouds from the detection of emission lines above the microwave background, and indeed to use observations of, say, CO, HCN and CS to draw density contours of a cloud, although between contours one only has a lower limit to the density. The very fact that lines from molecules with high critical densities seem to come from the denser parts of clouds is itself confirmation that ionization is low in dense clouds and that charged particles do not play a significant role in excitation. It should be noted that the A_{ji} values scale as $(J+1)^4/(2J+3)$, whereas C_{ji} does not vary much with J, so transitions with higher values of J require larger densities to be thermalized.

The argument so far has assumed that radiation in the line itself does not affect the excitation temperature, and this is no longer true for an optically thick line. Absorbed line photons will tend to populate the upper level and hence to increase the excitation temperature. This is called *trapping*. Consider the two-level case where the radiation field is dominated by the line radiation itself. Suppose also that line broadening is dominated by velocity gradients rather than by turbulence, so the observed line width is much greater than the Doppler width. Suppose also that the line optical depth stays constant over much of the line profile, as does the mean intensity J_ν, so we can write $\int \phi_\nu J_\nu \, d\nu = J$. Then

$$S = \frac{J + \varepsilon B_\nu}{1 + \varepsilon}$$

Using

$$S_\nu = \frac{\dfrac{2h\nu^3}{c^2}}{e^{T_0/T_{ex}} - 1}$$

$$\frac{T_{ex}}{T_0} = \frac{1}{\ln_e\left(\dfrac{2h\nu^3}{c^2 S} + 1\right)}$$

If we write β as the escape probability, then $J = S(1-\beta)$ and $S = (\varepsilon B_\nu)/(\beta + \varepsilon)$. Substituting for S_ν and putting $\varepsilon = C_{ji}/A_{ji}(1 - \exp(-T_0/T_K))$, $B_\nu = 2h\nu^3/c^2 \cdot 1/[\exp(T_0/T_K) - 1]$

$$\varepsilon B_\nu = \frac{2h\nu^3}{c^2}\frac{C_{ji}}{A_{ji}}e^{-T_0/T_K}$$

$$\frac{T_{ex}}{T_0} = \frac{1}{\ln_e\left[\dfrac{(1-e^{-T_0/T_K}) + \dfrac{A_{ji}}{C_{ji}}\beta}{e^{-T_0/T_K}} + 1\right]}$$

$$= \frac{T_K}{T_0}\frac{1}{1 + \dfrac{T_K}{T_0}\ln_e\left[1 + \dfrac{A_{ji}}{C_{ji}}\beta\right]} \tag{8.42}$$

The line optical depth is given by

$$\tau_\nu = \frac{c^2}{8\pi\nu^2}\frac{g_j A_{ji}}{g_i}(1 - e^{-T_0/T_{ex}})\int \phi[\Delta\nu(z)]n_i(z)\mathrm{d}z$$

where

$$\Delta\nu = \nu - \nu_0 - \Delta\nu_D(z)$$

Here ν_0 is the line centre rest frequency and $\Delta\nu_D$ is the Doppler shift in the line centre frequency due to motion at velocity V along the line of sight, so $\Delta\nu$ is the distance from the Doppler shifted line centre. n_i is the density of molecules in the lower level of the transition at a point a distance z along the line of sight. The Doppler shift is due to a velocity gradient through the cloud which we will assume to be approximately linear and equal to $\mathrm{d}V/\mathrm{d}z$.

Then

$$\mathrm{d}z = \frac{\mathrm{d}V}{\left(\dfrac{\mathrm{d}V}{\mathrm{d}z}\right)} = \frac{c}{\nu_0}\frac{\mathrm{d}(\Delta\nu_D)}{\left(\dfrac{\mathrm{d}V}{\mathrm{d}z}\right)}$$

$$\tau_\nu = \frac{c^3}{8\pi\nu^3}\frac{g_j A_{ji}}{g_i\left(\dfrac{\mathrm{d}V}{\mathrm{d}z}\right)}(1 - e^{-T_0/T_{ex}})\int \phi[\Delta\nu(\Delta\nu_D)]n_i(\Delta\nu_D)\mathrm{d}(\Delta\nu_D)$$

If n_i = constant throughout the cloud and the frequency at which τ is estimated is not too near the edges of the line, so ϕ is negligible at the values of $\Delta\nu_D$ at which n_i falls to zero, then n_i can be taken outside the integral. If the variable is changed to

$\Delta\nu$, the integral can be seen to be 1.0 since ϕ is normalized. Then

$$\tau_\nu = \frac{c^3}{8\pi\nu^3} \frac{g_j A_{ji}}{g_i\left(\frac{dV}{dz}\right)} n_i(1 - e^{-T_0/T_{ex}})$$

$$= \alpha A_{ji}(1 - e^{-T_0/T_{ex}})n_i \tag{8.43}$$

where

$$\alpha = \frac{c^3}{8\pi\nu^3} \frac{g_j}{g_i\left(\frac{dV}{dz}\right)}$$

Hence if $\tau \gg 1$, so $\beta \sim 1/(3\tau_\nu)$ according to (4.47):

$$T_{ex} = \frac{T_K}{1 + \frac{T_K}{T_0} \ln_e \left[1 + \frac{1}{3C_{ji}\alpha(1 - e^{-T_0/T_{ex}})n_i}\right]} \tag{8.44}$$

Thus for large optical depths the excitation temperature is independent of A_{ji} and depends only on $n_i C_{ji}$. According to Scoville and Solomon [1] for $\tau C_{ji} \gg A_{ji}$ and $T_K > T_{ex}$, the excitation temperature rises as roughly the 1/3 power of $(\alpha n_i C_{ji})$, unit it eventually reaches T_K.

There is considerable controversy about whether the widths of CO lines are due to systematic velocities (expansion, contraction, rotation) as assumed in the preceding discussion, or are due to microturbulence. In the former case one sees contributions to the observed line from all over the cloud, while in the latter an observed optically thick line comes only from the outer regions of the cloud. In some cases the observed line widths are large and if entirely due to turbulence would seem to imply supersonic turbulence velocities which, it is argued, should lead to rapid dissipation of the turbulence. The outer layers of clouds may be cooler than the central regions, and it has been suggested that if the lines are broadened by turbulence, the centres of very strong CO lines will be formed in cooler regions than points on the line profile on either side of the centre, and one should obtain a self-reversed profile (an apparent absorption line at the centre of the main emission line). On the other hand, some models of the heating and cooling of clouds without embedded stars suggest that the outer layers are hotter than the centres (because of the absorption of interstellar radiation by dust in the outer regions of the cloud and the consequent release of photoelectrons). Clouds containing embedded stars will presumably be hotter in the inner regions near the stars. Arguments have also been advanced about the degree of correlation in velocity between lines that can be formed all over a cloud and those only excited in the denser core regions of the cloud and about the degree of correlation between lines of large optical thickness and small optical thickness, where the former in a turbulence picture come only from the outer regions whereas the latter come from the whole cloud with a large contribution from the central regions.

Another important question is the adequacy of a two-level model. The levels $J = 2$ and $J = 3$ have comparable LTE populations to $J = 0$ and $J = 1$ even at 20 K and so cannot be neglected. Multi-level calculations are needed (since LTE cannot be assumed, especially for the higher levels with large radiative transition probabilities) to predict T_{ex}, τ, and the partition function for given densities and kinetic temperatures. $A_{ji} \propto (J + 1)^4/(2J + 3)$ and so is about 9.6 times larger for $J = 2$ to $J = 1$ than it is for $J = 1$ to $J = 0$. On the other hand the collisional excitation rate to $J = 2$ from $J = 0$ is comparable to that to $J = 1$ from $J = 0$. Hence collisional excitations from $J = 0$ to $J = 2$ will rapidly decay radiatively to $J = 1$, which will decay much more slowly to the ground state. This will lead to a build up of population in $J = 1$, raising $\tau(J = 1)$ and $T_{ex}(1\text{–}0)$, and lowering $T_{ex}(2\text{–}1)$. These effects will disappear at large optical depths when radiative absorption will remove the excess population but can be significant at optical depths less than 1.

The interpretation of the lines of CS is similar to that of the lines of CO.

We now turn to rather different approaches that can be used with some other molecules. An important tool in the determination of optical depth is the use of hyperfine structure. This is closely analogous to the use of isotopic lines in that one employs the ratio of the observed strengths of two transitions of a molecule where in the weak line limit the ratio can be predicted. In the case of hyperfine structure one is concerned with the ratio of lines differing only in the total angular momentum quantum number F (including nuclear spin) of the levels concerned, all other quantum numbers being the same. Exact formulae for the ratios of the transition probabilities are available. The hyperfine levels are very close together in energy so the relative populations are likely to be determined by collisions and to be given by Boltzmann's formula. In fact the levels are so close together that the relative populations are proportional to the statistical weights of the hyperfine levels $2F + 1$. Now equation (8.31) shows that the optical depth of a line $\propto g_j A_{ji}$ so for hyperfine split transitions $j \to i$ and $l \to m$:

$$\tau_{ji}/\tau_{lm} = (g_j A_{ji})/(g_l A_{lm}) = c$$

For a homogeneous gas where the background can be ignored:

$$\frac{T_B(ji)}{T_B(lm)} = \frac{T_{ex}(ji)}{T_{ex}(lm)} \frac{(1 - e^{-\tau_{ji}})}{(1 - e^{-\tau_{lm}})}$$

$$= \frac{(1 - e^{-c\tau_{lm}})}{(1 - e^{-\tau_{lm}})} \tag{8.45}$$

since collisional transitions between j and l and between i and m will keep the two excitation temperatures equal. This equation can be solved for τ_{lm} and then the individual equation for $T_B(lm)$ can be used to obtain T_{ex}.

The method has a number of advantages over that using isotope ratios, the principal one being that no assumption has to be made about the isotope ratio being solar. Isotope ratios are usually large, leading to very different optical depths and degrees of trapping, and hence to the possibility of differences in the excitation temperatures of the isotopic variants. This difficulty does not arise in the case of hyperfine structure (where in any case the ratios of optical depths are fairly modest). Finally the hyperfine lines are closer together in frequency than the lines of isotopic variants, and this has technical advantages in that questions of the variation of receiver characteristics (beam efficiency) with frequency do not need to be taken into account.

The hyperfine splitting must be greater than the line width (the latter is usually of the order of a few kilometres per second) to be observed as a resolved structure, that is the separation $\Delta\nu$ of the hyperfine lines must satisfy $\Delta\nu/\nu > v_{\text{turb}}/c$. Molecules containing only nuclei with $I = 0$ (e.g. ^{12}C, ^{16}O, ^{32}S) have no hyperfine structure and molecules containing only nuclei with $I = = 1/2$ (e.g. hydrogen) in addition to those with $I = 0$ have only magnetic hyperfine structure where the frequency shifts are small. However, the magnetic hyperfine shifts can sometimes be resolved if the turbulence velocities are small (as in dark clouds with no embedded stars) and if the transition energy is small as in Λ doubling or inversion doubling than $\Delta\nu/\nu$ is relatively large. Molecules containing at least one nucleus with $I = 1$ or greater (for interstellar lines this means ^{14}N with $I = 1$) exhibit the much larger quadrupole splitting of the line.

The shift in energy produced by a quadrupole interaction is proportional to the quadrupole moment of the nucleus and to a quantum number dependent factor:

$$\frac{3/4C(C+1) - I(I+1)J(J+1)}{2I(2I-1)(2J-1)(2J+3)} \tag{8.46}$$

with $C = F(F+1) - I(I+1) - J(J+1)$. I is, of course, fixed for a given molecule. For $I = 1$, F can take on the values $J+1$, J, $J-1$. The selection rules for transitions are $\Delta F = +1, 0,$ or -1 with $F = 0$ to $F = 0$ excluded. Transitions with the same value of J in the upper and lower levels (as in inversion doubling, Λ doubling) will have upper and lower hyperfine shifts depending only on F and hence transitions with $\Delta F = 0$ form the central or *main* line and transitions with $\Delta F = +1$ and $\Delta F = -1$ will produce two satellite lines on each side of the central line (Figure 8.4).

The relative values of $g_j A_{ji}$ for the case $\Delta J = 0$ are given by

$$g_j A_{ji}(\Delta F = 0) \propto \frac{[J(J+1) + F(F+1) - I(I+1)]^2 (2F+1)}{F(F+1)}$$

$$g_j A_{ji}(\Delta F = +1, -1) \propto -\frac{(J+F+I+1)(J+F-1)(J-F+I+1)(J-F-I)}{F}$$

where in the second equation the F value is the larger of the F values involved in the transition. Thus for a $J = 1 \rightarrow J = 1, I = 1$, quadrupole transition, there will be a central line with 50% of the intensity ($\Delta F = 0$), satellites on either side representing $F = 2$ to $F = 1$ and F = 1 to $F = 2$, each with 14% of the intensity, and two satellites further out on either side ($F = 0$ to $F = 1$ and $F = 1$ to $F = 0$) with about 11% of the intensity each. Higher J values will have a larger proportion of the intensity in the central line. Similar sets of relative $g_j A_{ji}$ values are available for the cases where $\Delta J = +1$ or -1 are available [2]. The relative transition probabilities apply to both quadrupole and magnetic hyperfine cases, and with them the relative optical depths of the lines can be calculated.

The OH molecule shows magnetic hyperfine splitting in the transitions between Λ doubled levels, the very long wavelength (18 cm) of the transition allowing the hyperfine structure to be easily resolved. With $I = 1/2$ each level is split into two with $I = J + 1/2$ and $I = J - 1/2$, giving 4 transitions, since the central $\Delta F = 0$ transition is slightly split, with for the $^2\Pi_{3/2} J = 3/2$ ground state a frequency of 1665 MHz for $F = 1$ to $F = 1$, and a frequency of 1667 MHz for $F = 2$ to $F = 2$ (figure 8.5a). The satellites are at 1612 MHz for F(upper) = 1 to F(lower) = 2, and at 1720 MHz for F(upper) = 2 to F(lower) = 1. The formulae above show that the $g_j A_{ji}$ values for the lines and hence the optical depths and intensities in the optically thin case should be in the ratio for 1612:1665:1667:1720 of 1:5:9:1. The lines should be resolved since even the separation of the main lines is 350 km s^{-1}. However OH often shows anomalous excitation when the simple optical depth analysis becomes invalid, and this transition will be discussed again later in Chapter 9. CH similarly shows Λ doubling and magnetic hyperfine structure, and formaldehyde H_2CO shows K doubling (it is a slightly asymmetric rotor so each (J, K) level is split and since the parity of the split levels is different, there can be transitions between them) with magnetic hyperfine structure. The 1_{10} to 1_{11} transition of formaldehyde is at a wavelength of 6 cm. However, formaldehyde sometimes shows anomalous excitation and occasionally maser emission.

Turning now to nitrogen-containing molecules, $H^{12}C^{14}N$ has a resolvable quadrupole hyperfine structure in its ordinary rotational transitions (it is a linear molecule). This hyperfine structure has been used to obtain optical depths to compare with those obtained by finding the ratio of the brightnesses of lines belonging to isotopic variants of HCN ($H^{13}C^{14}N$ and $H^{12}C^{15}N$) and hence to check the isotope ratios assumed in the latter method.

The hyperfine structure of ammonia NH_3 has been extensively used to obtain optical depths. Ammonia is a symmetric top but has its levels split by inversion doubling . The transitions *within* the inversion doublets of the various levels labelled by J and K lie at frequencies around 23 MHz (wavelengths around 1.3 cm), and in dark clouds the splitting into additional components due to magnetic hyperfine interaction with the hydrogen spin can be observed as well as the normal larger quadrupole effect due to the nitrogen nucleus (Figure 8.5b).

The ammonia spectrum shows another important feature. Each inversion doublet is characterized by the rotational quantum number J and the quantum

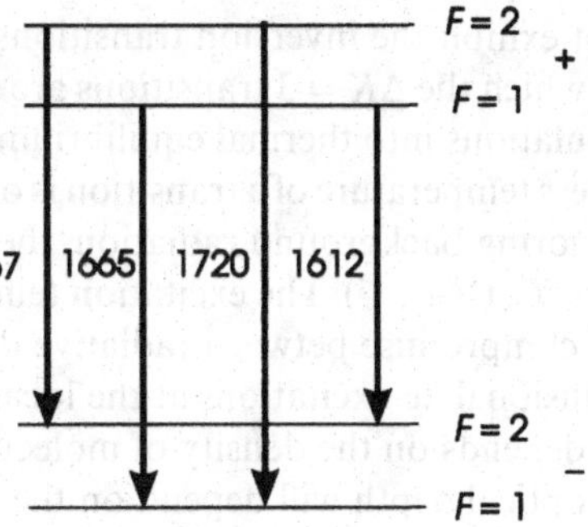

(a)

Λ doubled levels of $^2\Pi_{3/2}$ showing hyperfine splitting

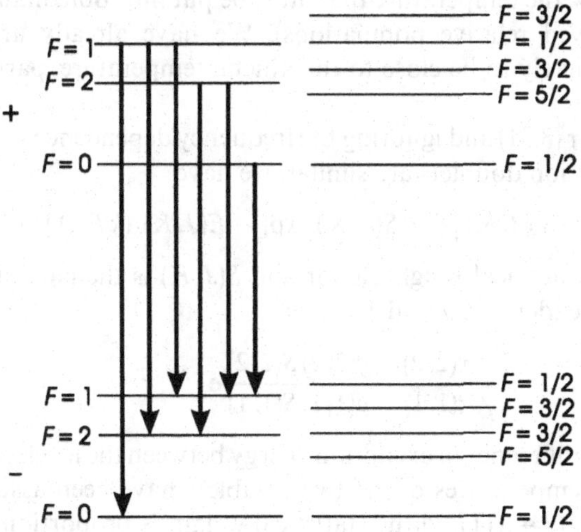

(b)

Figure 8.5. Inversion doubled levels of $J = 1$, $K = 1$ of NH_3 showing (left) quadrupole hyperfine splitting and (right) magnetic hyperfine splitting

number K representing the projection of the rotation on the molecular axis of symmetry, so $K \leqslant J$. The selection rule on K is $\Delta K = 0$, so the set of rotation levels belonging to a given K value ($J \geqslant K$), called a K ladder, does not link radiatively to other K values. The lowest inversion doublet in each K ladder cannot decay radiatively and is therefore metastable, while higher doublets in each K ladder can decay radiatively downwards to the metastable doublet at the bottom of the ladder (it should be noted that the J levels in the $K = 0$ ladder are single and therefore do not exhibit the inversion transitions). Collisions *between* the metastable levels, of which the $\Delta K = 3$ transitions are strongest, will tend to bring the metastable populations into thermal equilibrium at the kinetic temperature T_K.

If the brightness temperature of a transition is observed and the optical depth is known, then ignoring background radiation, the excitation temperature can be found from $T_B = T_{ex}(1 - e^{-\tau})$. The excitation temperature of an inversion doublet represents a compromise between radiative decays between the levels of the doublet and collisional de-excitations at the local kinetic temperature. Since the collisional rate depends on the density of molecular hydrogen, the latter can be estimated. The optical depth will depend on the population of the doublet, and here the special feature of the energy level structure of ammonia comes into play. Comparison of the optical depths in two metastable doublets will give an estimate of their relative populations and hence of the temperature T_{MS} connecting them (that is, the temperature that must be put into Boltzmann's equation to give the observed relative populations). We have already argued that this temperature is likely to be close to the kinetic temperature, particularly at low temperatures.

From (8.28) or (8.31) and ignoring the frequency dependence since the frequencies of the inversion doublets are similar, we have

$$\tau \propto g(J, K)\,[d_p]^2 S(J, K) \exp[-E(J, K)/(kT_{MS})]/T_{ex}$$

where g is the statistical weight factor and $S(J, K)$ is the part of the transition probability dependent on J and K. Hence

$$\frac{\tau(2,2)}{\tau(1,1)} = \frac{g(2,2)\,S(2,2)}{g(1,1)\,S(1,1)} e^{-\Delta E/kT_{MS}}$$

where ΔE is the difference in excitation energy between the levels 2, 2 and 1, 1, and the excitation temperatures of the two doublets have been assumed to be the same. $S \propto K^2/[J(J+1)]$ and the statistical weight is proportional to $(2J+1)$ if K is not a multiple of 3 (with an extra factor 2 if it is) so the ratio of the gS factors is 20/9 in this case.

Hence

$$\frac{T_B(2,2)}{T_B(1,1)} = \frac{(1 - e^{-\tau(1,1)[\tau(2,2)/\tau(1,1)]})}{(1 - e^{-\tau(1,1)})}$$

Then

$$\frac{\tau(2,2)}{\tau(1,1)} = -\frac{1}{\tau(1,1)} \ln_e\left[1 - \frac{T_B(2,2)}{T_B(1,1)}(1 - e^{-\tau(1,1)})\right]$$

$$= \frac{20}{9} e^{-\Delta E/kT_{MS}} \qquad (8.47)$$

which can be solved for T_{MS} if the brightness temperature ratio is known and $\tau(1,1)$ has been found from the hyperfine structure.

A rather similar approach can be adopted to find the kinetic temperature from observations of symmetric top molecules like CH_3CN and CH_3CCH. Again we have K ladders with radiative transitions forbidden between different ladders and with the bottom level of each ladder ($J = K$) metastable and hence probably populated mainly by collisions. In this case there is no inversion or K doubling, so the observed transitions are simply the rotational transitions $J + 1 \rightarrow J$ within each ladder. We can specify the temperature T_L characterizing the relative populations of the J levels of each ladder, and the temperature T_{MS} characterizing the relative populations of the metastable states, and approximating the kinetic temperature. For CH_3CCH the rotational transitions are probably optically thin. Then

$$\int T_B \mathrm{d}v \propto \frac{g_j}{g_i} A_{ji} g_i \mathrm{e}^{-E_1/kT_L} \mathrm{e}^{-E_2/kT_{MS}} N \tag{8.48}$$

where $E_1 = E(J, K) - E(K, K)$, the energy difference between the state concerned and the bottom of the ladder, and E_2 is the energy difference between the bottom of the ladder and the ground state (0,0). N is the population of the ground state. By observing the ratio of two or more brightness temperatures within a particular ladder, the temperature T_L can be determined since E_2/kT_{MS} will be the same for each transition. Observations of the ratio of the brightness temperatures from two ladders will then enable T_{MS} to be determined since E_1/kT_L is now known.

Cold Molecular Hydrogen

The largest component of molecular clouds, the H_2 molecule, is not directly detectable as an emission line at long wavelengths in the same way as, say, CO, because as a homonuclear molecule with no dipole moment, H_2 has neither a permitted dipole rotational spectrum nor a permitted dipole rotational–vibrational spectrum. However electronic transitions with appropriate vibrational–rotational structure are allowed, and such transitions from the ground electronic state should be observable in absorption in a similar way to atomic interstellar lines, namely in absorption against the background of a bright star. The ground electronic state, $X\,^1\Sigma_g$, which is the only electronic state likely to be populated in the cold interstellar medium, is linked by the permitted Lyman electronic transition at around 111 nm to the lowest excited level $B\,^1\Sigma_u$ and by the Werner transition at around 101 nm to the $C\,^1\Pi_u$ excited state, as we have seen in our discussion of PDRs in Chapter 6. Transitions with wavelengths less than 912 nm, the atomic hydrogen Lyman limit, will be obscured by absorption by cold atomic interstellar hydrogen in the ground state. The Lyman and Werner transitions in the ultraviolet must be observed from above the atmosphere, and suffer from the fact that molecular clouds are associated with dust, which of course extinguishes ultraviolet wavelengths strongly. Hence only moderate H_2 column densities with equally moderate dust extinctions can be observed. Nevertheless the measurements are very important as they give the only direct estimates for the density of the most abundant species in molecular clouds, and are clearly significant when it comes to estimating dust to gas ratios and the relation between gas column density and $E(B-V)$.

The electronic transitions have vibrational bands specified by the vibrational quantum numbers of the upper and lower levels v_U and v_L with $v_L = 0$ for absorption lines from the cold interstellar medium. For the Lyman transition the lines can have $J_U - J_L = \Delta J = +1$ for the R branch starting with R(0) for $J_L = 0$, and $\Delta J = -1$ for the P branch starting with P(1) with $J_L = 1$. For the Werner transition, with a $\prod$ upper state, the lowest rotational level of the upper state has $J = \Lambda = 1$ so the R branch starts with R(0) as before but the P branch starts with P(2). In this case, transitions with $\Delta J = 0$ are also allowed, giving a Q branch starting with Q(1). Frequencies can be obtained from

$$\nu = \nu_{\text{elec}} + [G(v_U) - G(v_L) + F(J_U) - F(J_L)]c$$

where we use traditional symbols so G and F are in wavenumber units and are given by

$$G(v) = \omega_e - \omega_e x_e (v + 1/2) + \omega_e y_e (v + 1/2)^2 - \cdots$$

and

$$F_v(J) = B_v J(J+1) - D_v J^2 (J+1)^2, \quad \text{with } B_v = B_e - \alpha_e (v + 1/2)$$

Here ω_e, $\omega_e x_e$ and $\omega_e y_e$ are vibrational constants and B_e, α_e and D_v are rotational constants to be found in tables of molecular constants.

The B coefficient for the ground state is much larger than those for the excited states ($B_e(\text{X}) = 6080\,\text{m}^{-1}$ as against $B_e(\text{B}) = 2001.59\,\text{m}^{-1}$ and $B_e(\text{C}) = 3134.0\,\text{m}^{-1}$ so the P branch in these cases runs to higher wavelengths with increasing J as the R branch always does. Thus the Lyman (0–0) R(0) line comes at 110.81 nm with R(1) at 110.86 nm, R(5) at 112.02 nm and P(1) at 111.01 nm, P(5) at 112.55 nm. The (5–0) band with some of the strongest lines has R(1) at 103.71 nm and P(1) at 103.82 nm. In the case of the Werner bands, the (0–0) R(0) line comes at 100.85 nm, the P(2) line at 101.22 nm and the Q(1) line at 100.97 nm. The oscillator strength for each line will be proportional to a factor depending on which vibrational transition is involved, increasing at first with v_U and then decreasing, and to a rotational line strength factor which is $(J+1)/(2J+1)$ for the Lyman R branch and $J/(2J+1)$ for the Lyman P branch, $(J+2)/(4J+2)$ for the Werner R branch, 1/2 for the Werner Q branch and $(J-1)/(4J+2)$ for the Werner P branch and so varies slowly with J.

The population of the levels of the ground state is in general a rather complicated question, involving both collisional processes and radiative excitation of excited electronic levels by ultraviolet radiation followed by downward radiative cascade. However the lowest levels $J = 0$ and $J = 1$ of the ground state are likely to be populated almost entirely by collisions so their population should be given by Boltzmann's equation. The statistical weights require a little further consideration, since with identical nuclei nuclear triplet states (with nuclear spins parallel) must combine with odd J rotational states if the overall wavefunction is to be antisymmetric to give *ortho*-hydrogen molecules and vice versa for the nuclear singlet states which must combine with even J levels in *para*-hydrogen molecules. Thus for odd J the statistical weight is $(2I+1)(2J+1)=3(2J+1)$ whereas for even J levels the statistical weight is $(2J+1)$. Thus the relative populations of the $J=1$ and $J=0$ levels should be $N(1)/N(0)=9\exp(-\Delta E/kT)=9\exp(-171/T)$. Transitions like Lyman 1–0, 2–0, 4–0 and 5–0 R(0), R(1) and P(1) and Werner (0–0),(1–0),(2–0) and (3–0) R(0), R(1), and Q(1) are strong and enable an average temperature of 80 K to be obtained for the molecular clouds.

Dust

Interstellar dust is important in its own right, and also has to be taken into account in every astronomical observation in the visible or shorter wavelengths. The terminology used in studies of dust will be introduced here and the spectrum of interstellar dust briefly discussed. Interstellar dust in circumstellar shells is considered in the next chapter.

We start by distinguishing between the properties of individual grains, which are 'optically thick' if the grain is larger than the wavelength of the radiation, but radiate less at some wavelengths than a blackbody at the same temperature, and ensembles of grains which can be described by emission and absorption coefficients per unit mass in the usual way and which may be optically thin even though the individual grains are optically thick at the same wavelength. The interactions of individual grains with radiation are described by an absorption cross-section which is written as the product of the geometrical cross-section σ and an *absorption efficiency* $Q_{abs}(\lambda)$ so $\sigma_{abs}=\sigma Q_{abs}(\lambda)$ and by the emittance (energy radiated per unit frequency interval per unit area) which is written as π times Planck's function $B_\nu(T)$ times the emissivity Q_ν. The *emissivity* is defined as the ratio of the emission from the grain surface to that from a perfect blackbody radiator, and a perfect blackbody radiator produces isotropic outward radiation with intensity B_ν and flux $=\pi^*$ intensity. Normally grains reach an equilibrium between the absorption and emission of radiation that can be characterized by a grain temperature T_G. Suppose for simplicity that the grains are spherical with radius a so the geometrical cross-section $\sigma=\pi a^2$ and is the same in all directions. Suppose further that the grain is immersed in isotropic radiation of blackbody intensity $B_\nu(T)$. Then the energy absorbed per unit frequency per second per unit solid angle is $\sigma B_\nu(T)Q_{abs}(\lambda)$ and summing over 4π, the total power absorbed per unit frequency range is $4\pi\sigma B_\nu(T)Q_{abs}(\lambda)$. The power emitted from the grain, which has a surface area $4\pi a^2$ and in full equilibrium must reach a temperature $T_G=T$, is $4\pi a^2\pi Q_\nu B_\nu(T)$. The power emitted integrated over all frequencies must equal the power absorbed integrated over all frequencies, but if this equality is to hold generally it must also be true at each frequency (detailed balance) so $Q_{abs}(\lambda)=Q_\nu$. This is true of any shape and is, of course, just an expression of Kirchoff's law.

The emission coefficient per unit volume if there are N grains per unit volume is given by

$$j_\nu\rho = 1/(4\pi)(\text{number of grains per unit volume})(\text{emission per grain})$$
$$= N\,4\pi a^2 \pi Q_\nu B_\nu(T)/(4\pi) = N\pi a^2 Q_\nu B_\nu(T)$$

while the absorption coefficient per unit volume is given by

$$\kappa_\nu\rho = N\pi a^2 Q_\nu$$

so the source function is $B_\nu(T)$, as one would expect. The absorption coefficient per unit grain mass if ρ_G is the density of the grain material is given by

$$\kappa_\nu = \frac{N\pi a^2 Q_\nu}{N\frac{4}{3}\pi a^3 \rho_G} = \frac{3Q_\nu}{4a\rho_G} \tag{8.49}$$

It follows that if Q_ν is constant with wavelength, the maximum absorption effect for a given mass of grains is produced when the grains are as small as possible. In fact Q_ν falls when the grain size becomes less than the wavelength, so the optimum absorption is produced when the grains are of the order of magnitude of the wavelength concerned. It is the presence in the interstellar medium of appreciable numbers of grains with sizes of the order of the wavelengths of visible and ultraviolet radiation that makes interstellar dust so efficient at extinguishing starlight—larger grains have too small cross-sections for their volumes, and smaller grains have cross-sections much less than their geometrical cross-sections.

In general grains do not find themselves in a blackbody radiation field, but in highly diluted starlight. Although collisions and chemical reactions also affect the energy balance, the balance is mainly one between the absorption and emission of radiation and it is this that determines the grain temperature. If the interstellar radiation field follows the Planck variation with frequency at some temperature T_{is}, we can write $I_\nu = WB_\nu(T_{is})$, where W is the dilution factor and will be very small, or alternatively $I_\nu = c/(4\pi)u_\nu$, where u_ν is the radiation energy density at frequency ν. Then the balance is represented by

$$4\pi\pi a^2 \int B_\nu(T_{is}) W Q_\nu \,\mathrm{d}\nu = 4\pi a^2 \pi \int Q_\nu B_\nu(T_G)\,\mathrm{d}\nu$$

$$W\int B_\nu(T_{is}) Q_\nu \,\mathrm{d}\nu = \int Q_\nu B_\nu(T_G)\,\mathrm{d}\nu \tag{8.50}$$

If we define an average efficiency $\langle Q_\nu \rangle$ by

$$\langle Q_\nu(T)\rangle = \frac{\int B_\nu(T) Q_\nu \,\mathrm{d}\nu}{\int B_\nu(T)\,\mathrm{d}\nu} = \frac{\pi}{\sigma_{SB} T^4}\int B_\nu(T) Q_\nu \,\mathrm{d}\nu$$

where is σ_{SB} the Stefan–Boltzmann constant, then (8.50) becomes

$$W\frac{\sigma_{SB}}{\pi}T_{is}^4\langle Q_\nu(T_{is})\rangle = \frac{\sigma_{SB}}{\pi}T_G^4\langle Q_\nu(T_G)\rangle$$

$$T_G = T_{is}W^{1/4}\left(\frac{\langle Q_\nu(T_{is})\rangle}{\langle Q_\nu(T_G)\rangle}\right)^{1/4} \tag{8.51}$$

where the efficiency corresponding to the interstellar radiation field is fixed and is close to 1.0, while the efficiency corresponding to the grain temperature is lower and dependent on the grain temperature. Grains absorb mainly at short wavelengths (UV) since the interstellar radiation field is dominated by hot stars with a frequency distribution corresponding to a temperature of around 10 000 K, and emit at long wavelengths (FIR for far infrared) since the grains have a very low temperature. Very crudely, if we replace the mean efficiencies by their average values in the ultraviolet and far infrared:

$$T_G \sim T_{is}W^{1/4}(\langle Q_{UV}\rangle/\langle Q_{FIR}\rangle)^{1/4} \tag{8.51a}$$

The dilution factor W has a value of about 10^{-14}, so if $\langle Q_{UV}\rangle$ and $\langle Q_{FIR}\rangle$ were the same the grain temperature would be about 3 K. As we have already remarked, the absorption efficiency is close to 1 at short wavelengths, but the absorption efficiency at long wavelengths is considerably less, making the grain temperature greater than 3 K. The final result depends on the nature of the grain material (which affects Q_{FIR}), but calculations evaluating the integrals $\int BQ_\nu\,d\nu$ for a weakly absorbing material give grain temperatures around 15 K, while those for strongly absorbing materials which have larger values of Q_{FIR} give grain temperatures a factor 2 higher. Inside an optically thick dust cloud there will be little ultraviolet radiation and absorption will be in the infrared only, leading to lower dust grain temperatures.

Dust grains scatter as well as absorb, particularly at short wavelengths. Scattered light is not 'lost' but is deflected out of the line of sight to the star. If we used an infinitely large receiving aperture, we would pick up light that was travelling away from the star in directions that missed the Earth but was scattered into our field of view, and scattering would have no effect on the signal received. Usually most scattering takes place far from the star, and since in practice we measure the flux from a star only in a one arc second or so radius around the star, the scattered radiation is almost entirely lost from the measurement and reduces the observed flux as effectively as absorption does. An exception to this occurs when there is denser dust than usual close to the star, in which case we may see a resolved halo of scattered light around the star. These *reflection nebulae* are most frequently associated with B type stars where the dust represents the remnants of the cloud from which the star was born. In other cases, the scattered light just contributes to the general background light in the sky. Particularly in the ultraviolet, if the airglow, zodiacal light and unresolved starlight are removed from the night sky intensity, one is left with a contribution linked with the Milky Way which can be attributed to dust scattered light in our galaxy.

The cross-section for scattering is written $\sigma_{sca} = \sigma Q_{sca}(\lambda)$, where $Q_{sca}(\lambda)$ is the scattering efficiency. The overall effect of scattering and absorption is referred to as *extinction*, with

$$\kappa_{ext}\rho = \kappa_{abs}\rho + \kappa_{sca}\rho = N\sigma Q_{ext}(\lambda) = N\sigma[(Q_{sca}(\lambda) + Q_{abs}(\lambda)]$$

The ratio $Q_{sca}/Q_{ext} = 1 - Q_{abs}/Q_{ext}$ is called the albedo , and since Q_{abs} and Q_{sca} vary differently with wavelength, the albedo is also a function of wavelength. It is the extinction that we measure when we observe the effect of dust on a background star, but as far as grain temperature and grain emission are concerned it is the absorption efficiency alone that enters into the calculation. However, radiative transfer calculations in a very optically thick cloud must take into account scattering, for multiple scattering increases the path length in the cloud and hence gives further opportunities for absorption.

We have so far largely ignored the question of grain size, which considerably complicates the issue since the efficiencies Q depend on the grain size, and real dust clouds contain a wide range of grain sizes. Thus all predictions have to be averaged over the grain size distribution $n(a)$. The crucial parameter is the ratio of grain size to wavelength, usually expressed as $x = 2\pi a/\lambda$.

In Mie scattering theory, Maxwell's equations are solved with the appropriate boundary conditions, the grain material being described by its complex refractive index $m = n - ik$, where the coefficient of the imaginary part, k, gives rise to absorption as opposed to scattering. In turn the refractive index can be written in terms of the complex dielectric constant. Weakly absorbing materials have $m \sim n$, with a weak dependence on wavelength, but the refractive index of a strongly absorbing material can show a considerable dependence on wavelength. Mie scattering theory gives $Q_{sca}(x)$ and $Q_{abs}(x)$ for various combinations of n and k. For large values of x, $Q_{sca} \sim Q_{abs} \sim 1$ which represents the situation at very short wavelengths. For small x (and therefore long wavelengths for a given grain size):

$$Q_{abs} \simeq 4x\Re\left[\frac{m^2-1}{m^2+1}\right]$$

$$Q_{sca} \simeq \frac{8}{3}x^4\left|\frac{m^2-1}{m^2+1}\right|^2 \tag{8.52}$$

with Q tending to zero at very long wavelengths. If the variation of m with λ is small, then absorption will clearly dominate at small x, and $Q_{ext} \propto Q_{abs} \propto 1/\lambda$ for a given grain size. The inverse proportionality of Q_{ext} with wavelength holds fairly generally for x around 1, with a rise of Q flattening out at $x \sim 5$. Including the effects of the wavelength variation of the refractive index, Q_{abs} is often written approximately as $Q =$ constant ν^{β}, with β having values between 1 and 2 for different materials. In addition there will be resonances and characteristic frequencies in the grain material which will give rise to enhanced cross-sections over certain wavelength ranges which are too broad to be called 'lines'. An example is the Si–O bond stretching frequency found in silicate materials at a wavelength of about 9.7 μm.

We now consider the observation of a star through a column of dust of optical depth τ_D at some wavelength. The optical depth is $\int \kappa_{ext}(\lambda)\rho\, dz = \int\int \pi a^2 Q_{ext}(a, \lambda)\, n(z, a)\, da\, dz$ if there are $n(a, z)\, da$ grains with radii between a and $a + da$ per unit volume at position z along the line of sight. The flux from the star will be reduced by $\exp(-\tau_D)$ so the magnitude of the star at the given wavelength will be increased by

$$\Delta m = A(\lambda) = 2.5 \log_{10}[\exp(-\tau_D)] = 1.086\, \tau_D = 1.086 < \pi a^2 Q_{ext}(a, \lambda) > N_{coll} \tag{8.53}$$

where the average is over particle size.

Now suppose two identical stars are compared, with star 2 being close enough to be unreddened and star 1 more distant. Then the difference in magnitude at some wavelength λ is

$$m_1(\lambda) - m_2(\lambda) = A(\lambda) + 5 \log d_1/d_2$$

with $A(\lambda)$ the dust extinction to star 1 since $A_2(\lambda) = 0$, and d_1 and d_2 the distances to the two stars. Observationally identical stars are selected on the basis of identical spectra, since the strengths of lines relative to the continuum are unaffected by dust extinction. If we observe at two wavelengths:

$$m_1(\lambda_1) - m_1(\lambda_2) - [m_2(\lambda_1) - m_2(\lambda_2)] = A(\lambda_1) - A(\lambda_2).$$

In particular if the two wavelengths are chosen to be those representing the B and V bands:

$$E(B - V) = m_1(B) - m_1(V) - [m_2(B) - m_2(V)] = A_B - A_V$$

where $E(B - V)$ is called the *colour excess*, the amount by which the $B - V$ colour of a star exceeds the star's intrinsic $B - V$ colour because of dust reddening. The colour excess is widely used as a measure of the amount of dust in the line of sight to a particular star. What we want to know when correcting observational data for the presence of dust is the A values, and in particular A_V, the total extinction in the V band. The quantity $A_V/E(B - V)$ is called R, the ratio of total to selective extinction.

The shape of the extinction curve is best expressed in terms of the relative extinction, normalized to $E(B - V)$:

$$\text{Relative extinction} = \frac{E(\lambda - V)}{E(B - V)} = \frac{Q_{ext}(\lambda) - Q_{ext}(V)}{Q_{ext}(B) - Q_{ext}(V)}$$

If the relative extinction is found from comparisons of identical stars of well-established spectral type (usually OB stars because of their luminosity and simple spectra) as a function of wavelength, and R is known, then $A(\lambda)$ can be found for any wavelength:

$$A(\lambda) = A_V\left[\frac{E(\lambda - V)}{E(B-V)}\frac{1}{R} + 1\right] \tag{8.54}$$

Determining R is more difficult. However in the limit of very long wavelengths the extinction predicted by Mie scattering goes to zero as a power law in $1/\lambda$. Extrapolation of the observed values of the relative extinction as a function of wavelength in the infrared gives a value of 3.05 , with a small real variation in different regions of about ± 0.2.

The relative extinction curve is normally plotted against $1/\lambda$ and is fairly linear in the visible and near ultraviolet with $A(\lambda)/A(V)$ roughly proportional to $1/\lambda$ (there is a slight bend at 440 nm) but in the ultraviolet there is a broad symmetric peak in the extinction at 217.5 nm, followed by a dip to 170 nm and a rise at shorter wavelengths. The slope in the near infrared (out to 5 μm) is given by selective extinction and $A(\lambda) \sim 1/\lambda^{1.84}$. The variation of A with wavelength should be proportional to the variation of Q_{ext} with wavelength. Extinction curves only determine directly the combined quantity $Q_{\text{ext}}(\lambda)a^2N$, but it is possible to obtain an overall estimate of the mass of dust per square metre corresponding to a given A_V as follows. Suppose the column number density of grains of radii between a and $a + da$ is $n_{\text{col}}(a)\,\text{d}$. Take equation (8.53) and integrate over all wavelengths:

$$\int A(\lambda)\,\text{d}\lambda = 1.086\int \pi a^2 n_{\text{col}}(a)\int Q_{\text{ext}}(a,\lambda)\,\text{d}\lambda\,\text{d}a$$

Mie scattering theory predicts that for constant m:

$$\int Q_{\text{ext}}(a,\lambda)\,\text{d}\lambda = 4\pi^2 a\frac{m^2-1}{m^2+2}$$

Hence if the density of the grain material is ρ_G and the mass of dust per unit area in the line of sight is M_{dust}:

$$\begin{aligned}\int A(\lambda)\,\text{d}\lambda &= 1.086 * 4\pi^3\frac{m^2-1}{m^2+2}\int a^3 n_{\text{col}}(a)\,\text{d}a\\ &= 1.086\times\frac{3\pi^2}{\rho_G}\frac{m^2-1}{m^2+2}\int m_G(a)n_{\text{col}}(a)\,\text{d}a, \quad \text{where } m_G = \frac{4}{3}\pi a^3\rho_G\\ &= 1.086\times\frac{3\pi^2}{\rho_G}\frac{m^2-1}{m^2+2}M_{\text{dust}}\end{aligned}$$

Hence:

$$M_{\text{dust}} = \frac{A_V\displaystyle\int\frac{A(\lambda)}{A_V}\,\text{d}\lambda}{1.086\times 3\pi^2}\rho_G\frac{m^2+2}{m^2-1} \tag{8.55}$$

Provided that a reasonable guess can be made at ρ_G (say 3000 kg m^{-3}), at m (say around 1.6), and the extinction curve $A(\lambda)/A_V$ is known over a wide range of wavelengths, a relationship can be found between A_V and column dust mass. Observations of H I and H_2 in the same line of sight (with an allowance for H II if necessary) then provide a calibration of the gas to dust ratio.

Additional information about the nature of the grains is also available from scattered light. Scattered light from the stars of the galaxy in general is seen as diffuse galactic light. The dust-scattering component of the diffuse galactic light must be separated from faint starlight and of course from solar system sources. The average intensity of starlight at any point in the interstellar medium can in principle be found from the known distribution of stars, the luminosity function, and the energy distributions in the spectra of stars of various types. The ratio of scattered light to light available to be scattered tells about the product of $Q_{sca}\pi a^2$ and the number of particles along a given line of sight. However extinction measures tell us about the product of $Q_{ext}\pi a^2$ and the number of particles along the line of sight, so we can obtain the ratio of Q_{sca} to Q_{ext}, the albedo. Of course this must be worked out using a model of the distribution of dust and of the distribution of stars in the galaxy. The fact that scattering is not in general isotropic but depends on the angle through which the light is scattered must also be taken into account. We can define a scattering function $f(\theta)$ such that intensity scattered through θ from incident intensity I_0 is given by $I_{sca} = I_0 f(\theta)\sigma_{sca}$ per grain, although other normalizations are sometimes used. It turns out that the diffuse galactic light intensity at galactic latitudes between 17° and 20° hardly depends on the scattering function and so determines the albedo. Satellite measurements eliminate airglow from the Earth's atmosphere and have been made in the visible from Pioneer 10 (eliminating also some of the zodiacal light) and in the ultraviolet. The albedo is quite high at 0.6 to 0.7 in the blue and the visible, with a dip for the 217.5 nm feature, rising again at shorter wavelengths. This demonstrates that the 217.5 nm feature is indeed an absorption feature, but that the agent responsible for the rest of the extinction in the visible must have a predominantly dielectric nature.

Scattered light observations have also been made of dark nebulae where the interstellar radiation field is scattered from the surface of a dense optically thick dust cloud. We shall consider here reflection nebulae where we are concerned with scattered light from a single star, usually embedded within the dust cloud in the most easily observed cases although many other geometries also occur, such as back-scattering from a dust layer behind the star as seen by the observer. The nebula emits scattered light, mainly in the ultraviolet with luminosity L_{sca}(UV) say, and far-infrared light which represents radiation absorbed and re-emitted thermally, with luminosity L_{abs}(FIR) say. The ratio $L_{sca}/(L_{sca} + L_{abs})$ is then a measure of the albedo ω, or more strictly of the albedo raised to the power of the mean number of scatterings (at each 'extinguishing encounter' with a grain the probability that a photon is scattered and therefore lives to fight again is given by the albedo).

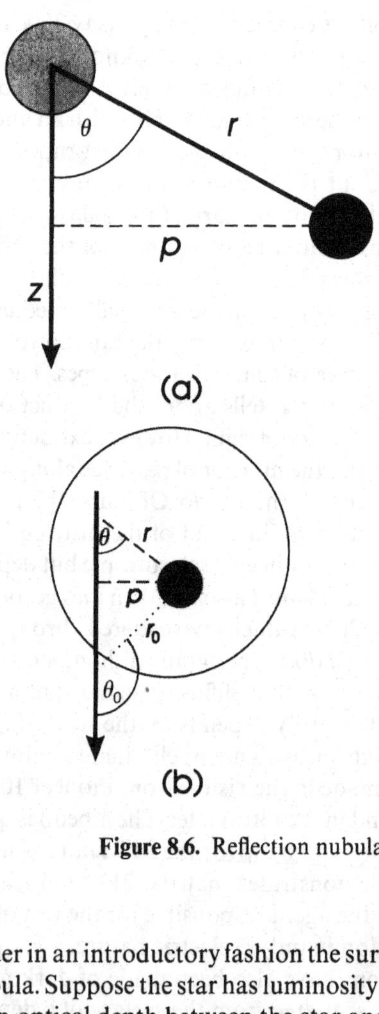

Figure 8.6. Reflection nubula

We now consider in an introductory fashion the surface brightness distribution in a reflection nebula. Suppose the star has luminosity per unit frequency range L_ν and the extinction optical depth between the star and a grain distant r from the star is $\tau_{\text{ext}}(r)$. Then the flux falling on the grain is $L_\nu/(4\pi r^2)(\exp(-\tau_{\text{ext}})$, the power scattered by the grain is $Q_{\text{sca}}\pi a^2 L_\nu/(4\pi r^2)\exp[-\tau_{\text{ext}}(r)]$ and the power scattered by volume $\mathrm{d}V$ into the solid angle $\mathrm{d}\Omega(\theta)$ if there are N grains per unit volume is

$$I_{\text{sca}}\mathrm{d}\Omega = \frac{L_\nu}{4\pi r^2}\mathrm{e}^{-\tau_{\text{ext}}}Q_{\text{sca}}\pi a^2 N f(\theta)\,\mathrm{d}V\mathrm{d}\Omega$$

where it should be noted that $L_\nu \sim \pi B_\nu(T^*)4\pi R^2$, where the star has radius R and effective temperature T^*, and that for isotropic scattering $f(\theta) = \text{constant} = 1/(4\pi)$. Let z measure the distance along the line of sight, p be the distance of closest approach to the star along the line of sight, and let $\tau_{\text{ext}}(z, p)$ be the extinction optical depth along the line of sight. Then the surface brightness of the reflection nebula (the observed intensity) at a distance from the star corresponding to p (Figure 8.6) is

$$I(p) = \int Q_{sca} \pi a^2 \frac{L_\nu}{4\pi r^2} N e^{-\tau_{ext}(r)} e^{-\tau_{ext}(z,p)} f(\theta)\, dz \tag{8.56}$$

$$= \omega \int \frac{L_\nu}{4\pi r^2} e^{-\tau_{ext}(r)} e^{-\tau_{ext}(z,p)} f(\theta)\, d\tau_{ext}(z) \tag{8.56a}$$

where $d\tau_{ext}(z) = 1/\omega\, d\tau_{sca} = 1/\omega\, N\, Q_{sca}\pi a^2\, dz$. If the cloud size is small compared with r so that r is nearly constant along the line of sight and scattering is isotropic:

$$I(p) = \omega \frac{L_\nu}{16\pi^2 r^2} e^{-\tau_{ext}(r)}(1 - e^{\tau_{ext}}) \tag{8.56b}$$

where τ_{ext} is the extinction optical depth of the dust cloud along the line of sight. The observed flux from the star itself is $F_\nu(*) = L_\nu/(4\pi d^2) \exp[-\tau_{ext}(SO)]$, where the optical depth is that between the star and the observer.

If on the other hand we are dealing with a uniform density spherically symmetrical dust cloud of radius r_0 with the star at the centre so $\tau_{ext}(r) = \kappa\rho\, p/\sin\theta$ and $\tau_{ext}(z, p) = \kappa\rho[\surd(r_0^2 - p^2) + p \cot\theta]$, with the scattering angle θ_0 at the edge of the cloud (Figure 8.6(b)), we can write

$$I(p) = \frac{\omega\kappa_{ext}\rho L_\nu}{4\pi}\int_{\theta_0}^{\pi-\theta_0} \frac{1}{r^2} f(\theta) \exp\left\{-\kappa_{ext}\rho\left[\frac{p}{\sin\theta} + \sqrt{r_0^2 - p^2} + p\cot\theta\right]\right\} d(p\cot\theta)$$

$$= \frac{\omega\tau_0 L_\nu}{4\pi r_0}\int_{\theta_0}^{\pi-\theta_0} f(\theta)\frac{\sin^2\theta}{p^2} \exp\left\{-\kappa_{ext}\rho\left[\frac{p}{\sin\theta} + \sqrt{r_0^2 - p^2} + p\cot\theta\right]\right\}\frac{-p}{\sin^2\theta} d\theta \tag{8.57}$$

since $dz = d(p\cot\theta)$ and writing $\tau_0 = \kappa_{ext}\rho r_0$.

If the optical depth is small, following Sobolev, we can further write

$$I(p) = \frac{\omega L_\nu \tau_0}{4\pi r_0 p}\int_{\theta_0}^{\pi-\theta_0} f(\theta) d\theta$$

$$\frac{d[pI(p)]}{dp} = \frac{\omega\tau_0 L_\nu}{4\pi r_0^2 \cos\theta_0}\frac{d}{d\theta_0}\int_{\theta_0}^{\pi-\theta_0} f(\theta)\, d\theta, \quad \text{since } dp = r_0 \cos\theta_0\, d\theta_0$$

$$= \frac{\omega\tau_0 L_\nu}{4\pi r_0^2}\frac{f(\theta_0) + f(\pi - \theta_0)}{\cos\theta_0} \tag{8.58}$$

which in principle enables information about the scattering function to be obtained from the run of surface brightness with distance from the star, with the surface brightness distribution being more centrally concentrated if the scattering function is not isotropic but favours forward scattering. The formulae (8.56)–(8.58) must be taken as purely indicative since the brighter nebulae are generally not optically thin, and when the optical thickness becomes much greater than one, multiple scattering must be taken into account, so a numerical calculation using a model becomes necessary. It seems that the scattering function favours forward scattering in the visible with $\langle \cos\theta \rangle$ averaged over the solid angle about 0.7 ($\langle \cos\theta \rangle = 0$ for isotropic scattering, 1.0 for forward only scattering), but is probably lower in the ultraviolet. Forward scattering happens when the grains have sizes of the order of the wavelength, whereas smaller grains produce isotropic scattering. More precisely the visible results suggest grains with radii between 0.1 and 0.3 μm, while the ultraviolet results suggest that there the scattering is due to much smaller particles.

Information is also available from the spectral dependence of the polarization of starlight that has been extinguished by interstellar dust. The fact that linear polarization of the light from reddened stars is observed, while starlight is in general unpolarized, shows that grains are not always spherical and that the grains are aligned by interstellar magnetic fields. Long grains perpendicular to the direction of the radiation will have a slightly larger extinction efficiency when the electric vector is parallel to the grain than when it is perpendicular to the grain. Thus if the light is initially unpolarized, the slightly greater efficiency of removal of the parallel component of the electrical field will leave a slight excess of the perpendicular component, so the resulting light is slightly polarized. The only plausible alignment mechanism is a magnetic field, so the polarization of interstellar light shows the orientation of the interstellar magnetic field. The differential extinction is at a maximum when $x = 2\pi a/\lambda = 1/(m-1)$. The degree of polarization of stars plotted against wavelength goes through a maximum at a wavelength which varies from star to star but usually lies in the range 0.2 to 0.8 μm with an average value of 0.55 μm. This suggests that the grains responsible for polarization have sizes around 0.15 μm.

This discussion of the continuous properties of interstellar extinction has necessarily been very brief, and the variations found from place to place in our galaxy and between galaxies have not been discussed, and questions of grain shape have only been touched on briefly.

We now turn to the 'lines' observed in the extinction curve. These are generally rather broad features, with in some case $\Delta\lambda/\lambda$ as great as 0.1, and are thus very different from the lines met in the spectroscopy of gases. This is expected theoretically, and indeed some features are found in these different guises both from solid grains and molecules in gases (e.g. vibration of the C—O bond). The breadth of the solid-state features and the absence of a rotational fine structure makes unique identification much more difficult.

The strong 217.5 nm absorption (it is much weaker in scattered light) is still not positively identified, although the most popular identification remains that of an electronic resonance in graphite. The problem is that the observed feature has little variation in its wavelength, whereas theory requires graphite particles of 0.02 μm to fit the observations. The presence of an appreciable number of larger particles would increase the peak wavelength. Hence, if the graphite identification is correct, it seems that the graphite particle size distribution must be the same in a variety of environments.

In the visible about 40 diffuse interstellar bands — broad absorption features whose presence and strength correlate with extinction — have been found, the most prominent being those at 443.0 nm and 628.2 nm. No consensus has been reached as to the source of these features — suggestions include polycyclic aromatic hydrocarbons and fulleranes with varying degrees of hydrogenation.

In the infrared, a number of absorption features have been found when looking at sources with very large extinctions such as the galactic centre (where there are 30 magnitudes of visual extinction) and a highly reddened B supergiant, number 12 in the Cygnus OB2 association. In these cases we are seeing cooler dust against a background of hotter dust or luminous cool stars which are much hotter than the absorbing dust. These features can be characterized by their peak optical depth, that is the optical depth τ_X from $I(\text{centre}) = I(\text{neighbouring continuum}) \exp(-\tau_X)$. The most prominent features are those at 9.7 μm and (with about 40% of the strength) at 18.5 μm. There can be little doubt that these are to be attributed to the frequencies of stretching of the Si—O bond and bending of the O—Si—O bond in silicates of various sorts ($MgSiO_3$, Mg_2SiO_4, $FeSiO_3$, Fe_2SiO_4, etc.). The same features are seen in the spectra of the circumstellar shells of cool oxygen-rich stars, where refractory dust grains are expected to condense, and silicates are amongst the most refractory of oxygen-rich solids. The correlation of the two features, together with their absence from circumstellar spectra of carbon-rich stars, supports the identification. The smoothness of the features suggests that the silicates are in amorphous rather than crystalline form. Laboratory data suggest the peak absorption coefficient for such materials at 9.7 μm is about 300 $m^2\,kg^{-1}$ so that from the peak optical depth of the 9.7 μm absorption feature in the direction of the galactic centre of 3.6 and noting that our distance from the centre is about 8 kpc, the average density of silicate dust in this direction comes out as about $5*10^{-23}\,kg\,m^{-3}$. Other features are sometimes found at 3.1 and 3.4 μm, the former usually being attributed to the stretching of the O—H bond in water ice, the latter perhaps being due to C—H bond stretching, possibly in hydrogenated amorphous carbon. In molecular clouds, silicate features are present but the strongest feature is the 3.05 μm water ice absorption, with a H—O—H bending absorption at 6.0 μm also present. These features are accompanied in molecular clouds by 4.67 and 6.85 μm absorptions. The former has been identified with vibrations of solid CO, the latter may be a C—H deformation in hydrocarbons of some kind.

There is as yet no consensus about the make-up of interstellar dust. The classic models have a size distribution that goes as $n(a) \propto a^{-3.5}$ with a characteristic size of about 0.15 μm, and graphite and silicate grains with (sometimes) ice coatings. It seems likely that there is more than one grain 'population' and while it is generally accepted that there is an important component of amorphous silicates, the form in which carbon is present is disputed. Much hinges on the identification of the 217.5 nm feature with a resonance in graphite, and on the slightly mysterious fact that an infrared feature often identified with SiC is widely seen in the circumstellar shells of carbon-rich stars, but is not seen at all in the interstellar absorption curve, while the proponents of PAHs and fulleranes find fits with a number of diffuse interstellar bands apparently requiring considerable quantities of one or other of these substances.

Finally, we turn to look at emission from interstellar grains. The emission from interstellar dust is usually optically thin at the long wavelengths involved. We have seen that grains in the general interstellar radiation field will be heated to temperatures of the order of 15 K, and hence will radiate in the millimetre region of the spectrum. The emission coefficient per unit volume, as was seen earlier, is $N\pi a^2 Q_\nu B_\nu(T)$ for N grains per unit volume. Hence the flux received from volume $\mathrm{d}V$ is

$$F_\nu = \frac{N\pi a^2 Q_\nu B_\nu(T)\,\mathrm{d}V}{d^2}$$

For spherical grains (other shapes do not make a large difference) with grain mass $= 4/3\pi a^3 \rho_G$:

$$\begin{aligned} \text{mass of dust} &= N\,\mathrm{d}V m_G \\ &= \frac{4}{3}\rho_G \frac{F_\nu d^2}{B_\nu}\frac{a}{Q_\nu} \end{aligned} \tag{8.59}$$

where if x is small, $Q_\nu \propto x \propto a$ for the given frequency so a/Q_ν is independent of a and since the flux is proportional to a^3, a volume weighted average value of Q_ν/a will give a dust mass independent of the value of a.

If $Q_\nu \propto \nu^\beta$, then the flux per unit frequency interval peaks at a frequency given by

$$\frac{\mathrm{d}F_\nu}{d\nu} = 0, \quad \text{i.e.}\ \frac{\mathrm{d}}{\mathrm{d}\nu}\left[\frac{\nu^{\beta+3}}{\mathrm{e}^{h\nu/kT}-1}\right] = 0$$

i.e. when

$$\mathrm{e}^{h\nu/kT}\left[(\beta+3) - \frac{h\nu}{kT}\right] = (\beta+3)$$

when

$$\frac{h\nu}{kT} \simeq (\beta+3)$$

so

$$\lambda \simeq \frac{0.0144}{\beta+3}\frac{1}{T}\,m \tag{8.60}$$

which for $\beta = 1.5$ and $T = 15$ K gives a peak F_ν at around 200 μm. If we were able to observe over a sufficiently wide range of wavelengths and all the dust observed was isothermal, we could use the long wavelengths to find β since B_ν in the Rayleigh-Jeans approximation is proportional to ν^2 so F_ν should be proportional to $\nu^{\beta+2}$, and then use the shape of the curve at shorter wavelengths or the position of the peak to find the temperature. In practice it is difficult to separate β and T, but studies of likely grain materials suggest that β lies between 1 and 2 in the far infrared. The density of the grain material is likely to be around $3000\,\mathrm{kg\,m^{-3}}$, being somewhat less for graphite and somewhat more for silicates.

The determination of Q_ν in the far infrared is difficult. However, $Q_{ext}(\mathrm{UV})$ can be estimated, being somewhat higher than the short wavelength Mie scattering limit of 2, say around 3. Now $Q_\nu(\mathrm{FIR})/Q_{ext}(\mathrm{UV}) = \tau_{abs}(\mathrm{FIR})/\tau_{ext}(\mathrm{UV})$. For a particular cloud, say a reflection nebula, the visible extinction can be found from the reddening of background stars or of the illuminating star, and the known shape of the extinction curve can be used to obtain the extinction in the ultraviolet. The far infrared intensity from the cloud $= F_\nu(\mathrm{FIR})/(\text{beam solid angle}) = B_\nu(1 - e^{-\tau}) = B_\nu(T)\,\tau_\nu(\mathrm{FIR})$ since the emission is optically thin. Hence the ratio of efficiencies can be found, with some error since the extinction measures refer to all the material between the observer and the background object, and the far-infrared emission only to the particular cloud observed. The mass of a cloud can now be found from (8.59), but only with quite large errors, mainly due to uncertainty in the efficiency Q. A total cloud mass can be found if the ratio $M(\mathrm{gas})/M(\mathrm{dust})$ can be calibrated. A cloud which is optically thin enough for A_V to be found from measurements of $E(B-V)$ of background objects and yet is optically thick enough to emit sufficient far infrared for $\tau_\nu(\mathrm{FIR})$ to be determined as above supplies a value for $\tau_\nu(\mathrm{FIR})/A_V$. A relation between $N(H + H_2)$ and A_V can be found from atomic and molecular line observations of a cloud whose extinction can be measured, so that finally a $\tau_\nu(\mathrm{FIR})/N(\mathrm{H} + \mathrm{H_2})$ calibration can be found if the individual calibrations made in different objects apply universally.

We have already pointed out that dark clouds are unlikely to be isothermal, even when heated by the general galactic interstellar radiation field, and that while the centres may have temperatures of the order of 7 to 10 K, the outer layers may reach 15 to 30 K. At 100 μm, the hotter outer layers will dominate the flux, and observations will underestimate the total mass of the cloud. Many clouds contain embedded luminous stars, all of whose visible and ultraviolet light will be absorbed in the very optically thick dust cloud and converted to infrared radiation. It will be this infrared radiation field that will heat the outer layers of the cloud and determine their temperature. More general surveys of far infrared emission show a peak around 100 μm indicating temperatures around 25 K as expected, but there is also a secondary peak in the continuous emission around 10 μm corresponding to a much higher temperature, which cannot represent an equilibrium temperature far from individual stars. This emission has been attributed to very small grains. Such grains do not reach an equilibrium temperature with the radiation field, but are either very cold or raised to quite high temperatures by the absorption of a single photon. For further discussion see the book by Whittet [3].

References

[1] N. Z. Scoville and P. M. Solomon, *Astrophysical Journal Letters*, **187**, L67, 1974.
[2] C. H. Townes and A. L. Schawlow, *Microwave Spectroscopy*, Dover, 1955.
[3] D. C. B. Whittet, *Dust in the Galactic Environment*, Institute of Physics, 1990.

9 MASERS

Introduction

The essential feature of an astronomical maser is a millimetre or centimetre transition whose upper level population exceeds the lower level population. More precisely, the absorption coefficient is given by

$$\kappa_\nu \rho = N_i h\nu B_{ij}(1 - g_i/g_j N_j/N_i)$$

and masering occurs if this is negative, that is if $g_i/g_j N_j/N_i > 1$. One calls such a state of affairs a *population inversion.* A negative absorption coefficient means a negative optical depth and a negative source function since the emission coefficient is always positive. Strong lines, instead of saturating (as in the case of a positive absorption coefficient) grow in strength increasingly rapidly with increasing optical depth, leading to very high surface brightnesses at line frequencies and large luminosities from small volumes.

Such a population inversion is clearly a non-LTE situation. Of course non-LTE situations are common in interstellar clouds, but they most commonly produce sub-Boltzmann populations. To produce a population inversion requires some kind of pumping mechanism which has the net effect of transferring molecules from the lower to the upper state via some third state or states. A possible example is where the masering transition occurs in the ground state which is connected to higher vibrational or rotational levels by infrared transitions. Infrared radiation from a cool star can excite molecules out of the ground state, and the excited states can then decay radiatively back to the ground state. If selection rules or transition probabilities mean that more decays take place to the upper level of the ground state than to the lower level, then an inversion can be produced. Collisional effects could also produce such an inversion. We shall refer to such mechanisms as *pumps.*

First let us consider the consequences of masering for a homogeneous sphere of radius R in which the masering is of low enough intensity not to affect the population inversion significantly (of course at large maser intensities the population inversion must be reduced) and we simply characterize the system by a constant negative absorption coefficient and source function.

Let the optical thickness along some line of sight be τ and the constant source function be S. Then $I = S(1 - e^{-\tau})$, which saturates to $I = S$ for large positive τ, but becomes $I \sim |S|\exp(|\tau|)$ for large negative τ.

The flux emerging from unit area of sphere is given by

$$F = 2\pi \int_0^{+1} I(\mu)\mu \, d\mu$$

where μ is $\cos\theta$ and θ is the angle of the line of sight with a radial direction. The optical depth along such a line of sight is $\tau_0\mu$, where τ_0 is the optical depth through the diameter of the sphere (Figure 9.1). Then

$$\begin{aligned} L &= 4\pi R^2 F \\ &= 8\pi^2 R^2 |S| \int_0^{+1} (e^{|\tau_0|\mu} - 1)\mu \, d\mu \end{aligned}$$

and integrating

$$\begin{aligned} L &= 8\pi^2 R^2 |S| \left[\frac{e^{|\tau_0|}}{|\tau_0|} - \frac{e^{|\tau_0|}}{\tau_0^2} + \frac{1}{\tau_0^2} - \frac{1}{2} \right] \\ &\simeq 8\pi^2 R^2 |S| \frac{e^{|\tau_0|}}{|\tau_0|} \quad \text{for large } |\tau_0| \end{aligned} \tag{9.1}$$

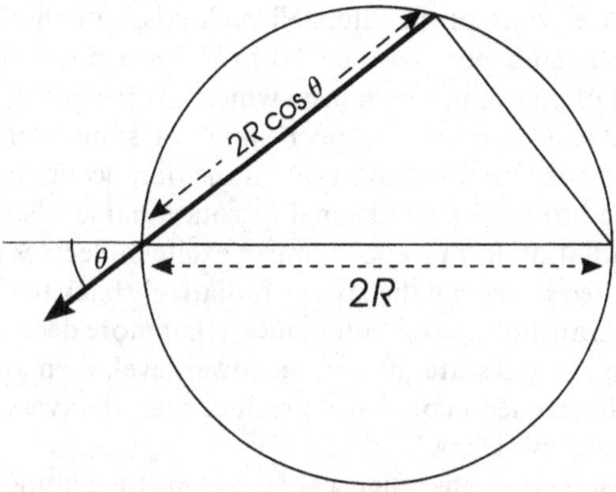

Figure 9.1. Spherical maser

This is to be compared with a (positively) optically thick system with $L = 4\pi^2 R^2 |S|$. Clearly for large negative τ_0, the maser luminosity can greatly exceed that of a blackbody.

We would normally use as an estimate of the size of a body the distance from the centre at which the intensity fell to half of its maximum value. A line of sight through a sphere passing (at closest approach) a distance $R\sin\theta$ from the centre has an optical thickness of magnitude $|\tau_0|\cos\theta$. The intensity at the centre is $|S|\exp(|\tau_0|)$ for large $|\tau_0|$ and the intensity emerging from the line of sight passing $R\sin\theta$ from the centre, whose length is $2R\cos\theta$, is $|S|\exp(|\tau_0|\cos\theta)$. Hence the half-maximum point occurs when $\tau_0\cos\theta = \tau_0 - \ln_e 2$, i.e. when

$$\sin\theta = \sqrt{\frac{2\ln_e 2}{|\tau_0|} - \frac{(\ln_e 2)^2}{|\tau_0|^2}}$$

$$\simeq \sqrt{\frac{2\ln_e 2}{|\tau_0|}} \quad \text{for large } |\tau_0| \tag{9.2}$$

This gives the half maximum projected distance from the centre of $R\sqrt{([2\ln_e 2]/|\tau_0|)}$, which is much smaller for large τ_0 than the value R found for a (positively) optically thick sphere. An optically thin cloud will have $I \propto \tau$ and the half-maximum intensity point will occur when $\cos\theta = 1/2$, $\sin\theta = \sqrt{(3/4)}$, so the projected distance from the centre is $0.866R$. The large luminosity from a small apparent size gives a very high surface brightness and brightness temperature.

Finally in this introductory review we consider the effects of masering on the line profile. Suppose the profile is Gaussian, as one might expect from Doppler broadening, so for Doppler width $\Delta\nu_D$ we have $\phi_\nu \propto \exp(-[\Delta\nu/\Delta\nu_D]^2)$. If $\Delta\nu < \Delta\nu_D$, then $\phi_\nu \sim 1 - [\Delta\nu/\Delta\nu_D]^2$ and

$$\tau = \tau_c(1 - [\Delta\nu/\Delta\nu_D]^2)$$

where τ_c is the line centre optical depth. Hence for large $|\tau|$:

$$I_\nu = |S|e^{|\tau|} = |S|e^{|\tau_c|}e^{-(\Delta\nu/\Delta\nu_D)^2|\tau_c|}$$

$I_\nu = 1/2 I_\nu(\max)$ when $1/2 = \exp(-[\Delta\nu/\Delta\nu_D]^2|\tau_c|)$, i.e. when $\Delta\nu = \Delta\nu_D\sqrt{(\ln_e 2/|\tau_c|)}$. The full width at half height is then less than the value of $1.66\Delta\nu_D$ normally assumed for a Gaussian profile.

Population inversion then leads to high brightness temperatures (10^{14} K has been observed), small apparent sizes and narrow lines. Interstellar conditions provide the low densities needed for departures from thermodynamic equilibrium and the large column densities needed to give big amplifications. The masers will tend to be beamed since amplification depends on path-length, but are unlikely to show the coherent output found from laboratory systems since the cavities and mirrors used in the latter systems to produce parallel beams are of course absent in the astrophysical environment.

Maser Properties in More Detail

We must now consider the population inversion in a little more detail. Consider a two-level system with upper level j and lower level i, with pump rates into the two levels from some reservoir of P_j and P_i respectively, and with loss rates from the two levels Γ_j and Γ_i. Then in the steady state:

$$\frac{\mathrm{d}N_j}{\mathrm{d}t}=0=P_j-\Gamma_j N_j-A_{ji}N_j-4\pi N_j\int\phi_\nu J_\nu B_{ji}\,\mathrm{d}\nu+N_i\int\phi_\nu J_\nu B_{ij}\,\mathrm{d}\nu-N_jC_{ji}+N_iC_{ji}$$

$$0=P_j-\Gamma_j N_j-A_{ji}N_j-4\pi B_{ji}\int\phi_\nu J_\nu\left(N_j-\frac{g_j}{g_i}N_i\right)\mathrm{d}\nu-C_{ji}\left[N_j-\frac{g_j}{g_i}\mathrm{e}^{-\Delta E/kT}\right] \tag{9.3a}$$

using the standard relationships between Einstein coefficients and collisional excitation and de-excitation rates (Figure 9.2). Similarly for the lower level:

$$0=P_i-\Gamma_i N_i+A_{ji}N_j+4\pi B_{ji}\int\phi_\nu J_\nu\left(N_j-\frac{g_j}{g_i}N_i\right)\mathrm{d}\nu+C_{ji}\left(N_j-\frac{g_j}{g_i}\mathrm{e}^{-\Delta E/kt}\right) \tag{9.3b}$$

These can then be solved for N_j and N_i to obtain the population inversion.

The collisional terms must be much less than the radiative terms, for otherwise collisions would drive the levels into thermodynamic equilibrium and there would be no population inversion. We will therefore neglect them in what follows. Write $\int\phi_\nu J_\nu\,\mathrm{d}\nu=J_{\mathrm{av}}$ and assume for simplicity that $g_j=g_i$ and $\Gamma_j=\Gamma_i$. Subtract

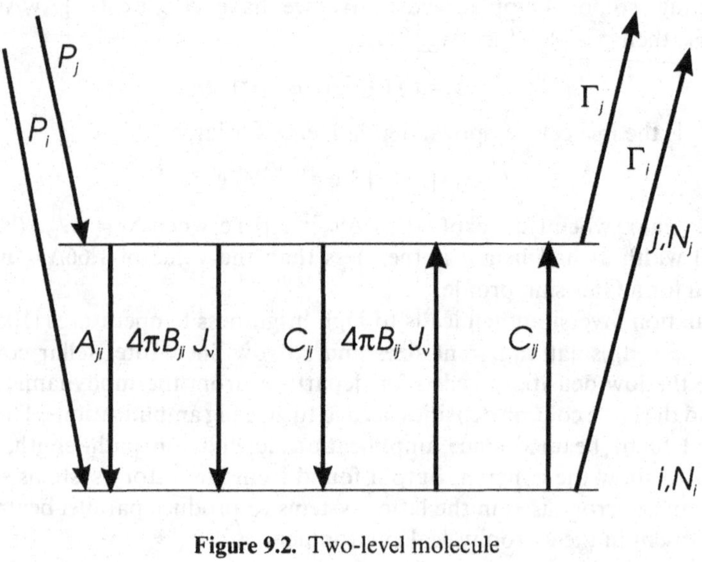

Figure 9.2. Two-level molecule

(9.3a) from (9.3b). Then

$$P_j - P_i = 8\pi B J_{av}(N_j - N_i) + \Gamma(N_j - N_i) + A_{ji}(N_j + N_i)$$

and adding

$$P_j + P_i = \Gamma(N_j + N_i)$$

The spontaneous emission term must also be much less than the stimulated term once masering amplification is taking place and so we will omit it in calculating the population inversion, although it cannot be neglected in finding the source function. Then

$$N_j - N_i = \frac{P_j - P_i}{\Gamma + 8\pi B_{ji} J_{av}}$$

$$= \frac{\Delta N_0}{1 + \dfrac{J_{av}}{J_s}} \tag{9.4}$$

where

$$\Delta N_0 = \frac{P_j - P_i}{\Gamma} \quad \text{and} \quad J_s = \frac{\Gamma}{8\pi B_{ji}}$$

It can be seen that ΔN_0 is the population inversion for very low maser intensity and J_s is the mean intensity, averaged over the line, at which the maser intensity starts to affect the population inversion.

It is normally assumed that the emission line profile is the same as the absorption line profile ϕ_ν, so the absorption coefficient is $\kappa_\nu \rho = -B_{ji} h\nu(N_j - N_i)\phi_\nu$ and the emission coefficient is

$$j_\nu \rho = A_{ji} N_j h_\nu/(4\pi)\phi_\nu$$
$$= A_{ji}[(N_j - N_i)/2 + (P_j + P_i)/(2\Gamma)]h\nu/(4\pi)\phi_\nu$$

where we have written

$$2N_j = (N_j - N_i) + (N_j + N_i) = (N_j - N_i) + (P_j + P_i)/\Gamma$$

If $J_{av} \ll J_s$, then

$$N_j - N_i = \Delta N_0 = (P_j - P_i)/\Gamma$$

and the source function is given by

$$S = 2h\nu^3/c^2(\Delta N_0/2 + [P_j + P_i]/[2\Gamma])/\Delta N_0$$

or writing the efficiency of the maser as

$$\eta = (P_j - P_i)/(P_j + P_i)$$

we have

$$S = 2h\nu^3/c^2(\eta + 1)/(2\eta) \tag{9.5}$$

However, the emission line profile ψ_ν and the absorption line profile ϕ_ν are not the same for large maser intensities. Collisions are not important in line broadening, which is dominated by Doppler broadening. Normally, Doppler broadening leads to redistribution in frequency, but the beaming effect means that we can treat the stimulated emission as being coherent with the stimulating radiation, and hence can write our statistical equilibrium equations for a particular frequency. The pump rates, if they represent downward radiative transitions from some reservoir state or states, will have a frequency profile representing the Doppler distribution of the originating states, ϕ_ν, so giving terms like $P_i\phi_\nu$ in the statistical equilibrium balance at a particular frequency. The upper level population at a particular Doppler shift is $N_{j\nu} = N_j\psi_{j\nu}$ and the lower level population is $N_{i\nu} = N_i\phi_{i\nu}$, so for the upper level statistical equilibrium equation we have

$$0 = P_j\phi_\nu - \Gamma N_{j\nu} - A_{ji}N_{j\nu} - 4\pi B_{ji}J_\nu(N_{j\nu} - N_{i\nu})$$

and similarly for the lower level.

Soving in the same way as before we obtain instead of (9.4):

$$N_{j\nu} - N_{i\nu} = \frac{\Delta N_0}{1 + \dfrac{J_\nu}{J_s}}\phi_\nu \tag{9.4a}$$

where $\Delta N_0 = (P_j - P_i)/\Gamma$ and J_s have the same values as before, but the frequency dependence of the populations can be different from the Doppler profile if $J_\nu > J_s$, when the maser emission can selectively lower the upper level population at line centre frequencies.

In the most general case, if we allow both g_i and g_j to be different and Γ_i and Γ_j to be different:

$$\frac{P_j\phi_\nu}{\Gamma_j g_j} - \frac{N_{j\nu}}{g_j} - \frac{4\pi J_\nu B_{ji}}{\Gamma_j}\left(\frac{N_{j\nu}}{g_j} - \frac{N_{i\nu}}{g_i}\right) = 0$$

$$\frac{P_i\phi_\nu}{\Gamma_i g_i} - \frac{N_{i\nu}}{g_i} + \frac{4\pi J_\nu B_{ji}\,g_j}{\Gamma_i\;\; g_i}\left(\frac{N_{j\nu}}{g_j} - \frac{N_{i\nu}}{g_i}\right) = 0$$

and adding

$$\frac{P_j\phi_\nu}{\Gamma_j g_j} + \frac{P_i\phi_\nu}{\Gamma_i g_i} = \frac{1}{g_j}\left(N_{j\nu} + \frac{g_j}{g_i}N_{i\nu}\right)$$

Subtracting and rearranging:

$$\Delta N_\nu = N_{j\nu} - \frac{g_j}{g_i}N_{i\nu} = g_j\left(\frac{N_{j\nu}}{g_j} - \frac{N_{i\nu}}{g_i}\right)$$

$$= g_j\frac{\left[\dfrac{P_j}{\Gamma_j g_j} - \dfrac{P_i}{\Gamma_i g_i}\right]\phi_\nu}{\left[1 + 4\pi g_j J_\nu B_{ji}\left(\dfrac{1}{g_i\Gamma_i} + \dfrac{1}{g_j\Gamma_j}\right)\right]}$$

so the population inversion is still given by

$$\Delta N_\nu = \frac{\Delta N_0}{1+\dfrac{J_\nu}{J_s}}\phi_\nu \tag{9.6}$$

with

$$\Delta N_0 = g_j\left[\frac{P_j}{g_j\Gamma_j} - \frac{P_i}{g_i\Gamma_i}\right]$$

and

$$J_s = \frac{1}{4\pi g_j B_{ji}\left[\dfrac{1}{g_j\Gamma_j} + \dfrac{1}{g_i\Gamma_i}\right]}$$

and the source function is still given by (9.5) in the unsaturated case with

$$\eta = \left[\frac{P_j}{\Gamma_j g_j} - \frac{P_i}{\Gamma_i g_i}\right] \Big/ \left[\frac{P_j}{\Gamma_j g_j} + \frac{P_i}{\Gamma_i g_i}\right]$$

although for large J_ν, η must be divided by $(1 + J_\nu/J_s)$.

If we take the *unsaturated case*, where $J_\nu \ll J_s$, and so for a homogeneous gas the emission and absorption coefficients are constant, the radiative transfer equation

$$\frac{dI_\nu}{dx} = -\kappa_\nu \rho I_\nu + j_\nu \rho$$

has the solution

$$I_\nu(x) = \frac{j_\nu}{\kappa_\nu}(1 - e^{-\kappa\nu\rho x}) + I_o e^{-\kappa\nu\rho x}$$

$$= |S|(e^{|\kappa\nu|\rho x} - 1) + I_0 e^{|\kappa\nu|\rho x} \tag{9.7}$$

where S is the source function and I_0 is the intensity at $x = 0$, i.e. an external input into the maser system. This solution is independent of geometry and was used in our initial discussion of masering. The *gain*, that is the exponent in (9.7), is proportional to the path length times the population inversion, and the latter is proportional to ΔN_0, which depends only on the pump rate (and loss rate).

If $J_\nu \gg J_s$, we have the rather different *saturated* case. The equation of radiative transfer can be written $dI_\nu/d\tau_\nu = -I_\nu + S_\nu$ and since the saturated case implies large intensities, we can neglect the source function in comparison with the intensity and write $dI_\nu/dx = -\kappa_\nu \rho I_\nu$. If the beam can be approximately represented as being of constant intensity in the solid angle Ω, then $J_\nu = I_\nu \Omega/(4\pi)$ and

$$\frac{\mathrm{d}I_\nu}{\mathrm{d}x} = \frac{B_{ji} h\nu \Delta N_0}{1 + \dfrac{I_\nu \Omega}{4\pi J_s}} \phi_\nu I_\nu$$

$$= \frac{4\pi B_{ji} h\nu \Delta N_0 J_s}{\Omega} \phi_\nu, \quad J_\nu \gg J_s$$

Hence

$$I_\nu = \frac{4\pi B_{ji} h\nu \Delta N_0 J_s}{\Omega} \phi_\nu (x - x_s) + I_\nu(x_s) \tag{9.8}$$

where x_s is the point along the ray where saturation sets in, $I_\nu(x_s)$ being given approximately by the unsaturated formula.

It will be noted that the intensity of the saturated part of the maser increases only linearly rather than exponentially with path length. This means that properties like beaming, decrease of apparent source size and line narrowing that are characteristic of unsaturated masers will not be further enhanced in the saturated region. Saturation will be reached at the end of a large path length of unsaturated masering gas, that is in the *outer* regions of a masering object. The value of J_s (for Γ and g equal for the two levels) is $\Gamma/(8\pi B_{ji})$, which in the constant intensity within a beam picture corresponds to an intensity of

$$I_s = \Gamma/(2B_{ji}\Omega) \tag{9.9}$$

or a brightness temperature of

$$T_B = c^2/(4k) \cdot 1/\nu^2 \cdot \Gamma/(B_{ji}\Omega) = (h\nu)/(2k) \cdot \Gamma/A_{ji} \cdot (4\pi/\Omega) \tag{9.9a}$$

We have made an approximation in jumping from unsaturated to completely saturated, whereas there must be a region of partial saturation, and we have also assumed a simple relation between I and J which is quite a good approximation for a thin cylinder, but may not work for some other geometries.

In the fully saturated case an expression can be obtained for the luminosity of the saturated region directly. In this case the spontaneous emission can be neglected, and so the luminosity is the excess of stimulated emissions over absorptions, i.e. for volume V:

$$L_\nu = h\nu V 4\pi J_\nu B_{ji} \Delta N_\nu$$

$$= h\nu V J_\nu 4\pi B_{ji} \phi_\nu \frac{\Delta N_0}{1 + \dfrac{J_\nu}{J_s}}$$

$$= h\nu V 4\pi B_{ji} \phi_\nu J_s \Delta N_0, \quad \text{if } J_\nu \gg J_s$$

$$= h\nu \phi_\nu V \frac{P_j - P_i}{2} \tag{9.10}$$

where in the last step Γ and g for the two energy levels have been assumed to be the same.

Now consider the particular case of a spherical maser. The observed brightness temperatures of many masers indicate that they are saturated. A typical model of a spherical maser will therefore have a saturated shell between radii R_s and R surrounding an unsaturated core (Figure 9.3). Exponential amplification takes place in the core, so only rays passing through the core are likely to achieve large intensities. We have essentially already dealt with the case of an unsaturated sphere. At any point a distance r from the centre, the longest path length will be the radial ray at $2r$. Rays making an angle θ with the radial ray will have path length $2r\cos\theta$, and since θ must be small for large amplification, we can write $\cos\theta = 1 - 1/2\theta^2$. Then

$$I(r, \theta = 0) = |S| e^{2r|\kappa_\nu \rho|}$$

$$I(r_\theta) = |S| e^{2r|\kappa_\nu \rho|\cos\theta} = |S| e^{2r|\kappa_\nu \rho|(1 - \theta^2/2)}$$

$$\frac{I(r, \theta)}{I(R, 0)} = e^{-r|\kappa_\nu \rho|\theta^2}$$

where both κ_ν and S are in their unsaturated forms and hence independent of the mean intensity.

The beam size, taken as the angle at which the beam falls to $1/e$ of its maximum intensity, is then

$$\theta_s = 1/\sqrt{(\kappa_\nu \rho r)} \tag{9.11}$$

as, indeed, we found before. The mean intensity is then

$$\begin{aligned} J_\nu &= \frac{1}{2}\int I_\nu \mathrm{d}(\cos\theta) \\ &= \frac{1}{2}|S| e^{2r|\kappa_\nu \rho|}\int e^{-r|\kappa_\nu \rho|\theta^2}\theta \mathrm{d}\theta \quad \text{for small } \theta \\ &= \frac{1}{2}|S| \frac{e^{2r|\kappa_\nu \rho|}}{2r|\kappa_\nu \rho|} \end{aligned}$$

which reaches J_s when $r = R_s$. Thus

$$\frac{e^{2R_s|\kappa_\nu \rho|}}{2R_s|\kappa_\nu \rho|} = \frac{2\Gamma}{8\pi B_{ji}|S|} \simeq \frac{2\Gamma\eta}{A_{ji}}, \quad \eta \ll 1 \tag{9.12}$$

An approximate solution for large optical depth is

$$R_s \sim 1/(2\kappa_\nu \rho)\ln_e[2\Gamma\eta/A_{ji}] \tag{9.13}$$

but if there is a saturated shell surrounding the unsaturated core there will be a contribution to the mean intensity in the core from inward flowing radiation,

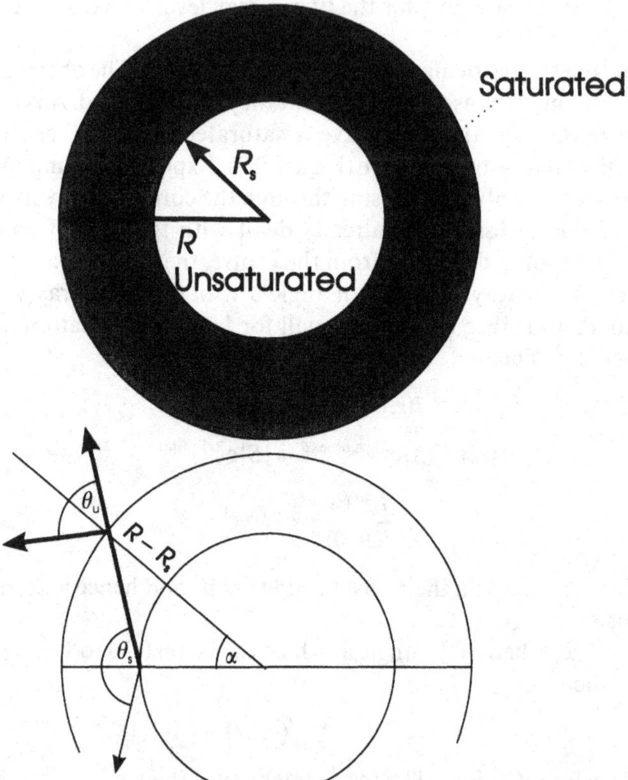

Figure 9.3. Saturated spherical maser

and R_s will be less than calculated above. Only rays that have been exponentially amplified in the core will have large intensities after being linearly amplified in the saturated shell. Hence the final beam emerging from the saturated shell is determined by the range of angles at the outer radius R which can be covered by the beam from the core. Let the final beam angle be θ_u and the beam angle from the core be θ_s, as before, with α the angle between the radius from the centre to the point on the surface being considered and the radius from the centre to the point on the edge of the core from which the core beam can just reach the surface point being considered (Figure 9.3). Then $\alpha = \theta_s - \theta_u$, $R_s \sin\alpha = (R - R_s)\tan\theta_u$ or $R_s \alpha = (R - R_s)\,\theta_u$ since the angles are small. Hence

$$\theta_u = R_s/R\theta_s = \sqrt{(R_s/|\kappa_\nu \rho|)}1/R \tag{9.14}$$

Now the luminosity of the shell can be written $L_\nu = h\nu(4/3)\pi(R^3 - R_s^3)4\pi B_{ji}\Delta N_0 J_s \phi_\nu$ from (9.10), and $L = 4\pi R^2 F = 16\pi^2 R^2 J$ for the mean intensity J at R since the radiation is beamed and $F = 2\pi \int I\mu d\mu$ and $J = (1/2)\int I d\mu$, so for strong beaming $F = 4\pi J$. Hence for $R \gg R_s$ we have

$$J_\nu = (1/3)RJ_s h\nu B_{ji}\,\Delta N_0 \phi_\nu$$

and with $I\pi\theta^2/(4\pi) = J$, the emergent intensity can be written

$$I_\nu(R) = \frac{4J_\nu}{\theta_u^2} = \frac{4J_s h\nu B_{ji}\Delta N_0 \phi_\nu}{3R_s^2\theta_s^2}R^3$$

$$= \frac{4J_s(|\kappa_\nu \rho|)^2}{3R_s}R^3 = \frac{4J_s(h\nu B_{ji}\Delta N_0\phi_\nu)^2}{3R_s}R^3 \tag{9.15}$$

so the brightness temperature

$$T_B = \frac{c^2 h}{12\pi k}(P_j - P_j)\frac{|\kappa_\nu \rho|}{\nu R_s}R^3 \tag{9.15a}$$

where substitution has been made for κ_ν (which throughout this argument refers to the unsaturated case) and for J_s. It must be remembered that R_s is also frequency dependent, and that the brightness temperature is usually quoted at the line centre, where the line profile factor ϕ_ν is $1/(\sqrt{\pi}\Delta\nu_D)$. If R_s is independentof R, the intensity and the brightness temperature increase with the cube of the radius. However, the radius of the saturated core should be calculated including the contribution to the mean intensity from the saturated shell, and this gives (see, e.g., Elitzur [1])

$$R_s\kappa_\nu = 1/2\log_e\left[\frac{48\eta\Gamma}{A_{ji}}\frac{1}{(\kappa_\nu R)^4}\right] \tag{9.16}$$

which means that R_s decreases very slowly with R. The flux at the surface of such a maser is simply found from the luminosity since a sphere must radiate isotropically so $F_\nu = L_\nu/(4\pi R^2)$, or using (9.10):

$$F_\nu = (h\nu)/6\cdot(P_j - P_i)R\cdot\phi_\nu$$

and hence is proportional to R.

The case of a cylindrical maser is similar, but it should be noted that a considerable proportion of the luminosity of a saturated cylindrical maser comes from the sides of the cylinder (see again Elitzur [1] for a discussion of this point and other details). The geometry does not, in the end, greatly alter our conclusions about masers. Naturally a given pumping rate can produce a larger flux from a cylinder than a sphere if it is pointing at the observer, but such an orientation can only hold for a small proportion of cases. The actual shape is determined not by the cloud shape, but by the velocity distribution. Systematic velocities in a cloud mean that the variously moving components of the cloud are Doppler shifted to different frequencies, and hence cannot amplify radiation from other parts. The shape of the maser is the shape of the surface enclosing gas moving at the same velocity.

In the case of OH masers, polarization is sometimes observed, with left and right circularly polarized components separated in apparent velocity. This is due to the Zeeman effect in a magnetic field. The OH radical has an electronic angular momentum which acts as a magnet, and the magnetic interaction causes the levels of OH to be split by an amount proportional to the magnetic field, the sub-levels thus distinguished being labelled by m. It will be recalled from Chapter 5 that transitions with $\Delta m = \pm 1$ are called σ transitions and give rise to lines on either side in the frequency of the zero field transition, while transitions with $\Delta m = 0$ are called π transitions and lie close to the zero field frequency. If the magnetic field is parallel to the line of sight (longitudinal field), the σ transitions are left and right circularly polarized respectively, while the π component is missing. If the field is perpendicular to the line of sight (transverse field), then the σ and π components are present and linearly polarized. The observations show left and right circularly polarized lines from the same position but displaced in velocity, which is to be interpreted as a displacement in frequency due to the field. Linearly polarized features are not seen. This may be due to Faraday rotation of the plane of polarization in the maser, which would mean that linearly polarized light produced in one part of the maser would not be able to interact with another part and hence would not be amplified. Typical fields are of the order of 0.5 μT.

OH is paramagnetic because of its electronic angular momentum. The other maser molecules have closed shells and no net electronic angular momentum, so the Zeeman effect is due only to the much smaller nuclear magnetic moment. Linear polarization but not line splitting is observed in some H_2O masers and some SiO masers.

What observational information (apart from polarization) is actually available? One directly measures the flux received on Earth, F_ν, and (sometimes using VLBI techniques) the apparent angular size of the source, θ. The flux divided by the apparent solid angle subtended by the maser ($\pi\theta^2$) gives the intensity and hence the brightness temperature. The distance to the maser, d, enables the apparent angular size to be turned into an apparent linear size. However the apparent radius of a spherical maser is considerably smaller than its physical radius R, as we have seen, while the apparent radius of a cylinder is close to its physical radius r but tells the observer nothing about its length l along the line of sight. If the maser is spherical and therefore an isotropic radiator, then the luminosity $L_\nu = 4\pi d^2 F_\nu$, but if the maser is cylindrical and elongated and only radiates into beams of solid angle Ω_b at either end, then $L_\nu = 2I_\nu\Omega_b = 2\Omega_b/(4\pi)4\pi d^2 F_\nu$, and even if the maser is saturated and some radiation is emitted along the sides of the cylinder, the luminosity calculated assuming isotropy may still be much greater than the true luminosity.

However the criterion for saturation give values of similar order of magnitude for the gain factor, $\kappa_\nu\rho R$ for a sphere and $\kappa_\nu\rho r$ for a cylinder of radius $r \ll$ length l, at which saturation sets in (the expression for a cylinder contains a factor $\log_e(\kappa_{\nu r}r)$. The unsaturated output beam intensity at which saturation is about to set in in the outer regions of the maser is therefore similar in different geometric arrangements, with a value that depends on $\eta\Gamma/A_{ji}$, with A known, Γ roughly known, and η certainly being less than 1. We can therefore certainly conclude that the brightness temperatures of the more powerful masers indicate that they are saturated. This enables us to set some lower limits on the required pump rate.

A saturated maser has a total luminosity given by (9.10) as

$$L = \int L_\nu \mathrm{d}\nu = Vh\nu(P_j - P_i)/2 = \eta(h\nu)(P_j + P_i)/2$$

Even assuming 100% efficiency, the average pump rate $(P_j + P_i)/2$ must exceed the number of maser photons emitted, $L/h\nu$, i.e.

$$\frac{(P_j + P_i)}{2} > \frac{F_0 \Delta\nu}{h\nu} 4\pi d^2 \frac{2\Omega_b}{4\pi}$$

where we have written $\int F_\nu \mathrm{d}\nu \sim F_0 \Delta\nu$ for the line centre observed flux F_0 and line width $\Delta\nu$, and put the beam solid angle as Ω_b for each end of a cylinder with $2\Omega_b = 4\pi$ for a sphere. If the pumping is radiative by some line or lines with a frequency of around ν_p and with width $\Delta\nu_p$, then the number of pumping photons available is $L_{\nu p}\Delta\nu_p A/(4\pi d^2_{pm} h\nu_p)$, where $L_{\nu p}$ is the luminosity of the pump per unit frequency interval, and the pump is a distance d_{mp} from the maser, with maser cross-sectional area A. Now $A/(4\pi d^2_{pm}) = \Omega_{pm}/(4\pi)$, where Ω_{pm} is the solid angle subtended by the maser at the pump. If $F_{p\nu}$ is the pump flux per unit frequency interval received on Earth, then the number of pumping photons $= F_{p\nu}\Delta\nu_p d^2/(h\nu_p)(\Omega_{pm}/4\pi)$. Hence we require

$$F_{p\nu} > F_0(\Delta\nu/\Delta\nu_p)(\nu_p/\nu)(2\Omega_b/\Omega_{pm}) \qquad (9.17)$$

The beam width is in general unknown, and the maser solid angle from the pump source will often be less than 4π if the source is a star some distance from the maser, but $\nu_p > \nu$, since the pump will be infrared or even ultraviolet, whereas the maser frequency will correspond to a centimetre or millimetre wavelength. Hence the pump flux and luminosity must be large, and the requisite luminosities from some pump models are greater than any stellar luminosity.

Examples of Astronomical Masers

We now turn to particular examples of masers, considering first the OH radical. We start by considering the energy level structure of OH. The ground electronic state of OH, unlike those of most of the molecules discussed so far, has an electronic orbital angular momentum, which couples strongly with the internuclear axis, so it is the component along this axis, specified by the quantum number

Λ, which is significant (see Chapter 8). The ground electronic state of OH has $\Lambda = 1$ and is called a Π state. The ground electronic state of OH also has a spin of 1/2, and if this couples strongly to the internuclear axis again it will be the component along this axis, called Σ, which will be significant. In this case Σ can take the values $+1/2$ or $-1/2$, giving a total electronic angular momentum along the axis of Ω, for OH taking the values 3/2 or 1/2. These states are designated $^2\Pi_{3/2}$ and $^2\Pi_{1/2}$, where the superscript 2 indicates the multiplicity in the same way as for atomic states. The total electronic angular momentum then interacts with the rotation of the molecule, which is here specified by quantum number N, to give a total angular momentum with quantum number J. In OH this gives rise to two rotational ladders, $^2\Pi_{3/2}(J = 3/2, 5/2, 7/2, \ldots)$ and $^2\Pi_{1/2}(J = 1/2, 3/2, 5/2\ldots)$. This coupling scheme is called Hundt's case (a) and has the selection rule $\Delta\Sigma = 0$, so radiative transitions between ladders are forbidden, but $\Delta J = 1$ transitions within a ladder are permitted. However, other coupling schemes are possible, the relevant one here being Hund's case (b), where the electronic spin couples weakly to the internuclear axis, so the component of the orbital angular momentum specified by Λ couples with the rotation specified by N to give a resultant angular momentum specified by K (Λ, $\Lambda + 1$, $\Lambda + 2\ldots$). Each level is slightly split by interaction with the spin to give levels specified by the total angular momentum quantum number J, so for $S = 1/2$ one obtains for each K two levels, $J = K + 1/2$ and $J = K - 1/2$. The selection rule here is $\Delta K = 0, \pm 1$. Now OH lies between the two cases, so we can describe the levels by the terminology of case (a) with its two rotational ladders, but transitions between ladders are permitted although with a lower transition probability than a similar transition within a ladder. The ground state is $^2\Pi_{3/2}$ $(J = 3/2)$ and the lowest level of the $^2\Pi_{1/2}$ ladder with $J = 1/2$ lies above the second level $J = 5/2$ of the $^2\Pi_{3/2}$ ladder. Infrared transitions connect these rotational levels so the ground state $^2\Pi_{3/2}(3/2)$ is linked to $^2\Pi_{1/2}(1/2)$, (3/2, and (5/2) by transitions at 79, 53, and 35 μm respectively, while the ground state is linked to the next higher level in its own ladder by a transition at 119 μm (Figure 9.4).

Now we turn to the sub-structure of each level, already briefly discussed. Each level is split into an upper and lower level by Λ doubling, the upper and lower levels being distinguished here by their parity ($+$ or $-$), which is always different, so that radiative transitions between the Λ doubled sub-levels of a given J level are permitted. It is such transitions that show evidence of the maser phenomenon. In the ground state $^2\Pi_{3/2}(3/2)$, $3/2^-$ lies lowest, whereas in the lowest level of the $^2\Pi_{1/2}$ ladder, $1/2^+$ lies lowest. The energy order of the parities alternates as one goes up each ladder. Since parity must change in a radiative transition, in a downward cascade if one starts with the upper sub-level the cascade will always link one to upper sub-levels and vice-versa. Each Λ doubling sub-level is further split into hyperfine structure by the nuclear spin $I = 1/2$, so for a given value of J the total angular momentum quantum number taking nuclear spin into account, F, can take on the values $J + 1/2$ and $J - 1/2$ and each value of J gives

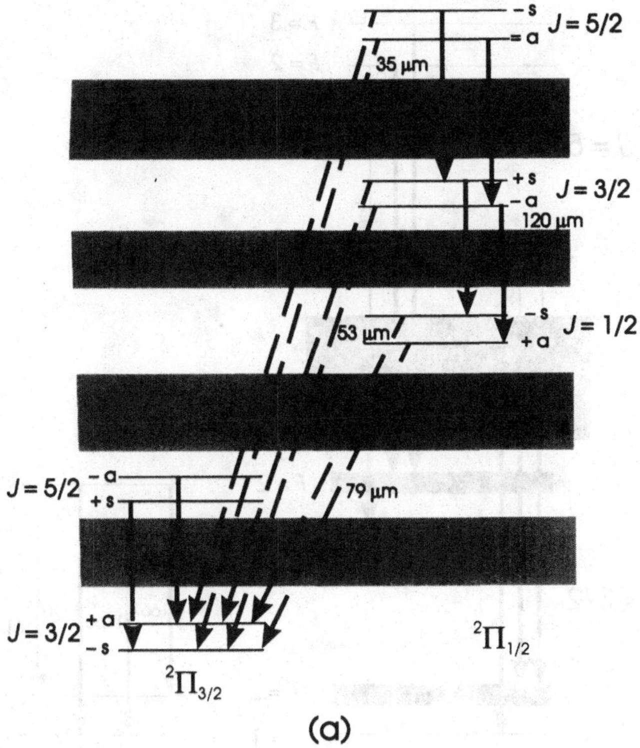

Figure 9.4. (a) OH lower levels; hyperfine structure not shown, parity + or −. (b) OH levels showing hyperfine splitting, $F = 2$ inversion, and on the left maser transitions. (c) OH maser levels showing inversion of $F = 1$ hyperfine levels

rise to 4 levels. The selection rule for F is $\Delta F = 0, 1$, so normally one obtains 4 lines with two main lines with $\Delta F = 0$, and two satellite lines with $\Delta F = \pm 1$. However $F = 0$ to $F = 0$ is forbidden, so there are only 3 lines in the $^2\Pi_{1/2}(1/2)$ transition.

The most extensive observations of maser activity have been made with the ground state transition $^2\Pi_{3/2}(3/2)$ at around the 18 cm wavelength, with the two main lines at frequencies of 1667 MHz ($F = 2$ to $F = 2$) and 1665 MHz ($F = 1$ to $F = 1$), and satellite lines at 1720 MHz ($F = 2$ (upper) to $F = 1$ (lower)) and 1612 MHz ($F = 1$ (upper) to $F = 2$ (lower)). We have already shown that the line strengths should be in the ratio 1:5:9:1 in order of increasing frequency for the optically thin case, and of course equal intensity if all are optically thick. A wide variety of patterns of anomalous strengths are observed, sometimes the main lines being anomalously strong, sometimes the 1612 MHz line appearing very strong and 1720 MHz very weak and sometimes 1720 MHz appearing very

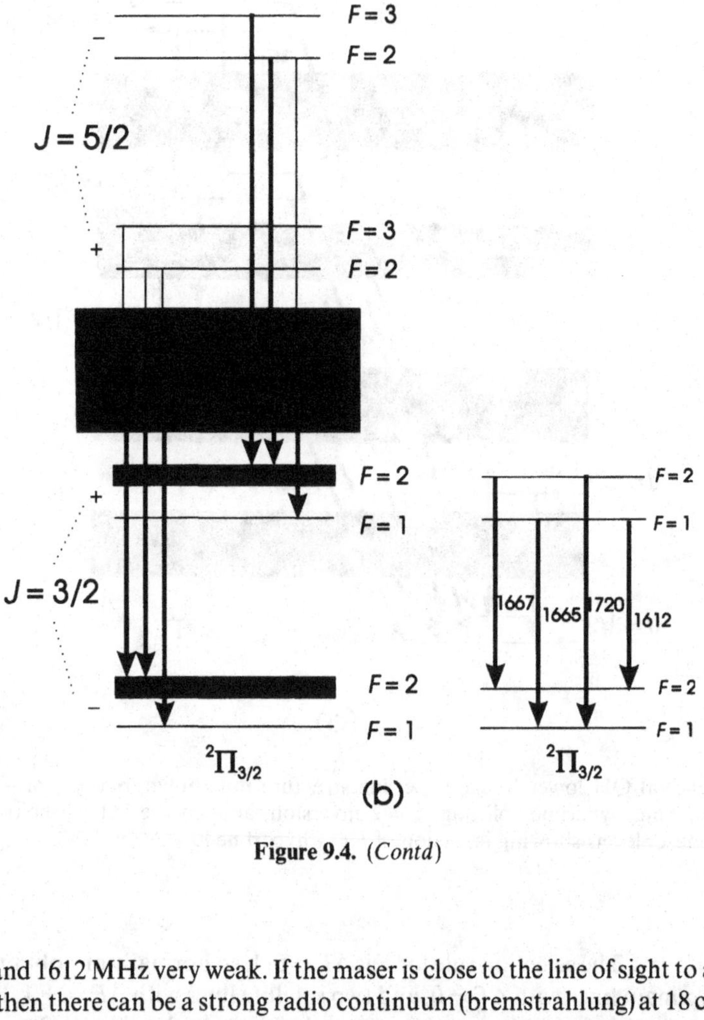

Figure 9.4. *(Contd)*

strong and 1612 MHz very weak. If the maser is close to the line of sight to an H II region, then there can be a strong radio continuum (bremstrahlung) at 18 cm, and sometimes one of the satellite lines may appear in absorption while the other lines are in emission. Maser activity has also been found in the three lines of the $^2\Pi_{1/2}(1/2)$ with the satellite line at 4765 MHz from $F = 1$ to $F = 0$ anomalously strong, and in the $^2\Pi_{3/2}(5/2)$ quadruplet at around 6030 MHz. Detections have also been made of OH lines from several higher rotational levels in each ladder.

How have these patterns been produced? It is generally accepted that the population of the masering levels proceeds from the radiative decay of excited rotational levels, the excitation being radiative, collisional, or through the formation of the molecule in the first place. The easiest question to answer is: Why does one satellite or the other seem to be preferentially produced in some cases? Consider first a decay cascade down the $^2\Pi_{3/2}$ ladder. The $J = 3/2^+$ $F = 2$ and

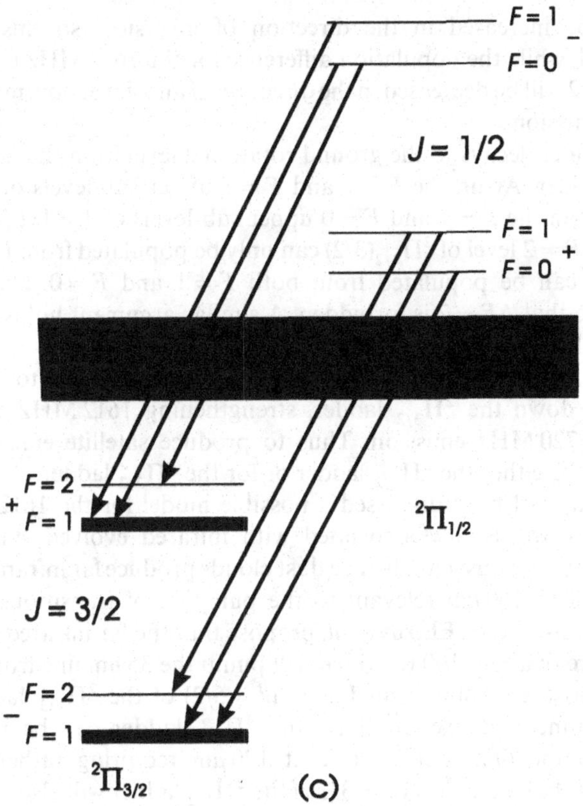

Figure 9.4. (*Contd*)

$F = 1$ upper sub-levels of the ground rotational state can only be produced by radiative decay from the upper $J = 5/2^-$ $F = 3$ and $F = 2$ sub-levels of the next higher rotational state. Now $J = 3/2^+$ $F = 2$ can be produced from $J = 5/2^-$ $F = 3$ and $F = 2$, while $J = 3/2^+$ $F = 1$ can only be produced from $J = 5/2^-$ $F = 2$ by the selection rules on F. Similarly, the lower sub-levels of the ground rotational state are populated from the lower levels of $J = 5/2$, but $F = 1$ can only receive decays from $J = 5/2$ $F = 2$, whereas $F = 2$ can receive decays from $J = 5/2$ $F = 2$ and $F = 3$. The net effect is to increase the population of the $F = 2$ sub-levels of the ground rotational level at the expense of the $F = 1$ sub-levels of the ground rotational level. The population difference between the upper and lower $F = 2$ sub-levels will be unaffected, as will the population difference between the upper and lower $F = 1$ sub-levels, so there will be no effect on main line strengths. However, the population difference for the 1720 MHz transition from $F = 2$ to

$F = 1$ will be increased in the direction of inversion, so this line will be strengthened, while the population difference for the 1612 MHz transition from $F = 1$ to $F = 2$ will be decreased in the direction of anti-inversion, making this line weaker in emission.

Now consider decays to the ground rotational level from the bottom level of the $^2\Pi_{1/2}$ ladder. Again the $F = 2$ and $F = 1$ upper sub-levels of $J = 3/2^+$ are populated from the $F = 1$ and $F = 0$ upper sub-levels of $J = 1/2^-$. However, in this case the $F = 2$ level of $^2\Pi_{3/2}(3/2)$ can only be populated from $F = 1$ of $^2\Pi_{1/2}$, while $F = 1$ can be populated from both $F = 1$ and $F = 0$, again using the selection rule that $\Delta F = 2$ is forbidden. A similar argument holds for the lower sub-levels. The net result is to transfer population from the $F = 2$ sub-levels of $^2\Pi_{3/2}$ to the $F = 1$ sub-levels. The overall effect is the opposite to that produced by cascades down the $^2\Pi_{3/2}$ ladder, strengthening 1612 MHz emission, and weakening 1720 MHz emission. Thus to produce satellite emission we need a preference for either the $^2\Pi_{1/2}$ ladder or for the $^2\Pi_{3/2}$ ladder.

Elitzur *et al.* [2] have discussed a possible model for the 1612 MHz maser. These masers are usually associated with infrared evolved stars which are surrounded by cool dust clouds. The dust clouds produce far infrared emission in the region 30 to 160 μm relevant to the pumping of transitions between the rotational levels of OH. Elitzur *et al.* propose that the far infrared radiation has a temperature of about 100 K, which will pump the 35 μm line from the ground state $^2\Pi_{3/2}$ to the second excited state ($J = 5/2$) of the $^2\Pi_{1/2}$ ladder. For the moment assume that excitation up the $^2\Pi_{3/2}$ ladder can be neglected, the transition to the first excited state at 120 μm requiring rather cooler dust (excitation to the second level $J = 3/2$ of the $^2\Pi_{1/2}$ ladder will also take place, but does not alter the main conclusions). We will consider only one half of the Λ doubling, say the upper half, since the two halves perform almost identical cycles. The proposed cycle is then from the ground state $^2\Pi_{3/2}$ with hyperfine levels $F = 2$ and $F = 1$ having populations N_{G2} and N_{G1} which are pumped by 35 μm radiation from the dust up to the excited level $^2\Pi_{1/2}(J = 5/2)$ with hyperfine levels with $F = 3$ and $F = 2$ and populations N_{H3} and N_{H2}. These then decay radiatively to $^2\Pi_{1/2}(J = 3/2)$ with hyperfine levels with $F = 2$ and $F = 1$ and populations N_{M2} and N_{M1}, which in turn decay to $^2\Pi_{1/2}(J = 1/2)$ with hyperfine levels with $F = 1$ and $F = 0$ and populations N_{L1} and N_{L0}, since transitions within a ladder have larger probabilities than those between ladders. $^2\Pi_{1/2}(J = 1/2)$ then decays radiatively back to the ground state. Let the transition probability between level j with $F = F_j$ and level i with $F = F_i$ be written as $A_{ji}(F_j, F_i)$ and suppose all the transitions between rotational levels are optically thin so the rate of decay per unit volume is $N_j A_{ji}$. The excitation rate is $N_i B_{ij} J_{av} = N_i A_{ji} g_j / g_i \, J'_{av}$, where $J'_{av} = c^2/(8\pi h \nu^3) \, J_{av}$ and J_{av} is the mean radiation intensity from the dust at the line frequency. Then in equilibrium we have

$$N_{G1} g_2 / g_1 A_{HG}(2,1) J'_{av} = N_{L0} A_{LG}(0,1) + N_{L1} A_{LG}(1,1)$$

and similarly for the other transitions. The transition probabilities for $A_{3/2,1/2}$ are in the ratio 1/2:1/3:1/6 for (2,1), (1,0) and (1,1) and similarly those for $A_{1/2,3/2}$ are in the ratio 1/2:5/12:1/12 for (0,1), (1,2) and (1,1) and those for $A_{5/2,3/2}$ are in the ratio 1/2:9/20:1/20 for (3,2), (2,1) and (2,2).

Solving and substituting:

$$\frac{N_{L1}}{N_{L0}} = \left[\frac{N_{M2}}{N_{M1}} + \frac{A_{ML}(1,1}{A_{ML}(2,1)}\right]\frac{A_{ML}(2,1)}{A_{ML}(1,0)}\frac{A_{LG}(0,1)}{A_{LG}(1,2)+A_{LG}(1,1)}$$

$$= \frac{3}{2}\left[\frac{N_{M2}}{N_{M1}} + \frac{1}{3}\right]$$

Solving similarly for the other population ratios, we obtain $N_{G1}/N_{G2} = 3/5$, the statistical equilibrium value with no population inversion.

If, however, we assume the infrared excitation transition and the interladder transition (the transitions with the ground state as the lower level) are both optically thick, then the three transitions from ${}^2\Pi_{3/2}(J = 3/2)$ to ${}^2\Pi_{1/2}(J = 5/2)$ will all have the same number of transitions per second per unit volume, say k, since they will be optically thick saturated absorption lines of the external infrared radiation. Hence we can write $B_{ij}N_iJ_{av}$ as k for all three lines. Following through this result to the ${}^2\Pi_{3/2}$ $J = 5/2$, $J = 3/2$ and $J = 1/2$ levels as before means that the ratio of number of photons populating ${}^2\Pi_{1/2}(J = 1/2)$ $F = 1$ to the number populating $F = 0$ is 3:2 and the ratio of the numbers leaving the two levels must be the same in equilibrium. Optical thickness in the ${}^2\Pi_{1/2}(J = 1/2)$ to ${}^2\Pi_{3/2}(J = 3/2)$ emission transitions means that the radiative decay rate is reduced by re-absorptions and is given by N_jA_{ji} (escape probability) which for large optical depth $\sim N_j A_{ji}$/(optical depth) using the results of our previous discussion of escape probability. Now the OH molecules are likely to be predominantly in the ground rotational state so the optical depth will be proportional to N_iB_{ij} which in turn is proportional to N_ig_j/g_iA_{ji}. Hence if the hyperfine levels are populated with their equilibrium populations which are proportional to go (ratio 5:3), the optical depths will be proportional to g_jA_{ji} and the radiative decay rate will be independent of A_{ji}, with the result that the two decays from $F = 1$ will have equal rates. The result of these considerations is that the upper $F = 2$ sub-level of the ground rotational level receives from $F = 1$ at a rate 3/7 of that with which the $F = 1$ level receives from $F = 1$ and $F = 0$. Similarly the $F = 2$ level loses population to ${}^2\Pi_{1/2}(5/2)$ at twice the rate that $F = 1$ loses population. Both these effects transfer population from $F = 2$ to $F = 1$, and although the ground level sub-populations are no longer in the ratio of their statistical weights as assumed above, the effect will still be to produce a population inversion in the 1612 MHz line.

Of course a similar cycle of far infrared radiative excitation followed by a downward cascade could still operate within the ${}^2\Pi_{3/2}$ ladder, and if the transition to the ground level is optically thick an inversion in favour of the $F = 1$

sub-level is produced, thus enhancing the 1720 MHz line. However Elitzur argues that of the two competing cascades, the $^2\Pi_{1/2}$ cascade will normally predominate in deciding which level will be inverted and so one should normally obtain maser activity in the 1612 MHz line rather than the 1720 MHz line. Elitzur points out that if the $^2\Pi_{1/2}(J = 1/2)$ to $^2\Pi_{3/2}(J = 3/2)$ transitions were optically thin but the $^2\Pi_{3/2}(J = 5/2)$ to $^2\Pi_{3/2}(J = 3/2)$ transitions optically thick, then the overpopulation would be of the $F = 2$ level producing the 1720 MHz line, but the OH column density required would only produce a moderate inversion and hence the resulting maser would be weak. Collisional excitation at low temperatures might populate the second lowest rotational level only, namely the $^2\Pi_{3/2}(J = 5/2)$ level, and hence produce 1720 MHz. These mechanisms would not produce the overall inversion of the Λ doublets needed to produce main line maser emission. Elitzur [3] has pointed out that the far infrared radiative pump transition to the upper of a doubled pair has a shorter wavelength than the lower of such a pair since the separation of the Λ doubled levels increases as one goes to higher rotational levels and hence if the flux from the dust decreases with increasing wavelength, the upper level will be slightly more strongly pumped than the lower level. Upper levels cascade downwards to upper levels so the end result will be a ground level with the upper half of the Λ doublet over populated with respect to the lower level, leading to main line maser activity.

Andresen [4] has considered the inversion of OH in terms of the reflection symmetry of the electronic wavefunction with respect to the plane defined by the two nuclei. In the $^2\Pi_{3/2}$ ladder the upper component of each Λ doubled level is always antisymmetric, whereas in the $^2\Pi_{1/2}$ ladder the upper component of each Λ doubled level is always symmetric. Hence any process which leads to the formation of excited levels of OH in a symmetric state will result in a cascade down the upper levels (the upper component always decays to the upper component of the next lower rotational level) of the $^2\Pi_{1/2}$ ladder, the lowest level of the ladder then decaying to the upper component of the ground state with a change in symmetry (this is an electronic transition). The upper component of the ground state will become overpopulated, leading to a main line inversion. For the same reasons as before, the $F = 1$ hyperfine sub-level of the ground state is preferentially populated, leading to 1612 MHz emission. A number of different processes give rise to OH in excited symmetric levels—Andresen favours collisional excitation by H_2 molecules from the ground state. However one process leads to the preferential formation of OH in antisymmetric excited states, and that is the photodissociation of H_2O. This will then populate the upper component of each Λ doublet in the $^2\Pi_{3/2}$ ladder, which will cascade down to overpopulate the upper component of the ground state, leading to main line maser emission, with a preference for the $F = 2$ hyperfine sub-level which will give a 1720 MHz maser.

Observationally, OH masers divide into two categories: those found in close association with evolved giants, and those found in association with H II regions. The former are termed *stellar masers*, and the latter *interstellar masers*.

The stellar masers are found in conjunction with stars with infrared excesses, where the infrared excess is due to a circumstellar dust shell, produced by mass loss from the star which is also evidenced in many cases by circumstellar lines in the visible. In some instances the dust shell is so thick that the star is not detected in the visible, but where they can be observed the underlying stars include both red giants and red supergiants. In nearly all cases the star is variable, usually with a long period of the order of years. The far infrared radiation from the dust is a possible pump. The maser activity round the star lies in a number of spots with different Doppler shifts, often grouped in a shell-like arrangement, and it will be seen later that the arrangement in three dimensions is an expanding shell. The radius of the shell is usually around 10^{14} m to 10^{15} m and the brightness temperature of the masers is typically 10^8 K with an output of 10^{42} photons per second if the radiation is assumed to be isotropic, although there are more luminous examples. The maser spectra are classed into type II in which the strongest emission comes from the 1612 MHz line and type I in which the strongest emission comes from the main lines, although type I masers normally show maser action in the 1612 MHz line as well. Stellar masers do not exhibit maser activity in the 1720 MHz line, strongly suggesting that the $^2\Pi_{1/2}$ ladder is dominant. The main line emitting region is typically smaller and less luminous than that producing 1612 MHz emission. The type II masers are found to vary in brightness with the infrared flux from the underlying variable star with little time lag, so these masers must be radiatively pumped. The type I masers also vary in step with the star, although the link is not so clear as with type IIs, and radiative pumping is likely here too. The infrared flux seems capable of meeting the requirement for the number of pump photons to equal or exceed the number of maser photons. It has been suggested that stars with large mass loss rates and hence thick dust shells develop 1612 MHz regions outside the main line emitting regions.

The interstellar masers would be better described as star-formation linked masers. OH emission is nearly always associated with a compact H II region, that is one smaller than $3*10^{15}$ m in size, and hence with very young massive stars that have just turned on and started emitting the ultraviolet photons that ionize the H II region. The interstellar masers often have higher brightness temperatures than stellar masers of around 10^{12} K with higher photon outputs of around 10^{44} photons per second if isotropic, and again are arrangements of maser-emitting spots with different Doppler shifts roughly spread over the projected area of the H II region. However the maser spots themselves are smaller in apparent size than those found in stellar masers, perhaps 10^{12} m as opposed to 10^{14} m. The interstellar masers most often have either the main lines or the 1720 MHz line as the strongest, although there are a number of examples of 1612 MHz masers. The 1665 MHz main line is usually stronger than the 1667 MHz line, which is not the case for stellar masers. The central stars of the H II regions are not variable in most cases, so the test for radiative pumping used with the stellar masers is not applicable. The pattern of lines suggests that in the majority of cases pumping is through the $^2\Pi_{3/2}$ ladder. The high luminosities present difficulties for many pumping models. The observations would seem to be consistent with a model in which dissociation of H_2O in the interstellar clouds by ultraviolet radiation from the hot star forms OH and drives the maser pump.

The other intensively studied maser is H_2O. The water molecule is an asymmetric top, so the energy level structure is complex. Each level of rotational quantum number J is split into $2J+1$ levels, which can be specified by the quantum numbers K_{-1} and K_1 which would represent the projection of the angular momentum on the symmetry axis if the molecule were a symmetric top in the prolate and oblate cases respectively. K_{-1} and K_1 can take values from 0 to J, and for a given J and K_{-1}, K_1 takes the values $J-K_{-1}$ and $J-K_{-1}+1$, except for $K_{-1}=0$ for which K_1 only takes the value $J-K_{-1}$. Thus in the case of water, each value of J gives rise to a short ladder of $K_{-1}K_1$ values, with the smallest K_{-1} lying lowest of the K_{-1} values. The ladders for neighbouring J values overlap (Figure 9.5). Now the selection rules require $\Delta J = 0, \pm 1$ as usual, and that K_{-1} and K_1 both change by 1 (strictly both change their evenness, so that changes by 3, etc. are also allowed). If it is also noted that the two hydrogen nuclear spins can combine to give $I=0$ or $I=1$ with appropriate symmetry, we can divide water levels into *ortho*-H_2O with $I=1$ for which it can be shown that one of K_{-1} and K_1 must be odd and the other even and *para*-H_2O with $I=0$ for which both of K_{-1} and K_1 must be even or both odd. Since they are unconnected by radiative transitions, *ortho*- and *para*- H_2O can for some purposes be treated as separate species.

Consider the case of *ortho*-H_2O. Transitions are permitted between adjacent levels within a ladder and between neighbouring ladders. Transitions within a ladder will have wavelengths in the infrared typically of the order of 100 μm and so will many of the transitions between ladders. However it will sometimes happen that levels in adjacent ladders nearly coincide in energy, and if the selection rules on the change of evenness in the K quantum numbers are obeyed, we will obtain a transition at much longer wavelengths. Such is the case of the transition from 6_{16} to 5_{23} which is at 1.35 cm or 22 GHz, and is the maser H_2O line by far the most commonly observed. Other maser transitions of *ortho*-H_2O have been observed at 380 GHz (4_{14} to 3_{21}) and 321 GHz (10_{29} to 9_{36}), and of *para*-H_2O at 183 GHz (3_{13} to 2_{20}), and 325 GHz (5_{15} to 4_{22}). It seems likely that there are a number of other maser transitions, and the lines actually observed are constrained by observational limitations such as terrestrial absorption (by H_2O lines!).

De Jong [5] argues that what he calls the backbone of the ortho energy level diagram, the bottom levels of each ladder, 1_{01}, 2_{12}, 3_{03}, 4_{14}, 5_{05}, 6_{16}, 7_{07}, etc. are connected by strong radiative transitions which will therefore tend to be optically thick. Furthermore in most cases the only radiative transition out of these levels is to the next level down in the backbone (the exceptions being maser transitions

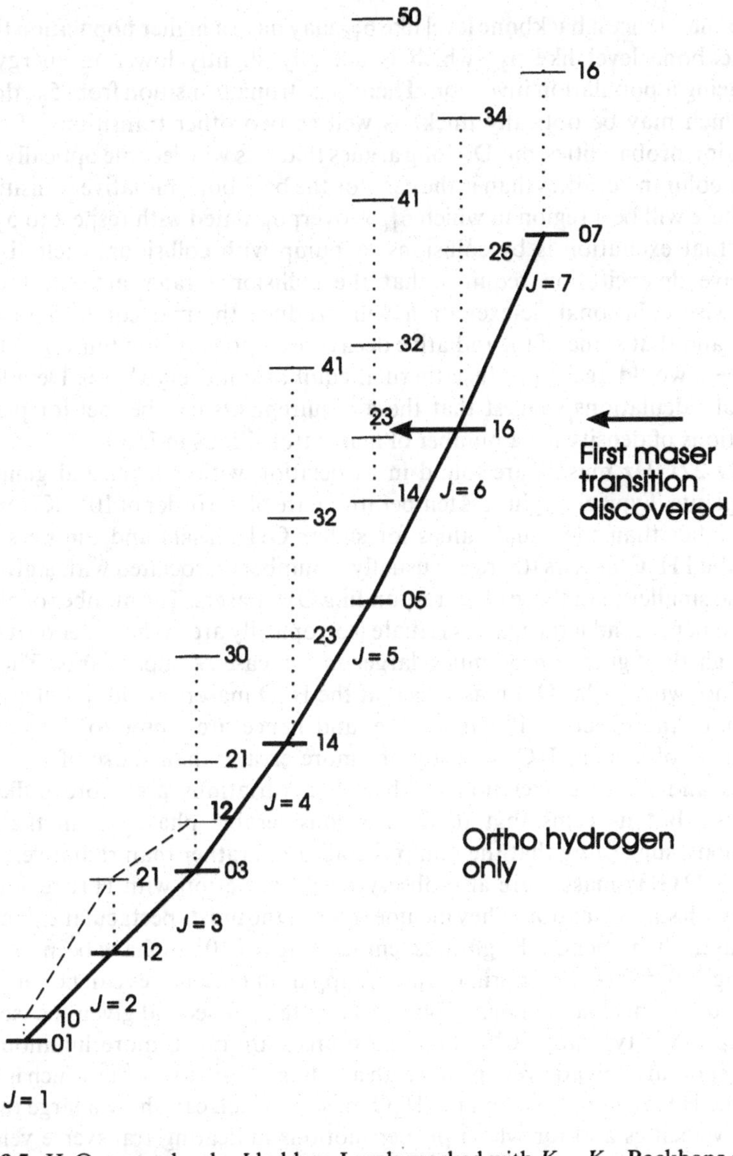

Figure 9.5. H_2O energy levels: J ladders. Levels marked with $K_{-1}K_1$. Backbone and its transitions heavily marked. Some other transitions dotted

that have small transition probabilities. The net result of these two factors will be to maintain the backbone level populations in thermal equilibrium. Other levels have several radiative downward transitions, some of them optically thin because of small transition probabilities and will tend to cascade downwards to the backbone. Hence a backbone level like 6_{16} may have a higher population than an off-backbone level like 5_{23} which is actually slightly lower in energy, thus producing a population inversion. There is a strong transition from 5_{23} down to 4_{14} which may be optically thick, as well as two other transitions of smaller transition probabilities, but De Jong argues that this will become optically thin at higher column densities than is the case for the backbone radiative transitions so that there will be a region in which 6_{16} is overpopulated with respect to 5_{23}. It is likely that excitation is by collisions. A pump with collisional excitation and radiative de-excitation requires that the collisional rates are not too high (otherwise collisional de-excitation will produce thermal equilibrium for all levels) and that some of the radiative decays are optically thin (universal optical thickness would again produce thermal equilibrium everywhere). Detailed numerical calculations suggest that these requirements can be met for plausible conditions of density for a number of maser transitions in H_2O.

H_2O 22 GHz masers are found in association with evolved red giants and supergiants. Typical brightness temperatures are of the order of 10^{11} K, considerably higher than the usual values for stellar OH masers and the sizes of the individual H_2O masers (there are usually a number) associated with a given star are also smaller than those of corresponding OH masers. The number of photons emitted per second if the masers radiate isotropically are of the order of 10^{43} s^{-1}, although the figure can be much larger in the case of supergiants. The other difference with stellar OH masers is that the H_2O masers are found in a smaller region, of the order of 10^{13} m in size, and hence are closer to the star. The variations of stellar H_2O intensity are more erratic than those of stellar OH masers and hence correlations with stellar variations are more difficult to establish, but it seems that there is a considerable phase lag in the maser variations, suggesting that the pump is collisional rather than radiative.

H_2O 22 GHz masers are also observed in association with H II regions and hence with star formation. They include some of the most spectacular examples of the maser phenomenon. Brightness temperatures of 10^{14} K have been observed, although 10^{12} K is more normal, and the apparent maser sizes can be very small, down to 10^{11} m. The number of photons emitted per second given the isotropic assumption is typically 10^{45} but some sources are much more luminous. The maser spots are spread over an area with a radius of 10^{14} to 10^{15} m which is bigger than the H II region itself, and the H_2O masers, which can show a large range of radial velocities and for which proper motions indicating transverse velocities can be measured, appear to be an expanding system and are normally taken to be associated with a high velocity outflow from the young star. It has been suggested that shocks created by the outflow both produce the H_2O and provide the energy to drive the collisional pumps.

The third commonly observed masering molecule is SiO. SiO is necessarily a linear molecule with a simple level structure. The maser transitions come from excited *vibrational* states, with observations in the $v = 1$ vibrational state of the

$J = 1$ to 0, 2 to 1, 3 to 2, 4 to 3, 5 to 4 and 6 to 5 transitions, of the same transitions in the $v = 2$ vibrational state and of $J = 1$ to 0 in the $v = 3$ vibrational state. The first maser transition discovered was the $v = 1$ $J = 2$ to $J = 1$ 86 GHz (3.49 mm) transition. Similar numbers of SiO and H_2O masers are known, but the SiO transitions are in observationally difficult regions so they may be the most numerous class of masers. Nearly all are associated with evolved stars.

Rotational levels in excited vibrational states decay almost entirely by radiative transitions to the next lower vibrational state, since the transition probabilities for purely rotational transitions within a vibrational state are much less. These decays, of the form v, J to $v - 1, J \pm 1$, may be optically thick, in which case, as has already been discussed, the net radiative decay rate becomes

$$(A_{ji})(\text{escape probability}) \sim A_{ji}/\tau \text{ if the optical depth } \tau \text{ is large}$$

Take the optical depth averaged over the line:

$$\tau = \kappa\rho\, x = (N_i/g_i - N_j/g_j)c^2/(8\pi h v^3) g_j A_{ji} h v$$

Now

$$N_i/g_i = N(v-1, J-1)/g(J-1)$$
$$= N(v-1, 0)/g(0) * N(v-1, J-1)/N(v-1, 0) * g(0)/g(J-1)$$

for a transition from J to $J - 1$, and if the rotational levels were populated by collisions, their relative populations would be in the ratio of their statistical weights so $N_i/g_i = N(v-1, 0)/g(0)$. Watson *et al.* [6] show that the collisional rates from the ground state are to a first approximation proportional to the statistical weight of the rotational level. Similarly $N_j/g_j = N(v, 1)/g(1)$ giving

$$\tau = \tau(v, J = 1; v - 1, J = 0) * A(J, J-1)/A(1, 0) * g(J)/g(1)$$

In the same way, for a transition from $v - 1, J + 1$ to v, J

$$\tau = \tau(v, J = 1; v - 1, J = 0) * A(J, J+1)/A(1, 0) * g(J)/g(1)$$

so the net decay rate is

$$2/\tau(v, J = 1;\ v - 1, J = 0) * [g(1)A(1, 0)]/g(J)$$

which is inversely proportional to $g(J) = 2J + 1$. Hence decay rates for higher rotational levels are less, with the result that population by collisions followed by optically thick radiative decay leads to overpopulation of higher rotational levels compared with the proportionality to the statistical weight assumed (the latter being an essentially Boltzmann distribution among rotational levels since $\Delta E \ll kT$). Thus a population inversion is produced among the rotational levels of excited vibrational states. Elitzur [7] argues that collisional pumping is much more likely than radiative pumping.

Stellar SiO masers have typically fairly modest brightness temperatures of about 10^{10} K, and are situated very close to the parent star at radii from $2 * 10^{11}$

to $4*10^{12}$ m, which correspond to only a few stellar radii. The velocity components of SiO masers show an irregular pattern, and the time variations are also irregular. If the star varies in the visible, the maser variation seems to lag far enough behind to rule out optical pumping by the star, whereas temperature variations controlling the collisional pumping rates and produced by the stellar variation would be expected to lag.

Weak maser emission has also been detected from the CH radical at a wavelength of around 9 cm, the CH radical being closely analogous to the OH radical in having Λ doubling and hyperfine structure, and from the HCN linear molecule in the $J = 1$ to $J = 0$ transition of an excited vibrational state. Other molecules sometimes showing maser emission include the asymmetric rotor, formaldehyde H_2CO, at 6 cm, and ammonia NH_3 in the inversion transition at 1.6 cm (a frequency of 18.5 GHz). CH_3OH, methanol, shows the richest maser spectrum of all with more than 20 different maser transitions detected.

References

[1] M. Elitzur, *Astronomical Masers*, Kluwer, 1992.
[2] M. Elitzur, P. Goldreich, and N. Scoville, *Astrophysical Journal*, **205**, 384, 1976.
[3] M. Elitzur, *Ast. Ap.*, **62**, 305, 1978.
[4] P. Andresen, *Ast. Ap.*, **154**, 42, 1986.
[5] T. De Jong, *Ast. Ap.*, **26**, 297, 1973.
[6] W. Watson, M. Elitzur and R. Bienick, *Astrophysical Journal*, **240**, 547, 1980.
[7] M. Elitzur, *Astrophysical Journal*, **240**, 553, 1980.

10 WINDS AND CIRCUMSTELLAR SHELLS

Stars of many types have winds blowing outwards from their surfaces which can lead to shells of material (dust, molecules, etc.) around the star. Main sequence stars of solar type and cooler have winds, but these are insignificant as far as mass loss is concerned and do not show up in the stellar spectrum. Main sequence stars of type B sometimes show broad emission lines superimposed on the photospheric spectrum correlated with fast rotation of the star, and these stars are called Be stars. The emission lines come from a rotating low density ring or disc around the star, and are associated with modest rates of mass loss. Main sequence stars of spectral type O have strong winds with terminal velocities of up to $2000\,\mathrm{km\,s^{-1}}$, which are not surprisingly photoionized by the central star and form part of the H II region surrounding the star. Wolf–Rayet stars are very hot stars lying a little above the main sequence, and having such strong winds that the whole spectrum is dominated by emission from the wind, and no static surface of the star can be seen.

It must be remembered that the wind region is usually not spatially resolved from the star in observations and so the spectrum of the wind region is seen superimposed on that of the star. The typical spectrum line seen in the O and Wolf–Rayet stars is a *P-Cygni* line profile with a broad emission line with a sharp absorption line at its violet short-wavelength edge. This is produced by a wind where the emission from most of the wind flow is not blocked so that one sees both the blue-shifted back side of the wind region and the red-shifted near side of the wind region, giving a nearly symmetrical broadened emission line since the wind region is not dense enough to emit appreciable continuum radiation (Figure 10.1). The part of the wind directly between the observer and the star gives an absorption line if the wind source function is less than that of the background star, the absorption line having a blue shift corresponding to the full wind velocity

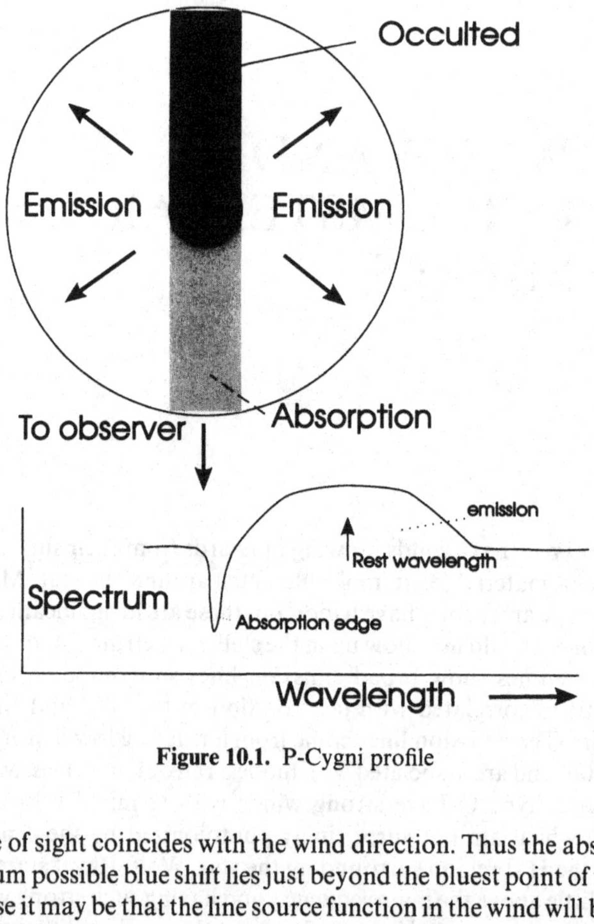

Figure 10.1. P-Cygni profile

since the line of sight coincides with the wind direction. Thus the absorption line with maximum possible blue shift lies just beyond the bluest point of the emission line. Of course it may be that the line source function in the wind will be too low to give appreciable emission, and then all that will be seen is the blue-shifted absorption line. Sometimes the wind spectral lines are also seen as absorption lines in the underlying photospheric spectrum (e.g. hydrogen lines seen in emission in the circumstellar spectrum of Be stars and as absorption lines in the spectrum of the star) and sometimes conditions of ionization/excitation in the photosphere mean that the circumstellar lines do not appear in the photospheric spectrum. P-Cygni line profiles are also characteristic of the spectra of supernovae, representing a kind of transient wind where the whole envelope of the star is expanding outwards and thinning. The velocities are large, from 10 000 to 20 000 km s^{-1}.

Red giants and supergiants all have winds giving appreciable mass loss rates, and for supergiants the spectral type range extends as hot as G stars. These winds are much slower than those found in hot stars, typically 15 km s^{-1}, and show the low degrees of ionization to be expected when the central star produces little

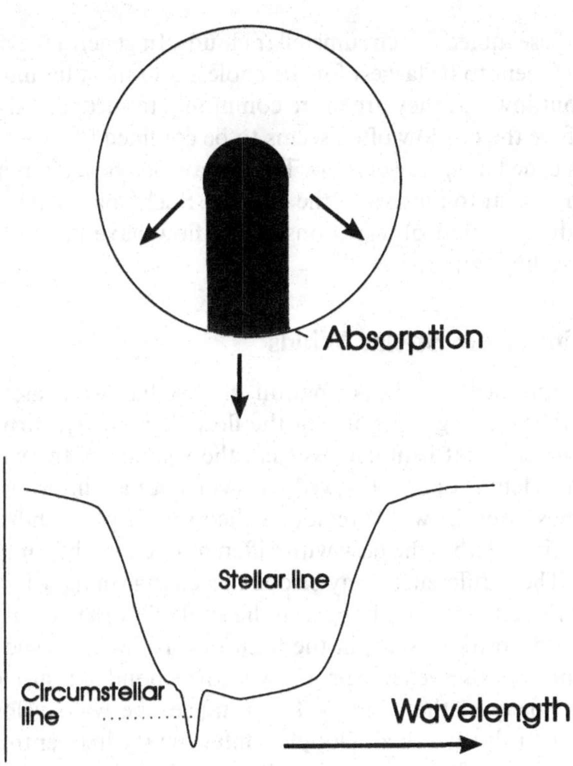

Figure 10.2. Circumstellar line in a cool star

ultraviolet radiation. The low temperatures in the circumstellar envelopes mean that emission is weak, and usually all that is seen in the optical part of the spectrum is the blue-shifted absorption component of a P-Cygni profile. These low excitation and ionization lines are usually also produced by the cool underlying photosphere, so a typical circumstellar line appears as a slightly blue-shifted sharp 'extra' absorption component to a low excitation photospheric line (Figure 10.2). The low temperatures involved mean that molecules and dust can form and can sometimes be seen as shells around the star. The dust absorbs photospheric radiation in the visible and reradiates it at longer wavelengths in the infrared, giving rise to a continuous spectrum that looks a little like the superimposition of two black bodies, one at the temperature of the photosphere and one at the lower temperature of the dust—one sometimes refers to the latter as a 'dust bump'. The molecules radiate in rotational transitions in the millimetre region where there is little photospheric radiation and hence give an emission line spectrum as though from a molecular cloud. Many of the most important maser sources lie in these molecular circumstellar clouds. In general the mass loss rates from cool stars seem to be largest for the coolest and most luminous stars.

Winds, or outflows, as they are more commonly termed, are also a feature of protostars, where the outflow often seems to be confined to two opposite cones, giving what is called a *bipolar outflow*. There may also be a circumstellar disc in a plane perpendicular to the axis of the outflow. Such objects are always heavily obscured by dust, so that observations of the flow have to be in the infrared, millimetre or radio regions.

Emission Lines from Stellar Winds

The Sobolev approach to line formation in a stellar wind assumes that the intrinsic line width at a given point in the flow due to, say, turbulence, corresponds to a velocity that is much less than the velocity of the wind. The main cause of the broadening of the observed line, which for an unresolved circumstellar region comes from the whole region, is then the Doppler shifts produced by various sub-regions within the flow with different line of sight components of the wind velocity. These differences may be due to a change in the wind velocity with position (usually radius) or to changes in the angle that the flow makes with the line of sight with position. Thus in the limit of zero intrinsic width of the line, a given point on the observed line profile will correspond to a surface of constant line of sight velocity in the wind. If, for instance, the wind velocity increases outwards, the intensity at a small Doppler shift from the line centre will originate from a surface starting from a point well in towards the centre of the star on the line of sight towards the centre, where the full wind velocity is directed along the line of sight, and moving outwards for lines of sight passing well away from the centre of the star where only a fraction of the larger wind velocity is along the line of sight.

Consider a spherically symmetric flow with polar coordinates (r, θ) of a point in the flow (strictly a ring) measured from the centre of the star where r is the radial distance from the centre and θ is the angle between the radial line from the centre to the point and the line of sight from the centre to the observer. Now take a line of sight from the point to the observer, which will be parallel to the line of sight from the centre to the observer, and suppose this line of sight passes the centre at closest approach at distance p. Measure the distance along this line of sight by z ($z = 0$ at closest approach) so $z = \pm\sqrt{(r^2 - p^2)}$. The angle between the line of sight through (r, θ) and the radial direction from the centre is also θ (Figure 10.3). If the flow velocity is $V(r)$, then the line of sight velocity at (r, θ) is $V_z = V(r)\cos\theta$. Suppose we are observing a line with a rest frequency ν_0. The observed frequency of radiation emitted at this point at the centre of the line will be $\nu'_0 = \nu_0(1 + V_z/c)$ if positive velocity is taken in the direction of the observer. Conversely an observed frequency ν'_0 for a line with no intrinsic broadening will be produced on a surface with r given by $V(r) = (\nu'_0 - \nu_0)/\nu_0 \cdot c/\cos\theta$.

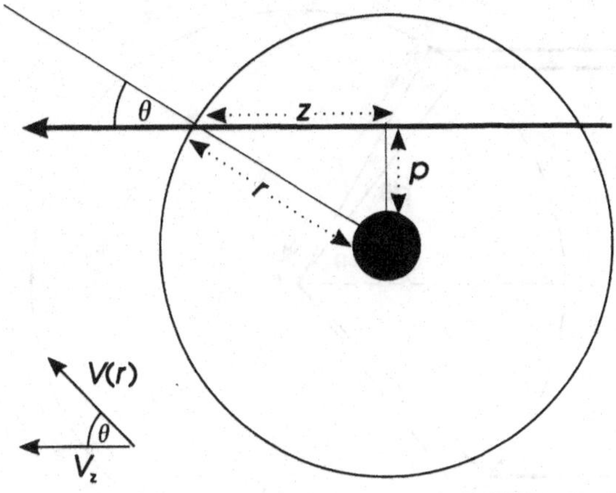

Figure 10.3.

We consider first the simple cases where the line is optically thin. We shall also ignore the fact that the line emitted at a point has an intrinsic width, which it would certainly be wrong to do if optical depth effects had to be considered. Suppose that $V(r)$ = constant and no part of the emitting region is occulted by the star. Then V_z depends only on $\cos\theta$ and at each value of r there will be a ring of material with the right value of θ to Doppler shift the emission to ν'_0. We wish to determine the intensity per unit frequency interval in the resulting line profile. For fixed r, a small range in frequency corresponds to a small range in θ, with $d\theta = d\nu/\nu'_0\, c/(V\sin\theta)$.

For fixed r again, the area of the ring corresponding to the frequency interval from ν'_0 to $\nu'_0 + d\nu$ is $2\pi r^2 \sin\theta d\theta = 2\pi r^2 c/V\, d\nu/\nu'_0$. Hence the intensity per unit frequency interval produced by each annulus of thickness dr is nearly proportional to $2\pi r^2\, dr\, c/V\ 1/\nu_0$, which is independent of frequency, as will be the result of summing a number of such annuli at different values of r, although the contribution from each r will be different as the density and hence the emission coefficient will be different. Thus the observed line profile will be flat-topped and rectangular in shape, with $I_{\nu'}$ = constant from $\nu'_0 = \nu_0(1 + V/c)$ to $\nu_0(1 - V/c)$ (Figure 10.4). This result was first obtained by Beals [1].

Of course the line will in reality have a finite intrinsic width, so the sides of the rectangular profile will not be vertical, and the star is bound to occult some of the emitting gas which lies directly behind the star, so that the red end of the profile may be truncated. More complicated geometries are certainly found for young stars where the outflow is often bipolar and in the form of a cone, and there may be occultation of part of the emission by an opaque equatorial disc around the star. The latter is relatively simply dealt with; see, for example, Appenzeller et al. [2]. Consider a point star surrounded by a spherical emitting region and an

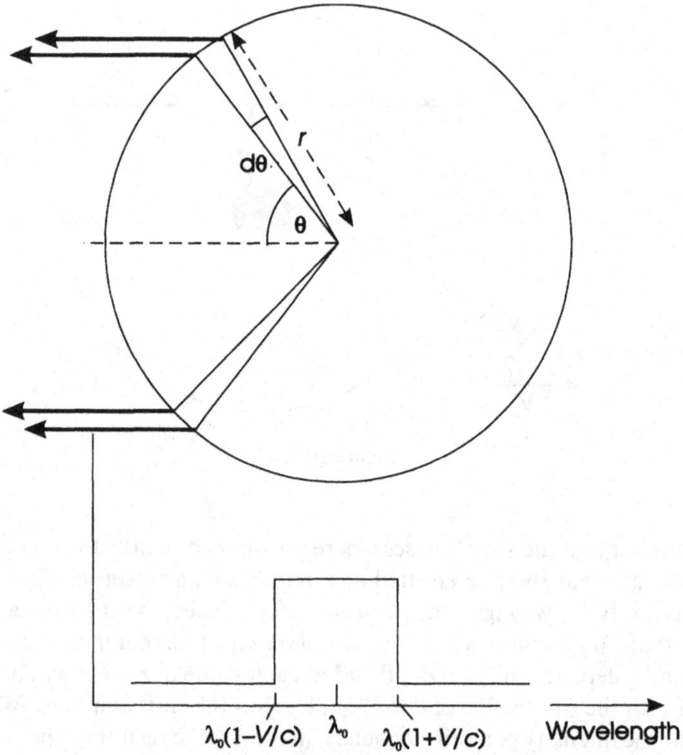

Figure 10.4. Constant velocity outflow

opaque disc, with the axis of the disc at an angle ψ to the line of sight (Figure 10.5). The maximum velocity observable is $-V$, since gas in front of the star moving towards the observer is not blocked, but the maximum positive velocity observable is $+V\sin\psi$, so there is a truncation of the red wing of the emission line, with a gradual decline in intensity in going from $+V$ to $-V\sin\psi$.

We have already seen that in the constant velocity case a particular observed frequency corresponds to a particular value of θ, which is also the angle subtended by the emitting ring at the centre. The side of the disc closest to the observer makes an angle $90-\psi$ with the line of sight so if $\theta < 90-\psi$ there is no occultation of any part of the ring (the degree of occultation is independent of r). Now this condition can be written as $\cos\theta = V_z/V < \cos(90-\psi) = -\sin\psi$. Hence the profile is flat-topped from the frequency equivalent of $-V$ to the frequency equivalent of $-V\sin\psi$ and then drops gradually to zero at $+\mathrm{V}\sin\psi$. In the partly occulted region and for a given r, instead of the full ring of emitting gas subtending an angle 2π at the ring's centre one has an arc of a ring subtending, say, $2\pi - 2x$ at the ring's centre (radius of ring $= r \sin\theta$). If the centre of the ring lies a distance h (in the plane of the ring) above the line where the occulting disc cuts the plane of the ring (Figure 10.5(b)), then $\cos x = h/(r \sin\theta)$. We can determine h by noting that the plane of the ring lies a distance $r\cos\theta$ from the centre of the system along the line of sight from the centre to the observer, and that the disc makes an angle $90 - \psi$ with the line of sight from the centre to the observer, so $h/(r\cos\theta) = \tan(90 - \psi)$. Hence $\cos x = (r\cos\theta\cot\psi)/(r\sin\theta) = \cot\theta\cot\psi$. The intensity at line shift $\Delta\nu$ will be proportional to the angle

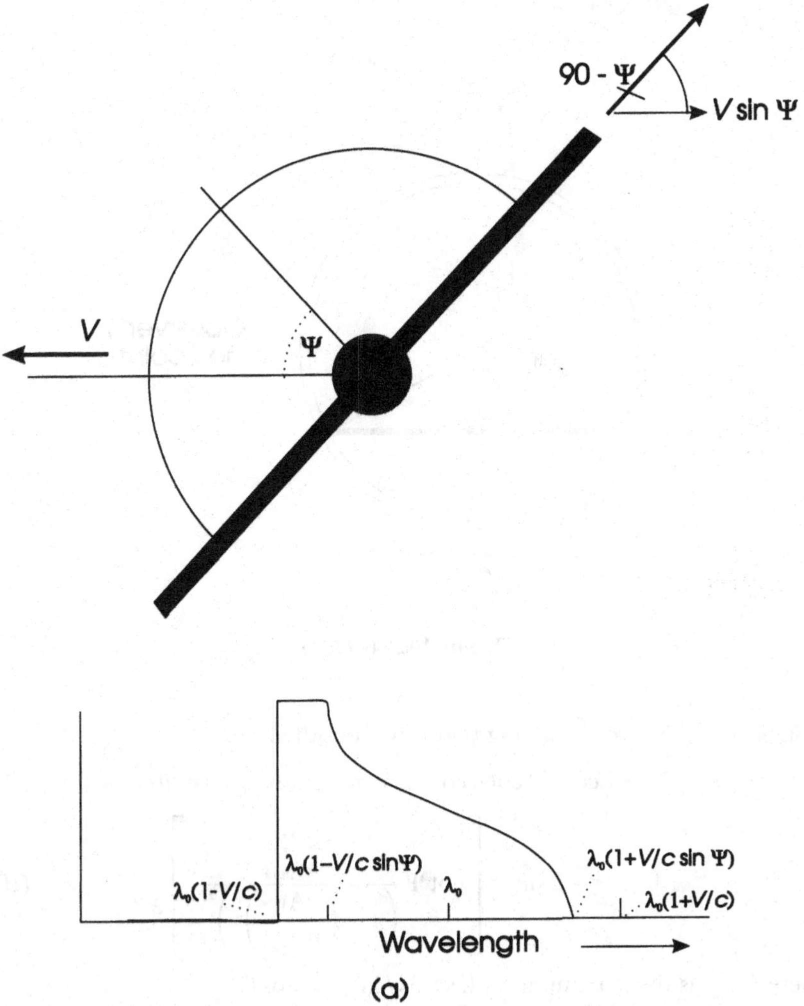

Figure 10.5. (a) Effect of occulting: $\Psi = 45°$. (b) Effect of occulting disc

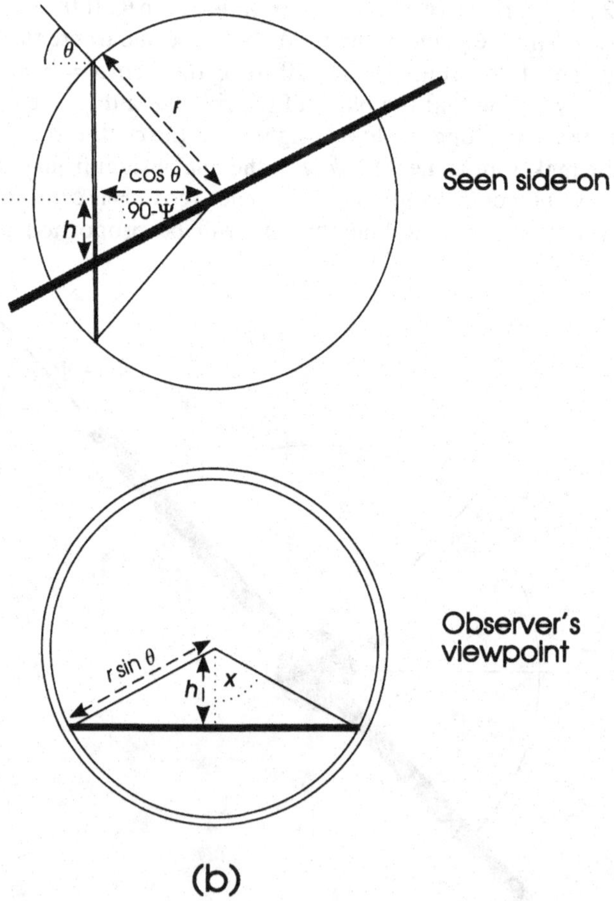

Figure 10.5. (*Contd*)

subtended by the arc of the ring that is not occulted, i.e.

$$I_\nu \propto 2\pi - 2\cos^{-1}[\cot\theta\cot\psi] \propto \pi - 2\sin^{-1}[-\cot\theta\cot\Psi]$$

$$\propto \pi - 2\sin^{-1}\left[\cot\Psi\,\frac{\dfrac{\Delta\nu}{\Delta\nu_{\max}}}{\left(1-\left(\dfrac{\Delta\nu}{\Delta\nu_{\max}}\right)^2\right)^{1/2}}\right] \tag{10.1}$$

where $\Delta\nu_{\max}$ is the maximum shift, so $\Delta\nu/\Delta\nu_{\max} = \cos\theta$.

Appenzeller et al. [2] also consider the case of a flow restricted to a cone of half angle ω where the axis of the cone makes an angle ψ with the line of sight (Figure

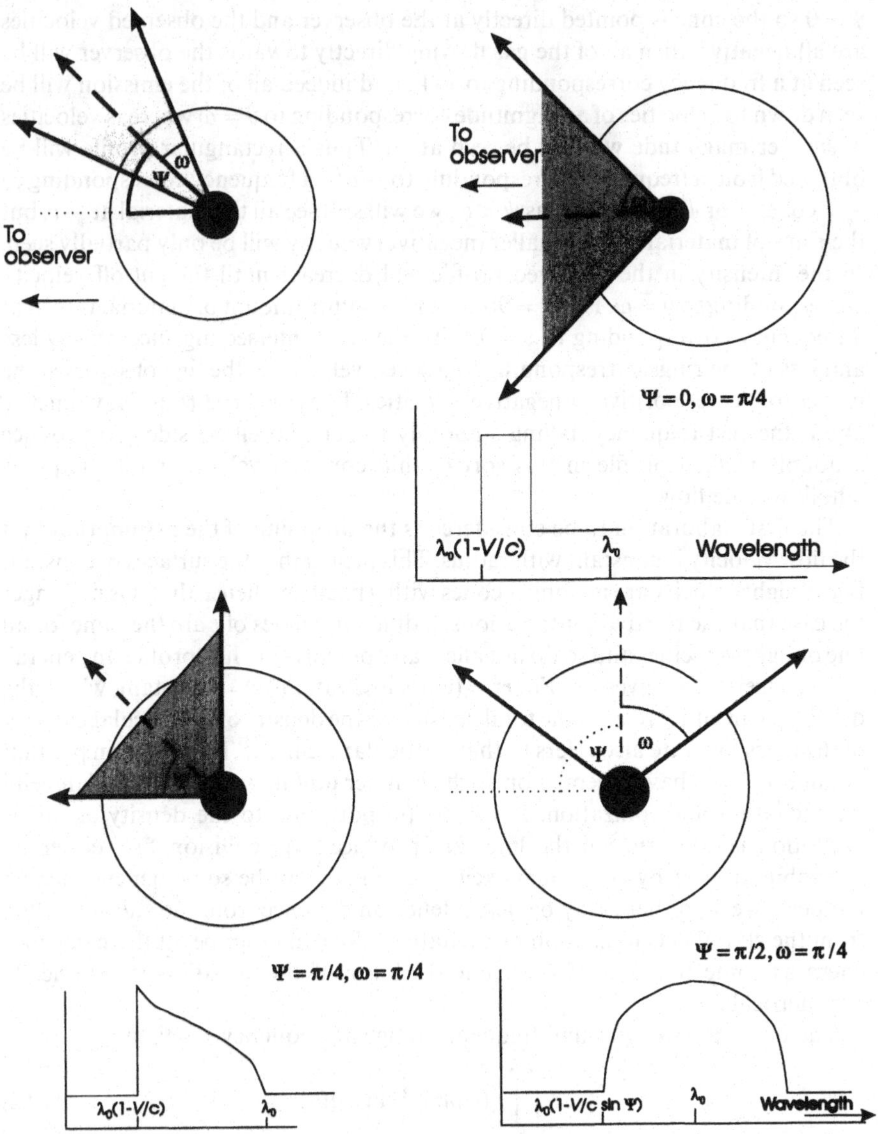

Figure 10.6. Emission from cone

10.6). This problem can also be dealt with by finding the (angular) length of the arc of the emitting ring which is intersected by the flow cone. This can be calculated in a similar way to that used above, but the trigonometry is more messy, and in general the answer has to be calculated numerically. It can however be seen that if

$\psi = 0$ so the cone is pointed directly at the observer and the observed velocities are all negative, then all of the gas flowing directly towards the observer will be seen at a frequency corresponding to $-V$, and indeed all of the emission will be seen down to velocities of a magnitude corresponding to $\theta = \omega$ whereas velocities of smaller magnitude will not be seen at all. Thus a rectangular profile will be obtained from a frequency corresponding to $-v$ to a frequency corresponding to $-V\cos\omega$. For $\psi > 0$, as long as $\psi < \omega$, we will still see all the material at $-v$, but the rings of material with a smaller (negative) velocity will be only partially seen, so the intensity in the observed profile will decrease until the cut-off velocity corresponding to $\theta = \omega$. For $\psi = 90°$, the maximum amount of material is seen at a frequency corresponding to $v = 0$, with the cone intersecting successively less and less of the rings corresponding to a given velocity in the line of sight as one moves to greater positive or negative velocities. Thus the line profile is symmetric about the rest frequency, falling smoothly to zero on either side. To produce a double peaked profile in this sort of thin constant velocity model requires a hollow cone flow.

The first elaboration to be considered is the dropping of the assumption that the flow velocity is constant with radius. This means that the surfaces of constant line of sight velocity are no longer cones with θ fixed and hence that it is no longer the case that the relative contributions of different values of r are the same for all line of sight velocities and so do not affect an optically thin line profile. In general, mass conservation gives $\rho(r)V(r)r^2 = (\text{mass loss rate})/(4\pi) =$ constant, where the density ρ can either refer to the total density, or the density of a particular element (if the mass loss rate also refers to that particular element). The emission per unit volume, $j_\nu\rho$, will have an emission coefficient per unit mass, j_ν, which will depend on excitation and ionization. It will be proportional to the density as far as excitation is concerned if the lines are produced via collisions, i.e. either by recombination or by collisional excitation. There will be some dependence on temperature, and there may be dependence on distance from the star if the flux from the star affects ionization or excitation. We can describe all these dependences as some function of r or indeed of $V(r)$ if the situation is spherically symmetrical.

The total emission per unit frequency range at frequency ν is then

$$2\pi \int\int j_\nu(r)\rho(r)r^2 \sin\theta \, d\theta \, dr \tag{10.2}$$

where the integral runs over the shell with the line of sight velocity giving frequency $\nu = \nu_0 + \Delta\nu$, with $\Delta\nu = \nu_0/c \, V(r)\cos\theta$ and $d(\Delta\nu) = \nu_0/c\cos\theta(dV/dr)dr$ at fixed θ. Hence

$$I_\nu = 2\pi \frac{c}{\nu_0} \int j_\nu(r)\rho(r)r^2 \frac{\sin\theta}{\dfrac{dV}{dr}\cos\theta} \, d\theta \tag{10.3}$$

where θ has a maximum value for a given $\Delta\nu$ in the line profile which can be determined by noting that the largest value of θ will be needed to reduce the line of sight component of the largest value of $V(r)$ to that appropriate for a Doppler shift $\Delta\nu$, so $\theta_{max} = \cos^{-1}(c/\nu_0 \Delta\nu / V_{max})$.

There will also be a minimum value of θ similarly corresponding to the minimum value of $V(r)$. The low frequency red wing of the line may have a θ_{min} depending on r because the star will occult material at small values of θ. The mass conservation formula above enables the density to be eliminated in terms of r and $v(r)$.

Chandrasekhar [3] takes $j_\nu \propto \rho V^\beta$ and discusses the cases of decelerating flows with $V = V_e\sqrt{(V_0^2/V_e^2 + r_0/r - 1)}$ with initial velocity V_0 at $r = r_0$ and accelerating flows with $V = V_t\sqrt{(1 - r_0/r)}$ with initial velocity zero at $r = r_0$, V_t being the terminal velocity in both cases. Substituting $r^2 \mathrm{d}V/\mathrm{d}r = \pm V_t^2 r_0/2V$, $V = c/\nu_0 \Delta\nu/\cos\theta$ and $\rho^2 \propto V^{-2} r^{-4}$ in

$$I_\nu \propto \int \rho^2 V^\beta \frac{r^2 \sin\theta}{\dfrac{\mathrm{d}V}{\mathrm{d}r} \cos\theta}\,\mathrm{d}\theta$$

$$\propto \int \frac{V^{\beta-2}}{r^2 \dfrac{\mathrm{d}V}{\mathrm{d}r}} \frac{\sin\theta}{\cos\theta}\,\mathrm{d}\theta$$

we obtain:

$$I_\nu \propto (\Delta\nu)^{\beta-1} [\sec^{\beta-1}\theta]_{\theta_{min}}^{\theta_{max}} \tag{10.4}$$

In the decelerating case $\theta_{max} = \cos^{-1}(c/\nu_0 \Delta\nu/V_0)$ and $\theta_{min} = \cos^{-1}(c/\nu_0 \Delta\nu/V_t)$ if we are observing escaping material and the velocity never falls to zero. Then

$$I_\nu \propto V_0^{\beta-1} - V_t^{\beta-1}$$

and is independent of $\Delta\nu$ so a flat-topped profile is obtained. In the accelerating case:

$$\theta_{max} = \cos^{-1}(c/\nu_0 \Delta\nu/V_t) \quad \text{and} \quad \theta_{min} = 0$$

but in practice there may be an inner bound to the radiating part of the flow at some radius r_1 and velocity V_1. If $\Delta\nu < \nu_0/c\, V_1$, then $\theta_{min} = \cos^{-1}(c/\nu \Delta\nu/ V_1)$ and $I_\nu \propto V_t^{\beta-1} - V_1^{\beta-1}$, which is independent of $\Delta\nu$ and therefore flat-topped, but for $\Delta\nu > \nu_0/c\, V_1$, $\theta_{min} = 0$, so

$$I_\nu \propto (\nu_0/c V_t)^{\beta-1} - (\Delta\nu)^{\beta-1}$$

which falls gradually with increasing distance from the line centre until it reaches zero intensity at a shift corresponding to the terminal velocity.

The next stage in elaborating the wind model is to allow for optical thickness in the line. This means evaluating the effect of absorption along the line of sight (it

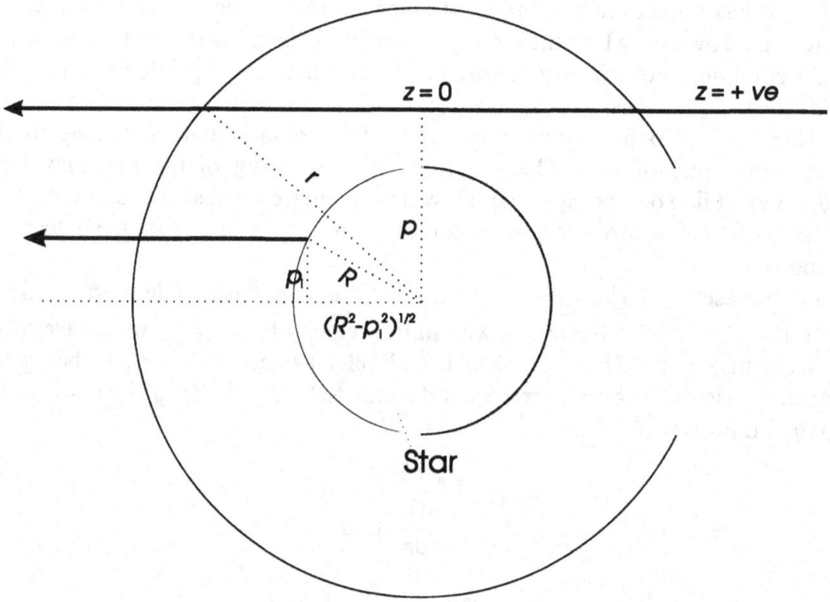

Figure 10.7. Emission geometry including the effects of the finite size of the star

will be recalled that z measures the distance along the line of sight). Here we adopt the convention, following Castor [4] that z is positive in the direction away from the observer (Figure 10.7) and $\mu = \cos\theta = -z/r$. The optical depth through the wind to the point z is given by

$$\tau_\nu = \int_{-\infty}^{z} \kappa_\nu(p, z')\rho \, dz' \tag{10.5}$$

where

$$\kappa_\nu = \kappa_0(r)\,\phi\left(\nu - \frac{\nu_0}{c}V(r)\mu - \nu_0\right) \tag{10.6}$$

since material moving towards the observer ($\mu > 0$) will appear blue shifted to higher frequencies and hence observations at a particular frequency will originate at a lower frequency in the rest frame of the emitting and absorbing material.

Writing $\Delta\nu' = \nu - \nu_0/cv(r')\mu - \nu_0$:

$$\tau_\nu(z) = \int_{-\infty}^{\Delta\nu(z)} \kappa_0(r')\rho(r')\frac{\phi(\Delta\nu')}{\left(\dfrac{\partial(\Delta\nu')}{\partial z'}\right)_p} d(\Delta\nu') \tag{10.7}$$

where r' is defined by

$$c[\Delta\nu' - (\nu - \nu_0)]/\nu_0 = V(r')\mu = V(r')\sqrt{(1 - p^2/r'^2)} \tag{10.8}$$

and $\Delta\nu'(z)$ by

$$\Delta\nu'(z) = \nu - \nu_0 + \nu_0/cV(\sqrt{(p^2+z^2)})z/r$$

Now

$$\left(\frac{\partial(\mu V)}{\partial z}\right)_p = V\left(\frac{\partial\mu}{\partial z}\right)_p + \mu\frac{\mathrm{d}V}{\mathrm{d}r}\left(\frac{\partial r}{\partial z}\right)_p$$

$$= -\left[\frac{V}{r}(1-\mu^2) + \mu^2\frac{\mathrm{d}V}{\mathrm{d}r}\right] \tag{10.9}$$

since

$$\left(\frac{\partial\mu}{\partial z}\right)_p = -\frac{1-\mu^2}{r}$$

and

$$\left(\frac{\partial r}{\partial z}\right)_p = \frac{z}{r}$$

using $r^2 = p^2 + z^2$ and $z = -\mu r$. Hence

$$\tau_\nu(z) = \int_{-\infty}^{\Delta\nu'(z)} \frac{\kappa_0(r')\rho(r')\phi(\Delta\nu')}{\dfrac{\nu_0}{c}\dfrac{V(r')}{r'}\left[1+\mu^2\left(\dfrac{r'}{V(r')}\dfrac{\mathrm{d}V(r')}{\mathrm{d}r'}-1\right)\right]}\mathrm{d}(\Delta\nu') \tag{10.10}$$

Then

$$\text{emergent intensity } I_\nu(p) = \int_{-\infty}^{\infty} S_\nu(z)\mathrm{e}^{-\tau(z)}\mathrm{d}\tau(z) \tag{10.11}$$

which should give an exact solution if the source function $S(r)$ is known.

The Sobolev approximation lies in assuming that the intrinsic width of the line is much less than $V(r)$ so that along a given line of sight the line originates over a relatively small range of z around z_m, which is given by $c(\nu-\nu_0)/\nu_0 = -z_\mathrm{m}/r_\mathrm{m}V(r_\mathrm{m})$ with $r_\mathrm{m} = \sqrt{(z_\mathrm{m}^2 + p^2)}$. Then $\kappa_0(z)$ may be treated as constant over this small range in z and be put equal to $\kappa_0(z_\mathrm{m})$ and removed from the integral for τ. Similarly $\delta(V\mu)/\partial z$ can be evaluated for $z = z_\mathrm{m}$ and removed from the integral. Finally the source function can be taken as having its value at $r = r_\mathrm{m}$ and removed from the integral for the emergent intensity.

Hence

$$\tau_\nu(z,p) = \frac{\kappa_0(r_\mathrm{m})\rho(r_\mathrm{m})}{\dfrac{\nu_0}{c}\dfrac{vV(r_\mathrm{m})}{r_\mathrm{m}}\left[1+\mu^2\left(\dfrac{r_\mathrm{m}}{V(r_\mathrm{m})}\dfrac{\mathrm{d}V(r_\mathrm{m})}{\mathrm{d}r}-1\right)\right]}\int_{-\infty}^{\Delta\nu'(z)}\phi(\Delta\nu')\mathrm{d}(\Delta\nu') \tag{10.12}$$

$$\tau_\nu(\infty,p) = \frac{\kappa_0\rho}{\dfrac{\nu_0}{c}\dfrac{V}{r}\left[1+\mu^2\left(\dfrac{r}{V}\dfrac{\mathrm{d}V}{\mathrm{d}r}-1\right)\right]} = \frac{\tau_0}{1+\mu^2\left(\dfrac{r}{V}\dfrac{\mathrm{d}V}{\mathrm{d}r}-1\right)} \tag{10.13}$$

$$\tau_\nu(z,p) = \tau_\nu(\infty,p)\Phi(z) \tag{10.14}$$

where

$$\Phi(\Delta\nu'(z)) = \int_{-\infty}^{\Delta\nu'} \phi(\Delta\nu')\mathrm{d}(\Delta\nu')$$

where all quantities are evaluated at r_m and we have used $\Phi(\infty) = 1$ (the line profile is normalized).

A similar answer can be obtained by arguing that along a given line of sight the line will be formed between z_1 and z_2 where $[V(z_1)\mu(z_1) - V(z_2)\mu(z_2)]\nu_0/c =$ frequency width of line $= 2\Delta\nu_D$ say, where we approximate the intrinsic line profile by a rectangular shape with width $2\Delta\nu_D$ and absorption coefficient κ_D within the profile ($2\kappa_D\Delta\nu_D = \kappa_0$). Then $\mathrm{d}(V\mu)/\mathrm{d}z\ (z_2 - z_1) = c/\nu_0 2\Delta\nu_D$ and $\tau = \kappa_D\rho(z_2 - z_1)$ gives essentially the same result as above.

The emergent intensity is then given by

$$I_\nu(p) = S_\nu(r_m)(1 - \mathrm{e}^{-\tau_\nu(\infty,p)}) \tag{10.15}$$

The luminosity of the wind at frequency ν is given by $L_\nu = 4\pi \int I_\nu(p) 2\pi p\,\mathrm{d}p$ so the flux received at distance d is $F_\nu = 1/d^2 \int I_\nu(p) 2\pi p\,\mathrm{d}p$. A particularly simple case occurs where $V =$ constant, so $\theta =$ constant in the integral over p, and since $p = r\sin\theta$, $\mathrm{d}p = \mathrm{d}r\sin\theta$. Furthermore $\mathrm{d}(\mu v)/\mathrm{d}z = v/r\sin^2\theta$. Hence

$$F_\nu = \frac{2\pi}{d^2}\sin^2\theta \int S_\nu[1 - \mathrm{e}^{-c\kappa_0\rho r/\nu_0 V\sin^2\theta}]\, r\,\mathrm{d}r$$

If the wind is optically thin, then

$$F_\nu = \frac{2\pi}{d^2}\sin^2\theta \int S_\nu(r_m)\frac{c\kappa_0\rho}{\nu_0 V\sin^2\theta} r^2\,\mathrm{d}r$$

$$= \frac{2\pi}{\nu_0 V d^2}\int S\kappa_0\rho r^2\,\mathrm{d}r$$

which is independent of frequency since S hardly varies over a line profile. Hence the observed line will have a rectangular profile, as we deduced earlier.

If, on the other hand, the wind is optically thick, then

$$F_\nu = \frac{2\pi}{d^2}\sin^2\theta \int Sr\,\mathrm{d}r$$

$$\propto \sin^2\theta \propto \left[1 - \frac{c^2}{V^2}\left(\frac{\nu - \nu_0}{\nu_0}\right)^2\right] \tag{10.16}$$

which gives a parabolic-shaped profile.

We have so far modelled the emission from the wind, ignoring the presence of the central star, which has two effects. Firstly, the portion of the envelope behind the star is occulted. Secondly, the star emits a continuum of intensity I_c, and the portion of the envelope in front of the star will absorb this continuum. Equation

(10.15) still holds for the intensity at projected distance p from the centre, but only for $p > R$, the radius of the star. For smaller p, (10.11) must be modified to

$$\text{emergent intensity } I_\nu(p) = \int_{-\infty}^{-\sqrt{R^2-p^2}} S_\nu(z) e^{-\tau(z)} \, d\tau(z) + I_c e^{-\tau(-\sqrt{R^2-p^2})}$$

and using the Sobolev approximation, we find for $p < R$:

$$I_\nu(p < R) = S(r_m)[1 - e^{-\tau_\nu(\infty,p)\Phi(Z_{min})}] + I_c e^{-\tau_\nu(\infty,p)\Phi(Z_{min})},$$

$$\text{where } Z_{min} = -\sqrt{R^2 - p^2} \tag{10.17}$$

$$I_\nu(p > R) = S(r_m)[1 - e^{\tau_\nu(\infty,p)}]$$

$$F_\nu = \frac{2\pi}{d^2}\left[\int_0^R I(p < R)\, p dp + \int_R^\infty I(p > R)\, p dp\right] \tag{10.18}$$

where of course the absorbed continuum only occurs at blue shifts.

Finally one needs to calculate the source function, which is very unlikely to be given by its LTE value in an envelope. For a two-level atom, $S = (1 - \varepsilon)J + \varepsilon B$ for the Planck function B and mean intensity J, where ε measures the relative importance of collisional and radiative de-excitations with rates C_{ji} and A_{ji}. The mean intensity (ignoring the presence of continuum radiation from the star) can be written as $(1 - \beta)S$ according to Chapter 4, where β is the escape probability. The continuum radiation from the star will also contribute to the mean intensity, diluted by distance and partially absorbed. This will give a contribution to the mean intensity proportional to the continuum intensity at the surface of the star I_c, say γI_c. Thus $S = (1 - \varepsilon)((1 - \beta)S + \gamma I_c) + \varepsilon B$, or

$$S = \frac{(1 - \varepsilon)\gamma I_c + \varepsilon B}{(1 - \varepsilon)\beta + \varepsilon} \tag{10.19}$$

The escape probability along a particular line of sight in the Sobolev approximation is given by (4.38) of Chapter 4 as

$$p_e = \frac{1 - e^{-\tau_s}}{\tau_s}$$

where

$$\tau_s = \frac{\kappa_0 \rho c}{\nu_0} \frac{1}{\dfrac{\partial(v\mu)}{\partial z}}$$

The escape probability is then given by averaging this expression over angle:

$$\beta = \frac{1}{2}\int_{-1}^{+1} \frac{1 - e^{-\tau_s}}{\tau_s} \, d\mu \tag{10.20}$$

In the particular case that $V = Ar$, τ_s is independent of μ and $\beta = [1 - \exp(-\tau_s)]/\tau_s$.

The mean intensity due to the stellar continuum at some point in the wind is found by integrating I_c times the probability of transmission of stellar continuum to that point over the range of values of μ subtended by the star at that point, namely from $\mu = \sqrt{(1 - R^2/r^2)}$ to $\mu = 1$. Absorption will happen in the vicinity of the point because of the velocity gradient and so the transmission will be the same as the escape probability in the particular direction involved. Hence

$$\gamma = \frac{1}{2}\int_{\sqrt{1-R^2/r^2}}^{1} \frac{1 - e^{-\tau_s}}{\tau_s}\, d\mu \tag{10.21}$$

Far from the star a small range of angles is involved and if the escape probability is independent of angle, we can write $\gamma \approx \beta W$, where W is the dilution factor $(1/2)(1 - \sqrt{(1 - R^2/r^2)}]$. Hence

$$S \simeq \frac{(1-\varepsilon)\beta W I_c + \varepsilon B}{(1-\varepsilon)\beta + \varepsilon} \tag{10.19a}$$

ε is usually small, so $S \approx (\beta W I_c + \varepsilon B)/(\beta + \varepsilon)$. At large optical depths (specified by τ_0—see (10.13), the escape probability becomes very small and so $S = B$. At very small optical depths $\beta = 1$ and since I_c and B are likely to be of comparable size, close to the star where $W \sim 1/2$, $S \approx WI_c$, and far from the star $S = \varepsilon B/\beta$. If the optical depth is fairly large, $\beta \sim 1/\tau_0$ and $S \sim \varepsilon B \tau_0$.

Now $F_c = \pi R^2/d^2 I_c$, so (10.18) finally gives

$$\begin{aligned} \frac{F_\nu - F_c}{F_c} &= \frac{1}{R^2}\int_0^\infty \frac{S(r_m)}{I_c}[1 - e^{-\tau(\infty,p)}]\,p dp \\ &- \frac{1}{R^2}\int_0^R \frac{S(r_m)}{I_c}[e^{-\tau(\infty,p)\Phi(Z_{min})} - e^{-\tau(\infty,p)}]\,p dp \\ &- \frac{1}{R^2}\int_0^R [1 - e^{-\tau(\infty,p)\Phi(Z_{min})}]\,p dp \end{aligned} \tag{10.22}$$

The first integral can be transformed into an integral over r using $p^2 = r^2 - z^2 = r^2 - (\Delta\nu/\nu_0 \cdot c/V)^2 r^2$ since the line at frequency $\Delta\nu$ from the centre comes from around z given by $V\mu = -V \cdot z/r = \Delta\nu/\nu_0 \cdot c$. Here one must note that there will be a minimum value of r for a given position in the line profile for an accelerating flow corresponding to $V(r_{min}) = \Delta\nu/\nu_0 c$. Hence

$$\begin{aligned} 2p\,dp &= 2r\,dr[1 - (\Delta\nu/\nu_0 \cdot c/V)^2\{1 - r/V \cdot dV/dr\}] \\ &= 2r\,dr[1 + z^2/r^2\{r/V \cdot dV/dV/dr - 1\}] \\ &= 2r\,dr\,\tau_0/\tau_s(\infty, p) \end{aligned}$$

Note that for $V = \text{constant}$, $r/V \cdot dV/dr - 1 = -1$ so the first term of (10.22) for the optically thick case becomes $1/(I_c R^2)\int S(1 - z^2/r^2) r dr$ with $z/r = \Delta\nu/\nu_0 c/V(r)$. Each shell contributes an amount proportional to $[1 - (\Delta\nu/\nu_0 c/V)^2]$, giving a parabolic profile, as we have already seen more directly. Such a velocity law cannot hold for the whole of the flow, but a rapid acceleration followed by a velocity asymptotically tending to the terminal velocity, which would produce a similar profile, is quite plausible. As was seen before, optical thinness tends to produce a more flat-topped profile.

As for the absorption edge, $z/r \sim 1$ for gas in front of the star so for a constant velocity flow $\tau = \tau_0/(1 - z^2/r^2)$ which can be much greater than τ_0, and the absorption edge can be deep even for the optically thin case. In the optically thick case, the prominence of the absorption component relative to the emission component increases as τ_0 is increased and ε is decreased.

In some cases (including Wolf–Rayet stars) it appears that the turbulence velocity is not negligible compared with the flow velocity. In the Sobolev Exact Integral approach, the source function is calculated using the Sobolev approximation (10.19a), but the emergent intensity is calculated exactly using (10.11) and analogous expressions for the occulted gas and the gas in front of the star (see, for example, Lamers *et al.* [5]). An exact solution for the whole problem can be obtained by working in the comoving frame as suggested by Mihalas *et al.* [6]. It turns out that the exact answers are close to those found from the Sobolev exact integral method in most cases. Problems can arise with the latter method because there is an implicit assumption that the situation can be divided into a 'core' and a 'halo', that is that we can talk about a photosphere illuminating an envelope in which the wind is blowing, and that one can meaningfully talk about the photospheric intensity I_c. In the case of Wolf–Rayet stars there is no photosphere—the spectrum that we see is entirely due to the wind. In such cases the comoving frame approach should be used.

Mass Loss from Hot Stars

The mass loss rate from hot stars is usually determined from the strength of the absorption component of the P-Cygni profile of ultraviolet lines (or from the whole line profile), or from the free–free emission from the wind in the infrared or radio regions. In the latter case the wind velocity has to be determined from the short wavelength side of the absorption component of a line profile.

In the spectra of O stars and OB supergiants, strong low excitation potential absorption lines show modest blue shifts indicating velocities of the order of $25\,\mathrm{km\,s^{-1}}$ in layers roughly 0.5 stellar radii above the continuum-forming photosphere. Hα is in emission in OB supergiants and Hα and He II 468.6 nm are seen in emission in Of stars, which are O stars with strong winds. The Hα line sometimes has a P-Cygni profile whose violet absorption edge indicates velocities of the order of $300\,\mathrm{km\,s^{-1}}$ and is formed perhaps 3 stellar radii above the photosphere. This velocity is still below the escape velocity at this distance from the centre of a massive star, and one has to observe material still farther out moving at the terminal velocity of the wind to be sure that the wind material is being lost. Farther out from the star the photoionizing flux from the star will be diluted, but the density and the accompanying recombination rate will have fallen even more, so the degree of ionization will increase. On the other hand the excitation may be fairly low. The upshot is that resonance lines of several times ionized species would appear to be the most likely spectral features to originate in the outer parts of the wind. Such lines occur in the ultraviolet. Ultraviolet observations of C IV, Si IV and N V lines show P-Cygni profiles with violet edges indicating velocities in the range 1500 to 3000 km s^{-1}, which are clearly greater than the escape velocity, unequivocally demonstrating mass loss. In the case of Wolf–Rayet stars, all the lines are very broad, although in some cases the emission is dominant and the violet absorption component is weak. The velocities indicated are of the order of 3000 km s^{-1}, and it is not necessary to use ultraviolet lines to see material moving at the terminal velocity in these very thick winds. Turbulence makes the definition of the violet edge difficult in Wolf-Rayet lines, and there are advantages in using infrared lines.

One of the first analyses of ultraviolet lines from a stellar wind was that performed by Morton [7] on observations of 3 O supergiants. Hydrogen and helium were assumed to be fully ionized throughout the wind, so that with a normal hydrogen to helium abundance of 9:1 and with two electrons per helium and one per hydrogen, $N_e = 1.22\, N_H$ for a total hydrogen number density of N_H. The mass density is almost entirely made up of hydrogen and helium, so

$$\rho(r) = 1.44\, m_H N_H(r) = 1.18\, m_H N_e(r)$$

The mass loss rate is then given by

$$dM/dt = -4\pi r^2 \rho(r)\, V(r) = -4\pi r^2 N_e(r)\, 1.18\, m_H V(r) \qquad (10.23)$$

The observations enable the column density $n_{col}(X)$ of some ion X to be found. Suppose the number density of the ion is $N_{ion}(r)$, the number density of the corresponding element is $N_{TOT}(r)$, and the abundance of the element with respect to hydrogen is N_{TOT}/N_H. Then

$$N_{ion} = N_{ion}/N_{TOT} \cdot N_{TOT}/N_H \cdot N_H = N_{ion}/N_{TOT} \cdot N_{TOT}/N_H \cdot N_e/1.2$$

$$n_{col} = \int_R^\infty N_{ion}(r)\, dr = \frac{N_{TOT}}{1.2 N_H} \int_R^\infty \frac{N_{ion}}{N_{TOT}}(r) N_e(r)\, dr \qquad (10.24)$$

where R is the inner boundary of the wind for the ion concerned.

Now it has already been shown in the Stromgren approximation for photoionization (2.48) that the ratio of the number density in a higher stage of ionization k to that in a lower stage of ionization i is

$$\frac{N_k}{N_i} = \frac{R_*^2}{4r^2} \frac{1}{N_e(r)} K_{ki} T_e^{1/2} T_R\, e^{-I/kT_R}$$

where

$$K_{ki} = \left(\frac{2\pi mk}{h^2}\right)^{3/2} \frac{2U_k}{U_i}$$

T_e is the electron kinetic temperature, T_R is the 'radiation' temperature characterizing the photoionizing flux from the star, and R_* is the radius of the star. For an ion where most of the element is in the next lower stage of ionization to that observed, so 'ion' = k and 'total' = i:

$$n_{\rm col} = \frac{N_{\rm TOT}}{4.8N_{\rm H}} R_*^2 K_{ki} T_e^{1/2} T_R e^{-I/kT_R} \int_R^\infty \frac{dr}{r^2}$$

$$= \frac{N_{\rm TOT}}{4.8N_{\rm H}} K_{ki} R T_e^{1/2} T_R e^{-I/kT_R} \qquad (10.25)$$

if $R_* = R$, that is if the region containing the relevant stage of ionization extends almost to the surface of the star. Morton assumed R, and calculated T_R, which is determined by the difficult to observe short wavelength flux.

On the other hand, for an ion where most of the element is in the next higher stage of ionization to that observed, 'ion' = i and 'total' = k, and

$$n_{\rm col} = \frac{N_{\rm TOT}}{1.2N_{\rm H}} \frac{1}{K_{ki}} \frac{4e^{I/kT_R}}{R_*^2 T_e^{1/2} T_R} \int_R^\infty r^2 N_e^2(r)\,dr \qquad (10.26)$$

Morton assumed a constant velocity flow so $\rho(r)r^2 =$ constant and since hydrogen and helium make up most of the density and provide most of the electrons, $N_e(r)r^2 = \text{constant} = N_e(R)R^2$. Hence

$$n_{\rm col} = \frac{N_{\rm TOT}}{1.2N_{\rm H}} \frac{1}{K_{ki}} \frac{4e^{I/kT_R}}{R_*^2 T_e^{1/2} T_R} N_e(R)^2 R^4 \int_R^\infty \frac{dr}{r^2}$$

$$= \frac{4N_{\rm TOT}}{1.2N_{\rm H}} \frac{1}{K_{ki}} \frac{e^{I/kT_R}}{T_e^{1/2} T_R} N_e(R)^2 R \qquad (10.27)$$

which, if T_R has been determined from (10.25), enables $N_e(R)$ and hence dM/dt (from (10.23) to be found, V being known from the Doppler shift of the violet edge.

The ultraviolet spectra used by Morton showed strong resonance lines of C IV (154.9 nm doublet), Si IV (140.3 nm doublet) and N V (124.0 nm doublet), all of which have similar electronic structures. The C IV absorption components were very strong and 100% deep, suggesting a very small source function, or in curve of growth terms, $R_C = 1.0$. The N V absorption components were 30% deep and hence weak in curve of growth terms, while the Si IV lines were rather deeper and hence on the 'bend' of the curve of growth.

The equivalent widths of the absorption components of the P-Cygni line profiles could then be interpreted using a Van der Held curve of growth (continuum source and uniform absorbing column). In the weak line limit applicable to N V:

$$W = (\pi e^2)/(4\pi\varepsilon_0 mc) f \lambda n_{col}(N\,V)$$

to give for these particular observations an estimate for the column density of N V ions of about 10^{19} m^2. The nitrogen is likely to be mainly in the form of NIV, so equation (10.25) can be used to find T_R if R can be estimated. The kinetic temperature was taken to have the typical value for a photoionized gas of 10^4 K, confirmed by the absence of lines with lower levels above the ground state, except for those of C III where the lower excitation energy is small. For $R_c = 1.0$, the central depth is $\exp(-\tau_c)$, where τ_c is the line centre optical depth, and the equivalent width of lines on the bend of the curve of growth is given by

$$W = \sqrt{\pi}\Delta\lambda_D \tau_C \left(1 - \frac{\tau_c}{2!\sqrt{2}} + \cdots\right)$$

where $\Delta\lambda_D$ is the Doppler width of the line. The central depth gives τ_c and the equivalent width then gives the Doppler width, so the column density of Si IV is determined by

$$\tau_c = \frac{\pi e^2}{4\pi\varepsilon_0 mc} \frac{f\lambda^2}{\sqrt{\pi}\Delta\lambda_D} n_c(\mathrm{Si\,IV})$$

Silicon should be mainly in the form of Si V so using (10.27), $N_e(R)$ can be determined and hence the mass loss rate found.

In this early work, Morton found mass loss rates for his supergiants of 1 to 2 times 10^{-6} solar masses per year, and shortly afterwards an even larger mass loss rate of $7*10^{-6}$ solar masses per year was found for the O5f star ξ Pup. Of course a number of simplifications were made in this analysis, including the assumption that V = constant, which cannot be true for all the flow, although it is quite plausible that V approaches the terminal velocity over much of the emitting region. It was also assumed that silicon is mainly in the form of Si V with a little Si IV all the way down to the surface of the star, whereas nitrogen is mainly in the form of N IV with a little in the form of N V all the way down to the surface of the star. Clearly, more realistic ionization models are called for, and indeed subsequent analyses have used models of the flow and information from the whole line profile. The ionization models present problems, as lines of O VI have been seen in the ultraviolet spectra of hot star winds, and such a high state of ionization was not predicted by simple models. Warm winds with kinetic temperatures of $T = 2*10^5$ K were suggested to produce O VI by collisional ionization, a thin corona at 10^6 K was postulated to produce O VI by photoionization (the whole wind could not be at such temperatures because this would produce X-rays at a level not observed), and a shock model seemed capable of fitting all the observations. However it now seems that more detailed calculations, without introducing any of the exotic circumstances postulated above, can produce strong O VI lines.

We now consider estimates of mass loss rate using free–free emission. When considering the spectrum of free–free emission, it must be remembered that the changing optical depth with wavelength means that different parts of the flow are 'seen' at different wavelengths. Assume a uniform source function, which will be blackbody since free–free is a collisional process, so $S_\nu = B_\nu(T_{ex})$, and consider a line of sight that passes a distance p from the centre of the star at closest approach.

Then the emergent intensity is

$$I_\nu(p) = B_\nu(T_{ex})(1 - e^{-\tau(p)})$$

$$\text{flux received} = \frac{1}{d^2}\int_0^q I\, 2\pi p\, dp \quad \text{at distance } d$$

$$= \frac{2\pi B_\nu(T_{ex})}{d^2}\int_0^q (1 - e^{-\tau(p)})p\, dp \qquad (10.28)$$

where q represents the outside radius of the emitting region.

For a typical temperature for an ionized gas of 10^4 K, the approximation $[1 - \exp(-h\nu/kT)] = h\nu/kT$ will be a good one for wavelengths of 10 μm and longer. The free–free absorption coefficient corrected for stimulated emission is then given by (2.22b):

$$\kappa_\nu \rho = 1.77 * 10^{-12}\frac{N_e(N_{ion}Z^2)g(\nu)}{T^{3/2}\nu^2}\, m^{-1}$$

where $g(\nu)$ is the Gaunt factor, a weak function of frequency and temperature, and Z is the net charge of the ion. For a pure hydrogen gas $N_e = N_{ion}$ and $Z = 1$, for a 90% hydrogen and 10% helium gas with the helium doubly ionized, $N_e = 1.1\, N_H$ so $\Sigma N_{ion}Z^2 = 1.4 * N_e/1.1$, and in general we can write $\Sigma N_{ion}Z^2 = f_i N_e$. It must be remembered that in the case of Wolf–Rayet stars, the wind will be hydrogen free but will have a high proportion of helium, and in some cases may have an appreciable percentage of carbon. We also have $dM/dt = 4\pi r^2 V \mu_e m_H N_e$, where μ_e is the mean molecular weight per electron = 1.3/1.1 for the 10% fully ionized helium case. Then

$$\kappa_\nu \rho = 1.77 * 10^{-12}\frac{f_i N_e^2 g(\nu)}{T^{3/2}\nu^2} = \kappa' N_e^2$$

$$\tau_\nu(p) = 2\kappa' \int_0^\infty N_e^2(z)\, dz$$

where $z^2 = r^2 - p^2$ for a distance z along the line of sight and a distance r from the centre of the system, and we have assumed that the flow extends to a very large distance from the star.

Substituting from the mass loss equation for N_e:

$$\tau_\nu(p) = \left(\frac{\frac{dM}{dt}}{4\pi m_H \mu_e} \right)^2 2\kappa' \int_0^\infty \frac{dz}{r^4 V^2} \tag{10.29}$$

The simplest case is a constant velocity flow when

$$\tau_\nu(p) = \left(\frac{\frac{dM}{dt}}{4\pi \mu_e m_H V} \right)^2 2\kappa' \int_0^\infty \frac{dz}{(z^2+p^2)^2}$$

$$= \left(\frac{\frac{dM}{dt}}{4\pi \mu_e m_H V} \right)^2 \frac{2\kappa'}{p^3} \int_0^\infty \frac{dx}{(1+x^2)^2}, \quad x = z/p$$

$$= \left(\frac{\frac{dM}{dt}}{4\pi \mu_e m_H V} \right)^2 \frac{2\kappa' \pi}{p^3\ 4} \tag{10.30}$$

since

$$\int \frac{dx}{(1+x^2)^2} = \tfrac{1}{2}[\theta + \sin\theta\cos\theta] \text{ with } \theta = \tan^{-1} x$$

Define t by $\tau = 1/t^3$, so

$$t = p \left(\frac{(4\pi\mu_e m_H V)^2}{\left(\frac{dM}{dt}\right)^2 2\kappa' \frac{\pi}{4}} \right)^{1/3}$$

and substituting in (10.28) for the flux, we obtain

$$\text{Flux} = \frac{2\pi B_\nu(T_e)}{d^2} \left(\frac{\pi\kappa'}{2} \right)^{2/3} \left(\frac{\frac{dM}{dt}}{4\pi\mu_e M_H V} \right)^{4/3} \int_0^\infty (1 - e^{-1/t^3}) t\, dt$$

where it has been assumed for the upper limit of the integral that the flow extends to large distances from the star. The value of the integral is 1.33. Using $B_\nu = 2kT/c^2 \cdot \nu^2$ at long wavelengths:

$$\text{flux} = \frac{4\pi k}{c^2} \left(\frac{\pi}{2} \right)^{2/3} \frac{(1.77 * 10^{-12})^{2/3}}{(4\pi m_H)^{4/3}} 1.333 \frac{1}{d^2} \left[\frac{\frac{dM}{dt}}{V} \right]^{4/3} \frac{f^{2/3}}{\mu_e^{4/3}} (g(\nu)\nu)^{2/3}$$

$$= 8.8 * 10^{-13} \frac{1}{d^2} \frac{f^{2/3}}{\mu_e^{4/3}} \left[\frac{\frac{dM}{dt}}{V} \right]^{4/3} (g(\nu)\nu)^{2/3} \tag{10.31}$$

If the mass loss rate is expressed in solar masses per year, the distance in kiloparsecs, the velocity in kilometres per seconds and the flux in Janskys, the mass loss rate is

$$dM/dt = 0.095\mu_e f^{-1/2} d^{3/2} v F_\nu^{3/4} (\nu g(\nu))^{-1/2} \qquad (10.32)$$

In the infrared region the frequency dependence of the Gaunt factor is small, so that the spectrum in flux per unit frequency units has a slope of 2/3. In the radio region the frequency dependence of the Gaunt factor has to be taken into account, giving a slope of 0.6 instead of 0.67. The optical depth increases with wavelength so that at long wavelengths one can only see down to the outer parts of the flow. Optically thick blackbody emission gives an intensity rising with ν^2, but at higher frequencies the emitting area is less (inner parts of flow) so the flux only rises as the two-thirds power of the frequency. There will be a frequency so high that the emission comes from the inner edge of the flow (in the limit, the surface of the star), and at higher frequencies the spectrum will flatten. Thus, in using (10.25) to find the mass loss rate, it must be ascertained that the slope of the spectrum has not fallen below 0.67 in the spectral region from which F_ν is measured. Of course some of the wind on the opposite side of the star to the observer is occulted by the star, and for small p the lower limit to the integral for the optical depth is not 0 but $\sqrt{(R^2 - p^2)}$, where R is the radius of the star, an effect that may be significant at high frequencies in the near infrared where much of the emission comes from the inner part of the wind. These points are discussed in Wright and Barlow [8].

Winds in hot stars are generally believed to be driven by radiation pressure. The subject of the driving force is beyond the area of interest of this book, but we include here a brief introduction. We can express the driving force either as a pressure gradient or as an acceleration:

$$\text{Acceleration} = g_{\text{rad}} = -\frac{1}{\rho}\frac{dP_{\text{rad}}}{dr}$$

$$= \text{rate of absorption of momentum per unit mass}$$

$$= \frac{4\pi}{\rho c}\int_0^\infty \kappa_\nu \rho H_\nu \, d\nu \qquad (10.33)$$

where H_ν is the (flux per unit frequency interval)/(4π) and the last line follows from equation (1.41). Deflection of radiation due to electron scattering is always present and produces a gradient of radiation pressure in the same way as absorption. The absorption coefficient for electron scattering is $\kappa_\nu = \sigma_T N'_e$, where σ_T is the Thomson cross-section for electron scattering (which is independent of frequency) and N'_e is the number of electrons per unit mass. N'_e can be written f''/m_H, where $f'' = 1$ for a pure hydrogen, completely ionized gas. In general $f'' = [1 + z'N(\text{He})/N(\text{H})]/[1 + 4N(\text{He})/N(\text{H})]$, where z' is the number of electrons released per helium ($z' = 2$ for completely doubly ionized helium). Thus $f'' = 0.85$ for a completely ionized 90% hydrogen, 10% helium mixture. Then the radiation pressure due to electron scattering is

$$g_{\text{rad}} = \frac{L}{4\pi r^2} \frac{\sigma_{\text{T}}}{c} \frac{f''}{m_{\text{H}}} \tag{10.34}$$

This acceleration is often quoted as a fraction of the gravitational acceleration ($g = GM/r^2$):

$$\frac{g_{\text{rad}}}{g} = \Gamma \frac{\sigma_{\text{T}} f''}{4\pi G c m_{\text{H}}} \frac{L}{M}$$

$$= 3 * 10^{-5} f'' \frac{L}{M} \quad L \text{ and } M \text{ in solar units} \tag{10.35}$$

However, electron scattering usually is not sufficient on its own to produce wind velocities of the magnitude observed fairly close to the star, and the acceleration due to line absorption must be added to that due to electron scattering. For simplicity, we take rectangular intrinsic line profiles of width $\Delta\nu_{\text{D}}$ corresponding to a thermal velocity V_{th}. Consider a shell of thickness dr, containing mass $4\pi r^2 \rho\, dr$. The range of velocities across the shell is $\Delta V = \mathrm{d}V/dr\, dr$ and the range of frequencies absorbed by a line of rest frequency ν (ignoring the intrinsic width for the moment) is

$$\Delta\nu = \Delta V \cdot \nu/c = \mathrm{d}V/\mathrm{d}r \cdot \nu/c\, \mathrm{d}r$$

Let the optical depth at a frequency in the line profile through the shell be τ, so the fraction of the stellar radiation absorbed at that frequency is $(1 - \exp(-\tau))$, and the fraction of the total luminosity L absorbed by that line in the shell is $[1 - \mathrm{e}^{-\tau}]\Delta\nu L_\nu/L$, where L_ν is the luminosity per unit frequency interval at that frequency. Hence the acceleration is

$$g_{\text{rad}}(\text{line}) = \frac{\frac{L}{c}\frac{L_\nu}{L}\Delta\nu(1 - \mathrm{e}^{-\tau})}{4\pi r^2 \rho \Delta r}$$

$$= \frac{L}{c^2}\frac{L_\nu}{L}\nu(1 - \mathrm{e}^{-\tau})\frac{1}{4\pi r^2 \rho}\frac{\mathrm{d}V}{\mathrm{d}r} \tag{10.36}$$

For strong lines the factor in parentheses is 1, which gives an acceleration:

$$g_{\text{rad}}(\text{line}) = \frac{L}{c^2 4\pi r^2 \rho}\frac{\mathrm{d}V}{\mathrm{d}r}\sum_{\text{strong}}\left(\frac{L_\nu \nu}{L}\right) \tag{10.37a}$$

For weak lines $(1 - \mathrm{e}^{-\tau}) \sim \tau$, and if N is the number of atoms per unit volume capable of absorbing the line, and we assume the line has an intrinsic width appropriate to the thermal velocity V_{th}:

$$\tau = \frac{\pi e^2}{4\pi\varepsilon_0 mc} \frac{f}{\Delta\nu_{\text{D}}} N\Delta r$$

$$= \frac{\pi e^2}{4\pi\varepsilon_0 mc} \frac{f}{\nu} c \frac{N}{\dfrac{dV}{dr}} = \frac{\pi e^2}{4\pi\varepsilon_0 mc} \frac{f}{\Delta\nu_{\text{D}}} N \frac{V_{\text{th}}}{\dfrac{dV}{dr}}$$

which on substitution gives a weak line acceleration:

$$g_{\text{rad}}(\text{line}) = \frac{L}{4\pi r^2 c^2} \frac{\pi e^2}{4\pi\varepsilon_0 m} \sum_{\text{weak}} f \frac{N}{\rho} \frac{L_\nu}{L} \tag{10.37b}$$

In general both weak and strong lines will contribute to the radiative acceleration. Following Castor *et al.* [9] and Abbott [10], define the ratio of line to electron scattering opacity as η, so

$$\eta = \frac{\pi e^2}{4\pi\varepsilon_0 mc} \frac{f}{\Delta\nu_{\text{D}}} \frac{N}{\rho} \frac{1}{\sigma_{\text{T}} N'_{\text{e}}}$$

Let $\tau = t\eta$ with

$$t = \frac{\sigma_{\text{T}} N'_{\text{e}} \rho V_{\text{th}}}{\dfrac{dV}{dr}}$$

Suppose the number of lines with values of the relative line absorption coefficient between η and $\eta + d\eta$ and frequencies between ν and $\nu + d\nu$ is $n'(\eta, \nu)\, d\nu\, d\eta$, and

$$dn = \int_0^\infty \frac{L_\nu \nu}{L} n'(\eta, \nu)\, d\nu\, d\eta$$

is a suitably frequency weighted quantity to represent the distribution of line strengths, then (10.36) becomes

$$g_{\text{rad}}(\text{line}) = \frac{L}{c^2} \frac{1}{4\pi r^2 \rho} \frac{dV}{dr} \int (1 - e^{-\tau(\eta)})\, dn(\eta) \tag{10.38}$$

Now it turns out that dN can be approximately represented by a power law:

$$dn = n_0(1 - \alpha)\eta^{\alpha - 2}\, d\eta \tag{10.39}$$

where n_0 represents the total 'number', and α is empirically determined by lists of line oscillator strengths to be about 0.7.

Inserting (10.39) in (10.38):

$$g_{\text{rad}}(\text{lines}) = \frac{L}{c^2}\frac{1}{4\pi r^2\rho}\frac{\mathrm{d}V}{\mathrm{d}r}n_0(1-\alpha)\int_0^{n_0}(1-\mathrm{e}^{-t\eta})\eta^{\alpha-2}\,\mathrm{d}\eta$$

$$= \frac{L}{c^2}\frac{n_0(1-\alpha)}{4\pi r^2\rho}\frac{\mathrm{d}V}{\mathrm{d}r}t^{-(\alpha-1)}\int\tau^{\alpha-2}(1-\mathrm{e}^{-\tau})\,\mathrm{d}\tau \text{ using } \tau = t\eta$$

$$= \frac{L}{c^2}\frac{V_{\text{th}}\sigma_{\text{T}}N'_{\text{e}}}{4\pi r^2}n_0(1-\alpha)t^{-\alpha}\int\tau^{\alpha-2}(1-\mathrm{e}^{-\tau})\,\mathrm{d}\tau$$

$$\simeq \frac{L}{c^2}\frac{1}{4\pi r^2}V_{\text{th}}\sigma_{\text{T}}N'_{\text{e}}n_0t^{-\alpha}\Gamma'(\alpha) \tag{10.40}$$

where, noting that t is small, the integral over τ has been extended from 0 to infinity, and after integration by parts twice, has been expressed as a gamma function, Γ' (not to be confused with the ratio of accelerations introduced earlier), which has a value of about 1.3 for $\alpha = 0.7$. In the papers by Abbott *et al.*, this result is expressed in terms of the force multiplier $M(t)$, which gives the ratio of the line acceleration to the acceleration due to electron scattering, so $M(t)$ is given by (10.40) divided by $\sigma_{\text{T}}N'_{\text{e}}(L/4\pi r^2c)$.

Mass Loss from Cool Stars: Evidence from Atomic Lines

The visible spectra of cool giants and supergiants show relatively strong violet absorption cores superimposed on the stronger photospheric resonance lines. Typical displacements correspond to a Doppler shift produced by movement relative to the star of 10 to 25 km s^{-1} towards the observer. The strongest violet displaced cores are always Ca II 393.3 and 396.8 nm, with Ca I 422.6 nm, the Na D 589.2 nm doublet and Sr II 407.7 nm fairly strong, but on high dispersion spectra many other lines, including those of Al I, Ti I and II, Cr I, Mn I, Fe I, Sc II and Ba II, can be observed. The violet displaced cores are seen in M giants, K and M stars of luminosity class II, and G, K and M supergiants, with the line strengths increasing as one goes to lower effective temperatures.

These violet displaced components were discovered in 1935 and tentatively attributed to expanding, circumstellar envelopes. The temperatures in envelopes surrounding cool stars would be expected to be modest and the excitation weak, so that little emission would be produced and only the violet absorption components of P-Cygni profiles would be seen. For the same reason, one would only expect to see lines from the ground state (resonance lines) or from low lying excited levels, and indeed the stronger observed lines are all resonance lines and only lines with lower excitation potentials of less than 1.5 eV are detected. Furthermore there will be no pressure broadening in such envelopes, leading to relatively sharp lines.

In one case the oscillations of the photosphere shown by a varying Doppler shift of the photospheric lines are not followed by corresponding variations in the Doppler shift of the violet displaced cores, and in several cases where the cool star is a member of a spectroscopic binary, the violet cores do not follow the Doppler shifts of photospheric lines produced by the cool star's orbital motion. These results show that the violet cores do not arise in the stellar atmospheres of the stars. On the other hand, the strengths of the displaced cores correlate with the surface temperature of the cool star so they cannot have an interstellar origin. Thus they must be *circumstellar lines*.

However, a velocity of a few tens of kilometres per second is less than the escape velocity at the surface of the stars concerned, so that it could not immediately be deduced that mass is being lost from the star. The crucial evidence was the observation of circumstellar lines in the spectra of *both* members of certain visual binaries, even though only one member of these binaries was cool enough for circumstellar lines to be expected. The conclusion has to be that the cool star circumstellar envelopes extend far enough out to engulf the other member of the binary system, so that cool star circumstellar absorption lines are seen in the spectrum of the hotter star. The classic example is α Her, composed of a G0 II–III star and an M5 star separated by 4.7″ in the sky, which corresponds to a distance of $1.1*10^{14}$ m. Circumstellar lines are never seen in the spectra of single giants as hot as G0, yet in this case strong circumstellar lines are seen in the spectrum of the G0 star, as well as in the spectrum of the M star. Now the distance apart of the G0 star and the M star must be at least as great as their projected separation in the sky, so the circumstellar envelope must extend out to at least $1.1*10^{14}$ m from the M star, at which distance 15 km s^{-1} certainly exceeds the escape velocity. In this case mass is certainly being lost, and as the strength of the circumstellar lines in the M star spectrum of α Her is similar to that found in single stars of the same spectral type and luminosity, it seems safe to assume that luminous cool star circumstellar lines are in general associated with mass loss.

Now the mass loss rate

$$\begin{aligned} dM/dt &\doteq 4\pi r^2 \rho(r) V(r) \\ &= 4\pi r^2 N_H(r) \mu m_H V(r) \end{aligned}$$

where N_H is the hydrogen number density and μ is the mean molecular weight.

If the fraction of element X in the stage of ionization i is f_{ion}, the abundance of element X relative to hydrogen is N_x/N_H, and the number density of ion i of element X is N_{Xi}:

$$\frac{dM}{dt} = 4\pi r^2 \frac{N_{Xi}(r)}{f_{ion}} \frac{N_H}{N_X} \mu m_H V(r) \tag{10.41}$$

Observations of circumstellar lines enable the column density of ion i of element

X, n_{Xi}, to be determined:

$$n_{Xi} = \int N_{Xi}(r)\,\mathrm{d}r$$

$$= \frac{N_{Xi}(r_{\min})r_{\min}^2 V(r_{\min})}{f_{\mathrm{ion}}(r_{\min})} \int_{r_{\min}}^{r_{\max}} \frac{f_{\mathrm{ion}}}{V(r)r^2}\,\mathrm{d}r \tag{10.42}$$

where $r_{\max}$ and $r_{\min}$ are the outer and inner boundaries of the region containing X_i and we have used $N_x(r)V(r)r^2 = \text{constant}$. If $V = \text{constant}$, then it may be reasonable to assume that f_i is roughly constant, at least between $r_{\min}$ and $r_{\max}$, since the degree of ionization will depend on the balance between photoionizations where the photoionizing flux is falling off as $1/r^2$ in the optically thin case, and recombinations which are proportional to the electron density, with the total density also falling of as $1/r^2$ in the constant velocity case. Then for

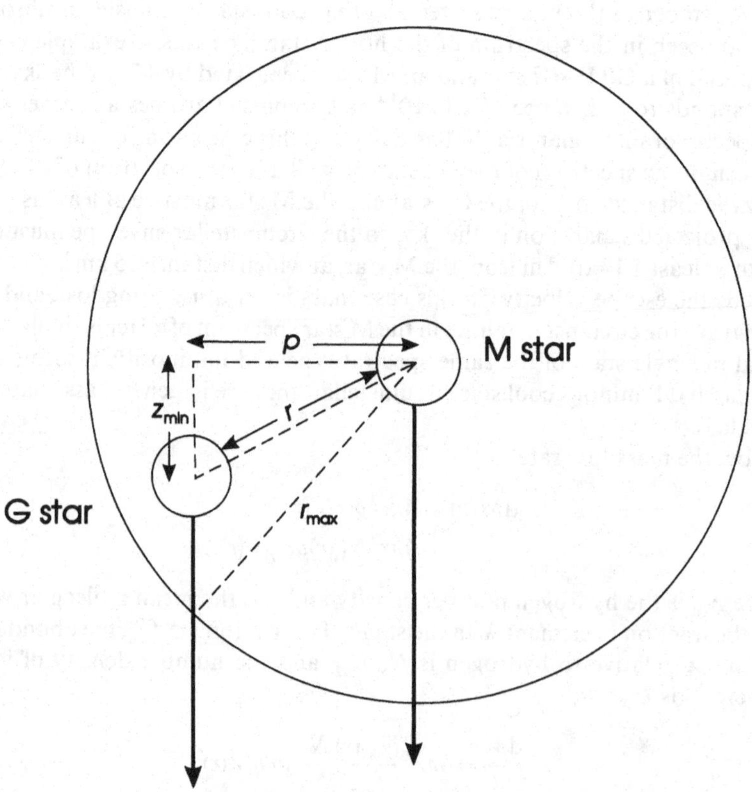

Figure 10.8. α Her system

$V = \text{constant}$:

$$n_{Xi} = N_{Xi}(r_{\min}) r_{\min}^2 \int_{r_{\min}}^{r_{\max}} \frac{1}{r^2} \mathrm{d}r$$

$$= N_{Xi}(r_{\min}) r_{\min}, \quad \text{if } r_{\max} > r_{\min} \qquad (10.43)$$

A major problem arises in fixing $r_{\min}$ for it is unlikely that the zone of constant ioniztion extends right down to the surface of the star, and the velocity must certainly drop close to the star.

It is interesting to consider briefly two of the first studies of mass loss from cool stars, that of Deutsch on the binary system α Her which contains an M giant, and that by Weymann of the single M supergiant α Ori. It must be remembered that circumstellar lines are usually superimposed on the same line originating in the photosphere, that there will also be weak emission from the circumstellar line, and that the Ca II photospheric lines have complicated profiles because of chromospheric emission and self-reversal near the line centre. Measuring accurately the strength of the circumstellar absorption component is not therefore always straightforward.

Deutsch's study of α Her (Deutsch [11]; see also Deutsch [12]), assumes $V = \text{constant}$ and also $f_{\text{ion}} = \text{constant}$, and avoids the problem of specifying $r_{\min}$ by using the strength of the Ca II H and K and the Ca I 422.6 nm circumstellar lines as seen in the spectrum of the G star. Suppose the G star is a distance p from the M star perpendicular to the line of sight (which can be calculated from the angular separation if the distance of the binary is known), and z represents distance along the line of sight measured form the point of closest approach to the M star. Let $z_{\min}$ be the z coordinate of the G star, positive if the G star is nearer to the observer than the M star (Figure 10.8).

Noting that $z^2 = r^2 - p^2$:

$$n_{Xi}(G) = N_{Xi}(p) p^2 \int_{z_{\min}}^{\infty} \frac{1}{r^2} \mathrm{d}z$$

$$= N_{Xi}(p) p^2 \int_{z_{\min}}^{\infty} \frac{1}{z^2 + p^2} \mathrm{d}z$$

$$= N_{Xi}(p) p \left[\frac{\pi}{2} - \tan^{-1}\left(\frac{z_{\min}}{p} \right) \right] \qquad (10.44)$$

If $z_{\min} = 0$, $n_{Xi}(p) = N_{Xi}(p) p\pi/2$, so

$$\frac{\mathrm{d}M}{\mathrm{d}t} = 8p \frac{n_{Xi}(G)}{f_{\text{ion}}} \left(\frac{N_{\text{H}}}{N_X} \right) \mu m_{\text{H}} V \qquad (10.45)$$

although, of course, we do not know the value of z_{min}. Deutsch writes:

$$\frac{dM}{dt} = 16pVn_{Xi}(G)\gamma\frac{N_H}{N_X}\mu m_H$$

where

$$\gamma = \frac{\frac{\pi}{4}}{\left[\tan^{-1}\left(\frac{p}{z_{min}}\right) - \sin^{-1}\left(\frac{p}{r_{max}}\right)\right]}$$

which can be derived by replacing infinity as the upper limit of the integral leading to (10.44) by z_{max} and noting that $\tan^{-1}(z_{max}/p) = \pi/2 - \sin^{-1}(p/r_{max})$ and $\tan^{-1}(z_{min}/p) = \pi/2 - \tan^{-1}(p/z_{min})$. Deutsch argues that although the value of γ is unknown, a value very different from 1/2 is unlikely. Since both Ca and Ca II are observed, the column densities can simply be added to give the total column density of calcium without calculating f_i explicitly, *provided* there is no second ionization of calcium.

A problem arises because the Ca II lines are clearly strongly saturated, as indicated by the near equality in equivalent width of the two circumstellar lines despite the fact that the oscillator strengths are in the ratio 2:1. It follows that a curve of growth is needed to derive a column density from these lines. An observational curve of growth cannot be constructed from just two lines so Deutsch assumed that the Ca II lines were Doppler broadened and that the Ca I line was unsaturated, the full width at half height of the latter line giving the Doppler width which could be used in deriving a column density from he Ca II lines ($R_c = 1.0$ seems a reasonable guess for a circumstellar line and the excitation is not needed for a ground state line). The column density derived from such saturated lines is likely to be very inacccurate, and Deutsch himself discusses the possibility that the Ca I line is beginning to become saturated, in which case the Doppler width will be overestimated which will have a strong effect on the column density estimates. There is also the question of second ionization of calcium, for calcium has a relatively low second ionization potential of 11.9 eV (compared with the 47 eV needed to produce Na III or the 66 eV needed to produce Fe III). Now cool stars are not expected to produce much ultraviolet flux, so in general second ionization can be ignored but Ca III is produced by radiation below 104.5 nm, and an appreciable flux at this wavelength could not be ruled out in the 1960s in the absence of observational data and in view of the likely departure of M star spectra from blackbody form and the possibility of variable chromospheric emission.

Weymann [13] used a high dispersion spectrum of the single M supergiant α Ori. One hundred circumstellar lines were identified, coming from all the species listed earlier, and column densities were estimated. In some cases, e.g. Fe I and Ti II, it was possible to draw a curve of growth, in others only one or two lines were observed. Now in the circumstellar shell continuous absorption is negligible and line thermal emission is likely to be very small, so it is reasonable to treat line formation as a pure scattering process, with a pure scattering curve of growth in which the source function is the mean intensity and the maximum line depth is 100% ($R_c = 1$). Weymann interpreted his observations using pure scattering curves of growth derived for stellar atmospheres but was hesitant about the correctness of applying these curves of growth to the situation of an expanding extended shell. Problems arise because if the angular resolution of the observations is sufficient to resolve the star from the shell, then when looking at the star light scattered out of the line of sight is lost, whereas if the shell is unresolved, scattered light is included, at least partially, in the spectrum as emission from the shell. In the stellar atmosphere case, in calculating the mean intensity it can be assumed that photons scattered back towards the star are absorbed and lost, whereas with a shell which may be many stellar radii in size, many back scattered photons will not strike the star at all. Furthermore, for an expanding shell line photons scattered by one part of the shell will have the wrong Doppler shift to be scattered by another part of the shell.

Column densities were calculated for 12 atoms and ions. In two cases, those of titanium and calcium, both the neutral and once ionized forms were observed, and on the assumption of negligible second ionization could be added together to give the total column densities of calcium and titanium. These quantities when multiplied by the abundances of hydrogen relative to titanium or calcium (taken to be the solar values) gave two estimates of the column density of hydrogen. There appeared to be 40 times as much Ti II as neutral titanium so that it seemed reasonable to assume that elements like scandium, barium, and strontium, with lower ionization potentials than titanium, were almost entirely in once ionized form, and hence to use the observed column densities of Sc II, Ba II and Sr II as representing the total column densities of those elements, which when multiplied by the solar abundances gave further estimates of the hydrogen column density. The average hydrogen column density thus obtained, n_H, then gave the mass loss rate as

$$dM/dt = -4\pi R^2 x_0^2 V m_H N_H(x_0) = -4\pi R x_0 v m_H n_H$$

for $x_0 = r_{min}/R$ and stellar radius R, if the flow is constant velocity and constant ionization from r_{min} to infinity. The main difficulty is in determining x_0.

Weymann used the fact that Ca II has a level $3d^2D_{3/2,5/2}$ about 1.7 eV above the ground state, which is metastable since the transition to the ground state is forbidden and which is too high to be populated appreciably by collisions at the likely temperature and density of the circumstellar shell. However, radiative population is possible at a rate depending on the diluted photospheric flux whose mean intensity will fall with distance from the star. The (near) infrared triplet 849.8 nm, 855.2 nm and 866.2 nm originates from this level (Figure 10.9). Weymann found no sign of a circumstellar core from the strongest member of this

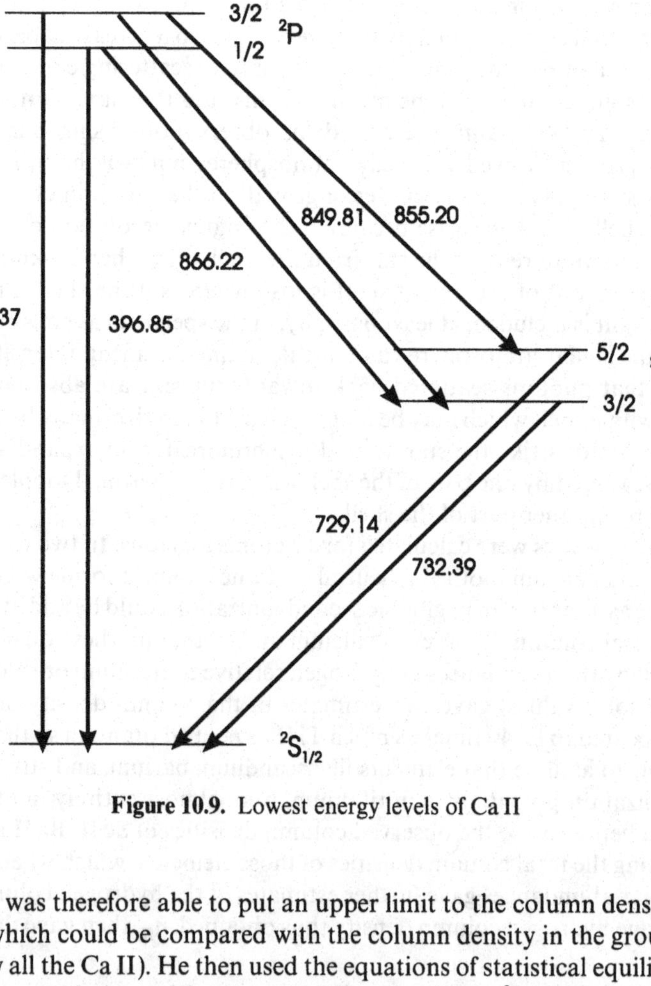

Figure 10.9. Lowest energy levels of Ca II

triplet and was therefore able to put an upper limit to the column density of the ^{2}D term, which could be compared with the column density in the ground state (essentially all the Ca II). He then used the equations of statistical equilibrium to solve for the populations of the three significant levels: the ^{2}S ground state, the ^{2}D first excited state and the ^{2}P second excited state, where ^{2}S and ^{2}D are joined by forbidden lines, ^{2}S and ^{2}P are joined by the very strong H and K lines, and ^{2}D and ^{2}P are joined by the optically thin infrared triplet. We can write:

$$n(S)[R_{SD}+R_{SP}]=n(D)R_{DS}+n(P)R_{PS} \quad \text{and} \quad n(\mathrm{D})[R_{DP}+R_{DS}]=n(\mathrm{S})R_{SD}+n(\mathrm{P})R_{PD}$$

since collisions can be neglected.

These solve to give

$$\frac{n(\mathrm{D})}{n(\mathrm{S})}=\frac{R_{SD}[R_{PD}+R_{PS}]+R_{SP}R_{PD}}{R_{DS}[R_{PS}+R_{PD}]+R_{PS}R_{DP}}$$

The upward rates from S to D and from D to P are just given by $4\pi J_\nu B_{ij} W$, where

J_ν is the mean intensity at the surface of the star at the appropriate frequency and W is the dilution factor, or replacing B_{ij} by A_{ji} and taking account of the fine-structure levels labelled ik for the lower state and jl for the upper state:

$$R_{ij} = \frac{c^2}{2h\nu_{ij}^3} J_{ij} W \sum_k \frac{g_{ik}}{g_i} \left[\sum_l \frac{g_{jl}}{g_{ik}} A_{jl,ik} \right]$$

where the fine-structure levels are assumed populated according to their statistical weights. In practice the S lower state has only one level so there is no sum over k for the SP and SD transitions. The corresponding downward rates are given by A_{ji} or again taking into account the fine structure we have a sum over l and k of $g_{jl}/g_j A_{jl,ik}$. The transition S to P is optically thick and therefore the radiation field is not the diluted photospheric intensity. A solution of the equation of radiative transfer must be used to find the rates, but Weymann argues that since the line is very optically thick the upward and downward rates in the P to S transition will come into detailed balance and the mean intensity in this line will come to equal the source function derived ignoring the scattering in the P to S transition itself, that is the source function produced by the optically thin transitions S to D and D to P which can be determined by balancing the upward and downward radiative transitions in these lines. In the end an expression for $n(\mathrm{D})/n(\mathrm{S})$ is found as a function of W, which in this case requires a dilution factor less than 10^{-3} or $x_0 > 14$.

It is also possible to derive a value of x_0 by comparing the strengths of the circumstellar lines in the spectra of the G star and the M star in α Her, where we know that the G strengths correspond to r greater than p, which suggests that $x_0 > 10$ for the M star. Finally the degree of ionization in the spectrum of the circumstellar shell of α Ori reflects the value of n_e/W, and if we suppose that we know the photoionizing radiation field and use the estimate of n_H found above, we can estimate a sort of average dilution factor for the various ions detected—this suggests a much smaller value for x_0. Weymann adopted $x_0 = 10$ and used this to find a mass loss rate of $4*10^{-6}$ M☉/year, but the large uncertainty in x_0 is reflected in a correspondingly large uncertainty in the mass loss rate.

An attempt to model the flow in more detail was made by Reimers [14]. The strengths of circumstellar lines in general increase with increasing luminosity and increase with decreasing effective temperature. In more detail, the Ca II H and K circumstellar lines in M giants increase in strength but decrease in velocity displacement as one goes to lower effective temperatures. In some cases the circumstellar lines are variable in strength and velocity displacement (in a way not correlated with oscillations of the star and so possibly due to a variation in chromospheric activity in the star) and in these cases the Ca II H and K velocity displacements vary inversely with the line strengths in a way that follows the relationship of velocity displacement and line strength for M stars of different effective temperatures. The chromospheric emission component of the Ca II K line (K_3) usually has a violet displacement compared with the photospheric absorption lines, but with a smaller displacement than the circumstellar lines. Finally, in the hotter M giants, the circumstellar lines other than those of Ca II have smaller velocity displacements than the Ca II H and K circumstellar cores, a difference which disappears or is even reversed for the cooler M stars.

To explain these observations, Reimers dropped the assumptions of constant velocity and no second ionization and developed models with an accelerating flow (Figure 10.10). The chromospheric displacement of a few kilometres per second represented the first stage in the acceleration. The region of double ionization of calcium falls in radius as one goes to lower effective temperatures of the central star, so that Ca II absorption comes from relatively large distances from the hotter M stars, distances at which the flow has reached its terminal velocity. On the other hand, in cooler M giants the Ca II region reaches farther down towards the star, giving stronger Ca II lines for the same mass loss rate, and

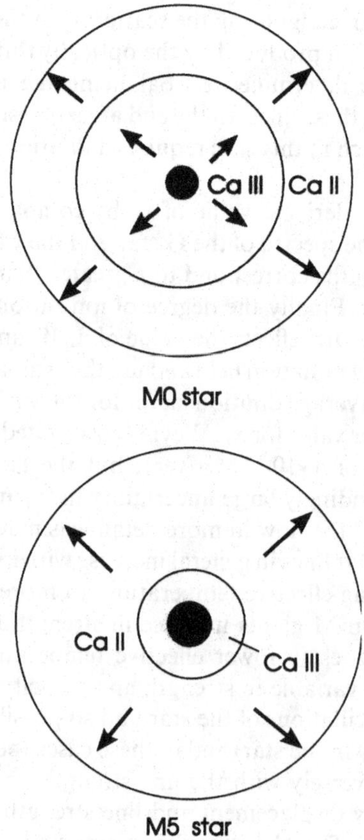

Figure 10.10. Effect of ionization

incorporating regions where the wind is still accelerating and hence giving a lower average velocity displacement, which will be close to that of other elements not affected by the problem of second ionization. Using these models with the observations, Reimers found that the mass loss rate increased with luminosity and decreased with increasing effective temperature in a way that could be summarized by

$$\frac{dM}{dt} = 4 \times 10^{-13} \eta \frac{L}{gR} \mathrm{M}_\odot \,\mathrm{year}^{-1} \qquad (10.46)$$

for stellar radius R, luminosity L and surface acceleration g, all in solar units. This gives a mass loss rate proportional to $L^{3/2}/(T_{\mathrm{eff}}^2 M)$. The calibration assumes that the Ca III zone shrinks to zero for the coolest M stars, and the parameter η allows for different calibrations. Although some aspects of Reimer's models have been disputed, in particular the empirical derivation of ionization structure from limited data, the formula is universally used in stellar evolution calculations.

Weymann (and Reimers) had used a plane-parallel atmosphere static curve of growth which essentially ignored the emission from the shell. Two investigations, those by Sanner [15] and Bernat [16], removed this limitation by using a proper expanding extended atmosphere analysis to predict line profiles to compare with observation. Both assumed a constant velocity flow, finding no significant velocity differences between the various lines used, and the models of both have no significant second ionization. The Ca II H and K lines are not used in these and subsequent investigations in view of the difficulties presented by their large optical depths and by chromospheric contributions. Sanner, for instance, divides his lines into three categories, those belonging to the dominant stage of ionization of an element for which the observed column density directly gives a total column density if an elemental abundance is assumed, the triplet lines from a metastable state of a dominant stage of ionization (the Ca II triplet) whose strength depends only on the dilution, and the trace lines belonging to a minority stage of ionization, which can give information on the degree of ionization. Examples of the dominant lines used are the resonance lines (corresponding to H and K of Ca II) of Sr II at 407.8 and 421.6 nm, and of Ba II at 455.4 and 493.4 nm which cover roughly an order of magnitude in strength, with the strongest line having an optical depth of 1 or less and line formation being predominantly via simple two-level scattering. The ionization potentials of Ba I and Ba II are 5.2 eV and 10.0 eV and the same quantities for Sr I and Sr II are 5.7 eV and 11.0 eV, so it would be expected that Sr II and Ba II are the dominant forms, the only problem being that deviations from solar abundances of these elements in red giants are possible through the convecting to the surface of the results of the s process. Examples of trace lines include the Na I D lines and the similar lines of K I at 766.5 and 769.9 nm.

The interpretation of the observations required that the circumstellar absorption and emission components and the photospheric profile are disentangled. For two-level scattering the total emission must equal the total absorption, so for observations in which the shell is unresolved, the emission component of the circumstellar line profile should equal (in equivalent width) the absorption component. The photospheric line should be symmetric, so in principle with high quality observations and a homogeneous shell at one velocity it should be possible to separate the components.

However there is a crucial difference between these two investigations in that Sanner claimed that he was able to make positive detections of the Ca II infrared triplet in a number of cases, whereas for α Ori Bernat put a much lower upper limit on the infrared triplet than Sanner's detection. Sanner, Bernat and Reimers find very different mass loss rates for the same star in some cases. A further complication is the discovery by Goldberg *et al.* [17], using high resolution observations, of the K I 769.9 nm line in α Ori that the shell is not homogeneous, but contains two components separated by about 7 km s^{-1}.

Further information about the structure of the shell has come from studies of the infrared CO vibrational–rotational lines, where the fundamental set of bands lie at around 4.6 μm, the first overtone bands at around 2.3 μm and the second overtone bands at around 1.6 μm. An enormous range of line strength and excitation energy is covered, enabling the shell structure to be probed. The circumstellar lines are seen superimposed on the very strong photospheric lines of CO. The first study of circumstellar absorption lines was that of Bernat *et al.* [18], who found in the 4.6 μm band the double velocity structure seen by Goldberg *et al.*, with the weaker, faster expanding group having a lower excitation temperature (70 K against 200 K), the 'temperature' being derived by comparing lines of different excitation energy and fitting a Boltzmann distribution to the relative column densities of the rotational levels found from the observations. The collisional cross-sections for excitation of the levels in the $v = 0$ ground vibrational level are large as the excitation energies are small, whereas the radiative transition probabilities are moderate, so that unlike the atomic circumstellar lines these CO lines are collisionally excited and hence the temperatures are kinetic temperatures.

An example of the complexity of the situation that can arise is given by the study of the Mira variable χ Cyg by Hinkle *et al.* [19]. This star differs very considerably from α Ori in being a cool giant and, more importantly, in being very strongly variable. Mira variables are believed to have pulsating photospheres in which shocks develop. Hinkle *et al.* used observations of a number of lines of different rotational quantum number and hence of different excitation energy in the fundamental set of bands at 4.6 μm, the first overtone set of bands at 2.3 μm, and the second overtone set of bands at 1.6 μms, thus sampling a wide variety of conditions in the photosphere and circumstellar envelope, and followed changes in the strengths and Doppler shifts of these lines with variation of the star. The lines often split into two or more components at different Doppler shifts. The second overtone lines are formed in the photosphere and show a large amplitude velocity curve (a range of $30\,km\,s^{-1}$) which shows discontinuous velocity jumps and excitation temperatures ranging from 2000 K to 4000 K, the latter being considerably higher than the effective temperature. These results can be interpreted in terms of an outwardly propagating shock wave. The first overtone lines come in part (the high excitation lines) from the photosphere and show the same properties as the second overtone lines, but also have a component exhibiting infall velocities at some phases, and a component characterized by an excitation temperature of 800 K and a constant velocity equal to the velocity of the star and so presumably represent an inner circumstellar non-expanding shell, perhaps at about 10 stellar radii from the star. Finally, the fundamental lines come from the circumstellar shell proper and appear to come from an expanding 300 K shell. It will be seen that the relation between the star and the expanding wind is complex, perhaps involving a stationary reservoir where grains form and are accelerated outwards.

It must be noted that the CO lines discussed here are *absorption lines* seen in the line of sight to the star, as opposed to the *emission lines* from the shell around the star, much more commonly observed but possibly coming from rather different parts of the circumstellar shell at lower temperatures.

More information on the structure of the circumstellar shell around α Ori has also become available from direct observations of the emission from the shell using scattered light from the K I 769.9 nm line. Descriptions can be found in Honeycutt *et al.* [20] and Mauron *et al.* [21] of observations out to 50–60″ away from the star. The observational problem is, of course, the removal of photospheric light scattered in the Earth's atmosphere into the apparent position of the shell. Suppose, following Bernat *et al.*, the number density of K I atoms is $n(r) = n_0(r/r_0)^\alpha$, where r_0 is the inner radius of the shell. Let the optical depth in K I in the line of sight to the star be τ_0. Then if σ is the scattering cross-section per atom:

$$\begin{aligned}\tau_0 &= \int \sigma n(r)\,dr \\ &= \int_{r_0}^{\infty} \sigma \frac{n_0}{r^\alpha} r_0^\alpha\,dr \\ &= -\left.\frac{\sigma n_0 r_0^\alpha}{(\alpha-1)r^{\alpha-1}}\right|_{r_0}^{\infty} \\ &= \frac{\sigma n_0}{\alpha-1} r_0 \qquad (10.47)\end{aligned}$$

Assume isotropic scattering and the by now familiar coordinate system with z along the line of sight and p the distance of closest approach of the line of sight to the star so $r = \sqrt{(z^2 + p^2)}$. Let the star have luminosity per unit frequency interval of L_ν. Then the intensity of the shell at p is

$$
\begin{aligned}
I_{\text{shell}}(p) &= \frac{1}{4\pi}\int \frac{\sigma n(r) L_\nu}{4\pi r^2}\,\mathrm{d}z \\
&= \frac{L_\nu}{16\pi^2}\sigma n_0 r_0^\alpha \int \frac{1}{r^{\alpha+2}}\,\mathrm{d}z \\
&= \frac{F_\nu}{4\pi}\mathrm{d}^2\sigma n_0 r_0^\alpha \int\left[\frac{1}{(p^2+z^2)^{(\alpha+2)/2}}\right]\mathrm{d}z \\
&= \frac{F_\nu}{4\pi}\mathrm{d}^2\tau_0(\alpha-1)\frac{r_0^{\alpha-1}}{p^{\alpha+1}}\int \frac{1}{(1+(z')^2)^{(\alpha+2)/2}}\,\mathrm{d}z'
\end{aligned}
\tag{10.48}
$$

where F_ν is the flux received from the star at distance d and $z' = z/p$.

If we put the angular size of the inner radius $= \theta = r_0/d$ and the angular radius of the position where the shell intensity is measured as $\phi = p/d$ and write the integral as $f(\alpha)$, with a value of order unity, then (10.48) can be rewritten as

$$
\frac{I_{\text{shell}}(\phi)}{F_\nu} = \frac{\tau_0(\alpha-1)}{4\pi}\frac{\theta^{\alpha-1}}{\phi^{\alpha+1}}f(\alpha)
\tag{10.49}
$$

where the left-hand side is observed, and τ_0 can be found from the absorption line in the stellar spectrum.

The slope of the surface brightness curve, $\alpha + 1$, gives α, and the absolute value of the left-hand side gives θ and hence r_0 if the distance is known. The values of τ_0 and r_0 then determine n_0, which with the outflow velocity in turn gives a mass loss rate of neutral potassium atoms. The abundance of potassium is almost certainly solar, but unfortunately potassium is mainly ionized so that an adequate ionization model is required before a mass loss rate can be determined. The two investigations found rather different values for α, Honeycutt *et al.* finding 1.65 ± 0.2, which would imply a changing degree of ionization in a constant velocity flow, and Mauron *et al.* finding 2.5 ± 0.8, consistent with the value 2 expected for a constant velocity, constant ionization flow. The absolute value of the shell emission found by Mauron *et al.* was about a factor 5 less than that found by Honeycutt *et al.*, with a consequent reduction in the estimated mass loss rate.

Mass Loss from Cool Stars—Evidence from Molecular Lines

We now turn to the estimation of mass loss rates using CO emission from the shell in the form of millimetre rotational lines. Carbon monoxide can be picked up from stars with moderate mass loss rates like α Ori which can also be observed in the visible, and also from stars with large mass loss rates where dust obscuration makes visible observations impossible. We have already discussed the determination of masses from CO observations of interstellar clouds and many of the same considerations hold here. A number of different approaches have been used and we start with the simplest.

Knapp *et al.* [22], assuming a constant velocity, distinguish between optically thin CO flows which have lines with flat-topped profiles and optically thin flows which have lines with parabolic profiles (see the discussion of lines from outflows in the constant velocity case earlier in this chapter). In the optically thick case Knapp *et al.* assume a constant brightness temperature out to a radius r_c and zero brightness temperature at larger radii, the position of r_c being determined by the point at which the density of *hydrogen molecules* falls to the critical value n_c just needed to thermalize the CO lines. If the CO is not thermalized, its excitation temperature falls to that of the microwave background and the line is not seen above the background, although of course the cut-off is not as abrupt as is assumed in this simplified model. Optically thick thermalized CO lines will radiate with a brightness temperature which depends on the kinetic temperature and Knapp *et al.* adopt as a reasonable approximation a constant brightness temperature of 13 K for the 1–0 transition and of 28 K for the 2–1 transition on the basis of more detailed models. The telescope is then predicted to 'see' a disc of radius r_c and constant brightness T_B and to convolve this with the telescope beam. If the telescope beam is represented as a Gaussian of the form

$$P(\theta) = 1/(\pi\theta_0^2)\exp(-\theta^2/\theta_0^2)$$

which is normalized so $\int P(\theta)2\pi\theta\,\mathrm{d}\theta = 1$, then writing $r_0 = \theta_0 \mathrm{d}$ for an object at distance d, the convolution gives for the observed brightness temperature:

$$\begin{aligned} T_{obs} &= T_B \int_0^{r_c} \frac{2\pi r e^{-r^2/r_0^2}}{\pi r_0^2}\,dr \\ &= T_B[1 - e^{-r_c^2/r_0^2}] \end{aligned}$$

The full width at half height of the profile is $\theta_H = \theta_0 2\sqrt{(\ln_e 2)}$ so rearranging we obtain:

$$r_c^2 = \frac{\theta_H^2 \mathrm{d}^2}{4\ln_e 2}\ln_e\left[\frac{T_B}{T_B - T_{obs}}\right] \qquad (10.50)$$

$$\simeq \frac{\theta_H^2 \mathrm{d}^2}{4\ln_e 2}\frac{T_{obs}}{T_B} \quad \text{if } T_B \gg T_{obs} \qquad (10.51)$$

where the last condition holds if $r_c \ll r_0$. Finally the mass loss rate is given by

$$\begin{aligned} \frac{\mathrm{d}N}{\mathrm{d}t} &= 4\pi r_c^2 n_c V \\ &= \frac{\pi\theta_H^2}{\ln_e 2}\mathrm{d}^2 V n_c \ln_e\left[\frac{T_B}{T_B - T_{obs}}\right] \\ &\simeq \frac{\pi\theta_H^2}{\ln_e 2}\mathrm{d}^2 V n_c \frac{T_{obs}}{T_B} \end{aligned} \qquad (10.52)$$

Knapp *et al.* checked that the envelopes concerned were really optically thick by comparing the observations of the ratio of the brightness temperatures of the 2–1 and 1–0 transitions with the predictions of models with the different T_B and n_c values for the two transitions.

In the optically thin case for a constant velocity V flow the profile should be rectangular from $+V$ to $-V$ (Figure 10.4). Under these circumstances, we can estimate the density of CO, so the total mass outflow rate can be found using an assumed ratio of H_2 to CO. The mass loss rate of CO molecules, dN_{CO}/dt, is given by $N_{CO}(r)/(4\pi r^2 V)$ and writing $n_{col} = dN(CO)/dV_z$, the column density of CO molecules $n_{col}\Delta V_z$ along some particular line of sight passing a distance p from the central star and with line of sight velocity between V_z and $V_z + \Delta V_z$ is given by $N_{CO}(r)\Delta z$, where Δz is the line of sight thickness corresponding to velocities in the range ΔV_z so $\Delta z = \Delta V_z/(dV_z/dz)$. From (10.10), in the constant velocity case:

$$\Delta z = \frac{\Delta V_z}{\dfrac{dV_z}{dz}} = \frac{\Delta V_z}{\dfrac{V}{r}(1-\mu^2)}$$

$$= \frac{\Delta V_z}{\dfrac{V}{r}\dfrac{r}{p}(1-\mu^2)^{3/2}} = \Delta V_z \frac{p}{V\left(1-\dfrac{V_z^2}{V^2}\right)^{3/2}}$$

Hence

$$n_{col}\Delta V_z = \frac{\dfrac{dN_{CO}}{dt}}{4\pi r^2 V} \frac{p\Delta V_z}{V\left(1-\dfrac{V_z^2}{V^2}\right)^{3/2}} \tag{10.53}$$

Now from (8.32) taking a relatively high temperature so the exponential in the Boltzmann factor can be written as unity, the roughest approximation is adequate for the partition function and the long wavelength approximation can be used for the stimulated emission correction:

$$\int \tau_\nu dV = \frac{16\pi^3 h}{3k^2 4\pi\varepsilon_0} \frac{B^2(J+1)^2}{T_{ex}^2} d_p^2 n_{col}$$

In the optically thin case $T_B = T_{ex}\tau_\nu$, so

$$T_B \Delta V_z = (16\pi^3 h)/(3k^2 4\pi\varepsilon_0) B^2 (J+1)^2 d_p^2 / T_{ex} n_{col} \Delta V_z$$

and hence

$$T_B(V_z, p) = \frac{16\pi^3 h}{3k^2 4\pi\varepsilon_0} \frac{B^2(J+1)^2 d_p^2}{T_{ex}} \frac{\dfrac{dN_{CO}}{dt}}{4\pi r^2 V} \frac{p}{V\left(1-\dfrac{V_z^2}{V^2}\right)^{3/2}} \tag{10.54}$$

The observed brightness temperature is found by convoluting $T_B(v_z, p)$ with the normalized beam profile $P(p)$ at the distance of the star:

$$T_B(v_z, \text{observed}) = \int 2\pi p P(p) T_B(V_z, p)\mathrm{d}p \tag{10.55}$$

If the beam is much larger than the envelope, then $P(p) = P(0) = 4\ln_e 2/(\pi\theta_H^2 d^2)$. Now

$$V_z/V = z/r \quad \text{and} \quad p^2 = r^2 - z^2$$

so

$$(1 - v_z^2/V^2) = (1 - z^2/r^2) = p^2/r^2$$

Hence for constant V_z so $p \propto r$:

$$T_B(V_z, \text{observed}) = \frac{16\pi^3 h}{3k^2 4\pi\varepsilon_0} B^2(J+1)^2 \mathrm{d}_p^2 \frac{4\ln_e 2}{\pi\theta_H^2 d^2} \frac{\frac{\mathrm{d}N_{CO}}{\mathrm{d}t}}{4\pi V^2} \int \frac{2\pi p^2}{r^2 T_{ex} \frac{p^3}{r^3}} \mathrm{d}p$$

$$= \frac{16\pi^3 h}{3k^2 4\pi\varepsilon_0} B^2(J+1)^2 \mathrm{d}_p^2 \frac{2\ln_e 2}{\pi\theta_H^2 d^2} \frac{\frac{\mathrm{d}N_{CO}}{\mathrm{d}t}}{V^2} \int_0^R \frac{\mathrm{d}r}{T_{ex}(r)} \tag{10.56}$$

where the integral has been transformed into one over r with a maximum radius of the CO envelope of R. It will be noticed that $T_B(v_z, \text{observed})$ is independent of V_z.

If the envelope is partially resolved by the beam, the integral in (10.56) can be replaced, if the beam is centred on the star, by one proportional to

$$\left(1 - \frac{V_z^2}{V^2}\right)^{-3/2} \int_0^{p_{max}} \frac{p^2 e^{-p^2/(\theta_0 d)^2}}{r^2 T_{ex}} \mathrm{d}p$$

so assuming that T_{ex} is constant and converting to an integral over r:

$$T_B \propto \int_0^R \exp\left(-\frac{r^2}{\theta_0^2 d^2}\left(1 - \frac{V_z^2}{V^2}\right)\right)\mathrm{d}r$$

$$\propto \frac{1}{\left(1 - \frac{V_z^2}{V^2}\right)^{1/2}} \int_0^X e^{-x^2}\mathrm{d}x, \quad \text{with } x^2 = r^2 \frac{\left(1 - \frac{V_z^2}{V^2}\right)}{\theta_o^2 d^2}$$

$$\propto \frac{1}{\left(1 - \frac{V_z^2}{V^2}\right)^{1/2}} \operatorname{erf}\left(\frac{R}{\theta_0 d}\left(1 - \frac{V_z^2}{V^2}\right)^{1/2}\right) \tag{10.57}$$

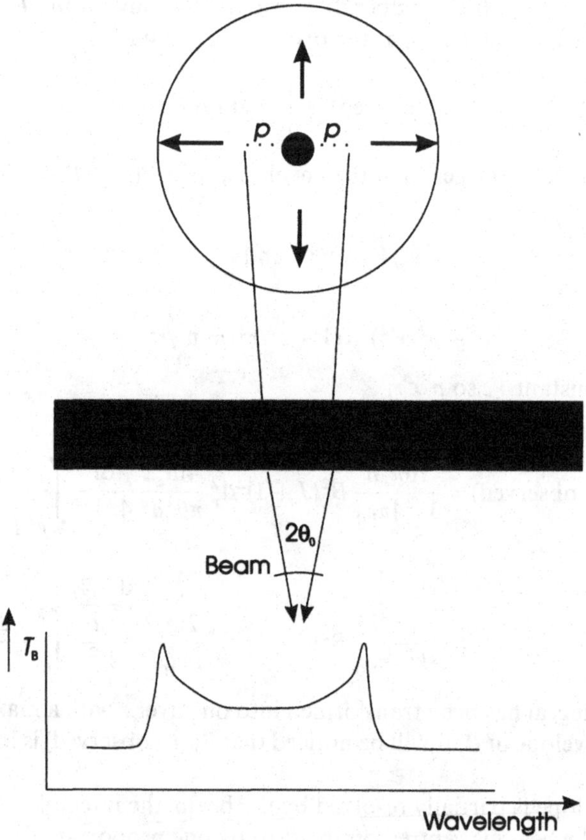

Figure 10.11. CO observations with resolution

where 'erf' stands for error function. This is no longer flat-topped and frequency independent, but has a characteristic horned-shape spectrum, rising to peaks at $v_z = \pm V$ as one might expect, for gas moving perpendicular to the line of sight now lies mainly outside the centred beam, which picks up best gas nearly in the line of sight to the star, namely gas moving directly towards or away from the observer (Figure 10.11). At the centre, $V_z = 0$, (10.57) has the value $\mathrm{erf}(R/\theta_0 d)$, whereas at the peaks, noting that erf $x \sim x$ for small x, (10.57) has the value $R/(\theta_0 d)$. Hence:

$$\frac{T_B(V)}{T_B(0)} = \frac{\sqrt{4\ln_e 2}\,\dfrac{R}{\theta_H d}}{\mathrm{erf}\left(\sqrt{4\ln_e 2}\,\dfrac{R}{\theta_H d}\right)} \tag{10.58}$$

Returning after this diversion to the more usual case of the unresolved envelope, models suggest that T_{ex} is proportional to $1/r$, so putting $T_{ex} = T_0R_0/r$ we obtain:

$$\frac{dN_{CO}}{dt} = \frac{3k^2 4\pi\varepsilon_0}{16\pi^3 h}\pi\frac{\theta_H^2 d^2}{\ln_e 2}\frac{1}{B^2(J+1)^2 d_p^2}\frac{T_0 R_0 V^2}{R^2}T_B(\text{observed}) \qquad (10.59)$$

Knapp *et al.* used the fact that their observations suggested incipient resolution in α Ori to make an estimate of R, and also employed the fact that the infrared observations in the absorption of fundamental CO lines of Lambert *et al.* of α Ori gave a value of 70 K for T_{ex}, then assuming that the value of T_0R_0/R^2 thus obtained for α Ori applied universally. Multiplication by $1/f*$(mass of hydrogen molecule), where f is the ratio of the CO number density to that of hydrogen molecules, then gives the mass loss rate. It will be noticed that there are considerable uncertainties in this procedure, which was only used by Knapp *et al.* for three stars.

Jura [23] has argued that the optically thick case of Knapp *et al.* where the emission cuts off completely at some critical radius where the excitation rate becomes too low is not realistic and that half or more of the emission will come from regions farther out with lower excitation temperatures.

Jura suggests an alternative method of finding mass loss rates for those cases where the envelope is resolved by using off-star observations of the 2–1 line, that is by finding the distance from the star on a map of 2–1 emission at which the brightness temperature falls to 1 K. Clearly the gas is optically thin along such a line of sight. Knapp had used an excitation temperature from a model, but Jura does not explicitly calculate an excitation temperature but instead estimates the optically thin emission from the outer envelope directly by assuming that every collisional excitation from the $J = 0$ or $J = 1$ levels of CO to $J = 2$ or higher gives rise to a $J = 2$ to $J = 1$ photon (which no longer holds if lines are optically thick). The collisional excitation rate is given by $N(\text{CO})\ N(\text{H}_2)\ \langle\sigma V_{th}\rangle$, where the quantity in brackets is the velocity averaged product of the collisional excitation cross-section of CO to rotational level 2 or higher and the *thermal* velocity. The thermal velocity is taken as that corresponding to a kinetic temperature of 20 K found for a detailed model of the envelope of IRC 10216, but fortunately the cross-section for such low excitation energies is not strongly dependent on the kinetic temperature. Assuming the CO molecules extend to infinity, the brightness temperature integrated over frequency at impact parameter p is given by

$$\begin{aligned}\int T_B(p)\,d\nu &= \frac{c^2}{2k\nu^2}\int I_\nu\,d\nu \\ &= \frac{c^2}{2k\nu^2}\int N(H_2)N(\text{CO})\langle\sigma V_{th}\rangle\frac{h\nu}{4\pi}\,dz \\ &= \frac{c^2 h}{8\pi k\nu}\left(\frac{dN(\text{H}_2)}{dt}\right)^2 f\langle\sigma V_{th}\rangle\frac{1}{V^2}\int\frac{1}{(4\pi r^2)^2}\,dz \\ &= \frac{c^2 h}{8\pi k\nu}\left(\frac{dN(\text{H}_2)}{dt}\right)^2 f\langle\sigma V_{th}\rangle\frac{1}{(4\pi V)^2}\,2\int_0^\infty\frac{dz}{(z^2+p^2)^2}\end{aligned}$$

The integral has the value $\pi/(4p^3)$ and if we take $\Delta v \sim v/c \cdot V$, we obtain:

$$T_B(p) \simeq \left(\frac{c}{4\pi}\right)^3 \frac{h}{k}\frac{\pi}{4}\frac{\langle \sigma V_{th} \rangle}{v^2}\frac{1}{(Vp)^3} f \left(\frac{dN(H_2)}{dt}\right)^2 \tag{10.60}$$

so setting $T_B(p) = 1$ we obtain a mass loss rate proportional to the distance away from the star at which this temperature is found to the power 3/2 (remembering that what is measured is an angular distance and hence that the mass loss rate is proportional to the assumed distance to the star to the power 3/2) and to $f^{-1/2}$.

There is no difficulty in obtaining predictions of brightness temperature profiles for given CO molecules loss rates taking into account optical thickness and beam profile *if* one can calculate the excitation temperature as a function of r or equivalently the excitation rate, and *if* the degree of dissociation of CO as a function of radius is known. The latter factor has not so far been considered, but ultraviolet radiation will dissociate CO molecules. The ultraviolet radiation field will be the interstellar field. Circumstellar shells are dusty and so CO molecules may be shielded from the interstellar UV field by dust. However for CO the continuum photodissociation rate is probably dominated by predissociation. Line absorption excites a molecule from the ground state to a vibrational level of an excited electronic state which is coupled to the vibrational continuum of a third electronic state. There can therefore be a radiationless transition to the third state, which is unbound (Figure 10.12). The net result is a line absorption giving rise to a dissociated molecule, and one can infer the existence of the process by the broadening of the line because of the reduced lifetime of the upper state.

The important point here is that while the CO envelope is unlikely to become optically thick to continuum ultraviolet radiation, the lines leading to predissociation may well become optically thick, preventing predissociation of molecules farther in. This is called *self-shielding*. Self-shielding will reduce the mean intensity at r by a factor equal to the probability of a photon of line frequency penetrating to r, which in turn is equal to the probability of such a photon escaping from r. Using (10.20) for the escape probability β, we obtain for the CO density:

$$\frac{d(r^2 N_{CO} V)}{dr} = -r^2 N_{CO} \frac{\pi e^2}{4\pi\varepsilon_0 m_e c} f \frac{4\pi J_{is}\beta}{hv} \frac{n_L}{N_{CO}}$$

where J_{is} is the mean interstellar intensity at the appropriate frequency, f is the oscillator strength of the line and n_L/N_{CO} is the fraction of CO molecules in the lower level of the transition. For large optical depth $\beta \sim 1/\tau \propto 1/(f n_L r)$, we find on substitution and integration that the radius at which the CO density falls to zero is given by

$$r = \sqrt{\frac{3c}{32\pi^2 m_H h J_{is}}} \sqrt{\frac{\frac{dM}{dt} f_{CO}}{V}} \tag{10.61}$$

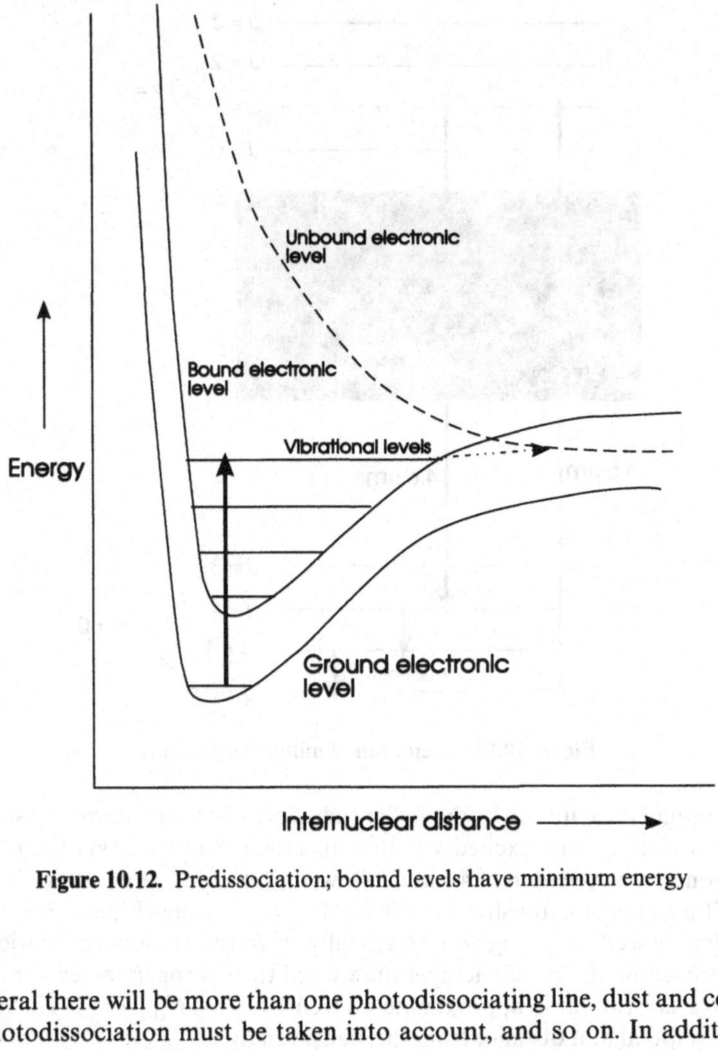

Figure 10.12. Predissociation; bound levels have minimum energy

In general there will be more than one photodissociating line, dust and continuum photodissociation must be taken into account, and so on. In addition H_2 molecules can shield the CO molecules. The detailed models of Mammon *et al.* [24] give similar results to (10.61) but predict rather larger radii fore the CO cut-off.

We next consider the excitation of the 1–0 and 2–1 lines. The microwave background can excite the millimetre transitions directly, but of course does not produce emission lines above the microwave continuous background. The star will produce negligible flux at millimetre wavelengths, but the millimetre lines are often optically thick, and absorption in these lines must also be taken into account in calculating level populations. However, the two important processes driving the excitation of the millimetre lines are (a) collisions with H_2 molecules

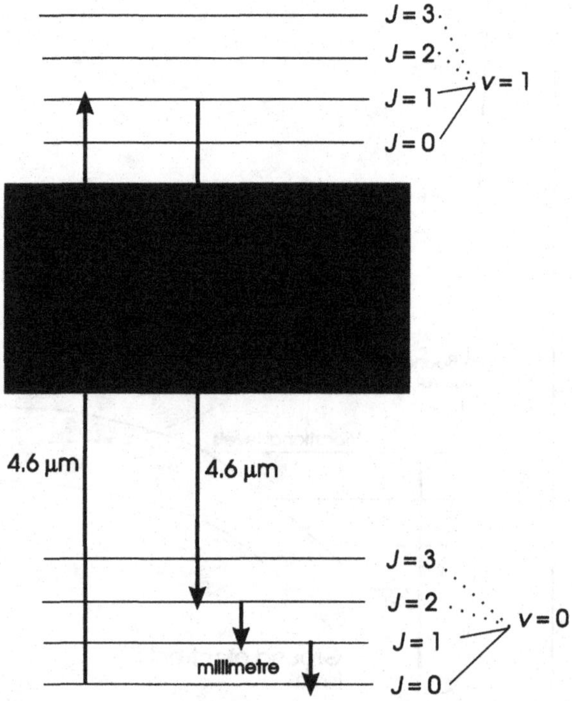

Figure 10.13. Excitation of millimetre transition

and (b) radiative excitation by the stellar radiation field from the ground state, say $v = 0$, $J = 0$ to the first excited vibrational state, $v = 1$, $J = 1$, via the radiative transition at a wavelength of 4.6 μm, followed by a radiative decay to the upper level of the millimetre transition, $v = 0, J = 1$ for the 1–0 line (Figure 10.13). As we have already seen, (a) acting alone gives fully effective collisional excitation with T_{ex} approaching the kinetic temperature when the hydrogen molecular density reaches a critical value approximately given by $\langle \sigma v \rangle N_{crit} \sim A_{ji}$. This critical density is found at a distance from the star given by

$$R_{crit} \simeq \left(\frac{\frac{dM}{dt}}{m(H_2)} \frac{1}{4\pi N_{crit} V} \right)^{1/2}$$

an expression which has nearly the same dependence on mass loss and velocity as the radius for photodissociation, and indeed for the 1–0 transition has nearly the same numerical value (the critical radius for the 2–1 transition is smaller).

If (b) acting alone is to produce excitation, the pump rate into the upper level of the transition per molecule must be equal to the spontaneous decay rate A_{ji} from that level, i.e.

$$B_\nu(T_*)\cdot R_*^2/4r^2\cdot B_{ij}(4.6\,\mu\text{m}) \sim A_{ji}, (\text{mm})$$

leaving aside many details, and so roughly the critical radius is given by

$$R \simeq \left(\frac{A_{ji}(4.6\,\mu\text{m})}{A_{ji}(\text{mm})} \frac{1}{e^{+h\nu(4.6\mu\text{m})/kT_*} - 1} \right)^{1/2} R_*$$

which is independent of wind conditions in this oversimplified picture, although we certainly should include the effects of dust absorption on the 4.6 μm emission from the star and the energy absorbed by the dust from shorter wavelengths and reradiated at 4.6 μm. Except for very large mass loss rates, we find that the critical radius for radiative excitation is larger than that for collisional excitation.

However, we have still to consider the effects of large optical depth, both for the vibrational transition (where the excitation rate will be reduced) and for the rotational millimetre transition (where the effect of trapping will be to move the critical radius farther out). The fact that we are dealing with a wind makes optical depth directionally dependent because in the radial direction all molecules in a constant velocity flow are moving with the same line of sight velocity, whereas in the tangential direction the molecules are moving with different line of sight velocities, so that radial optical depths can be much larger than tangential optical depths. The models of Morris [25] show that collisional excitation dominates radiative excitation if the radial vibrational optical depth becomes large (the radial optical depth being the relevant one for the absorption of stellar infrared radiation), which will be the case for large mass loss rates. The models also show that for low mass loss rates where the excitation is radiative, the envelope size is limited by photodissociation rather than excitation. Furthermore, the millimetre 1–0 transition becomes optically thick tangentially (and it is a depth averaged over direction that we are concerned with when considering radiative trapping) over most of the envelope when the vibrational radial optical depth becomes large, i.e. when collisional excitation dominates radiative excitation. Thus we obtain for high mass loss rate cases optically thick 1–0 lines which are collisionally excited and which are produced in a region limited by the critical radius for collisional excitation which is nearly the same as the radius for photodissociation, whereas for low mass cases we have optically thin 1–0 lines radiatively excited from a region limited by the radius for photodissociation. The dividing line between these two regimes occurs when the radial optical depth in the 4.6 μm transition is of the order of one, this optical depth being proportional to the column density of CO molecules from inner radius r_i ($\sim$ radius of star) divided by the intrinsic width of the line, ΔV. The column density for a constant density flow with density $n \propto 1/r^2$ is $n(r_i)r_i$ with (for a CO to H_2 ratio of f_{CO}):

$$n_i \propto dN/dt/(4\pi r_i^2 V) \propto dM/dt\, f_{CO}/(4\pi r_i^2 V)$$

so

$$\tau \propto dM/dt\, f_{CO}/(r_i V \Delta V)$$

Inserting numerical values, one finds the dividing line given by $\mathrm{d}M/\mathrm{d}t\, f_{\mathrm{CO}}/(V\Delta V) \sim 10^{-10}$ solar masses per year $\mathrm{km}^{-2}\mathrm{s}^2$ with $f_{\mathrm{CO}} \sim 10^{-3}$, $\Delta V \sim 1\,\mathrm{km\,s^{-1}}$ and $V \sim 15\,\mathrm{km\,s^{-1}}$.

An extensive investigation by Knapp and Morris [26], which obtained mass loss rates for 50 stars using observations of the 1–0 line of CO, found that the results of model predictions for optically thick models could be fitted by

$$T_{\mathrm{obs}} = 1.37 \times 10^{28} \frac{\dfrac{\mathrm{d}M}{\mathrm{d}t} f_{\mathrm{CO}}^{0.85}}{\mathrm{d}^2 V^2 \theta_{\mathrm{H}}^2} \tag{10.62}$$

where the units are mks, with d metres the distance of the star and θ_{H} radians the beam half-power width. The linear relation between brightness temperature and mass loss rate no longer holds for very large mass loss rates, but over a considerable range the increasing saturation of the millimetre lines with increasing mass loss rate is balanced by increasing excitation due to the greater density of H_2 molecules, giving (10.62). This equation also no longer holds when optical thinness and radiative excitation set in at low mass loss rates.

The brightness temperature of collisionally excited CO with optically thick lines depends on the kinetic temperature of the gas, which in turn depends on the balance between heating by the grains of dust drifting relative to the gas and cooling by radiation of molecular lines and adiabatic expansion. For carbon-rich stars the radiative cooling is mainly due to CO lines but for oxygen-rich stars H_2O lines may dominate the cooling. Kwan and Linke [26] have made models on the basis of this balance, and fitted them to the CO line observations of the particularly well-studied carbon-rich star IRC+10216. The resulting kinetic energy variation as a function of radius has been used in most attempts to deduce mass loss rates from CO observations. However stars with different chemical compositions, particularly those that are oxygen-rich, will have different dust properties (different values of Q for instance) and different rates of radiative cooling, and the extrapolation of the results of Kwan and Linke to stars of very different mass loss rates and compositions has been questioned.

It may also be noted that values for f_{CO} have to be assumed to obtain mass loss rates. It is customary to suppose that CO is completely associated in the atmosphere of the star, and that for oxygen-rich stars the number of CO molecules equals the total number of carbon atoms, whereas in carbon-rich stars the number of CO molecules equals the total number of oxygen atoms. Solar abundances are used, which in older work produced f_{CO}(oxygen rich) $= 7*10^{-4}$ and f_{CO}(carbon rich) $= 1.3*10^{-3}$, but these values may not be applicable to evolved stars. Laboratory measurements have suggested considerable revision of the photodissociation rates previously assumed for CO, and it may also be noted that the mass loss rates require a knowledge of the distance of the star, and evolved red giants and supergiants are poor standard candles in many cases.

Thus, despite the ubiquity of CO emission from mass-losing cool stars, and the apparently simple nature of thermal emission, there remain considerable uncertainties in the mass loss estimates made using this method.

We now turn to the use of the OH 18 cm lines from the envelopes of oxygen-rich evolved stars in determining mass loss rates. These lines from the circumstellar shells often show maser enhancement which gives them very high brightness temperatures, and the 1612 MHz line has proved particularly useful. The spectrum of this line has a characteristically double peaked form, the peaks lying at velocities of about $\pm 15\,\mathrm{km\,s^{-1}}$ on either side of the stellar velocity V_*. The use of interferometers to resolve the bright emission enables maps to be made at various velocities, and it is found that the maser emission tends to occur in small spots but the overall distribution at the stellar velocity is ring shaped about the star, whereas the distribution at the peak velocities is concentrated around the position of the star.

A plot of angular distance from the star, θ, against velocity, V_{obs}, shows a parabolic form:

$$\frac{\theta}{\theta_{\mathrm{max}}} = \left[1 - \frac{(V_{\mathrm{obs}} - V_*)^2}{V^2}\right]^{1/2} \tag{10.63}$$

where the maximum velocity relative to the star is V. These observations are consistent with the maser sources lying in a fairly thin shell of radius R through which the gas is flowing at velocity V. The material at impact parameter p then has a line of sight velocity $V_{\mathrm{obs}} - V_* = V\sqrt{(1 - p^2/R^2)}$. Now

$$\theta = \frac{p}{d} = \frac{R}{d}\sqrt{1 - \frac{(V_{\mathrm{obs}} - V_*)}{V^2}}$$

which agrees with (10.63) with $\theta_{\mathrm{max}} = R/d$ (Figure 10.14). The maser sources lying in the line of sight to the star have the maximum observed relative velocity with the positive sign if they lie on the opposite side of the star to the observer, while sources with zero line of sight velocity relative to the star will lie in a ring of radius R around the star. For maser emission, the amplification depends on the column density of OH molecules with the same velocity, and since the emission is beamed we are concerned with molecules with the same line of sight velocity component and lying along the line of sight to the observer. If the velocity gradient is small or indeed terminal velocity has been reached so V = constant, then all the OH molecules observed along the line of sight to the star, that is in the radial direction, will be moving with the same line of sight velocity component and hence the amplifying path length will be large and constrained only by the distribution of OH molecules with radius, whereas along a line of sight passing through the envelope far from the star at large p there will be a line of sight velocity component gradient $\mathrm{d}V_z/\mathrm{d}z = Vp^2/r^3$, leading to a smaller path length for maser amplification. Hence the maser spots in the shell that lie near the line of

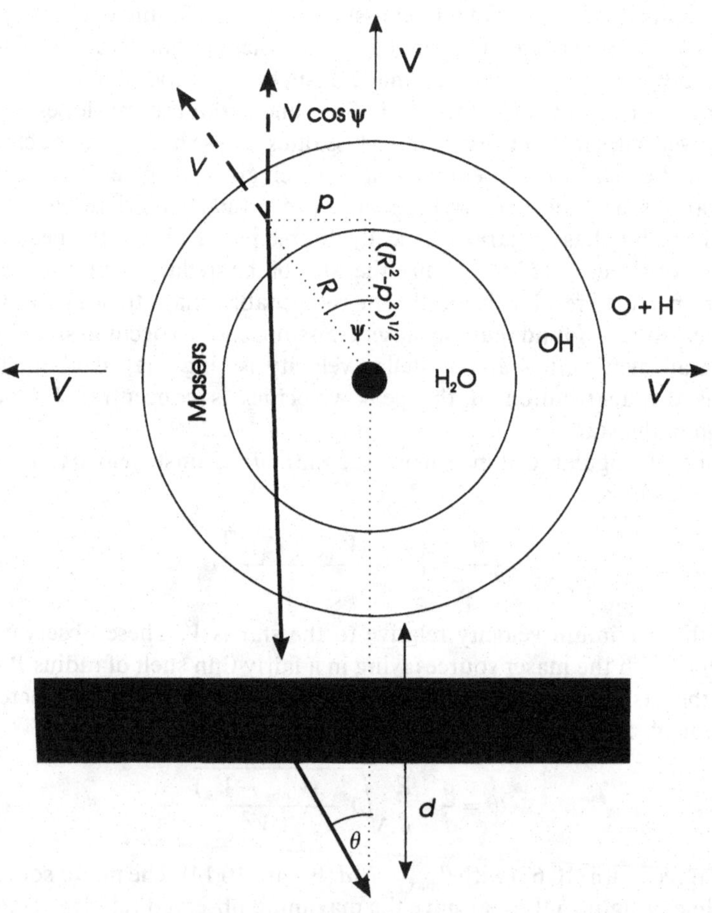

Figure 10.14. OH maser shell around cool star

sight to the star will show the largest amplifications and also the largest Doppler shifts, giving rise to the double peaked spectral line profile.

The radius of the shell of masers is determined by the balance between the production and destruction of OH molecules. In the outer parts of a cool star atmosphere the oxygen that is not bound in CO will be in the form of H_2O. As the stellar wind carries the molecules outwards with falling density and temperature this composition would remain 'frozen in' if it were not for the interstellar ultraviolet radiation field. This will dissociate H_2O into an OH radical and a hydrogen atom, and will eventually dissociate OH into an oxygen and a hydrogen atom. There will be a radius at which half the H_2O molecules are dissociated, representing the distance at which the dissociation rate per unit volume balances the number of molecules flowing through per unit time, and similarly there will be a radius at which half the OH molecules are dissociated, with most of the OH molecules lying between these two radii. However the overall density of a constant velocity flow falls off as $1/r^2$, so the peak density of OH molecules will lie close to the H_2O dissociation radius. The calculation of this radius follows much the same approach as has already been used for CO, with the exception that the dissociation of H_2O is mostly direct through continuum absorption as opposed to the dominating contribution of indirect predissociation through line absorption in the case of CO, so that self-shielding is less important. On the other hand, shielding by dust is just as important as for CO. The H_2O molecule is much less tightly bound than CO, so it is not surprising that the dissociation radius for CO is an order of magnitude greater than that for H_2O.

Netzer and Knapp [27] find for the peak OH density:

$$R_{\mathrm{OH}} = 2.95 \times 10^3 \left(\frac{\mathrm{d}M}{\mathrm{d}t}\right)^{0.7} V^{-0.4}\,\mathrm{m} \tag{10.64}$$

in mks units throughout.

The assumption is now made that the shell of OH masers lies close to this radius, which can be measured directly in a number of cases. The stars involved are long period variables, and the masers are pumped by the varying infrared flux from the star. An increase in stellar luminosity should be followed after the light-travel time across the shell radius by an increase in maser luminosity. However, the increase in brightness of the masers that lie on the far side of the star in the line of sight to the star should reach us $2R_{\mathrm{OH}}/c$ later than the increase in brightness of the masers that lie in the line of sight to the star but are on the near side of the star because of the light-travel time across the diameter of the shell, and these two groups of masers correspond to the red-shifted peak and the blue-shifted peak of the line profile, respectively (Figure 10.15). Thus the phase lag in the variation of the various maser sources enables the radius of the shell to be found. Not only does this then give the mass loss rate from (10.64) with V from the Doppler shifts of the peaks, but if the angular radius of the shell is known from interferometric measurements, the distance of the star can be determined from $\theta_{\mathrm{OH}} = R_{\mathrm{OH}}/d$.

This method is limited to oxygen-rich stars, and requires a number of observations to determine the phase lag, but has the enormous advantage that the result does not depend on uncertain distance estimates to the star, which are involved in all other mass loss rate estimates.

Mass Loss from Cool Stars: Evidence from Dust

We now turn to the estimates of mass loss rates using the emission of the dust in the circumstellar envelope. All such estimates directly measure the amount of dust in the circumstellar envelope, and hence must be multiplied by a gas to dust

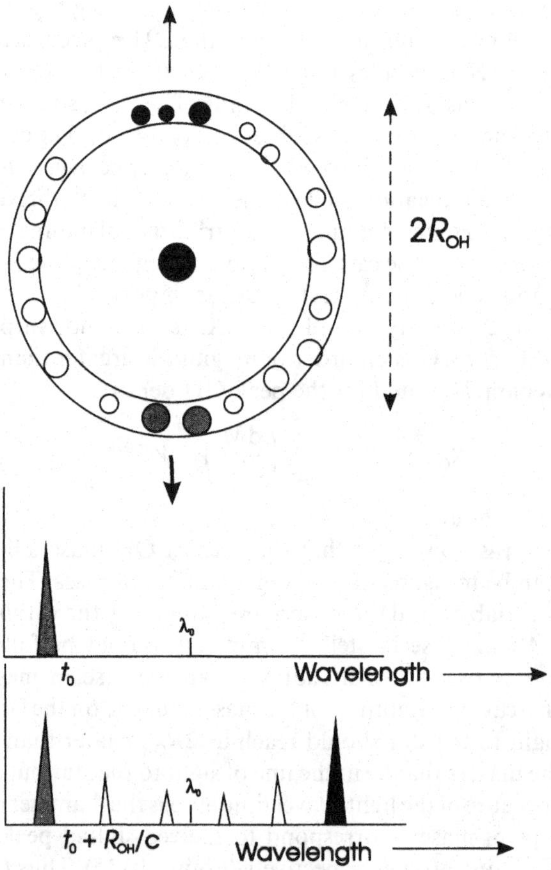

Figure 10.15. Time delay in variable maser spectra

mass ratio, which is normally taken from observations of the interstellar medium. The appropriate ratio for cool star winds may, of course, be rather different. We refer back to the previous chapter for a general study of interstellar dust, but clearly the uncertainties over the composition of both interstellar and circumstellar dust lead to uncertainties in the calculated mass loss rates. Dust mass loss rates have been estimated using the 9.7 μm silicate feature and using the general continuum emission of dust both at the IRAS satellite wavelengths (12, 25 and 60 μm) and at sub-millimetre wavelengths.

One of the first attempts to use dust emission to measure mass loss rates was that of Gehrz and Woolf [28] who used broad band measurements at 3.5, 8 and 11 μm of oxygen-rich cool evolved stars. The '11 μm' band included the peak of the 'dust bump', the feature attributed to silicates normally being taken as having a peak effect at 9.7 μm, and Gehrz and Woolf argued that this would often be optically thick, while at 3.5 μm the silicate feature would have no effect, and at 8 μm the silicate feature would be present but would be optically thin. Gehrz and Woolf estimated the optical depth in the 11 μm band as follows. They assumed the dust lay in a thin shell of radius R_S, temperature T_S and corresponding source function $S_\nu(T_S)$, with optical depth τ_ν, and that the star had radius R_* and source function $S_\nu(T_*)$, both shell and star having blackbody source functions.

The flux received from the star at distance d is

$$F_\nu(\text{star}) = \frac{\pi S_\nu(\text{star}) 4\pi R_*^2}{4\pi d^2} \mathrm{e}^{-\tau_\nu}$$

and the flux received from the shell is

$$F_\nu(\text{shell}) = \frac{\pi S_\nu(\text{shell}) 4\pi R_S^2 (1 - \mathrm{e}^{-\tau_\nu})}{4\pi d^2}$$

Hence

$$\begin{aligned} \text{total flux received from the system} &= F_\nu(\text{star}) + F_\nu(\text{shell}) \\ &= \pi S_\nu(\text{star}) \frac{R_*^2}{d^2} \left[\mathrm{e}^{-\tau_\nu} + \frac{R_S^2}{R_*^2} \frac{S_\nu(\text{shell})}{S_\nu(\text{star})} (1 - \mathrm{e}^{-\tau_\nu}) \right] \\ &= \pi S_\nu(\text{star}) \frac{R_*^2}{d^2} \left\{ \left[\left(\frac{R_S}{R_*} \right)^2 \frac{T_S}{T_*} - 1 \right] (1 - \mathrm{e}^{-\tau_\nu}) + 1 \right\} \end{aligned} \quad (10.65)$$

where it has been assumed that the source functions can be written in the Rayleigh–Jeans approximation $S_\nu = 2kT/c^2\nu^2$. Writing the flux from the star with no dust as $F_\nu(*)$:

$$\begin{aligned} \frac{F_{11}(\text{total})}{F_{3.5}(\text{total})} &= \frac{F_{11}(*)}{F_{3.5}(*)} \frac{\left[\left(\frac{R_S}{R_*} \right)^2 \frac{T_S}{T_*} - 1 \right] (1 - \mathrm{e}^{\tau_{11}}) + 1}{\left[\left(\frac{R_S}{R_*} \right)^2 \frac{T_S}{T_*} - 1 \right] (1 - \mathrm{e}^{\tau_{3.5}}) + 1} \\ &\simeq \frac{F_{11}(*)}{F_{3.5}(*)} \left[\left[\left(\frac{R_S}{R_*} \right)^2 \frac{T_S}{T_*} - 1 \right] (1 - \mathrm{e}^{-\tau_{11}}) + 1 \right] \end{aligned} \quad (10.66)$$

if $\tau_{3.5} = 0$.

If the shell is optically thin with $\tau_{11} \ll 1, (1 - \exp[-\tau_{11}]) \sim \tau_{11}$, and if the shell is optically thick at 11 μm, the 8.4 μm flux can be used instead in the comparison with the 3.5 μm flux. If we write the infrared excess (difference between the observed magnitude of the star at some wavelength and the intrinsic magnitude of the star from an extrapolated blackbody curve) as $[11] = m_{11} - m_{11}$ (intrinsic), and note $F_1/F_2 = 0.4(m_2 - m_1)$, then for optically thin shell emission at 11 μm we have

$$\tau_{11} = \frac{[\mathrm{antilog}_{10}\{0.4([3.5] - [11])\}] - 1}{\left(\frac{R_S}{R_*}\right)^2 \frac{T_S}{T_*} - 1} \tag{10.67}$$

and a corresponding relation for τ_8.

Now $\tau_{11} = \int \kappa_{11}\rho \, dz = \kappa_{11}\rho(r_{min})r_{min}$ if $\rho \propto 1/r^2$ and the shell extends from some minimum radius for dust formation r_{min} out to a large distance (although this is not consistent with the notion of a thin shell!). This gives the correct value for the radial optical depth absorbing light from the star but not for the optical depth in the expression for the emission which will average out over direction at a few times the radial optical depth. The mass loss rate is given as usual by

$$\begin{aligned} dM/dt &= 4\pi r_{min}^2 V \rho_{gas}(r_{min}) \\ &= 4\pi r_{min}^2 V \rho_{gas}/\rho_{dust} \tau_{11}/\kappa_{11} r_{min} \\ &= 4\pi/\kappa_{11}(\rho_{gas}/\rho_{dust}) V R_S \tau_{11} \end{aligned} \tag{10.68}$$

Gehrz and Woolf make a correction for the fact that the dust has a drift velocity with respect to the gas and hence ρ_{gas}/ρ_{dust} may be higher than the interstellar value. A plot of [8.4]–[11] versus [3.5]–[8.4] observations is compared with the predictions of (10.65) as τ_{11} is varied, using estimated ratios of the optical depth at 11, 8.4 and 3.5 μm. The form of the curve depends on the parameter $(R_S/R_*)^2$ (T_S/T_*) whose value can therefore be estimated for individual stars, with results lying between 100 and 4. Unfortunately we also need to know R_S separately, and this could be done if T_S were known. Silicate dust will not condense at temperatures much higher than 1200 K which sets an upper limit to T_S and indeed we would expect T_S to be close to this upper limit. Gehrz and Woolf also consider in their [8.4]–[11] versus [3.5]–[8.4] diagram the intersection of the curves (fixed R^2T parameter, varying τ_{11}) on which most of their stars lie with the curve followed by a blackbody, which represents the situation at very large optical depth when the shell would be optically thick at all wavelengths, and find temperatures of the order of 700 K, finally adopting a shell temperature of 900 K for all their stars. Clearly, this pioneering attack on the problem needed considerable elaboration if reliable mass loss rates were to be obtained. Real shells are not thin, and hence have a varying temperature with distance from the star. In optically thick shells, the radiative transfer problem needs to be solved.

Consider the temperature distribution $T_d(r)$ in the optically thin case with absorption efficiency Q_ν following a power law in frequency $Q_\nu = Q_0(\nu/\nu_0)^p$ so that $Q \propto \lambda^{-p}$ and with the central star radiating like a blackbody with effective temperature T_* and having radius R_*. Using a dilution factor $W = R^2/4r^2$ and balancing the heating and cooling of a grain of radius a:

$$4\pi \int \pi a^2 Q_\nu B_\nu(T_d) d\nu = \pi a^2 \int Q_\nu \frac{L_\nu}{4\pi r^2} d\nu$$
$$= \pi a^2 \frac{R_*^2}{r^2} \int Q_\nu \pi B_\nu(T_*) d\nu$$

Hence

$$\int Q_\nu B_\nu(T_d) d\nu = \frac{R_*^2}{4r^2} \int Q_\nu B_\nu(T_*) d\nu \tag{10.69}$$

Ignoring for the moment the fact that Q is in general a function of temperature, we can write

$$\int Q_\nu B_\nu d\nu = \frac{2h}{c^2} \frac{Q_0}{\nu_0^p} \int \frac{\nu^{3+p} d\nu}{e^{h\nu/kT} - 1}$$
$$= \frac{2h}{c^2} \frac{Q_0}{\nu_0^p} \left(\frac{kT}{h}\right)^{4+p} I_p \tag{10.70}$$

where

$$I_p = \int_0^\infty \frac{x^{3+p}}{e^x - 1} dx \simeq (p+3)!$$

where the last approximation holds in the Rayleigh–Jeans case. Hence

$$T_d = T_* \left(\frac{R_*}{2}\right)^{2/(4+p)} \frac{1}{r^{2/(4+p)}} \tag{10.71}$$

which for a grey efficiency independent of wavelength with $p = 0$ gives a dust temperature falling off as the inverse square root of the distance from the star. This simplest case of all produces a $T_d(r)$ function that starts at some distance r_0 where the equilibrium temperature T_0 is such that grains can condense, and then falls off as a power law in r. In oxygen-rich stars it is usual to presume that the grains are mainly silicates, and Ca–Al silicates can exist in solid form at 1400 K, with magnesium silicates (perhaps the bulk of the grain material) coming in at 1050 K. However the form of the silicate features near 10 and 20 μm suggests that the silicates may be amorphous rather than crystalline and may form at rather lower temperatures. Usually it is assumed that T_0 is around 1000 K. The efficiency Q is temperature dependent and drops when the temperature falls below 200 K because the 10 and 20 μm vibrational features are no longer excited. The resulting drop in emission from the colder outer parts of the dust envelope means that the temperature falls less fast with radius than it does in the inner envelope.

For such an optically thin envelope, the flux received from the whole envelope in the constant velocity V case, where $N(r) = N(r_0) r_0^2 / r^2$ and the dust mass loss rate is $4\pi r_0^2 N(r_0) m_{\rm grain} V$, is

$$F_\nu = \frac{1}{4\pi d^2}\int_{r0}^{\infty} 4\pi r^2 N_{\rm d}(r) 4\pi^2 a^2 Q_\nu B_\nu(T_{\rm d})\,{\rm d}r$$

$$= 4\pi^2 a^2 Q_\nu N_{\rm d}(r_0)\frac{r_0^2}{{\rm d}^2} r_0 \int B_\nu(T_{\rm d}(x))\,{\rm d}x, \quad x = r/r_0$$

$$= \frac{3}{4}\frac{{\rm d}M_{\rm d}}{{\rm d}t}\frac{1}{V\rho_{\rm g}}\frac{Q_\nu}{a}\frac{r_0}{d^2}\int_1^{\infty} B_\nu(T_{\rm d}(x))\,{\rm d}x \tag{10.72}$$

where ${\rm d}M_{\rm d}/{\rm d}t$ is the dust mass loss rate and $\rho_{\rm d}$ is the grain material density so the mass of a grain is $4/3\pi a^3 \rho_{\rm g}$. The integral has been taken to infinity but the exact value of the outer radius of the envelope makes little difference as long as it is much larger than r_0. Equation (10.71) gives $T_{\rm d}/T_0 = x^{(-2/[4+p])}$ which enables the integral in (10.72) to be calculated for given T_0, and r_0 is also fixed by (10.71) for given T_0 and stellar parameters. It will be noted that the resulting spectrum is considerably broader than a blackbody spectrum because of the range of temperatures in the envelope and the large volumes corresponding to a given temperature range at large r.

The determination of dust mass loss rates using sub-millimetre emission has the advantage that circumstellar envelopes are certainly optically thin at these wavelengths and that the emission is dominated by the cooler, outer parts of the envelope where the temperature is not strongly affected by radiative transfer effects at shorter wavelengths in the inner part of the cloud.

We follow here the discussion of Sopka *et al.* (Sopka, R. J., Hildebrand, R., Jaffe, D. T., Gatley, I., Roellig, T., Werner, M., Jura, M., and Zuckerman, B., Ap.J., *294*, 242, 1984) using observations at a wavelength of 400 μm. They argue that grains in the outer part of the envelope are mainly heated by the infrared flux from the inner part of the envelope, and since the outer part is fairly optically thin even at infrared wavelengths, the infrared flux in the outer envelope, F_ν, varies approximately as $1/r^2$ without changing its spectral distribution. Hence the observed infrared flux is $F_\nu({\rm obs}) = F_\nu(r) r^2/d^2$. We can modify (10.69) by replacing $L_\nu/4\pi r^2$ by $F_\nu(r)$:

$$\int Q_\nu B_\nu(T_{\rm d}){\rm d}\nu = \frac{{\rm d}^2}{4\pi r^2}\frac{Q_0}{\nu_0^p}\int \nu^p F_\nu({\rm obs})\,{\rm d}\nu$$

$$2\frac{h}{c^2}\left(\frac{kT_{\rm d}}{h}\right)^{4+p} I_p = \frac{{\rm d}^2}{4\pi r^2}\int \nu^p F_\nu({\rm obs})\,{\rm d}\nu$$

$$T_{\rm d}(r) = \left(\frac{r}{d}\right)^{-2/(4+p)}\frac{h}{k}\left[\frac{c^2}{8\pi h(p+3)!}\int \nu_p F_\nu({\rm obs})\,{\rm d}\nu\right]^{1/(4+p)} \tag{10.73}$$

The flux received from volume ${\rm d}W$ is, as before, $a^2/d^2 Q_\nu \pi B_\nu(T_{\rm d}) N_{\rm d}(r)\,{\rm d}W$, and writing ${\rm d}W$ as $2\pi r^2 \sin\theta\,{\rm d}\theta\,{\rm d}r$ and substituting for $N_{\rm d}$ in terms of

$\mathrm{d}M_\mathrm{d}/\mathrm{d}t/(4\pi r^2 Vm(\text{grain}))$, we obtain:

$$\mathrm{d}F_\nu(\text{obs}) = \frac{3Q_\nu \dfrac{\mathrm{d}M_\mathrm{d}}{\mathrm{d}t}}{8\mathrm{d}^2 V a\rho_\mathrm{g}} B_\nu(T_\mathrm{d})\sin\theta\,\mathrm{d}\theta\,\mathrm{d}r$$

and making allowance for the beam efficiency $P(r, \theta)$:

$$F_\nu(\text{obs}) = \frac{3}{8}\frac{Q_\nu}{a}\frac{\dfrac{\mathrm{d}M_\mathrm{d}}{\mathrm{d}t}}{\rho_\mathrm{g} V d^2}\iint B_\nu(T_\mathrm{d})P(r,\theta)\sin\theta\,\mathrm{d}\theta\,\mathrm{d}r \qquad (10.74)$$

where the beam efficiency is a function of angle as measured by the observer, and hence of r and θ in the circumstellar envelope coordinate system centred on the star.

Sopka *et al.* suggest that as a first approximation the beam efficiency is taken as 1 from the beam centre to the edge of the beam (at a projected distance b from the star at the distance of the star) and 0 outside, giving a rectangular profile, and that the emitting volume is taken as the sphere with $r = b$, the portions of the beam intersecting the envelope behind the star and in front of the star with $r > b$ being

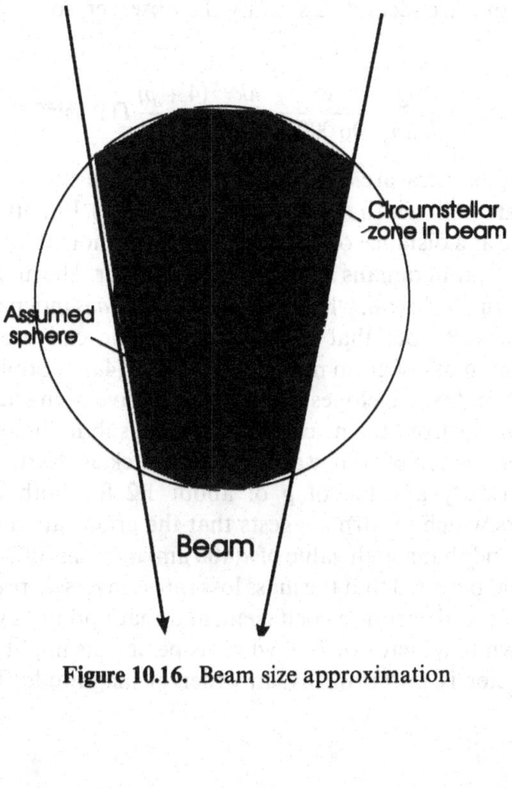

Figure 10.16. Beam size approximation

neglected (Figure 10.16), and show that the errors involved are fairly small. Using the Rayleigh–Jeans approximation for B_ν, and performing the integral over θ, (10.74) becomes:

$$F_\nu(\text{obs}) = \frac{3}{8}\frac{Q_\nu}{a\rho_g}\frac{\frac{dM_d}{dt}}{Vd^2}\frac{4k\nu^2}{c^2}\int_{r_0}^{b} T_d\,dr$$

Using the radial temperature dependence from (10.71), normalized so $T(r_1) = T_1$:

$$F_\nu(\text{obs}) = \frac{3}{8}\frac{Q_\nu}{a\rho_g}\frac{\frac{dM_d}{dt}}{Vd^2}\frac{4kT_1\nu^2}{c^2}\int_{r_0}^{b}\left(\frac{r}{r_1}\right)^{-2/(4+p)} dr$$

$$= \frac{3}{8}\frac{Q_\nu}{a\rho_g}\frac{\frac{dM}{dt}}{Vd^2}\frac{4k\nu^2 T_{1-i}}{c^2}\frac{\left(\frac{b}{r_1}\right)^{(2+p)/(4+p)}}{\frac{2+p}{4+p}} \qquad (10.74a)$$

where the lower limit of the integral (r_0/r_1) has been neglected compared with the upper limit.

Sopka *et al.* choose to put r_1 at a value corresponding to an angular distance from the star of one arc second as seen by the observer, so

$$F_\nu(\text{obs}) = \frac{3}{8}\frac{Q_\nu}{a\rho_g}\frac{\frac{dM_d}{dt}}{206\,000\,Vd}\frac{4k\nu^2\,(4+p)}{c^2\;(2+p)}T(1'')\phi^{(2+p)/(4+p)} \qquad (10.74b)$$

with T_1 the dust temperature one arc second from the star, $r_1 = d/206\,000$ and $b/r_1 = \phi$, the beam half- width measured in arc seconds. Finally, (10.73) gives the dust temperature at a distance of one arc second if we note that r/d is the angular distance from the star in radians as seen by the observer. The main problem lies in choosing values for p and Q/a, where we recall that Q/a is independent of a under some conditions and note that instead of Q/a one can use the absorption coefficient per unit mass of grain material, $\kappa_\nu = 3Q_\nu/(4a\rho_g)$. Sopka *et al.* compare observations of the dust envelopes of stars which have been studied over a wide range of wavelengths from the near infrared to the sub-millimetre and compare these with models (which of course are optically thick at shorter wavelengths) to determine empirically a value of p of about 1.2 for both carbon-rich and oxygen-rich stars which in turn suggests that the grains are amorphous rather than crystalline and that a high value of $\kappa(400\ \mu\text{m})$ microns) of 2 m^2/kg should be adopted. It should be noted that the mass loss rate is inversely proportional to the adopted value of the absorption coefficient, and that optimists would claim that this is now known to a factor of 2–3, whereas pessimists might reasonably hold that this parameter is uncertain by an order of magnitude. This uncertainty should only affect the absolute value of the mass loss rate and not the relative rates in stars of the same chemical type.

The same method has been applied to IRAS satellite observations at 60 μm, where again optical thinness can be assumed (e.g. Jura [29, 30]). The IRAS aperture size is large enough to assume that essentially all the envelope 60 μm flux is included. Then the flux received is given by

$$F_\nu(\text{obs}) = \frac{\kappa_\nu}{Vd^2}\frac{dM_d}{dt}\int \frac{\frac{2h\nu^3}{c^2}}{e^{h\nu/kT_D}-1}\,dr$$

Taking as before $T = T_1(r/r_1)^{-[2/(4+p)]}$ with $r_1 = d/206\,000$ and defining $x = h\nu/(kT)$,

$$F_\nu = \frac{\kappa_\nu}{Vd^2}\frac{dM_d}{dt}\frac{2h\nu^3}{c^2}\frac{4+p}{2}\frac{d}{206\,000}\left(\frac{kT_1}{h\nu}\right)^{(p/2+2)}\int_0^\infty \frac{x^{(p/2)+1}}{e^x-1}\,dx$$

Suppose the integral is $I(p)$ and take p as 1. Then

$$\frac{dM_d}{dt} = \frac{c^2}{h}\left(\frac{h}{k}\right)^{2.5}\frac{1}{2I(p)}\frac{2}{5}\nu^{-0.5}\frac{Vd*206\,000}{\kappa_\nu}\frac{F_\nu(\text{obs})}{T_1^{2.5}}$$

$$\simeq 2*10^{22}\frac{F_\nu(\text{obs})Vd}{\kappa_\nu T_1^{2.5}} \qquad (10.75)$$

where we have taken $\nu = 5*10^{12}\,\text{s}^{-1}$ corresponding to the wavelength of 60 μm and the integral $I(p)$ to give $I(p) \sim 2$, which has been used in the expression above.

Equation (10.73) determines T_1 for r_1 corresponding to one arc second, that is for $r_1/d = 1/20\,600$, and we can roughly replace $\int F_\nu \nu^p d\nu$ for $p = 1$ by $F_{TOT}\nu_{av}$, where F_{TOT} is the total observed flux and ν_{av} is the mean frequency of the observed flux, so $T_1 = 7.4*10^{-9}\,[2.25*10^{47}\nu_{av}F_{TOT}]^{0.2}$. If we also write $F_{TOT} = L_{TOT}/(4\pi d^2)$, we find

$$\frac{dM_d}{dt} \simeq \frac{5\times 10^{19}F_\nu(\text{obs}, 60\,\mu\text{m})\,V\,d^2}{\kappa_\nu \nu_{av}^{0.5}L_{TOT}^{0.5}} \qquad (10.76)$$

Observations at shorter wavelengths where the envelopes are in general optically thick need detailed modelling. The temperature is no longer simply given by (10.73), for the flux heating the dust is greater than that observed at shorter wavelengths. The 9.7 μm feature is an optical depth indicator, but in many cases is optically thick, and at large optical depths will turn from an emission feature into an absorption feature, because the observed flux comes from the cooler outer regions of the envelope. The transition from an emission feature to an absorption feature does not take place at the same optical depth as for a plane-parallel model (as was essentially used by Gehrz and Woolf in their pioneering work), because the spherical geometry means that the outer layers have larger volumes and therefore contribute more to the emission than inner layers, partly counterbalancing their lower temperatures. The models are complicated by the fact that at shorter wavelengths scattering is important and dust scattering is likely to be anisotropic.

All these methods of estimating the mass loss rate have assumed that mass loss is steady and isotropic. Observations in the few cases where the envelope can be resolved, such as α Ori, have shown that the envelope is not always symmetric about the star, and that the mass loss in these cases is anisotropic. The complicated structure observed in the outflow from χ Cyg suggests that the outflow is not always constant on short time scales, and the fact that some of the observed stars are evolving relatively rapidly would indicate that variations on long time scales might also occur. In principle, the different types of observation measure the mass loss rate at different distances from the star and hence at different epochs ($t \sim r/V$) so that CO for instance extends farther out than OH and hence on average refers to an earlier epoch of mass loss. Unfortunately the systematic errors that exist in all methods tend to be larger than all but the most dramatic variations in mass loss rate.

References

[1] C. Beals, *Monthly Notices of Royal Astronomical Society*, **91**, 966, 1931.
[2] I. Appenzeller *et al.*, *Astronomy and Astrophysics*, **141**, 108, 1984.
[3] S. Chandrasekhar, *Monthly Notices Royal Astronomical Society*, **94**, 534, 1934.
[4] J. I. Castor, *Monthly Notice Royal Astronomical Society*, **149**, 111, 1970.
[5] H. Lamers, M. Cerruti-Sola, and M. Perinotto, *Astrophysical Journal*, **314**, 726, 1987.
[6] D. Mihalas, P. Kumasz, and D. G. Hummer, *Astrophysical Journal*, **202**, 465, 1975.
[7] D. Morton, *Astrophysical Journal*, **150**, 535, 1967.
[8] A. E. Wright and M. Barlow, *Monthly Notices Royal Astronomical Society*, **170**, 41, 1975.
[9] J. Castor, D. Abbott, and R. Klein, *Astrophysical Journal*, **195**, 157, 1975.
[10] D. Abbott, *Astrophysical Journal*, **259**, 282, 1982.
[11] A. J. Deutsoch, *Astropphysical Journal*, **123**, 210, 1956.
[12] A. J. Deutsch, in J. L. Greenstein (ed.), *Stellar Atmospheres, Stan and Stellar Systems*, University of Chicago Press, 1960.
[13] R. Weymann, *Astrophysical Journal*, **136**, 844, 1960.
[14] D. Reimers, in B. Baschek, W. Kegel, and G. Traving (eds.), *Problems in Stellar Atmospheres and Envelopes*, Springer, 1971.
[15] F. Sanner, *Astrophysical Journal Supplement*, **32**, 115, 1976.
[16] A. P. Bernat, *Astrophysical Journal*, **213**, 756, 1977.
[17] L. Goldberg, L. Ramsey, L. Testerman, and Carbon, *Astrophysical Journal*, **199**, 427, 1975.
[18] A. P. Bernat, D. N. B. Hall, K. H. Hinkle, and S. T. Ridgway, *Astrophysical Journal*, **233**, L135, 1979.
[19] K. H. Hinkle, D. N. B. Hall, and S. T. Ridgway, *Astrophysical Journal*, **252**, 697, 1982.
[20] R. K. Honeycutt, A. P. Bernat, J. E. Kephart, C. E. Gow, M. T. Sandford II, and D. L. Lambert, *Astrophysical Journal*, **239**, 565, 1980.

[21] N. Mauron, B. Fort, F. Querci, T. Dreux, T. Fauconnier, and P. Lamy, *Astronomy and Astrophysics*, **130**, 341, 1984.
[22] G. R. Knapp, T. G. Phillips, R. B. Leighton, K. Y. Lo, P. G. Wannier, and H. A. Wootten, *Astrophysical Journal*, **252**, 616, 1982.
[23] M. Jura, *Astrophysical Journal*, **275**, 683, 1983.
[24] G. A. Mammon, A. E. Glassgold, and P. J. Huggins, *Astrophysical Journal*, **328**, 797, 1988.
[25] M. Morris, *Astrophysical Journal*, **236**, 823, 1980.
[26] J. Kwan and R. A. Linke, *Astrophysical Journal*, **254**, 587, 1982.
[27] N. Netzer and G. Knapp, *Astrophysical Journal*, **323**, 739, 1987.
[28] R. P. Gerhz and N. J. Woolf, *Astrophysical Journal*, **165**, 285, 1971.
[29] M. Jura, *Astrophysical Journal*, **303**, 327, 1986.
[30] M. Jura, *Astrophysical Journal*, **313**, 743, 1987.

11 VERY HOT THIN GASES AND CORONAE

In this chapter we discuss fairly briefly the radiation of high temperature, low density gas. Such a gas is found in the solar corona, the coronae of other stars, the high temperature component of the interstellar medium, and the intergalactic medium in clusters of galaxies. Most of these environments have temperatures of 500 000 K to 4 000 000 K, which are to be compared with the hottest stellar photospheres which have effective temperatures of 70 000 K to 100 000 K and relatively high densities, and photoionized gases which have low densities but temperatures in the range 7000 K to 25 000 K. Gas in the intermediate temperature range of 100 000 K to 500 000 K can be detected but is comparatively rare, and will be mentioned briefly in connection with the solar chromosphere and the chromosphere–corona transition. The energy source for the very hot gas to be discussed in this chapter is not radiation, but mass motions such as stellar winds and supernova shells impinging on the interstellar medium, and magnetic and acoustic waves.

The heating mechanisms listed above (mass motions, magnetic waves) together with absorption of ultraviolet radiation for photoionized regions, are largely independent of the temperature of the gas, but cooling by line and continuum radiation is strongly temperature dependent, since the degree of ionization is temperature dependent, and the reason for the temperature 'gap' lies in this temperature dependence of the cooling. Thus as far as cooling via the emission of continuum radiation is concerned, bound–free emission from hydrogen, which is very important at temperatures around 10 000 K, disappears at higher temperatures, leaving only free–free or bremstrahlung as a major source of cooling through continuum radiation. Now free–free is a collisional process, and under very hot gas conditions line emission is nearly always the result of collisional excitation from the ground state followed by radiative decay to the ground state, the main colliding particles being free electrons. At high temperatures hydrogen and helium are completely ionized so $n_e = n_H + 2n_{He} = 1.2n_p$. Thus the free–free emission is proportional to the density squared or, more precisely, to a sum of terms like $n_e n_{ion} Z^2$ and similarly the line emission is proportional to $n_{ion} n_e$. Since n_{ion} for a given temperature is proportional to the total density or to n_p, it is customary to write the overall radiative emission per unit volume integrated over frequency as

$$L = \int 4\pi j_\nu \rho \, d\nu = n_e n_p \Lambda(T) \, \mathrm{W\,m^{-3}} \tag{11.1}$$

where $\Lambda(T)$ is the *radiative cooling function*. For free–free alone, equation (6.23) gives

$$4\pi j_\nu \rho = 6.8 \times 10^{-51} \frac{N_e g_{ff} e^{-h\nu/kT}}{T^{1/2}} \sum_{ion} Z_{ion}^2 N_{ion}$$

$$\int_0^\infty 4\pi j_\nu \rho \, d\nu = \frac{k}{h} \times 6.8 \times 10^{+51} g_{ff}(\mathrm{av}) T^{1/2} N_e \int_0^\infty e^{-x} dx \sum_{ion} N_{ion} Z_{ion}^2$$

$$= 1.57 \times 10^{-40} T^{1/2} N_e N_H \sum \frac{N_{ion}}{N_H} Z^2 \tag{11.2}$$

where $g_{ff}(\mathrm{av})$ is an average of the Gaunt factor over frequency. It follows that for normal abundances:

$$\Lambda(T) \sim 2.5 * 10^{-40} \, T^{1/2} \, \mathrm{Wm^3} \tag{11.2a}$$

Bremstrahlung dominates the cooling at temperatures greater than 10^7 K, but at lower temperatures, line emission provides most of the cooling. The higher the temperature, the higher the mean kinetic energy available for excitation, but also the higher the ionization since ionization is mainly collisional, and in general, the higher the degree of ionization, the farther apart are the energy levels and the greater the excitation energy required for resonance lines. The balance between these two factors results in line radiation dominating bremstrahlung between a few times 10^4 K and a few times 10^6 K, with a peak in $\Lambda(T)$ roughly between 10^5 and $2.5 * 10^5$ K and with the cooling function on the high temperature side of this peak very roughly given by

$$\Lambda(T) \sim 6 * 10^{-32} T^{-0.6} \, \mathrm{Wm^3} \tag{11.3}$$

although in detail the cooling curve as a function of temperature between $2.5 * 10^5$ K and 10^7 K is quite complicated, at first falling rapidly at $T = 4 * 10^5$ K, then pausing as the temperature rises further before falling gradually for temperatures greater than $2 * 10^6$ K. The cooling is by strong ultraviolet lines between 91.2 and 12 nm with lines of carbon and oxygen playing a major role near $2 * 10^5$ K and lines of oxygen, silicon, sulphur and iron being important near $T = 10^6$ K. Gas heated by some mechanism to temperatures of a few million degrees and left to radiate will cool slowly at first, then rapidly as the cooling function rises to a maximum when the temperature passes through 10^6 K, and then will slow down as the temperature approaches 10^4 K. Thus we expect to see little gas with temperatures between a million and a few tens of thousand degrees.

Line Radiation from Very Hot, Thin Gases

In hot, thin gases, both excitation and ionization are usually collisional, since the large energies involved would require X-ray wavelength radiation if radiative excitation or ionization was to be important and stellar photospheres do not usually produce much radiation at such wavelengths. The situation may be different near AGN and indeed near some stellar accreting black holes. Collisionally excited lines are optically thin and can be treated in much the same way as collisionally excited lines from photoionized gas. In many cases, such as resonance lines, a two-level model is adequate. In general for a transition between level j and level i

$$I_\nu = \int j_\nu \rho \,\mathrm{d}x$$

$$= \int \frac{A_{ji}N_j}{4\pi} h\nu \,\mathrm{d}x$$

If the excitation can be modelled using only the two levels j and i, $N_i C_{ij} = N_{ji} A_{ji}$ and

$$I_\nu = \int \frac{C_{ij}N_i}{4\pi} h\nu \,\mathrm{d}x$$

$$= \frac{h\nu}{4\pi} \frac{N_{\mathrm{el}}}{N_{\mathrm{H}}} \frac{N_{\mathrm{H}}}{N_{\mathrm{e}}} 8.6 * 10^{-12} \frac{\Omega_{ij}}{g_i} \int \frac{N_{\mathrm{ion}}}{N_{\mathrm{el}}}(T) \frac{e^{-E_{ij}/kT}}{T^{1/2}} N_{\mathrm{e}}^2 \,\mathrm{d}x \tag{11.4}$$

where

$$C_{ij} = 8.6 * 10^{-12} \frac{\Omega_{ij} e^{-E_{ij}/kT}}{g_i T^{1/2}} N_{\mathrm{e}} \tag{2.28}$$

and

$$N_{\mathrm{ion}} = \left(\frac{N_{\mathrm{ion}}}{N_{\mathrm{el}}}\right)\left(\frac{N_{\mathrm{el}}}{N_{\mathrm{H}}}\right)\left(\frac{N_{\mathrm{H}}}{N_{\mathrm{e}}}\right) N_{\mathrm{e}}$$

and it has been assumed that Ω_{ij} is approximately independent of energy. For a completely ionized mixture of 90% hydrogen and 10% helium, $N_{\mathrm{e}} \sim 1.2 N_{\mathrm{H}}$, and $N_{\mathrm{el}}/N_{\mathrm{H}}$ is just the abundance of the element concerned. Ω_{ij} is the collision strength, which is of order 1 for forbidden transitions. For permitted lines in ions, if experimental or calculated values of Ω_{ij} are not available, Van Regemorter's formula is often used with $\Omega_{ij} = 2.32 * 10^{-18} g_i f_{ij}/E_{ij}$ times a factor that is around

0.2 for transitions where the quantum number n changes, and around 0.7 for those where n does not change (remembering that $E_{ij} = h\nu$ is $1.6*10^{-18}$ for an excitation energy difference of 10 eV). Formula (11.4) still applies in a system with more than two significant levels (provided that upper level j is only populated to any appreciable degree by collisions from the ground state) if (11.4) is multiplied by the fraction of downward transitions that go to the ground state, $A_{ji}/\Sigma A_{jk}$, where the sum is over all states k which are lower in energy than j.

The state of ionization is determined by the balance between collisional ionizations and radiative recombinations since the radiation field at very short wavelengths is too weak for photoionization and the density is too low for collisional three-body recombinations. Radiative recombination involves the meeting of an electron with an ion to produce the next lower stage of ionization and a continuum photon, so this too is a collisional process. Thus both ionization and recombination involve a collision, in the former case most usually with an electron and in the latter case always with an electron, so that the rates of both are proportional to the electron density.

Recombination can take place either directly to a bound single excited electron level of the recombined particle or to a level with two electrons excited which may be bound but lie above the single electron ionization limit. Such levels exist because if two electrons are excited, both will be less shielded by the remaining electrons from the nucleus than is the case if only one electron is excited. In the case of these levels, the excited level can transition back to the ionized state without input of radiative or collisional energy and is called an autoionizing level. Usually an autoionizing level rapidly ejects an electron, but occasionally manages to drop to a lower state of excitation before this can happen, leaving only one electron excited, which can decay at its leisure. This process is called *dielectronic* recombination. Both ordinary radiative recombination and dielectronic recombination rates are proportional to the electron density, but the former is only weakly dependent on the temperature, whereas the latter involves not only the incoming electron recombining but also collisionally exciting another electron with the usual $\exp(-E/kT)$ factor for collisional excitations, giving a very strong temperature dependence. Thus it is only at coronal temperatures that dielectronic recombination in some cases becomes the dominant mode of recombination.

In the steady state one can write $N_i I(i \rightarrow i+1) = N_{i+1} R(i+1 \rightarrow i)$, where I and R are the ionization and recombination rates, and the set of such equations can be solved for the relative populations N_i of the various stages of ionization of an element. Ionization rates are proportional to $T^{1/2}\exp(-I/kT)$, and since I increases with the degree of ionization, one would expect the predominant stage of ionization at any temperature to be the one for which $I \sim kT$. However, where an ion has a closed shell of electrons it is particularly difficult to remove an electron, and the ionization potential takes a large jump from that of the stage of ionization one short of a closed shell of electrons. On the other hand ionization potential of the ion with a closed shell of electrons plus one is only moderately higher than that of the closed shell ion. The result is that stages of ionization with closed shells are completely dominant over the appropriate temperature ranges and the temperature range over which such ions are present is much larger than for other ions. An example is oxygen, where the ionization potentials of O VI, O VII and O VIII are 138 eV, 739 eV and 871 eV, respectively, since O VII has a helium-like structure with two electrons in a closed shell. Oxygen is nearly entirely O VII from 220 000 K to 1 600 000 K whereas O VI never exceeds about 20% of all oxygen and only approaches this value in the temperature range 16 000 to 220 000 K. Autoionizing states of O VI are produced from O VII by recombination and excitation, but the excitation in a helium-like ion requires considerable energy and therefore relatively high temperatures. On the other hand, O VII is abundant over a large temperature range and so there is a high temperature tail to the distribution of O VI extending up to 10^6 K, although the abundance at this temperature is only about 0.3%.

In studies of the solar corona, it is customary to lump together the temperature dependent factors into $G(T)$ so

$$G(T) = \frac{N_{\rm ion}}{N_{\rm el}} \frac{1}{T^{1/2}} e^{-E_{ij}/kT}$$

$$I_\nu = 8.6 \times 10^{-12} \frac{h\nu}{4\pi} \frac{N_{\rm el}}{N_{\rm H}} \frac{N_{\rm H}}{N_{\rm e}} \frac{\Omega_{ij}}{g_i} \int N_{\rm e}^2 G(T) {\rm d}x \qquad (11.5)$$

from (11.4). The integral is now written as $\langle G(T) \rangle \int N_{\rm e}^2 {\rm d}x$, where $\langle G(T) \rangle$ is taken by Pottasch as 0.70 $G(T_{\rm max})$ where $G(T_{\rm max})$ is the maximum value of $G(T)$.

The observation of a line from a highly ionized species shows that gas at $T_{\rm max}$ for that line is present. It will be seen from equation (11.4) that the ratio of the intensities of two lines of a particular ion from levels of different excitation will just be in the ratio of the collisional excitation rates from the ground state times the photon energies, and if the excitation energies of the two levels are E_1 and E_2:

$$\frac{I_{2i}}{I_{1i}} = \frac{\Omega_{i2}}{\Omega_{i1}} \frac{\nu_{2i}}{\nu_{1i}} e^{-(E_2 - E_1)/kT} \qquad (11.6)$$

If there is a large energy difference, the intensity ratio will be very sensitive to temperature, although this equally means that the ratio is only usable over a small temperature range, and the lines will be far apart in wavelength which causes difficulties in calibration for the above-atmosphere detectors necessary for observing very hot gases. The strong temperature dependence means that a very small volume of very hot gas can change the ratio observed from a much larger volume of cooler gas appreciably. The method is similar in principle to the O III line ratios used in gaseous nebulae.

Sometimes *satellite* lines are observed in X-ray spectra on the long wavelength sides of resonance lines of hydrogen-like and helium-like ions. Take for example the first resonance line of a hydrogen-like ion, $2p \rightarrow 1s$, the equivalent of Lyman α. If an additional outer electron is present, say nl with $n \geqslant 2$, then the transition of the resulting helium-like ion, $2pnl \rightarrow 1snl$ will be similar to the $2p \rightarrow 1s$ transition of the hydrogen-like ion, but because of the small degree of shielding of the transitioning electron in the 2p state by the extra outer electron the energies involved will be slightly smaller and similarly the wavelength of the line produced will be slightly longer than that of the resonance line. Now consider the transition of a helium-like ion, $1s2p \rightarrow 1s^2$, and the equivalent transition of a lithium-like ion with an additional outer electron, $1s2pnl \rightarrow 1s^2nl$, which will again lie at a slightly longer wavelength than the helium-like transition, appearing as a satellite line. The satellites are only well separated from the parent lines for $n = 2$ and if the parent line is helium-like, the satellite may have several components—for instance $1s2p \rightarrow 1s^2$ gives rise to satellites $1s2p2s \rightarrow 1s^22s$ and $1s2p^2 \rightarrow 1s^22p$ (Figure 11.1).

The diagnostic significance of the satellite lines lies in the way that they are excited. The upper levels of satellite lines involve two excited electrons and therefore can be formed by dielectronic recombination to an autoionizing level. For example, 1s2p2s of a lithium-like ion can be formed from the ground state of a helium-like ion, $1s^2$, when an electron recombines in such a way that one of the 1s electrons is also collisionally excited to 2s. A two electron excited level of a lithium-like ion lies above the ionization limit for one electron excited levels ($1s^2$ nl as $n \rightarrow \infty$) and hence can undergo a radiationless autoionizing transition back to a helium-like ion and an electron. Indeed, such an autoionizing transition is the most likely fate for such a level, but there is a chance that the level will radiatively decay to a one electron excited state below the ionization limit with the emission of the satellite line.

We have for the satellite line:

$$4\pi j_\nu \rho = N(\text{He-like}) N_e \frac{A_s}{A_s + A_a} C h\nu \tag{11.7}$$

where $CN(\text{He-like})N_e$ is the dielectronic capture rate per unit volume, A_s is the transition probability for the satellite line and A_a is the probability of autoionization. This can be simplified by noting that in full equilibrium:

$$1s2s2p \rightleftarrows 1s^2 + e$$

with $N(1s2s2p)\ A_a = N(1s^2)N_eC$. In full equilibrium the population ratio of 1s2s2p to $1s^2$ can be written using Saha's equation:

$$\frac{C}{A_a} = \frac{N(1s2s2p)}{N(1s^2)N_e} = \left(\frac{h^2}{2\pi m_e kT}\right)^{3/2} \frac{1}{2} \frac{g(1s2s2p)}{g(1s^2)} e^{I_s/kT} \tag{11.8}$$

where I_s is the difference in energy between the two electron excited level and the ground state of the helium-like ion (note that I_s is negative since the 'ion' level lies

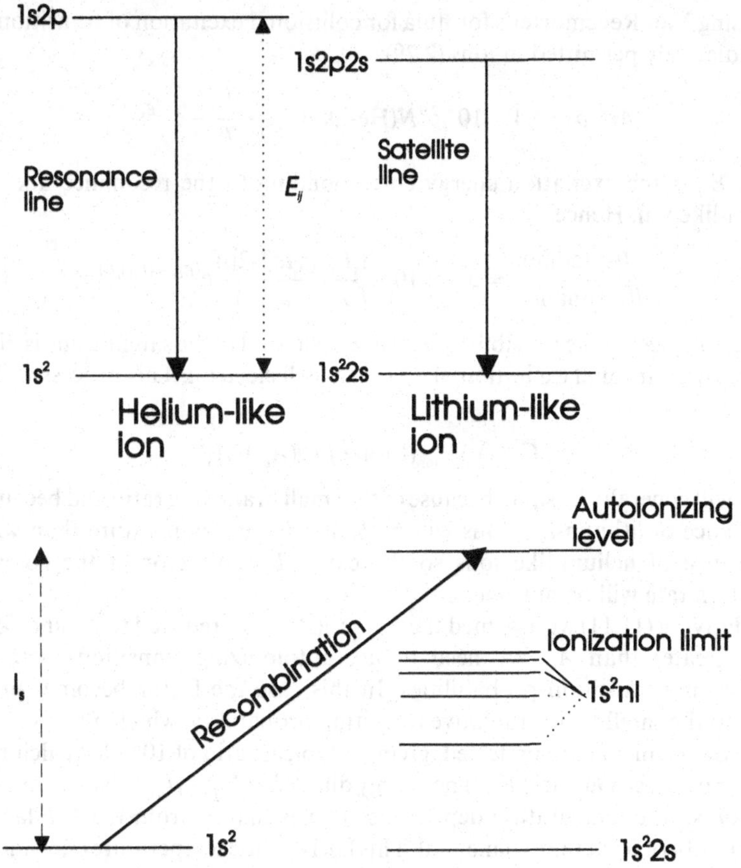

Figure 11.1. Satellite lines

lower in energy than the 'atom' level) and g is the statistical weight. Now $A_s \ll A_a$, so

$$4\pi j_\nu \rho = N(\text{He}-\text{like})N_e A_s \left(\frac{h^2}{2\pi m_e kT}\right)^{3/2} \frac{g(1s2s2p)}{2g(1s^2)} e^{-|I_s|/kT} \tag{11.9}$$

Now the emission coefficient for the resonance line of the helium-like ion is given by the number of collisional excitations per unit volume, neglecting a roughly 10% contribution from dielectronic recombinations to the upper level of the transition (that is, satellite line components that are not resolved from the resonance line since the shielding electron has a high quantum number), so

$$4\pi J_\nu \rho = C_{ij} N_{\text{ion}}(\text{He-like})$$

and using Van Regemorter's formula for collisional excitation of transitions that are radiatively permitted in ions (2.29):

$$4\pi j_\nu \rho = 5.4 \times 10^{-29} N(\text{He-like}) N_e \frac{f}{E_{ij} T^{1/2}} e^{-E_{ij}/kT} \tag{11.10}$$

where E_{ij} is the excitation energy corresponding to the resonance line of the helium-like ion. Hence

$$\frac{I(\text{satellite})}{I(\text{resonance})} = 3.8 \times 10^6 \frac{A_s E_{ij}}{f T} \frac{g(1s2s2p)}{g(2s^2)} e^{(E_{ij} - |I_s|)/kT} \tag{11.11}$$

We have neglected the possibility that the upper level of the satellite line is directly excited by collisional excitation of an inner shell electron. This process will have a rate:

$$C_{ij}(\text{sat}) N_{\text{ion}}(\text{Li-like}) A_s/(A_s + A_a)$$

which will generally be small because of the small branching ratio and because the abundance of lithium-like ions will peak at a lower temperature than will the abundance of helium-like ions so the $\exp(-E/kT)$ factor in the collisional excitation rate will be much less.

In deriving (11.11) we assumed that $A_a \cdot A_s/(A_a + A_s)$ reduced to A_s since A_a was much greater than A_s. We need to use autoionizing transitions with large autoionizing transition probabilities. In this case the factor becomes proportional to the satellite line radiative transition probability, which increases as Z^4, so heavier elements are preferred, giving a typical ratio of 10% for calcium and 50% for iron (see Gabriel [1]). The energy difference $E_{ij} - |I_s|$ is usually much less than kT so the temperature dependence comes mainly from the $1/T$ factor in (11.11) and not from the exponential. This fairly gentle temperature dependence is an advantage, giving a reasonably wide useful temperature range limited by the degree of ionization and not overly dependent on small volumes of very hot gas. The predicted line ratio is completely independent of density, whereas most other temperature diagnostics have a weak dependence on density. Observationally the fact that the resonance and satellite lines are close together in wavelength removes calibration problems. This method has been particularly applied to solar flares where the temperatures are high enough to produce helium-like ions of heavy elements, an example being the resonance line of helium-like Ca XIX at 0.3176 nm and its various satellites (different *nl*, different couplings to give different *L*, *S* and *J*) between 0.321 and 0.319 nm. Another example at still higher temperatures is the helium-like Fe XXV resonance line at 0.185 nm with its satellites at longer wavelengths. Calcium temperatures of around $1.5 * 10^7$ K and iron temperatures of around $2.2 * 10^7$ K have been recorded at some stages of a flare. It should be noted that in these transient events complete equilibrium may not be reached and the ionization may lag behind that appropriate to the kinetic temperatures found by the above method.

We now turn to the question of measuring densities (more strictly speaking electron densities) in very hot gases. The line diagnostics used here are similar in principle to those used for gaseous nebulae, and hinge on the competition between radiative de-excitation and collisional de-excitation of a level, the importance of the latter depending on the electron density. The level will be collisionally excited at a rate that will be proportional to the electron density and if the de-excitation is purely radiative the line strength per atom will just be proportional to the excitation rate and hence to the density, and the total line strength will go up as the density squared, but if collisional de-excitation becomes dominant, both excitation and de-excitation will increase with the electron density so the fraction of atoms excited will saturate at a value given by Boltzmann's equation and independent of the density and the total line intensity will be proportional to the density. The change in the way that line strength varies with density will occur when $C_{ji} \sim A_{ji}$ with $C_{ji} \propto N_e$ so that if we can observe a line ratio in the transition region we can determine the density. For gaseous nebulae we used the ratio of two such density sensitive forbidden lines of O II or S II, but we can also use the ratio of a line which is sensitive to collisional de-excitation to one which is not sensitive.

Now the electron density in very hot gases is low so collisional de-excitation can only compete with radiative de-excitation if the probabilities of downward radiative transitions from a level are all low, which usually means forbidden or semi-forbidden transitions. The chosen level must then be a metastable one. Thus the simplest arrangement is when one has a ground state and two excited levels 2 and 3 with 2 having permitted radiative decays to the ground state and 3 having only forbidden transitions downwards (Figure 11.2). Then $C_{21} \ll A_{21}$ and

$$N_1C_{12} = N_2A_{21} \quad \text{with} \quad 4\pi j_{12}\rho = N_2A_{21}h\nu_{21} = N_1C_{12}h\nu_{21}$$

$$N_1C_{13} = N_3(A_{31} + C_{31}), \text{ with } \quad 4\pi j_{31}\rho = N_3A_{31}h\nu_{31}$$

$$= N_1C_{13}/(A_{31} + C_{31})A_{31}h\nu_{31}$$

so

$$\frac{I_{31}}{I_{21}} = \frac{j_{31}}{j_{21}} = \frac{C_{13}\nu_{31}}{C_{12}\nu_{21}} \frac{A_{31}}{A_{31} + C_{31}} \tag{11.12}$$

where C_{31} may be replaced more generally by $C_{31} + C_{32}$ or indeed the sum of de-exciting collisions to all other levels and the system of equations may be elaborated in various ways.

Note that ideally levels 2 and 3 should be close in excitation energy since $C_{13}/C_{12} \propto \exp([E_1 - E_3]/[kT])$ and the smaller the sensitivity to temperature when density is being measured, the better. Reasonably close correspondence in energy also ensures that the wavelengths of the lines are close together in wavelength so that calibration problems are again minimized.

An example making an interesting contrast with the case of O II in gaseous nebulae is that of Fe IX where the ground state $3p^6\ ^1S$ (leaving out the inner

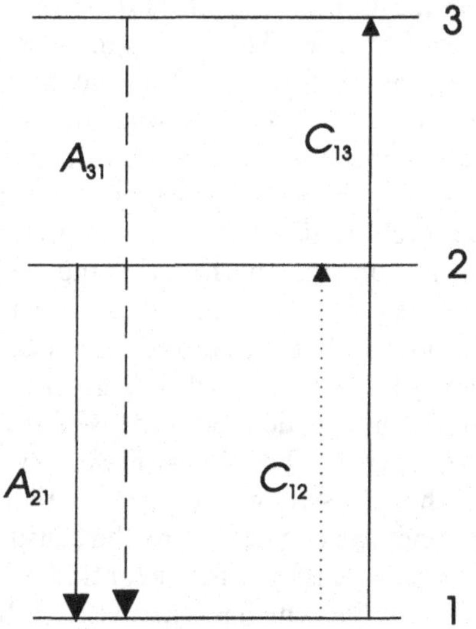

Figure 11.2.

electron configuration) connects to two fine structure levels of the first excited configuration $3p^5 3d\ ^3P_1$ and $^3P_2$ by the lines 24.491 nm and 24.174 nm, the first of these lines being permitted (strictly semi-forbidden since it is an intercombination transition) with a transition probability of $9.3 * 0^6\ s^{-1}$ and the second being forbidden since $\Delta J = 2$ and having a transition probability of $71\ s^{-1}$. Transitions from the third fine structure level $^3P_0$ to the ground state are very strongly forbidden since they involve a $J=0$ to $J=0$ transition, and the radiative transition probabilities between the fine structure levels are small so $^3P_2$ is metastable (Figure 11.3). Equation (11.12), modified to include collisional transitions between $^3P_2$ and $^3P_1$ and with other levels of the $3p^5 3d$ configuration, now holds. At low densities, collisional de-excitation is not important and the ratio of the two lines depends only on the relative collisional excitation rates, which since the excitation energies are nearly equal, is essentially the ratio of the collision strengths which for two such similar transitions should be approximately the ratio of the statistical weights of the final levels ~ 1.7. However at high densities collisional equilibrium will bring the populations of the fine-structure levels into the Boltzmann ratio, and the line intensity ratio 24.174/24.491 will be 5/3 $A(24.17)/A(24.49)$ which will be very small—unlike the case of the [O II] lines where the ratio fell to a constant value because the line transition probabilities were in that case similar. The density dependence here is a good diagnostic in the range 10^{16} to $10^{18}\ m^{-3}$, the ion existing at temperatures around 10^6 K.

The boron-like ions have a fairly complicated energy level structure, with a $2s^2 2p\ ^2P$ ground state with fine structure levels $J = 1/2$ (the lowest) and $J = 3/2$.

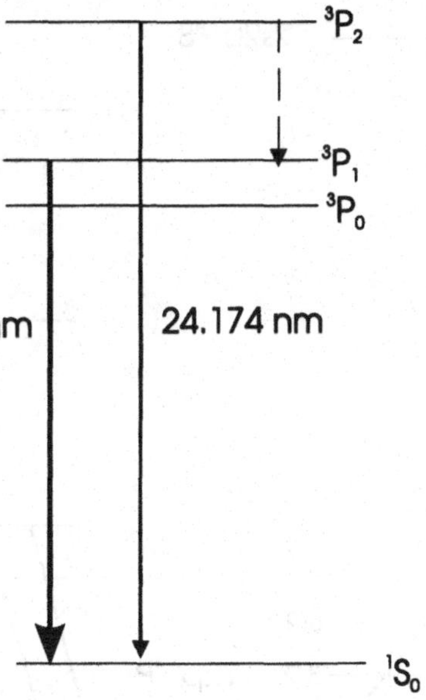

Figure 11.3. Fe IX lower levels

The $2s2p^2$ configuration gives rise to the doublets ^2D with $J = 5/2$ lower than $J = 3/2$, ^2P with $J = 1/2$ lower than $J = 3/2$ and ^2S, and the quadruplet ^4P with $J = 1/2$ (the lowest), $J = 3/2$ and $J = 5/2$. The next lowest quadruplet level is ^4S (Figure 11.4). A number of density dependent line ratios can be extracted from this system, where it will be noted that ^4P is metastable since the intercombination line to the ground state is semi-forbidden. Thus both ^2D and ^4P are collisionally excited, but the ^4P levels being metastable may also be collisionally de-excited at high densities and acquire a Boltzmann population relative to the ground state, weakening lines from this level relative to, say, a resonance line from ^2D. Thus the ratio $I(^4P \rightarrow {}^2P)/I(^2D \rightarrow {}^2P)$ should be a density indicator where for the $^4P \rightarrow {}^2P$ one could choose one of the 140.7/140.4/140.1/ 139.9 nm multiplet in O IV or at higher temperatures one would normally choose from the equivalent multiplet in Mg VIII the 79.4 or 78.3 nm lines, while from the $^2D \rightarrow {}^2P$ multiplet one might choose 79.0 nm for O IV and 43.0 nm for Mg VIII. It is also possible to use the relative strengths of lines within the $^2D \rightarrow {}^2P$ multiplet, say 43.04 nm connecting $^2D_{3/2}$ with $^2P_{1/2}$ and 436.6 nm connecting $^2D_{5/2}$ with $^2P_{3/2}$ in Mg VIII. The point here is that $^2D_{3/2}$ is fed by collisions from both $^2P_{1/2}$ and $^2P_{3/2}$ whereas $^2D_{5/2}$ is fed mainly by collisions from $^2P_{3/2}$ since $\Delta J = 2$ collisions are less common. Hence the relative populations of the ^2D levels are determined by the relative populations of the two ^2P levels. At low densities the population of the higher $^2P_{3/2}$ level will be very small, increasing as the density and collisional

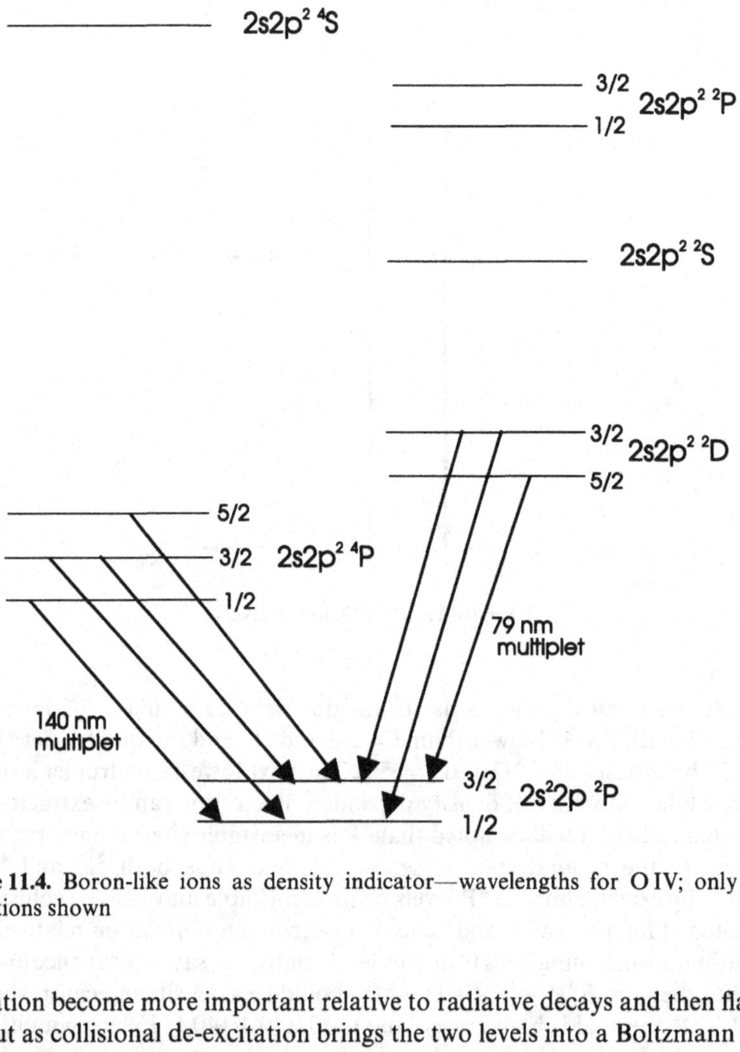

Figure 11.4. Boron-like ions as density indicator—wavelengths for O IV; only some transitions shown

excitation become more important relative to radiative decays and then flattening out as collisional de-excitation brings the two levels into a Boltzmann ratio. Hence $I(43.0)/I(43.66)$ will at first fall with increasing electron density and then remain constant.

Consider now a case where collisional excitation introduces a density dependence. O VII is helium-like and therefore has a $1s^2\,{}^1S$ ground state, singlet excited states 1s2s ${}^1S$ and 1s2p ${}^1P$, and triplet excited states 1s2s ${}^3S$ and 1s2p ${}^3P$. There is no line permitted transition from $1s2s\,{}^1S$ to 1s2s ${}^1S$ since no electron changes orbital angular momentum, but there are two photon continuum transitions like those of hydrogen. The 2.16 nm transition from ${}^1P$ to ${}^1S$ is permitted and is the resonance line. Transitions from the triplet to the singlet system are semi-forbidden ($\Delta S = 0$ is a weak selection rule) so the 2.18 nm line from ${}^3P$ to the

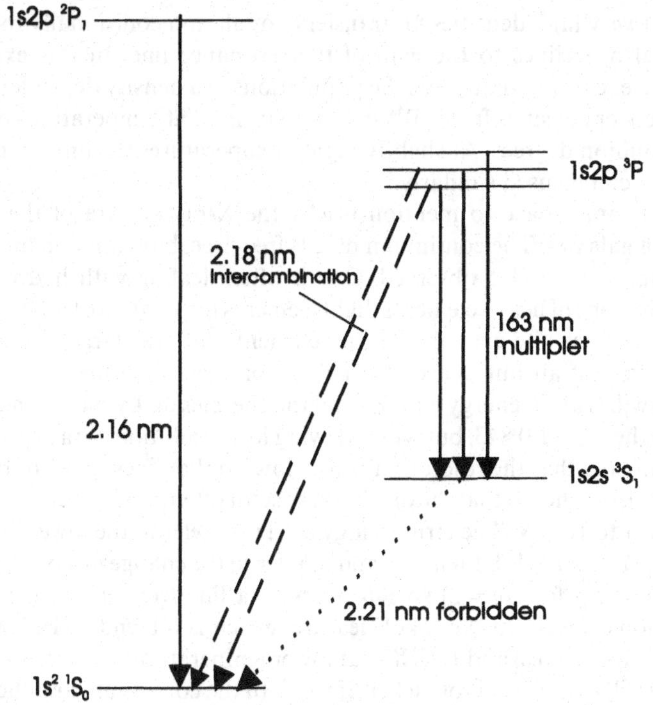

Figure 11.5. O VII levels

ground state is a semi-forbidden intercombination line, whereas the transition at 2.21 nm from ^{3}S to the ground state is not only an intercombination line but also breaks the selection rule $\Delta l = 1$ and hence is a forbidden line. Finally the radiative transfer from ^{3}P to ^{3}S is a permitted line at 163 nm (Figure 11.5). The transition probability for the two photon process lies in between that for an intercombination line and that for a forbidden line.

At low densities all four excited levels are collisionally excited and decay radiatively. The collisional excitation rate is higher for the lower ^{3}S level than for ^{3}P and so despite being more highly forbidden the 2.21 nm line is stronger than the 21.8 nm line. At higher densities collisional excitation from ^{3}S to ^{3}P competes with the forbidden line so that the population of ^{3}P increases relative to ^{3}S and the 2.18 nm line becomes much stronger than the 2.21 nm line (of course the 163 nm permitted line between these levels must be taken into account in numerical calculations). In effect, excitations to both ^{3}S and ^{3}P decay eventually through the 2.18 nm line because of the strong radiative and collisional coupling of the levels. At rather higher densities, 1s2s ^{1}S transfers to ^{1}P by collisions, increasing the strength of the resonance line at the expense of two photon decays.

Finally at very high densities ^3P transfers population collisionally to ^1P so that 2.18 nm also declines to the gain of the resonance line. In this example it is collisional excitation that alters the populations in a density dependent way. The changes cover densities from $10^{16}\,m^{-3}$ to $10^{24}\,m^{-3}$ at temperatures of the order of a few million degrees. At slightly higher temperatures the lines of the helium-like Ne IX can be used similarly.

There is only space to mention briefly the X-ray spectra of the hot gas in clusters of galaxies. The continuum here is free–free, but emission lines of highly ionized species are also observed. One is often dealing with hydrogen-like or helium-like ions of heavy elements (light elements may have lost all their electrons in the hottest gas and even the heavier elements with the large atomic numbers Z will have lost all but one or two). The corresponding lines of one electron elements will scale in energy with Z^2 and the energies of 'Lyman α' and 'Lyman β' will be in the ratio 1:0.843, but when viewing low resolution X-ray spectra it must be remembered that the resolution is very low, so that lines may not be resolved. It should also be noted that the displayed spectra often cover an enormous energy range, say 1 to 10 keV. The terminology used is to refer to the lower level of a line by a letter (K for $n = 1$, L for $n = 2$) and to refer to the change in n by a Greek letter (α for $\Delta n = 1$, β for $\Delta n = 2$) so Kβ refers to a line from $n = 3$ to $n = 1$. Early observations picked up the '7 keV feature' which is a blend of Fe XXVI (at the highest temperatures) and Fe XXV (at lower temperatures) K lines together with weaker similar transitions of nickel. Since both the continuum and the lines have strengths proportional to N_e^2, the equivalent width of the lines should depend on the abundance and not on the density. If the temperature is found from the continuum shape, it is possible to determine the iron abundance, which turns out to be surprisingly large. The resolution of Kβ from Kα of Fe XXV, and the relatively large strength of Kβ confirmed the thermal origin of these lines. At lower energies (longer wavelengths) the K lines of elements with lower atomic numbers like oxygen, silicon, magnesium, and sulphur are found, together with the L transitions of a variety of iron ions.

The Solar Corona

The Solar corona forms a special case in the observation of very hot and thin gas around stars because it is well resolved, and because the very low intensity visible light scattered from the photosphere can be detected. The corona has a temperature of the order of 10^6 K and a scale height (roughly the distance over which the density halves) of the order of $5 * 10^7$ m, or around one-tenth of a solar radius. Gas at this sort of temperature emits mainly in the X-ray region, where indeed it is easily observed since the photosphere emits no energy at these wavelengths. However, the electrons in the ionized gas will also scatter light from the photosphere, and since the photosphere emits mainly in the visible, this will be mainly visible light. The intensity of the scattered light is very low—only of the

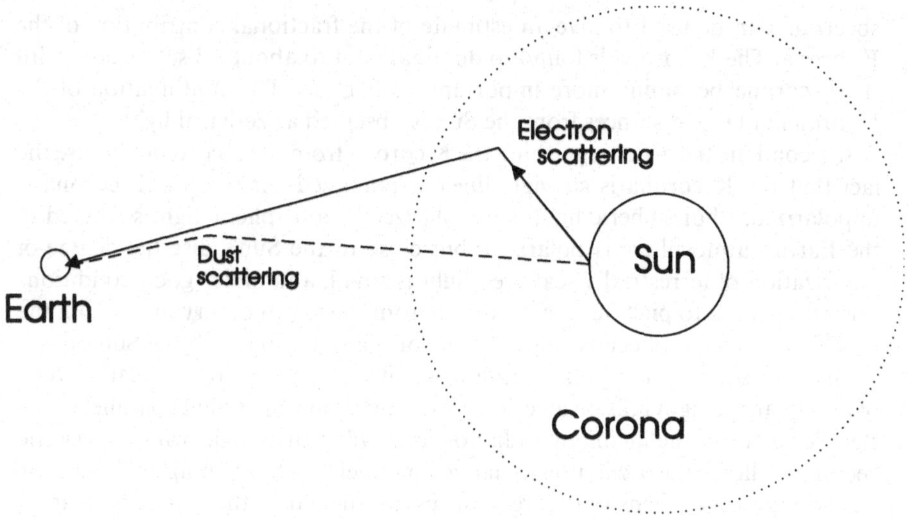

Figure 11.6. K and F coronae (not to scale)

order of 10^{-6} of the photospheric intensity even close to the Sun—and much less than the intensity of the sky background of sunlight scattered in the Earth's atmosphere close to the Sun, so that the visible corona can usually only be observed from the ground during eclipses.

A major difficulty is that the visible light corona has a major contribution from the scattering of photospheric radiation by dust particles in the interplanetary medium. Dust particles cannot exist closer to the Sun than about 4 solar radii, but when we look along a particular line of sight we will see both radiation scattered through fairly large angles by electrons close to the Sun and radiation scattered through very small angles by dust particles far from the Sun (Figure 11.6). The scattered spectrum will have the form but not the intensity of that from the photosphere but the electrons will be moving rapidly at velocities of the order of 5000 km s^{-1} so that light scattered by them will have large Doppler shifts and photospheric absorption lines will be broadened on scattering to widths of the order of 10 nm and hence will be almost totally washed out. On the other hand the slowly moving dust particles will not broaden lines on scattering and the dust scattered spectrum will show the photospheric absorption lines in full. The electron scattering contribution, the 'true' corona, is called the K corona, where K stands for 'kontiniuerlich' or 'continuum', while the dust scattered component is called the F corona, where F stands for Fraunhofer since the absorption lines in the Solar spectrum were originally called Fraunhofer lines. The strengths of the Fraunhofer lines in the observed coronal spectrum at any given projected distance from the Sun relative to the strength of the same lines in the photospheric spectrum can be used to give an estimate of the fractional contribution of the F corona. The K corona is found to dominate out to about 2.3 solar radii, with the F corona becoming more important farther out. The continuation of the F corona at large distances from the Sun is observed as zodiacal light.

A second means of distinguishing the K corona from the F corona is to use the fact that the K corona is strongly linearly polarized whereas the F corona is unpolarized. Photospheric light is unpolarized. Photospheric light scattered in the Earth's atmosphere is polarized, but close to the Sun's disc the degree of polarization of terrestrially scattered light is small, and under good conditions this can be used to pick out the K corona from the sky in observations from the Earth's surface out of eclipse but using a coronagraph to occult the Sun's disc.

One can see why the polarization arises if one notes that classically, when plane-polarized light falls on an electron, the electron will oscillate parallel to the electric vector of the incoming radiation, and will radiate light with the electric vector parallel to the oscillation (polarized parallel to the incoming polarization), and with greatest intensity in directions perpendicular to the axis of oscillation and with zero intensity along the axis of oscillation (Figure 11.7). Consider now radiation scattered through 90°, that is where the incident ray from the Sun, treated for the moment as a point source, is at right angles to the line of sight to the observer, so that we are looking at scattering electrons at the closest point of approach of the line of sight to the Sun. We can regard the unpolarized light coming from the Sun as made up of a component with the electric vector perpendicular to the scattering plane (the plane containing the Sun, the Earth and a scattering point) and of an equal component with the electric vector in the plane. The latter component will cause the electrons to oscillate along an axis which points along the line of sight, and the intensity scattered in this direction will be zero. On the other hand, the component with the electric vector perpendicular to the scattering plane will make the electron oscillate in this direction scattering maximum intensity in directions perpendicular to this axis, which include the line of sight, and the scattered radiation will be polarized perpendicular to the scattering plane. Hence if all scattering took place under these conditions, the observer would see pure tangential polarization with the electric vectors perpendicular to apparent radii from the Sun, and no radial polarization. However, at points along the line of sight nearer to the observer or farther from the observer than the point of closest approach to the Sun, the electric vector in the scattering plane will have a component perpendicular to the line of sight which will give a small radial component to the polarization. Furthermore, the Sun is not a point source and rays from the limb and the line of sight will define a slightly different scattering plane from that defined by rays from the centre and the line of sight.

The total scattering cross-section of an electron is the Thomson cross-section:

$$\sigma_T = 6.65 * 10^{-29}\,\mathrm{m}^2$$

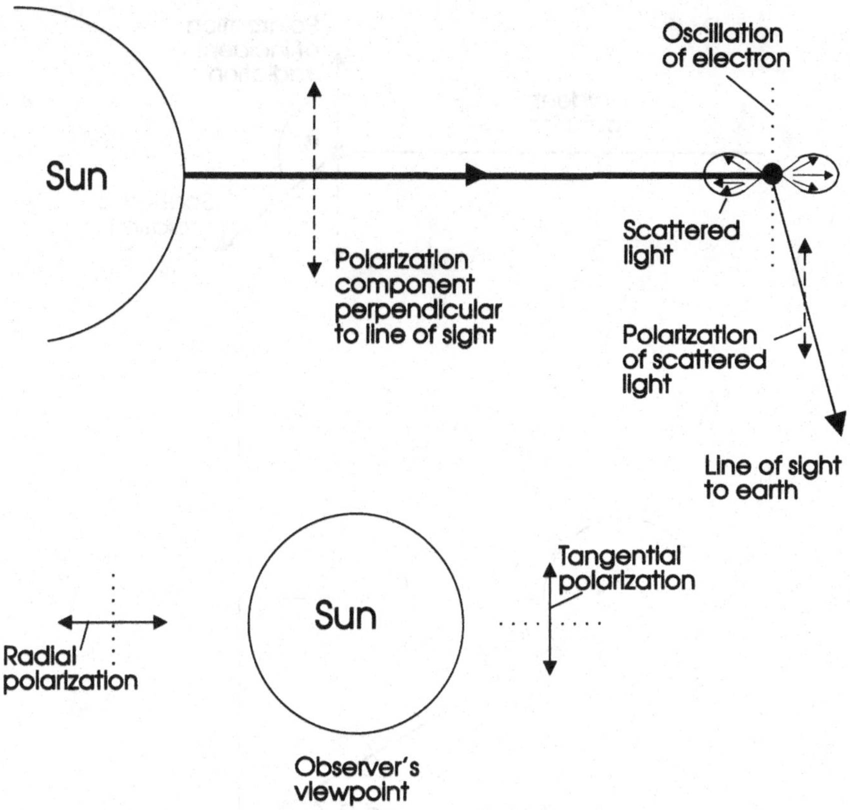

Figure 11.7. Polarization of electron scattered light from solar corona

The intensity scattered varies as the square of the sine of the angle α between the electric vector of the incident radiation and the scattered ray (Figure 11.8) so for linearly polarized radiation of intensity I'_0 in a solid angle $\Delta\Omega$, the scattered intensity is $3/(8\pi)I'_0\Delta\Omega\sigma_T \sin^2\alpha$. For initially unpolarized radiation of intensity I_0 in a small solid angle $\Delta\Omega$, the intensity scattered through an angle θ is $I_0\Delta\Omega\sigma_T$ $3/(16\pi)\,[1+\cos^2\theta]$. If we ignore limb-darkening, the solid angle subtended by the Sun (radius R) at a distance of r is:

$$\begin{aligned}\Delta\Omega &= \int_0^{\phi} 2\pi \sin\phi' \mathrm{d}\phi' \\ &= 2\pi[1-\cos\phi] \\ &= 2\pi\left[1-\sqrt{1-\frac{R^2}{r^2}}\right] \qquad (11.13)\end{aligned}$$

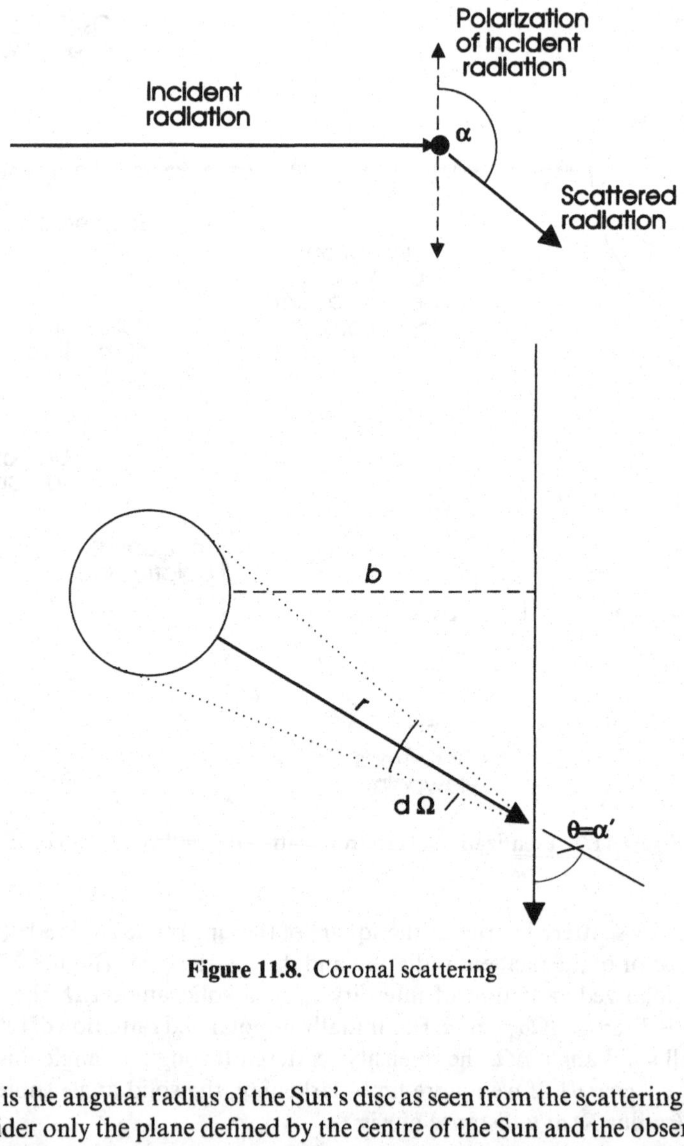

Figure 11.8. Coronal scattering

where ϕ is the angular radius of the Sun's disc as seen from the scattering point. If we consider only the plane defined by the centre of the Sun and the observed line of sight (ignoring the fact that rays from the limb and the line of sight define a slightly different plane) then we can represent the unpolarized radiation falling on the scattering electrons by equal plane polarized components with electric vectors perpendicular to the scattering and in the scattering plane, with power $I_0' = I_0 \Delta\Omega/2$, where I_0 is the mean intensity of the Sun's disc. The component perpendicular to the scattering plane will have an angle of 90° between the electric vector and the direction of the scattered radiation and will give the tangentially polarized component of the scattered radiation with intensity

$$3/(8\pi)(1/2)I'_0\Delta\Omega\sigma_T \text{ per electron}$$

The component in the scattering plane will have an angle $90° - \alpha'$ between the electric vector and the direction of the scattered ray, where α' is the angle between the incident ray and the line of sight, remembering that the electric vector is always perpendicular to the ray. This will give rise to the radially polarized component of the scattered radiation with intensity:

$$3/(8\pi)(1/2)I'_0\Delta\Omega\sigma_T \cos^2\alpha' \text{ per electron}$$

where $\cos\alpha' = \sqrt{(1 - b^2/r^2)}$ if b is the 'impact parameter'—the distance of closest approach of the line of sight to the Sun, and hence the apparent or projected distance from the observer's point of view of the point observed from the centre of the Sun.

Finally we must integrate the scattered intensity along the line of sight, with z as the distance along this line from the point of closest approach so $I_{obs} = \int I_{sca}\,dz$. Now $z^2 = r^2 - b^2$, so

$$dz = r\,dr/\sqrt{(r^2 - b^2)}$$

and hence

$$I_{tang}(b) = C\int_b^{\infty} N_e(r)\Delta\Omega(r)\frac{r\,dr}{\sqrt{r^2 - b^2}}$$

$$I_{rad}(b) = C\int_b^{\infty} N_e(r)\left(1 - \frac{b^2}{r^2}\right)\Delta\Omega(r)\frac{r\,dr}{\sqrt{r^2 - b^2}} \qquad (11.14)$$

with

$$C = \frac{3}{8\pi}I_0\sigma_T$$

and noting that at large r, $\Delta\Omega \sim \pi R^2/r^2$. For the modifications needed to take into account the slight variation of the scattering plane with the point of origin on the Sun and the limb-darkening, see Van der Hulst [2], noting that his r is scaled to the solar radius, and his function $A(r)$ corresponds in the limit of a point Sun to $\Delta\Omega/(2\pi)$. Three observations with a polarimeter at each point enable the F corona and the radial and tangential components to be separated, and either a simplified form of (11.14) to be inverted to give N_e or the parameters of a model density distribution to be adjusted until the predictions of (11.14) agree with observation. If we ignore the anisotropy of the scattering and polarization we have

$$I(b) = \sigma_T I_0\frac{\Delta\Omega}{4\pi}2\int_b^{\infty}\frac{N_e(r)r}{\sqrt{r^2 - b^2}}dr$$

which can be inverted using an Abel transform

$$N_e(r) = -\frac{4\pi}{\sigma_T I_0 \Delta\Omega}\frac{1}{\pi}\int_r^\infty \frac{\frac{dI(b)}{db}}{\sqrt{b^2 - r^2}}\,db \tag{11.15}$$

So far we have assumed a spherically symmetric corona, and pictures of the corona show not only that it is different at the Sun's magnetic poles from at the equator, but that the corona is far from homogeneous with holes, condensations and plumes. It is also important to note that the density of the corona varies very considerably with the stage in the solar cycle. The base of the corona has a density of a few times $10^{14}\,m^{-3}$ at the maximum of the Solar activity cycle, falling to about $10^{12}\,m^{-3}$ at 2.3 solar radii from the Sun's centre, but at the minimum of the cycle while the density at 2.3 solar radii from the centre will have fallen only slightly from the value at maximum in the equatorial plane, at the poles there will have been a fall by a factor of the order of 10. Thus at minimum the corona becomes highly flattened. The density halves roughly every 10^8 m in the inner corona, and a scale height $[N_e/(dN_e/dr)]$ of about $5*10^7$ m is sometimes used in approximate calculations although since the fall-off is not exponential, the scale height will not be independent of r. The scale height corresponds to a temperature of about 10^6 K if the corona is in hydrostatic equilibrium.

The corona has no permitted emission lines in the visible, because the high degree of ionization means that the energy levels are far apart and permitted lines are all down in the far ultraviolet or X-ray regions of the spectrum. However, transitions between levels with the same configuration (and therefore forbidden by the selection rule $\Delta l = 0$) but with different values of the total angular momentum J (fine structure splitting) are at much longer wavelengths. For low stages of ionization or neutral atoms these transitions lie at wavelengths in the far infrared—the familiar photospheric forbidden lines in the visible are transitions within a given configuration between levels with different values of the total orbital quantum number L, with the total spin quantum number S sometimes also changing, with for example the [O I] line at 557.7 nm being a transition within the $2p^4$ configuration between ^1D and ^1S. However the fine-structure splitting increases with the fourth power of the effective charge on the nucleus, so for high degrees of ionization the fine structure forbidden transitions move into the visible. Examples include the *red* coronal line at 637.4 nm which is a transition between the upper $^2P_{1/2}$ and lower $^2P_{3/2}$ fine structure levels of the ground state ($3p^5$) of iron Fe X, the similar *green* coronal line at 530.3 nm between the upper $^2P_{3/2}$ and lower $^2P_{1/2}$ fine-structure levels of the ground state (3p) of Fe XIV and in hot condensations the *yellow* line at 569.4 nm between the middle $^3P_1$ and lower $^3P_0$ fine-structure levels of the ground state ($2p^2$) of Ca XV, the $^3P_2$ to $^3P_1$ transition at 544.5 nm also being seen. Other fairly strong forbidden coronal lines in the visible include Fe XIII ($3p^2$) $^3P_2$ to $^3P_1$ and $^3P_1$ to $^3P_0$ at 1079.8 nm and

1074.7 nm, Fe XI ($3p^4$) $^3P_2$ to $^3P_1$ at 789.2 nm, Ni XV ($3p^2$) $^3P_1$ to $^3P_0$ at 670.2 nm , Ar XIV (2p) $^2P_{3/2}$ to $^2P_{1/2}$ at 441.2 nm and Ca XIII ($2p^4$) $^3P_1$ to $^3P_2$ at 408.6 nm. At shorter wavelengths in the ultraviolet starting with Fe XIII ($3p^2$) $^1D_2$ to $^3P_2$ at 338.8 nm and extending down to 110 nm and beyond, we find forbidden lines between different terms of the same configuration, that is levels differing in L and S.

The forbidden lines can be detected out of eclipse close to the Sun's disc using a coronagraph. Their study differs in two ways from that of the permitted lines at short wavelengths. Firstly, they are always observed against the background of the K corona and hence an emission line equivalent width can be found. Relative measurements at the same wavelength are easier than the absolute measurements of the intensities needed for the permitted lines. Secondly, the excitation of these lines may be by radiation, since the photospheric intensity at the appropriate visible wavelengths is quite high. The rate of radiative excitation is given by $B_{ij}I_0$ $\Delta\Omega$ and so we have

$$N_i(B_{ij}I_0\Delta\Omega + C_{ij}) = N_jA_{ji} \tag{11.16}$$

for the level populations. C_{ij} is of course proportional to N_e, which drops off fairly steeply with a scale height of the order of 100 000 km, whereas the dilution factor falls off as $1/r^2$ at large distances. Hence it is likely the radiative excitation will predominate far from the Sun, but collisional excitation may predominate in some cases in the inner corona. The strength of the line will be proportional to the density of the ion times the appropriate excitation rate integrated over the line of sight, and for constant temperature the density of the ion will be proportional to the overall number density of all ions, which will not be very different from the electron density. The strength of the K corona, as we have seen, is also proportional to the electron density integrated along the line of sight. Hence the strength of a forbidden line relative to the K corona will be roughly proportional to the excitation rate, in which case it should follow the fairly rapid decline in electron density close to the star where it is collisionally excited, and farther out should follow the more gentle inverse square decline of the radiative excitation (remembering always that we have to integrate the local collision coefficient along the line of sight). The changeover point should give some indication of electron density. However, the same problem that bedevils work on gaseous nebulae appears here also in the collisionally excited region. The line intensity is proportional to $\int N_e^2\,dz$, while the K corona intensity is proportional to $\int N_e\,dz$ but with steeply varying density this does not mean that the ratio is proportional to the mean electron density, for high density regions like condensations will contribute disproportionably to the line integral.

One should also notice that (11.16) and the subsequent discussion assume a two-level ion. In the case of a triplet fine structure, the upper level can not be populated by the absorption of radiation from the ground level but only by collisions, while the middle level is also populated by decays from the upper level

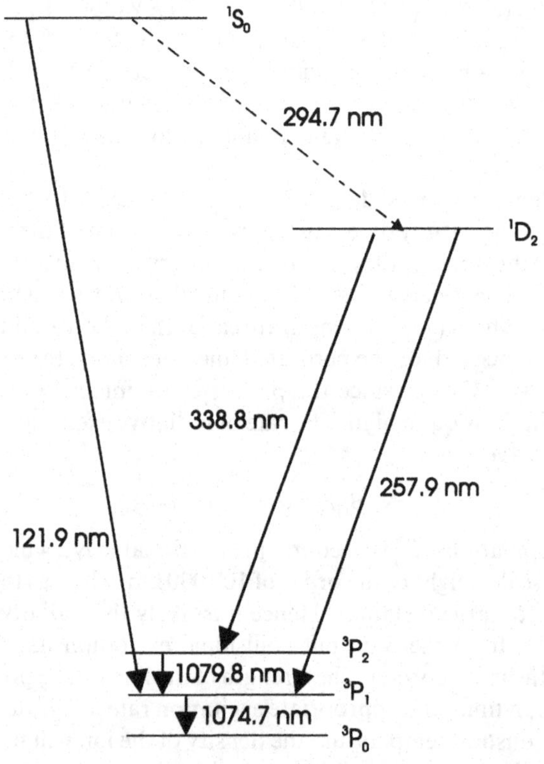

Figure 11.9. Lower levels of Fe XIII

which cannot decay directly to the lowest level for the same reason, namely because $\Delta J = 2$ is forbidden (Figure 11.9). Hence we must add to the direct excitation of the middle level the excitation rate to the upper level and radiative decay to the middle level. Thus the transition from the middle to the lower level, e.g. 569.4 nm in Ca XV or 1074.7 nm in Fe XIII must be stronger than the transition from the upper level to the middle level, e.g. the 544.5 nm transition in Ca XV or the 1079.8 nm in Fe XIII. At high densities, collisional de-excitation may become important, in which case the levels will assume Boltzmann populations and the upper to middle transition may be strongest, although densities high enough to achieve this are likely to be rare in the corona.

The photosphere of the Sun extends about 600 km above the point where continuum optical depth 1 is reached in the continuum at a wavelength of 500 nm. At this level the temperature goes through a minimum of about 4400 K and then starts to rise as one moves further out. The region between the temperature minimum and a value of 25 000 K is rather arbitrarily termed the chromosphere. On average one can estimate that the chromosphere is about 2000 km thick, but this average picture is not very meaningful because limb observations show that much of the chromospheric emission comes from discrete structures called spicules with very little from the regions between the spicules. From 25 000 K to several hundred thousand degrees, the temperature rises extremely rapidly as one moves outwards in the chromosphere–corona transition region before beginning to climb more slowly to the 10^6 K degree plateau of the corona proper, the observations of which were discussed above. The density falls equally rapidly in passing through the transition region. It must, however, be kept in mind that both chromosphere and corona are very inhomogeneous and perturbed by the phenomena of solar activity. The centres of certain very strong photospheric lines are formed in the chromosphere. The temperature rise there means that the central regions of these absorption lines will appear as emission lines (although not so bright as to return to continuum brightness). Non-LTE effects produce a dip in the line profile at the very centre of the line where the source function falls below the blackbody value due to surface effects. Thus the profiles of Ca II H and K at 396.9 and 393.3 nm drop in intensity as one moves in from the wings, this region being called for the K line K1V or K1R depending on whether one is talking about the short wavelength ('Violet') wing or the long wavelength ('Red') wing. Nearer the line centre, the intensity starts to rise again, and one has the K2V and K2R emission features. Finally at the very centre of the line the intensity drops again to give the K3 line. Similar but even more marked line reversals are seen in the corresponding doublet of Mg II at 280 nm. On the other hand the strong Balmer lines of hydrogen do not show emission components although their centres are formed at chromospheric heights—the source functions of these lines are photoelectrically controlled and do not respond to the chromospheric temperature rise (see the line formation discussion in Chapter 4). Pictures of the Sun made at the wavelength of K2 show a chromospheric network with most of the emission around the edges of cells perhaps 30 000 km or so across, and observations near the limb show that the spicules are emission features perhaps 1000 km thick, stretching upwards for 4000 km or so, and clustered around the edges of the chromospheric network.

We now consider wavelengths outside the visible range. In the visible, the corona is optically thin, emits no detectable continuum, and can only be seen away from the disc via the low intensity scattered photospheric continuum and forbidden line emission, but at long wavelengths the rising absorption coefficient of the free–free process can make the corona optically thick, the photosphere is thereby hidden, and coronal continuum emission is seen overlying the disc. Equation (6.26) for free–free absorption, adapted for $T = 10^6$ K and wavelengths of about 1 m, gives

$$\kappa_\nu \rho \sim 2.2 * 10^{-11} Z^2 / \nu^{2.05} (N_e N_{\mathrm{ion}}) / T^{1.43}$$

For completely ionized helium, summing over the contributions to the free-free opacity of hydrogen and helium we have

$$\Sigma Z^2 N_e N_{ion} = 1.3/1.1 N_e^2$$

so that the optical depth over thickness R with average electron density N_e(av) is

$$\tau_\nu = \int \kappa_\nu \rho \, dr = 2.2 \times 10^{-11} \frac{1}{\nu^{2.05}} \int \frac{N_e^2(r)}{T^{1.43}} \, dr$$

$$\simeq 2.2 \times 10^{-11} \frac{\langle N_e^2 \rangle R}{\nu^{2.05} T^{1.43}} \quad (11.17)$$

$$\simeq 3 \times 10^{-12} \frac{\langle N_e^2 \rangle}{\nu^{2.05}} \quad (11.17a)$$

where in the last equation we have taken $T = 10^6$ K and for R the electron density scale height of about $5 * 10^7$ m. Thus optical depth 1, which is the level from which we will observe radiation if free–free is the dominant opacity, will be the level where $N_e \sim 9 * 10^5 \nu$.

At wavelengths over 1.6 μm, free–free is the dominant source of opacity and by 1 mm, the free–free opacity is high enough to push the observed 'photosphere' into the chromosphere. The free–free process is an LTE one (even though chromospheric conditions are usually far from LTE) since the process is a collisional one, so the brightness temperature of the Sun from 1 mm up to a few centimetres is the chromospheric kinetic temperature, in the range 5000–10 000 K, with a slow rise with increasing wavelength reflecting a movement 'outwards' as the free–free opacity increases, although it must be remembered that the chromosphere is highly inhomogeneous with much of the material concentrated into spicules with emptier regions in between. By a wavelength of 10 cm, the much lower density corona is making an appreciable contribution (the transition from chromosphere to corona is very abrupt). If we treat the corona as being a constant temperature, we can use our simple model for a homogeneous cloud illuminated by a background source, where the background source is a blackbody at the temperature of the chromosphere and the cloud has optical thickness τ_{cor} and kinetic temperature T_{cor}:

$$T_B(\text{obs}) = T_{chr} \exp(-\tau_{cor}) + T_{cor}[1 - \exp(-\tau_{cor})]$$

With a coronal temperature of 10^6 K as opposed to a chromospheric temperature of 10^4 K, the corona can make an appreciable contribution even when the coronal optical depth remains fairly small. In fact for small τ_{cor} the coronal contribution will approximate to $T_{cor}\tau_{cor}$, which is proportional to ν^{-2}. Finally as the wavelength approaches 1 m, the corona will become optically thick, and the brightness temperature will become the kinetic temperature of the base of the corona (the base because the density is highest there and hence most of the contribution to the optical depth comes from this region).

One might think that at even longer wavelengths, the free–free emission would come from, and reflect, the temperature at higher levels in the corona. However, we must take account of the fact that electromagnetic waves can only propagate in a plasma at frequencies higher than the plasma frequency. An electron with average separation r_0 from an ion of the order of $N_e^{-1/3}$ will suffer a restoring force of the order of $e^2/(4\pi\varepsilon_0)\Delta r/r_0^3$ for a displacement Δr and so will oscillate with a circular frequency of about $\sqrt{[e^2/(4\pi\varepsilon_0 m_e)]}N_e^{1/2}$. A more exact argument gives a plasma frequency of

$$\nu_p = \frac{1}{2\pi}\left(\frac{4\pi e^2 N_e}{4\pi\varepsilon_0 m_e}\right)^{1/2}$$

$$= 8.98\, N_e^{1/2}\ \text{Hz} \tag{11.18}$$

Oscillations slower than this will be damped and will not propagate. Thus for any observed frequency, emission will only reach the observer from regions of the Sun's atmosphere where the density is lower than $\nu^2/81\,\text{m}^{-3}$. At relatively high frequencies or short wavelengths, the density at which the plasma frequency is reached lies deeper in the atmosphere than the level at which free–free becomes optically thick, and the plasma frequency limit has no effect. However at a wavelength of 4 m according to our approximate estimate after (11.17), more accurately around 2 m, the plasma frequency at the depth where free–free absorption becomes optically thick equals the observed wavelength, and at longer wavelengths the plasma frequency limitation prevents observation down to the level at which the gas is optically thick, and the observed brightness temperature will fall below the kinetic temperature. One must also note that when observing along lines of sight away from the Sun's centre, the path of a ray is bent since the refractive index is given by $[1-(\nu_p/\nu)^2]^{1/2}$. Over active regions magnetic fields may produce absorption by gyroresonance, etc. but consideration of these effects and of radio bursts would take us beyond the scope of this book.

At wavelengths shorter than the Lyman limit, 91.2 nm, the continuum comes from the chromosphere and is mainly produced by the bound–free process from and to the ground state of hydrogen, with helium taking over the role of the main source of bound–free opacity at wavelengths shorter than 50.4 nm. However at wavelengths less than 20 nm and extending down into the soft X-ray region, free–free re-emerges as the main source of continuous emission. The emission is optically thin and now comes mainly from the corona, but can be seen on the disc of the Sun because there is essentially no photospheric or chromospheric emission at these very short wavelengths. The free–free emission coefficient is given by (6.23) as $j_\nu\rho = 5.44 * 10^{-52} N_{\text{ion}} N_e Z^2 T^{1/2} \exp(-h\nu/kT)\ J\,\text{m}^{-3}\,\text{s}^{-1}\,\text{sr}^{-1}$.

The disc spectrum of the Sun shows a change at about 170 nm between a predominance of absorption lines above 170 nm to a predominance of emission lines below 170 nm, with the emission lines increasing in strength relative to the continuum as one goes to shorter wavelengths, which represents the weakening in absolute terms of the continuum as the wavelength decreases. Thus ultraviolet emission lines from the chromosphere and corona can be seen on the disc, whereas visible coronal lines can only be seen off the disc. The same wavelength of 170 nm marks the point where limb-darkening (a decrease in photospheric continuum intensity as one moves from the centre of the disc to the limb) changes to limb-brightening. The reason in both cases is the chromospheric temperature rise, for 170 nm in the continuum roughly comes from the temperature minimum at the top of the photosphere, with at shorter wavelengths continuum observations near the limb coming from higher and therefore hotter and brighter levels than those from the centre of the disc. Similarly at line wavelengths the line opacity forces the light which reaches us to come from higher and therefore hotter layers than those giving rise to the neighbouring continuum, and hence we obtain an emission line. It must, however, be remembered that line formation in the chromosphere is far from being in LTE, although all that is required for an emission line is that the line source function is greater than that of the continuum. It should also be recalled again that the chromosphere is very far from being homogeneous. At the shortest wavelengths we are looking at the corona, and the continuum is optically thin.

The lines below 170 nm can be categorized by the isoelectronic sequence. Thus one has lines that are lithium-like in that they originate in an ion that has one electron outside a closed $1s^2$ shell. Such ions have a spectrum that resembles sodium or Ca II in general form, and so has a resonance doublet (the 'D' lines) from $1s^22p\ ^2P$ to $1s^22s\ ^2S$, with the wavelengths scaling down from those of the corresponding lithium doublet at 670.8 nm as the nuclear charge Z increases. Thus we have the C IV doublet at 154.8 and 155.1 nm, the N V doublet at 123.8 and 124.3 nm, the O VI doublet at 103.2 and 103.8 nm, the Ne VIII doublet at 77.0 and 78.0 nm and the Mg X doublet at 61.0 and 62.5 nm (Fluorine and sodium have relatively low abundances). It will be noted that the spin–orbit splitting increases with Z. These lines peak at temperatures ranging from 10^5 to 10^6 K as higher degrees of ionization and higher excitation energies are required. The corresponding sodium isoelectronic sequence includes Si IV with doublet wavelengths shortwards of the C IV wavelengths at 139.4 nm and 140.3 nm. Similarly there is a beryllium isoelectronic sequence of ions with ground state $1s^22s^2$ with a resonance line in the singlet system $1s^22s2p\ ^1P$ to $1s^22s^2\ ^1S$ which for beryllium has a wavelength lying at 234.8 nm (beryllium has a very low abundance in the Sun), for C III at 97.70 nm, for N IV at 76.54 nm, for O V at 62.97 nm and for Ne VII at 46.52 nm. These isoelectronic sequences are examples of what are sometimes called 'screening doublets' because there is no change of shell (of quantum number n) involved, only of l, so that all that changes in the transition is the degree of screening of the nucleus, leading to a relatively small change in energy, a relatively long wavelength and a relatively low excitation energy for the degree of ionization involved, giving strong lines, especially for the lithium-like ions which result from recombination onto abundant helium-like ions. In the same wavelength region are the two strongest emission lines of all, Lyman α at 121.6 nm and Lyman α of He II at 30.4 nm, with the He I line at 58.4 nm also strong. Various transitions of iron ions which have 11 to 18 electrons (Fe XVI to IX) and involving no change in n lie shortwards of 40 nm, but at wavelengths shorter still we come to the n changing transitions and indeed in the soft X-ray region to the resonance lines of helium-like ions.

We have already discussed in the previous section how these lines can be used as diagnostics of temperature and density. It should be added that the highest stages of ionization and excitation are seen in solar flares, but here conditions may change rapidly with time and ionization may not be in equilibrium—considerations that would lead us beyond the scope of this book.

References

[1] A. H. Gabriel, in J. T. Schmulz and J. C. Brown (eds.), *The Sun: A Laboratory for Astrophysics*, Kluwer, 1992.
[2] Van der Hulst, Bulletin of the Astronomical Institute of the Netherlands, **111**, 131, 1950.

APPENDIX 1

Line Source Function at the Surface of an Atmosphere (Complete Redistribution)

The determination of the source function under non-LTE conditions is considerably more complicated for the case of complete redistribution than it is for the case of coherent scattering. One approach is to divide the frequency range covered by the line profile into N discrete frequencies labelled i to N, and to write

$$\int \phi_\nu J_\nu \, d\nu = \sum_i w_i J_i, \qquad \sum w_i = 1$$

where at the discrete frequencies the line absorption profile is represented by ϕ_i, the mean intensity by J_i and the condition $\sum w_i = 1$ normalizes the weights w_i assigned to each frequency. Solutions are of similar form to the coherent case but with the exponential in the optical depth replaced by a sum of N exponentials with different scales $1/z_k$ weighted by constants (that is, constants with depth) J_{ki} where k goes from l to N. Thus

$$\begin{aligned} J_i &= B + (1 - \varepsilon') \sum_k J_{ki} e^{-z_k \tau'} \\ &= B[1 + (1 - \varepsilon') \sum_k C_k f_i e^{-z_k \tau'}] \end{aligned}$$

Substituting in (4.17a):

$$\frac{d^2 J}{(d\tau')^2} = 3\phi_i^2 [J_i - (1 - \varepsilon') \sum_i J_i w_i - \varepsilon' B]$$

Taking for simplicity B (and ε') as constant with optical depth, we find that

$$\sum_k C_k f_i \left(1 - \frac{z_k^2}{3\phi_i^2}\right) e^{-z_k \tau'} = \sum_k C_k e^{-z_k \tau'} (1 - \varepsilon') \sum_i w_i f_i$$

and note that the RHS is independent of i so that the f_i must have the form $1/(1 - z_k^2/3\phi_i^2)$ and hence

$$J_i = B \left[1 + (1 - \varepsilon') \sum_k \frac{C_k e^{-z_k \tau'}}{1 - \dfrac{z_k^2}{3\phi_i^2}} \right] \tag{A1}$$

with the z_k satisfying

$$1 = \sum_i \frac{w_i(1-\varepsilon')}{1 - \dfrac{z_k^2}{3\phi_i^2}} \tag{A2}$$

The source function then becomes

$$S = \left[1 + (1-\varepsilon') \sum_i w_i \sum_k \frac{C_k e^{-z_k\tau'}}{1 - \dfrac{z_k^2}{3\phi_i^2}} \right]$$
$$= B[1 + (1-\varepsilon') \sum_k C_k e^{-z_k\tau'}] \text{ using (A2)}$$

Hence

$$S(0) = B\left[1 + (1-\varepsilon') \sum_k C_k \right]$$

If we now substitute (A1) in the boundary condition

$$\frac{dJ_i}{d\tau'}(\tau' = 0) = \sqrt{3}\phi_i J_i(\tau' = 0)$$

we find

$$1 + (1-\varepsilon') \sum_k \frac{C_k}{1 - \dfrac{z_k}{\sqrt{3}\phi_i}} = 0 \tag{A3}$$

In principle, the N equations (A2) (one for each value of z_k) and the N equations (A3) (one for each value of i) can be solved numerically for the given ϕ_i and w_i to obtain the C_k and the z_k. However, there is a slightly involved argument that gives $\sum C_k$ and hence $S(0)$ directly.

We first define a function F(q):

$$F(q) = 1 + (1-\varepsilon') \sum_k \frac{C_k}{1 - z_k q} \tag{A4}$$

so the source function becomes

$$S(0) = BF(0)$$

and the boundary condition becomes

$$F\left(\frac{1}{\sqrt{\phi_i}}\right) = 0$$

Multiplying through (A4):

$$F(q)\prod_k^N (1 - z_k q) = \prod_k^N (1 - z_k q) + (1-\varepsilon') \sum_k^N C_k \prod_{m \neq k}^N (1 - z_m q) \tag{A5}$$
$$= 0 \quad (q = 1/(\sqrt{3}\phi_i), \text{ boundary condition})$$

Now the left-hand side of (A5) is an algebraic equation in q and we know that $q = 1/(\sqrt{3}\phi_i)$ has N values. Hence it must be possible to write the LHS of (A5) in terms of these roots q_i:

$$C'(q - q_i)(q - q_2)\ldots(q - q_i)\ldots(q - q_N) = 0$$

where C' is a constant. If we compare this with the right-hand side of (4.25), and equate the coefficients of q^N, we find $C' = (-1)^N \prod z_k$. Hence

$$F(q) = \prod_k z_k \frac{\prod_i (q_i - q)}{\prod_k (1 - z_k q)}$$

$$F(0) = \sum_k z_k q_k \tag{A6a}$$

One can now apply a similar argument to (A2), defining $G(r)$ at the boundary:

$$G(r) = 1 - (1 - \varepsilon') \sum_i \frac{w_i}{1 - \frac{q_i^2}{r}}$$

$$= 0, \quad \text{if } r = \frac{1}{z_k^2}$$

Hence

$$G(r) = \sum_i \frac{w_i q_i^2}{q_i^2 - r} - \varepsilon' r \sum_i \frac{w_i}{q_i^2 - r} = 0 \tag{A6b}$$

where in the last equation we have used $\sum p_i = 1$. Multiplying through by $\prod (q_i^2 - r)$ and noting that since N values of r satisfy $G(r) = 0$, it must be possible to write the left-hand side of (4.26b) as

$$G(r) \prod_i^N (q_i^2 - r) = C'' \prod_s^N (r - r_s)$$

for the N roots r_s. On comparing the coefficients of r^N on both sides, we find that $C'' = \varepsilon'$ $(-1)^N$. Hence

$$G(r) = C'' \frac{\prod_s^N (r - r_s)}{\prod_i^N (q_i^2 - r)}$$

$$G(0) = \varepsilon' \prod_s^N \frac{r_s}{\prod_i q_i^2}$$

But (A6b) shows that $G(0) = 1$, so $\prod z_i q_i = \sqrt{\varepsilon'}$.

Finally $S(0) = BF(0) = B\prod z_k q_k = \sqrt{(\varepsilon')}B$, a result that holds exactly for the depth independent source function even when the Eddington approximation is dropped.

APPENDIX 2: Escape Probabilities

We follow the method of Canfield *et al.* [1]. Equation (4.34) gives the escape probability as

$$P_e = \int_0^\infty \phi_x E_2(\phi_x \tau_{OP})\,dx$$

Write $z = 1/\phi_x$ and then $y = \tau_{OP}/z$:

$$P_e = \int_1^\infty \frac{dx}{dz}\frac{1}{z} E_2\left(\frac{\tau_{OP}}{z}\right) dz$$

$$= \int_0^{\tau_{OP}} \frac{E_2(y)}{y} \frac{dx\left(\frac{\tau_{OP}}{y}\right)}{dz} dy$$

$$= \frac{dx(\tau_{OP})}{dz} \int_0^{\tau_{OP}} \frac{E_2(y)}{y} \frac{\dfrac{dx\left(\frac{\tau_{OP}}{y}\right)}{dz}}{\dfrac{dx(\tau_{OP})}{dz}} dy$$

and note that it is usually possible to write

$$\text{limit as } \tau_{OP} \Rightarrow \infty \left[\frac{\dfrac{dx\left(\frac{\tau_{OP}}{y}\right)}{dz}}{\dfrac{dx(\tau_{OP})}{dz}}\right] = y^{2\delta}$$

Thus for large optical depth, the escape probability becomes

$$P_e(\tau_0 P) = \frac{dx(\tau_{OP})}{dz} \int_0^\infty y^{2\delta-1} E_2(y)\,dy$$

$$= \frac{dx(\tau_{OP})}{dz} \int_0^\infty y^{2\delta-1} \int_1^\infty \frac{e^{-ty}}{t^2}\,dt\,dy$$

$$= \frac{dx(\tau_{OP})}{dz} \int_1^\infty \frac{1}{t^2} \int_0^\infty y^{2\delta-1} e^{-ty}\,dy\,dt$$

where the functionE_2 has been written out and the order of integration changed.

Writing $u = ty$:

$$P_e(\tau_{OP}) = \frac{dx(\tau_{OP})}{dz} \int_1^{\infty} \frac{1}{t^{2\delta+2}} \int_0^{\infty} u^{2\delta-1} e^{-u} du$$

$$= \frac{dx(\tau_{OP})}{dz} \frac{1}{2\delta+1} \Gamma(2\delta)$$

where the integral over u has been recognized as the gamma function, whose value in this case is independent of t so the integral over t can be performed immediately.

Consider first the application of this approach to the Doppler case where $\phi_x = 1/\sqrt{\pi}\exp(-x^2)$, so $z = \sqrt{\pi}\exp(x^2)$ and $x = \sqrt{(\ln_e[z/\sqrt{\pi}])}$. Then

$$\frac{dx(z)}{dz} = \frac{1}{2} \frac{1}{z\sqrt{\ln_e\left(\frac{z}{\sqrt{\pi}}\right)}}$$

$$\frac{dx(z = \tau_{OP})}{dz} = \frac{1}{2} \frac{1}{\tau_{OP}\sqrt{\ln_e\left(\frac{\tau_{OP}}{\sqrt{n}}\right)}}$$

$$\frac{dx\left(z = \frac{\tau_{OP}}{y}\right)}{dz} = \frac{1}{2} \frac{1}{\frac{\tau_{OP}}{y}\sqrt{\ln_e\left(\frac{\tau_{OP}}{\sqrt{\pi}}\right) - \ln_e y}}$$

Hence

$$\frac{\dfrac{dx\left(z = \frac{\tau_{OP}}{y}\right)}{dz}}{\dfrac{dx(z = \tau_{OP})}{dz}} \to y \text{ as } \tau_{OP} \to \infty$$

Hence $2\delta = 1$ and

$$P_e(\tau_{OP}) \frac{1}{2} \frac{\Gamma(1)}{2} \frac{1}{\tau_{OP}\sqrt{\ln_e\left(\frac{\tau_{OP}}{\sqrt{\pi}}\right)}}$$

$$= \frac{1}{4\tau_{OP}\sqrt{\ln_e\left(\frac{\tau_{OP}}{\sqrt{\pi}}\right)}} \tag{4.40a}$$

Finally, turning to the collision broadened case and writing generally $z = 1/\phi_x = z_0 x^k$, we have $x = (z/z_0)^{1/k}$. Hence

$$\frac{dx(z)}{dz} = \frac{1}{k} \frac{z^{(1/k)-1}}{z_0^{(1/k)}}$$

$$\frac{dx(z = \tau_{OP})}{dz} = \frac{1}{k} \frac{\tau_{OP}^{(1/k)-1}}{z_0^{(1/k)}}$$

$$\frac{\mathrm{d}x\left(z=\dfrac{\tau_{OP}}{y}\right)}{\mathrm{d}z}=\frac{1}{k}\frac{\left(\dfrac{\tau_{OP}}{y}\right)^{(1/k)-1}}{z_0^{1/k}}$$

$$\frac{\dfrac{\mathrm{d}x\left(z=\dfrac{\tau_{OP}}{y}\right)}{\mathrm{d}z}}{\dfrac{\mathrm{d}x(z=\tau_{OP})}{\mathrm{d}z}}=\left(\frac{1}{y}\right)^{(1/k)-1}$$

Thus $2\delta=(k-1)/k$ and

$$P_e(\tau_{OP})=\frac{1}{2k-1}\Gamma\left(\frac{k-1}{k}\right)\frac{1}{z_0^{1/k}}\frac{1}{\tau_{OP}^{(k-1)/k}}$$

where for instance in the case of Lorentzian wings with $\phi=a/(\pi x^2)$, $k=2$, $\Gamma(1/2)=\sqrt{(\pi)}$ and $z_0=\pi/a$, so

$$P_e(\tau_{OP})=\frac{1}{3}\sqrt{\frac{a}{\tau_{OP}}} \tag{4.41a}$$

Reference

[1] R. C. Canfield, R. C. Puetter et al., in W. Kalkofen (ed.), *Methods in Radative Transfer*, Cambridge University Press, 1985

SELECT BIBLIOGRAPHY

This bibliography is far from being complete, and is confined largely to textbooks and review articles. A number of detailed references are given in the text.

General

The following collections each contain a number of relevant articles, the most important of which are referred to under specialist headings:

Baschek, B., Kegel, W. H., and Traving, G., (eds.), *Problems in Stellar Atmospheres and Envelopes*, Springer, 1971.
Burbidge, G., Layzer, D., and Phillips, J. G., *Annual Reviews of Astronomy and Astrophysics*, Annual Reviews Inc. (Highly condensed articles with extensive bibliography).
Dalgarno, A. and Layzer, D., (eds.), *Spectroscopy of Astrophysical Plasmas*, Cambridge University Press, 1987. (All the articles are relevant).
Hartquist, T., (ed.), *Molecular Astrophysics*, Cambridge University Press, 1990.
Verschuur, G. L. and Kellerman, R. I., (eds.), *Galactic and Extragalactic Radio Astronomy*, 2nd edn, Springer, 1988.

Stars and Stellar Atmospheres

The fundamental reference here, particularly for model atmospheres but also for line formation is:
Mihalas, D., *Stellar Atmospheres*, 2nd edn., Freeman, 1978.

Two older but still useful books are:
Aller, L. H., *Astrophysics: The Atmospheres of the Sun and Stars*, 2nd edn, Ronald Press, 1963.
Unsöld, A., *Physik der Sternatmosphären*, 2nd edn., Springer, 1962.

See also:
Böhm-Vitense, E., 'Model Stellar Atmospheres and Heavy Element Abundances', in *Problems in Stellar Atmospheres and Envelopes*.

For details on non-LTE line formation see:
Jefferies, J. T., *Spectral Line Formation*, Blaisdell, 1968.
Athay, R. G., *Radiation Transport in Spectral Lines*, Reidel, 1971.
Kalkofen, W. (ed.), *Methods in Radiative Transfer*, Cambridge University Press, 1985.

For the analysis of line profiles see:
Gray, D. F., *The Observation and Analysis of Stellar Photospheres*, Wiley, 1976.
Gray, D. F., *Lectures on Spectral-Line Analysis of F, G, and K Stars*, 1988.

For hot stars see:

Kudritski, R. P., 'The Atmospheres of Hot Stars: Modern Theory and Observation', in Kudritski, R. P. *et al.*, *Radiation in Moving Gaseous Media*, 18th Advanced Course, Swiss Society for Astrophysics, Saas-Fee, 1988.

Underhill, A. and Doazan, V. (eds.) *B Stars With and Without Emission Lines*, NASA, 1982.

For cool stars see:

Gustaffson, B., 'Chemical Analysis of Cool stars', *Annual Reviews of Astronomy and Astrophysics*, **27**, 1989

Johnson, H. R. and Zuckerman, B., *Evolution of Peculiar Red Giant Stars*, Astronomical Union Symposium, **106**, Cambridge University Press, 1985.

For oscillator strengths see:

Parkinson, W. H., 'Laboratory Astrophysics', in *Spectroscopy of Astrophysical Plasmas.*

Gaseous Nebulae

The fundamental reference is:

Osterbrock, D., *Astrophysics of Gaseous Nebulae and Active Galactic Nuclei*, University Science Books, 1984.

Also very useful:

Aller, L. H., *Physics of Thermal Gaseous Nebulae*, D. Reidel, 1984.

See also:

Pottasch, S., *Planetary Nebulae*, D. Reidel, 1984.

For H II regions and recombination lines see:

Brown, R. L., 'Radio Observation of H II regions', in *Spectroscopy of Astrophysical Plasmas.*

Gordon, M., 'H II Regions and Recombination Lines', in "*Galactic and Extragalactic Radio Astronomy.*

For photodissociation regions see:

Hollenbach, D. J., 'Photodissociation Regions', in Blitz, L. (ed.), *Evolution of the Interstellar Medium*, Proceedings of the Astronomical Society of the Pacific Meeting, 1990.

Sternberg, A., 'IR Molecular H_2 Emission from Interstellar Photodissociation Regions', in *Molecular Astrophysics.*

For AGN see:

Bregmann, J. N., 'Continuum Radiation from AGN', *The Astronomy and Astrophysics Review* **2**, 1990.

Netzer, H. in Blandford, R., *et al.*, *Active Galactic Nuclei*, Saas-Fee Advanced Course 20, Springer, 1990.

Osterbrock, D. E., 'Quasars, Seyfert Galaxies and AGN', in *Spectroscopy of Astrophysical Plasmas.*

The Cold Interstellar Medium

For molecular spectroscopy see:

Herzberg, G., *Spectroscopy of Diatomic Molecules*, Van Nostrand, 1950.

Townes, G. H., Schawlow, A. L., *Microwave Spectroscopy*, Dover, 1955, 1975.

For molecular dissociation see:
Kirby, K. T., 'Molecular Photoabsorption Processes', in *Molecular Astrophysics.*

For absorption lines see:
Black, J. H., 'Diffuse Interstellar Clouds', in *Spectroscopy of Astrophysical Plasma.*

For molecular emission lines see:
Genzel, R., Physics and Chemistry of Molecular Clouds, in Burton, W. L., *et al.*, *The Galactic Interstellar Medium*, Saas-fee advanced course 21, Springer, 1992.
Turner, B. E., 'Molecules as Probes of the Interstellar Medium', in *Galactic and Extragalactic Radioastronomy.*
Turner, B. E. and Ziurys, L., 'Interstellar Molecules, Astrochemistry', in *Galactic and Extragalactic Radioastronomy.*

See also:
Geballe, T., 'Infrared Observations of Line Emission from Molecular H_2', in *Molecular Astrophysics.*
Ho, P. T. P. and Townes, C. H., 'Interstellar Ammonia', in *Annual Reviews of Astronomy and Astrophysics*, **21**, 1983.
Shull, J. M. and Beckwith, S., 'Interstellar Molecular Hydrogen', in *Annual Reviews of Astronomy and Astrophysics*, **20**, 1982.

For an introduction to interstellar dust see:
Evans, A., *The Dusty Universe*, Howarth, 1993.
Whittet, D. C. B., *Dust in the Galactic Environment*, Institute of Physics, 1990.

Masers

The fundamental reference is:
Elitzur, M., *Astronomical Masers*, Kluwer, 1992.

See also:
Kegel, W. H., 'Cosmic Masers', in *Problems in Stellar Atmospheres.*
Moran, J. M., 'Masers in the Envelopes of Young and Old Stars', in *Molecular Astrophysics.*
Reid, M. J. and Moran, J. M., 'Astronomical Masers', in *Galactic and Extragalactic Radio Astronomy.*

Circumstellar Shells and Mass Loss

For hot stars see:
References under 'hot stars'.
Kitchin, C. R., *Early Emission Line Stars*, Hilger, 1982.

For cool stars see:
Reimers, D., 'Circumstellar Envelopes and Mass Loss in Red Giant Stars', in *Problems in Stellar Atmospheres.*
Zuckermann, B., 'Spectroscopy of Circumstellar Shells', in *Spectroscopy of Astrophysical Plasmas.*
Morris, M. and Zuckermann, B., (eds.), *Mass Loss from Red Giants*, D. Reidel, 1985.

The Solar Corona and Chromosphere

For a good general reference see:
Zirin, H., *Astrophysics of the Sun*, Cambridge University Press, 1988.

See also:

Gabriel, A. H., 'Spectroscopic Diagnostics', in Schmetz, J. T. and Brown, J. C., *The Sun: A Laboratory for Astrophysics*, Kluwer, 1992.

Marasca, J. T., *The Solar Transition Region*, Cambridge University Press, 1992.

Zirker, J. B., 'Spectroscopy of the Solar Corona', in *Spectroscopy of Astrophysical Plasmas.*

Other Subjects

Amongst the subjects that there has not been room to discuss in this book,

For shock waves see:

Chernoff, D. F. and McKee, C. F., 'Shocks in Dense molecular Clouds', in *Molecular Astrophysics.*

McKee, C. F., 'Astrophysical Shocks in Diffuse Gas', in *Spectroscopy of Astrophysical Plasmas.*

For X-rays from hot gas, see:

Treves, A., *et al.* (eds.), *Iron Line Diagnostics*, Springer, 1991.

For techniques of observation, see:

Kitchin, C. R., *Astrophysical Techniques*, Institute of Physics, 1991.

Stellar Spectral Atlases

Jacoby, G. H., Hunter, D. A., and Christian, C. A. 'A Library of Stellar Spectra', *Astrophysical Journal Supplement*, **56**, 257, 1984.

Pickles, A. J., 'Differential Population Synthesis of Early Type Galaxies. I Spectrophotometric Atlas of Synthesis Standard Stars', *Astrophysical Journal Supplement*, **59**, 33, 1985.

Danks, A. C. and Dennefeld, M., 'An Atlas of Southern MK Standards from 5800 to 10200 A' *Publications of the Astronomical Society of the Pacific* **106**, 382, 1994.

Fanelli, M. N., O'Connell, R. W., Burstein, D., and Wu, C-C., 'Spectral Synthesis in the UV. A Library of Mean Stellar Groups', *Astrophysical Journal Supplement*, **82**, 197, 1992.

Index

9 780471 976790